Nanostructure Chemistry

Nanostructure Chemistry

Edited by **Lindy Bowman**

Willford Press
New York

Published by Willford Press,
118-35 Queens Blvd., Suite 400,
Forest Hills, NY 11375, USA
www.willfordpress.com

Nanostructure Chemistry
Edited by Lindy Bowman

International Standard Book Number: 978-1-68285-290-3 (Hardback)

Printed in China.

Contents

	Preface	**IX**
Chapter 1	**Simulation of SnO_2/WO_3 nanofilms for alcohol of gas sensor based on metal dioxides: MC and LD studies** Leila Mahdavian	**1**
Chapter 2	**Synthesis and nucleation-growth mechanism of almost catalyst-free carbon nanotubes grown from Fe-filled sphere-like graphene-shell surface** Muhammad Salleh Shamsudin, Maryam Mohammad, Syazwan Afif Mohd Zobir, Noor Asnida Asli, Suriani Abu Bakar, Saifollah Abdullah, Syed Yusainee Syed Yahya and Mohamad Rusop Mahmood	**10**
Chapter 3	**Effect of catalyst on growth of diamond by plasma-enhanced chemical vapor deposition** Karim Zare and Morasa Amani Malkeshi	**22**
Chapter 4	**Fabrication of various electrical resistances producing Zn nanowires and subsequent oxidation fabricating ZnO nanowires in PAA template by periodic and pulsed electrochemical deposition** Mahmood Moradi, Mina Zamanian and Mohammad Noormohammadi	**27**
Chapter 5	**Adsorption kinetics, thermodynamic studies, and high performance of CdO cauliflower-like nanostructure on the removal of Congo red from aqueous solution** Azadeh Tadjarodi, Mina Imani and Hamed Kerdari	**33**
Chapter 6	**Catalytic effect of commercial nano-CuO and nano-Fe_2O_3 on thermal decomposition of ammonium perchlorate** Mostafa Mahinroosta	**41**
Chapter 7	**A potentiometric solid state copper electrode based on nanostructure polypyrrole conducting polymer film doped with 5-sulfosalicylic acid** Reza Ansari, Zahra Mosayebzadeh, Majid Arvand and Ali Mohammad-khah	**47**
Chapter 8	**Superior photocatalytic degradation of azo dyes in aqueous solutions using $TiO_2/SrTiO_3$ nanocomposite** Loghman Karimi and Salar Zohoori	**56**
Chapter 9	**Preparation of transparent, conductive ZnO:Co and ZnO:In thin films by ultrasonic spray method** Said Benramache, Boubaker Benhaoua and Hamza Bentrah	**61**

Chapter 10 **Decoration of multi-walled carbon nanotubes (MWCNTs) with different ferrite nanoparticles and its use as an adsorbent** 68
Ahmed A Farghali, Mohamed Bahgat, Waleed M A ElRouby and Mohamed H Khedr

Chapter 11 **Synthesis of clay-CNTs nanocomposite** 80
Sajjad Sedaghat

Chapter 12 **One-step electrosynthesis of multi-walled carbon nanotube/poly-ortho-aminophenol composite film and investigation of its electrocatalytic properties** 84
Narges Ajami and Niloufar Bahrami Panah

Chapter 13 **A facile one-step synthesis of PTSA-doped tetraaniline nanostructure/magnetite nanoparticles via self-assembly method** 89
Keloth Basavaiah, Yarramsetti Pavankumar and Alamanda Vara Prasada Rao

Chapter 14 **Comparison of drug delivery systems: Nanotube and *p*- Sulphonatocalix[4]arene, by Density Functional Theory** 97
Karim Zare and Nasim Shadmani

Chapter 15 **DFT/NBO study of Nanotube and Calixarene with anti-cancer drug** 103
Karim Zare, Nasim Shadmani and Elham Pournamdari

Chapter 16 **Annealing temperature effect on the mechanical and tribological properties of molybdenum nitride thin films** 109
Kaykhosrow Khojier, Mohammad Reza Karami Mehr and Hadi Savaloni

Chapter 17 **Size-dependent *in vivo/in vitro* results of homoeopathic herbal extracts** 116
Rajesh Barve and Ramesh Chaughule

Chapter 18 **Investigation of a QM/MM study on interaction of a carbon nanotube with cytarabine drug in various solvents and temperatures** 121
Fahimeh Najafi

Chapter 19 **Nanocrystalline titania-based sulfonic acid (TiO_2-Pr-SO_3H) as a new, highly efficient, and recyclable solid acid catalyst for preparation of quinoxaline derivatives** 132
Seyyed Vahid Atghia and Somayyeh Sarvi Beigbaghlou

Chapter 20 ***Ab initio* simulations of p-type porous silicon nanostructures** 140
Emilye Rosas Landa Loustau, Jesús A del Río, Julia Tagüeña-Martínez, Luis E Sansores and Rocío Nava

Chapter 21 **Phytosynthesis of silver nanoparticles using *Pterocarpus santalinus* leaf extract and their antibacterial properties** 146
Kasi Gopinath, Shanmugam Gowri and Ayyakannu Arumugam

Chapter 22 **Preparation and study of vitamin A palmitate microemulsion drug delivery system and investigation of co-surfactant effect** 153
Solmaz Maleki Dizaj

Chapter 23 **Room temperature aqueous phase synthesis and characterization of novel nano-sized coordination polymers composed of copper(II), nickel(II), and zinc(II) metal ions with *p*-phenylenediamine (PPD) as the bridging ligand** 159
Rimpy Gupta, Sumit Sanotra, Haq Nawaz Sheikh and Bansi Lal Kalsotra

Chapter 24 **Synthesis, characterization, and experimental investigation of surface activity of SERS substrates using neodymium oxide (Nd_2O_3)** 168
Fatemeh Sadeghi, Farzad Khani, Abbas Hemati Azandaryani, Yadollah Mansouri, Zahra Shaykhi Mehrabadi and Amir Nikjou

Chapter 25 **Studies of *ab initio* and Monte Carlo simulation on interaction of fluorouracil anticancer drug with carbon nanotube** 174
Karim Zare, Fahimeh Najafi, Hamidreza Sadegh and Ramin Shahryari ghoshekandi

Chapter 26 **X-ray imaging technique using colloid solution of Au/silica core-shell nanoparticles** 182
Yoshio Kobayashi, Hiromitsu Inose, Tomohiko Nakagawa, Yohsuke Kubota, Kohsuke Gonda and Noriaki Ohuchi

Chapter 27 **Fabrication of ZnO nanoparticles by solution-combustion method for the photocatalytic degradation of organic dye** 188
Mohammad Amdad Ali, Mahmudur Rahman Idris and Md Emran Quayum

Chapter 28 **Optimization of ball milling parameters to produce *Centella asiatica* herbal nanopowders** 194
Muhammad Zuhairi Borhan, Rohaya Ahmad, Mohamad Rusop and Saifollah Abdullah

Chapter 29 **An experimental investigation of the enhanced oil recovery and improved performance of drilling fluids using titanium dioxide and fumed silica nanoparticles** 202
Goshtasp Cheraghian, Mahmood Hemmati, Mohsen Masihi and Saeed Bazgir

Chapter 30 **Study of influential factors in synthesis and characterization of cobalt oxide nanoparticles** 211
Ghazaleh Allaedini and Abubakar Muhammad

Chapter 31 **Simulation of nanodrug by theoretical approach** 227
Saeideh Ghorbaninezhad and Maryam Ghorbaninezhad

Chapter 32 **Functionalized multi-walled carbon nanotubes for enhanced photocurrent in dye-sensitized solar cells** 233
Chandra Sekharan Nair Omana Amma Sreekala, Jinchu Indiramma, Kandala Bala Subramanya Pavan Kumar, Karyaveettil Savithriamma Sreelatha and Mahesh Saran Roy

Chapter 33 **Bioengineered silver nanobowls using *Trichoderma viride* and its antibacterial activity against gram-positive and gram-negative bacteria** 241
Kethirabalan Chitra and Gurusamy Annadurai

Chapter 34 **The interaction between carbon nanotube and skin anti-cancer drugs: a DFT and NBO approach** **248**
Maryam Hesabi and Masoumeh Hesabi

Chapter 35 **Biosynthesis of silver nanoparticles using *Salmonella typhirium*** **254**
Hamid Reza Ghorbani

Chapter 36 **Antibacterial properties of nano silver finish cellulose fabric** **258**
Mohammad Mirjalili, Niloofar Yaghmaei and Marjan Mirjalili

Permissions

List of Contributors

Preface

In my initial years as a student, I used to run to the library at every possible instance to grab a book and learn something new. Books were my primary source of knowledge and I would not have come such a long way without all that I learnt from them. Thus, when I was approached to edit this book; I became understandably nostalgic. It was an absolute honor to be considered worthy of guiding the current generation as well as those to come. I put all my knowledge and hard work into making this book most beneficial for its readers.

In recent years nano-scale science and technology have undergone rapid progress. Nano-chemistry in particular presents a unique approach in building devices with a molecular-scale precision. The book provides a detailed study on nanostructure chemistry and synthesis of nanostructures and nano-materials. The text focuses on synthesis and processing of organic, inorganic and hybrid nano-sized particles by physical, chemical and biological methods. The book elucidates conceptual methods and applications on nanostructures and nano-devices. The book brings forth complete understanding of practical uses of nanotechnologies and nanostructures. The researches in this book will benefit students and professionals alike.

I wish to thank my publisher for supporting me at every step. I would also like to thank all the authors who have contributed their researches in this book. I hope this book will be a valuable contribution to the progress of the field.

Editor

1

Simulation of SnO_2/WO_3 nanofilms for alcohol of gas sensor based on metal dioxides: MC and LD studies

Leila Mahdavian

Abstract

This work presents a study of the adsorption properties of nanostructures. The moving gas technique was employed to determine the transient and steady-state response behavior of nanocrystalline gas sensors. SnO_2 sensors have shown high sensitivity to low concentrations of ethanol at moderate temperature. Tin dioxide is the most used material for gas sensing because its three-dimensional nanofilms and properties are related to the large surface exposed to gas adsorption. This study proposes the use of SnO_2 nanofilms in interaction with ethanol; we used different percentages of SnO_2 and WO_3 in the adsorption of ethanol by nanofilms. The total energy, potential energy, and kinetic energy were calculated for the interaction between nanofilms and ethanol at different concentrations and at 300 K. The calculations were achieved by Langevin dynamics and Monte Carlo simulation methods. The total energy decreased with additional tungsten percentage in the nanofilms and increased with additional number of ethanol molecules and interactions between them are endothermic.

Keywords: Metal dioxides, Nanofilm, SnO_2/WO_3, Alcohol, Langevin dynamics, Monte Carlo simulation

Background

It has been suggested that nanometer-sized particles, if properly treated, can be used to form high-surface-area catalysts, efficiencies of which improve on existing catalytic function or sensor function [1,2]. This might further be improved through changes in molecular electronic structure and as a result of the development of confined regions that accompany the transition to the macroscopic regime for both the active elements and their substrate supports [3]. In this paper, we used tin dioxide because it presents great potential properties for applications such as optoelectronic devices, gas sensors, and solar energy or even for detecting leakages of reducing gases such as H, H_2S, CO, and others [4].

Solids with a grain size of a few nanometers are normally referred to as nanocrystalline materials: they exhibit properties that are significantly different from their coarse-grained polycrystalline counterparts. A first classification of the change in the properties of a material relies on either 'bulk' or 'surface' effects. Typically, bulk effects involve the confinement of electrons in nanometric cells, which causes quantum phenomena to take place [5]. Surface effects normally appear because of the magnification of the specific surface of nanostructures, leading to the enhancement of the properties related to that, such as catalytic activity or surface adsorption [6].

Developing new solid-state gas sensors with improved properties carries with it an obvious close relationship between the sensing performance of the active materials and their surface chemical activity. The theoretical study of such surface adsorption interactions provides a valuable tool to get superior performances that are unattainable using only a trial-and-error approach together with a powerful analytic methodology to explain the experimental data. Tin dioxide (SnO_2) plays a key role as one of the more representative sensing materials in solid-state gas sensors [7], presenting a significant surface reactivity with many important reducing (CO, NO) and oxidizing gases (O_2, NO_2) [8,9].

However, to explain the sensing behavior, it is necessary to keep in mind that there exist interfering processes poisoning the surface [10] and that these can dramatically change the effective adsorptions of the target species and therefore their eventual detection as in

Correspondence: mahdavian_leila@yahoo.com
Department of Chemistry, Islamic Azad University, Doroud Branch, P.O. Box: 133, Doroud, Iran

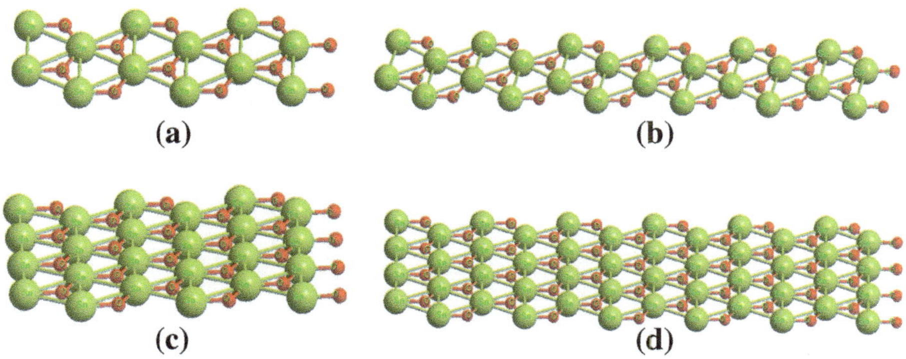

Figure 1 Optimized configuration (top view) of SnO_2 nanofilm replication count in (*i.j.k*) planes of the surface. **(a)** (1.1.2) with chirality (3,3), **(b)** (1.1.2) with chirality (6,6), **(c)** (1.1.4) with chirality (3,3), **(d)** (1.1.4) with chirality (6,6).

the case of the SnO_2 surface [11,12]. Thus, in the present analysis, their effects have also been studied in order to point up their consequences on the sensing mechanisms [13,14]. In this paper, we used SnO_2 nanofilms with different replication counts in the (*i.j.k*) plane that are shown in Figure 1. After structural characterizations, photoluminescence characteristics of the SnO_2 nanofilms were studied, so some tin atoms were replaced by tungsten atoms of different percentages in nanofilms (Figure 2).

Nowadays, first-principles methodologies are based on using Langevin dynamics to simulate the presence of a solvent during annealing and Monte Carlo simulation to sample configurations from a Boltzmann-weighted distribution. Langevin dynamics and Monte Carlo simulation can provide precise calculations of the total energy and potential energy of adsorption. Moreover, faster codes and new computational facilities allow dealing with numbers of surface adsorption configurations in moderate computing times. In this context, the aim of the present work is to provide theoretical hints for the development of improved C_2H_5OH gas sensors using SnO_2 and WO_3 as base sensing materials, and an interaction pattern between nanofilms and 45 molecules of ethanol in a mean distance of 13 Å is shown in Figure 3. The relevance of surface orientation is discussed, and the most significant adsorption sites of C_2H_5OH are identified.

Results and discussion

Semiconducting sensors offer an inexpensive and simple method for monitoring gases. The change of the electrical conductivity of semiconducting materials upon exposure to a reducing gas such as C_2H_5OH has been used for gas detection. Therefore, nanofilms have been utilized as a sensing material in pressure, flow, thermal, gas, optical, mass, position, stress, strain, chemical, and

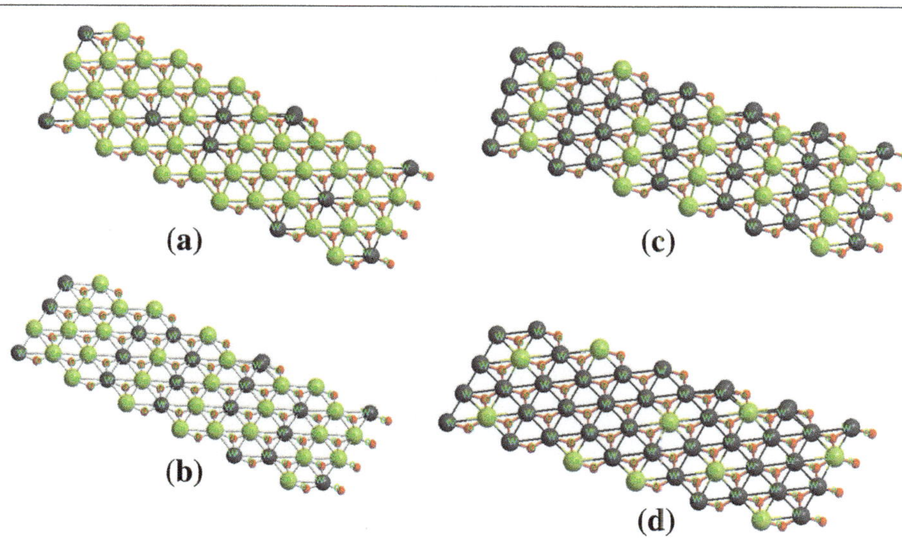

Figure 2 Sn and W dioxide nanofilms with different percentages for nanofilm (1.1.4) with chirality (6,6). **(a)** 20% (W/Sn), **(b)** 40% (W/Sn), **(c)** 60% (W/Sn), **(d)** 80% (W/Sn).

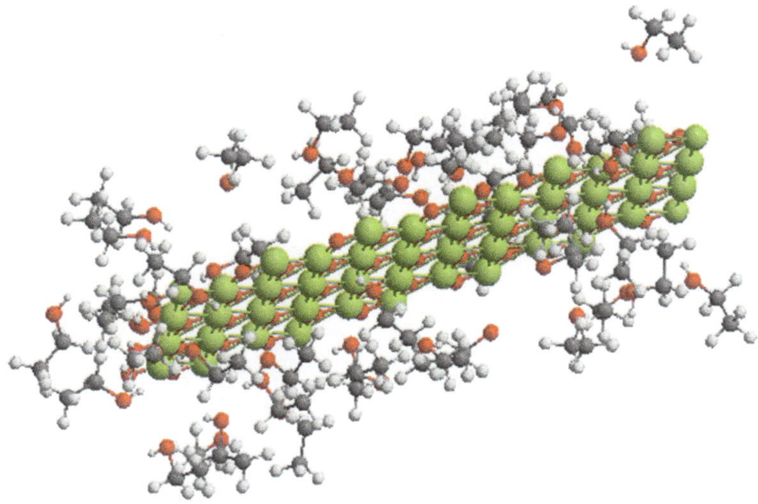

Figure 3 Interaction between SnO_2 nanofilm (1.1.4) with chirality (6,6) and 45 ethanol molecules in gas phase.

biological sensors. The sensor was based on the principle that the electronic properties of Sn and W nanofilms change when subjected to strains. The isotropic nature of nanofilms helps in measuring strains in multiple locations and in different directions. This manuscript demonstrated, using first-principles quantum transport calculations and Langevin dynamics (LD) and Monte Carlo (MC) simulation methods, that hydrostatic pressure can induce radial deformation and therefore electrical transition of Sn-W nanofilms. The measurements were carried out in a humid air atmosphere, which has the same condition as that of a general operation of a sensor. The total energy, potential energy, and kinetic energy were calculated for the interaction of Sn and W dioxide nanofilms and different numbers of ethanol molecules at 300 K. The calculations are shown in Tables 1, 2, 3, 4, 5, and 6, which show that with additional number of ethanol molecules, the energies increased. Figure 4 shows that the total energy increased with additional distance mean of ethanol molecules and its number is related to Sn-W nanofilms.

The characteristics of the sensor were investigated by changing of resistance in ethanol ambient with different products in air. Formation of acetaldehyde by the oxidative dehydrogenation of ethanol depends critically upon the reaction step that requires the oxide surface to acquire a negative charge: when ethanol is injected into the micro-reactor containing either a pure nanofilm compound or a large fraction of ethanol, it was found to be transformed to acetaldehyde. It then may be adsorbed in the form of ethoxide (CH_3CH_2O), producing an adsorbed hydrogen atom. On the contrary, this may be

Table 1 E_{total}, E_{pot}, and E_{kin} **calculated for** SnO_2 **nanofilms (1.1.2) with chirality (3,3) and chirality (6,6)**

Number (C_2H_5OH)	Nanofilm (1.1.2) with chirality (3,3)						Nanofilm (1.1.2) with chirality (6,6)					
	Langevin dynamics			Monte Carlo simulation			Langevin dynamics			Monte Carlo simulation		
	E_{kin}	E_{pot}	E_{total}	E_{kin}	E_{pot}	E_{total}	E_{kin}	E_{pot}	E_{total}	E_{kin}	E_{pot}	E_{total}
2	13,779.37	11,008.44	2,770.93	37.56	12,683.41	12,720.97	5,324.66	21,269.72	26,594.39	59.08	24,627.04	24,686.11
4	13,945.72	11,145.63	2,800.09	53.65	12,895.69	12,949.35	5,343.36	20,950.79	26,294.15	75.19	24,362.76	24,437.95
8	16,923.75	13,579.51	3,344.24	85.85	8,712.70	8,798.55	5,319.18	21,051.04	26,370.22	107.42	24,449.50	24,556.91
14	16,318.59	12,835.62	3,482.97	136.82	14,819.28	14,956.10	5,837.56	22,768.01	28,605.57	155.76	26,458.96	26,614.71
20	17,356.28	14,052.32	3,303.96	182.42	16,026.57	16,208.99	6,667.73	27,700.08	34,367.80	204.09	31,817.43	32,021.52
25	17,740.40	14,315.16	3,425.24	214.62	16,331.12	16,545.74	12,045.68	33,253.92	45,299.60	244.37	39,020.24	39,264.61
30	17,740.40	14,315.16	3,425.24	262.91	24,660.00	24,922.91	10,226.58	34,406.06	44,632.64	284.65	40,248.02	40,532.67
35	33,710.02	27,457.04	6,252.98	305.83	30,880.40	31,186.23	8,156.48	34,789.06	42,945.55	332.09	39,433.17	39,765.27
40	29,115.74	23,610.61	5,505.13	343.39	26,661.92	27,005.31	9,417.45	36,521.02	45,938.47	365.21	41,949.13	42,314.34
45	37,582.69	26,085.83	11,496.87	383.63	31,131.61	31,515.23	7,575.181	34,216.33	41,791.51	409.14	41,447.89	41,857.03

E_{kin}, kinetic energy; E_{pot}, potential energy; E_{total}, total energy.

Table 2 E_{total}, E_{pot}, and E_{kin} calculated for SnO_2 nanofilms (1.1.4) with chirality (3,3) and chirality (6,6)

Number (C_2H_5OH)	Nanofilm (1.1.4) with chirality (3,3)						Nanofilm (1.1.4) with chirality (6,6)					
	Langevin dynamics			Monte Carlo simulation			Langevin dynamics			Monte Carlo simulation		
	E_{kin}	E_{pot}	E_{total}	E_{kin}	E_{pot}	E_{total}	E_{kin}	E_{pot}	E_{total}	E_{kin}	E_{pot}	E_{total}
2	3,849.76	23,508.03	27,357.79	59.61	27,401.88	27,461.49	7,003.23	50,224.76	57,227.99	102.96	57,387.37	57,490.34
4	3,854.32	23,547.65	27,401.97	75.87	27,404.67	27,480.54	6,547.44	49,269.91	55,817.34	119.22	55,931.90	56,051.12
8	3,957.76	25,348.32	29,306.08	108.38	29,290.83	29,399.21	6,577.57	49,423.27	56,000.84	151.73	56,133.21	56,284.94
14	4,132.30	26,103.75	30,236.05	157.15	30,183.79	30,340.94	6,515.96	49,721.75	56,237.71	200.51	56,312.50	56,513.01
20	4,765.51	28,782.85	33,548.36	205.92	33,470.04	33,675.97	6,745.41	51,444.96	58,190.38	249.28	58,232.85	58,482.12
25	5,406.35	30,872.80	36,279.15	238.44	36,133.77	36,372.21	7,277.62	54,127.86	61,405.48	289.92	61,375.23	61,665.16
30	6,924.91	35,111.82	42,036.73	287.21	41,727.12	42,014.34	8,819.49	57,206.89	66,026.38	330.56	65,916.92	66,247.48
35	7,785.81	40,992.63	48,778.44	327.85	48,527.43	48,855.28	8,390.91	58,664.17	67,055.08	371.21	66,948.39	67,319.59
40	6,698.72	37,052.96	43,751.68	352.24	43,504.78	43,857.02	10,691.72	60,042.44	70,734.16	411.85	70,294.60	70,706.45
45	14,169.06	42,206.06	56,375.12	409.14	54,985.96	55,395.10	8,805.78	60,143.73	68,949.51	452.49	68,732.91	69,185.41

transformed to acetaldehyde (CH_3CHO), forming an adsorbed hydrogen atom:

$$\begin{aligned} &CH_3CH_2OH_{(g)} \rightarrow CH_3CH_2O_{ads} + H_{ads} \\ &CH_3CH_2O_{ads} \rightarrow CH_3CHO_{ads} + H_{ads} \\ &H_{ads} + H_{ads} \rightarrow 2H^+_{ads} + 2e^- \end{aligned}$$

The thermodynamic properties of this interaction are calculated by LD and MC simulation methods that are shown in Tables 1, 2, 3, 4, 5, and 6. Now, in the reaction environment, there are ethanol and acetaldehyde, from which methyl ethyl ketone represents the second most important reaction product in our calculations. This

Table 3 E_{total}, E_{pot}, and E_{kin} calculated for different percentages of SnO_2/WO_3 nanofilms (1.1.2) with chirality (3,3)

Number (C_2H_5OH)	80:20 (W/Sn)			60:40 (W/Sn)			40:60 (W/Sn)			20:80 (W/Sn)		
	E_{kin}	E_{pot}	E_{total}	E_{kin}	E_{pot}	E_{total}	E_{kin}	E_{pot}	E_{total}	E_{kin}	E_{pot}	E_{total}
Langevin dynamics												
2	15,54.04	8,165.71	9,719.75	1,822.83	8,564.03	10,386.86	2,111.08	9,380.21	11,491.29	2,291.73	9,585.20	11,876.93
4	1,610.63	8,497.87	10,108.50	1,886.40	8,935.71	10,822.11	2,168.96	9,341.16	11,510.12	2,462.70	10,241.20	12,703.91
8	606.60	6,004.59	6,611.19	692.07	6,283.64	6,975.71	724.37	6,445.04	7,169.41	860.46	7,039.84	7,900.30
14	2,376.31	10,044.76	12,421.06	2,821.23	11,139.06	13,960.29	2,901.86	11,317.48	14,219.35	31,77.64	12,005.30	15,182.93
20	2,062.18	10,732.53	12,794.71	2,295.12	11,177.52	13,472.64	2,575.76	12,020.05	14,595.81	2,825.58	12,633.59	15,459.17
25	2,137.36	11,033.01	13,170.36	2,414.31	11,442.94	13,857.25	2,694.22	12,268.44	14,962.66	2,968.88	12,751.69	15,720.57
30	3,518.02	17,802.29	21,320.30	3,769.88	18,575.42	22,345.29	4,045.41	19,496.10	23,541.51	4,350.71	20,361.66	24,712.37
35	4,940.62	23,470.34	28,410.96	5,228.05	24,372.44	29,600.49	5,511.25	25,205.43	30,716.68	5,788.76	25,976.53	31,765.29
40	4,256.95	20,159.47	24,416.42	4,504.25	20,863.25	25,367.50	4,775.13	21,383.31	26,158.44	5,091.34	22,269.85	27,361.19
45	10,020.35	22,086.69	32,107.04	10,724.36	23,433.19	34,157.56	10,927.98	24,008.3	34,936.28	11,154.76	25,011.04	36,165.80
Monte Carlo simulation												
2	37.93	9,699.82	9,737.75	37.93	10,365.47	10,403.40	37.93	11,473.01	11,510.94	37.93	11,853.55	11,891.49
4	54.19	10,074.46	10,128.65	54.19	10,789.42	10,843.61	54.19	11,459.26	11,513.45	54.19	12,663.02	12,717.21
8	86.71	6,563.80	6,650.51	86.71	2,638.45	2,725.16	86.71	7,117.41	7,204.12	86.71	7,853.92	7,940.63
14	137.28	12,266.40	12,403.68	137.28	12,945.73	13,083.01	137.28	1,4078.88	14,216.16	137.28	15,015.32	15,152.61
20	184.25	12,638.92	12,823.17	184.25	13,317.37	13,501.62	184.25	14,431.49	14,615.74	184.25	15,294.36	15,478.61
25	216.76	12,992.57	13,209.34	216.76	13,671.61	13,888.37	216.76	14,771.31	14,988.08	216.76	15,513.46	15,730.22
30	265.53	21,064.32	21,329.86	265.53	22,102.00	22,367.53	265.53	23,297.03	23,562.57	265.53	24,480.90	24,746.44
35	308.89	28,135.04	28,443.93	308.89	29,333.76	29,642.64	308.89	30,410.94	30,719.82	308.89	31,456.53	31,765.42
40	346.82	24,077.90	24,424.72	346.82	25,025.62	25,372.44	346.82	25,805.99	26,152.81	346.82	27,005.35	27,352.17
45	387.46	31,136.25	31,523.71	387.46	33,105.69	33,493.15	387.46	33,842.09	34,229.55	387.46	35,119.00	35,506.46

Table 4 E_{total}, E_{pot}, and E_{kin} calculated for different percentages of SnO_2/WO_3 nanofilms (1.1.2) with chirality (6,6)

Number (C_2H_5OH)	80:20 (W/Sn)			60:40 (W/Sn)			40:60 (W/Sn)			20:80 (W/Sn)		
	E_{kin}	E_{pot}	E_{total}	E_{kin}	E_{pot}	E_{total}	E_{kin}	E_{pot}	E_{total}	E_{kin}	E_{pot}	E_{total}
Langevin dynamics												
2	2,755.86	14,469.19	17,225.04	3,379.86	16,428.55	19,808.41	4,083.70	17,959.44	22,043.14	4,591.24	19,887.84	24,479.08
4	2,884.09	14,433.97	17,318.06	4,209.04	17,137.61	21,346.65	4,209.04	17,137.61	21,346.65	4,739.74	19,027.48	23,767.22
8	2,906.87	14,335.54	17,242.42	3,497.00	16,023.00	19,520.00	4,129.47	18,097.28	22,226.74	4,701.51	19,565.19	24,266.70
14	3,115.43	15,515.17	18,630.6	4,034.61	17,432.03	21,466.65	4,674.31	19,265.93	23,940.24	5,185.91	20,923.87	26,109.78
20	4,118.17	20,179.56	24,297.72	4,631.68	22,273.93	26,905.60	5,250.10	24,059.02	29,309.12	5,722.94	25,570.39	31,293.34
25	6,280.36	22,829.68	29,110.04	7,033.24	25,451.13	32,484.37	7,577.53	27,507.41	35,084.93	11,300.17	30,710.20	42,010.37
30	7,153.81	26,116.36	33,270.17	7,956.55	28,378.50	36,335.05	8,729.66	30,375.62	39,105.28	9,474.97	32,292.45	41,767.42
35	5,601.06	27,407.76	33,008.82	6,226.93	29,267.00	35,493.92	6,886.25	31,411.72	38,297.97	7,475.36	32,988.25	40,463.61
40	6,111.14	27,241.57	33,352.71	7,143.89	30,539.81	37,683.70	7,962.71	33,060.76	41,023.47	8,722.34	35,166.94	43,889.28
45	5,291.90	26,610.73	31,902.63	5,861.36	28,427.55	34,288.91	6,400.03	30,681.32	37,081.35	6,920.45	32,017.11	38,937.55
Monte Carlo simulation												
2	59.61	17,211.98	17,271.59	59.61	19,802.09	19,861.70	59.61	22,034.24	22,093.85	59.61	24,461.49	24,521.10
4	75.87	17,295.60	17,371.47	75.87	19,464.34	19,540.21	75.87	21,308.89	21,384.75	75.87	23,738.28	23,814.14
8	108.38	17,207.22	17,315.6	108.38	19,454.75	19,563.13	108.38	22,158.35	22,266.73	108.38	24,220.12	24,328.50
14	157.15	18,519.07	18,676.22	157.15	21,369.47	21,526.62	157.15	23,836.96	23,994.12	157.15	26,010.69	26,167.84
20	205.93	24,141.29	24,347.21	205.92	26,756.81	26,962.74	205.92	29,177.39	29,383.31	205.92	31,150.51	31,356.43
25	246.57	28,681.26	28,927.83	246.57	32,102.89	32,349.46	246.57	34,708.90	34,955.47	246.57	41,270.72	41,517.29
30	287.21	32,922.91	33,210.12	287.21	35,969.44	36,256.65	287.21	38,734.81	39,022.02	287.21	41,370.23	41,657.44
35	335.08	32,712.71	33,047.78	335.08	35,199.59	35,534.67	335.08	37,971.88	38,306.96	335.08	40,148.62	40,483.70
40	368.50	32,970.01	33,338.50	368.50	37,313.49	37,681.99	368.50	40,648.02	41,016.51	368.50	43,487.41	43,855.91
45	409.14	31,531.72	31,940.87	409.14	33,912.04	34,321.18	409.14	36,741.29	37,150.43	409.14	38,581.21	38,990.35

reaction requires H transfer from one adsorbed acetaldehyde, which becomes oxidized, to another adsorbed ethanol, which is reduced to alkoxide. Some Sn atoms are replaced by W atoms, making different nanofilms for the adsorption of ethanol molecules in gas phase. The energies of interaction are calculated by Langevin dynamics and Monte Carlo simulation that are shown in Tables 1, 2, 3, 4, 5, and 6. It was also found that the reduction in Sn atoms and length enhances the sensitivities of the sensors.

The trajectories obtained by applying this algorithm agree with those obtained by Langevin dynamics simulations. The efficiency of a Monte Carlo-based search engine depends on the interplay of the energy update protocol and the type of conformational movements used to modify a given conformation. The total energy and potential energy are approximating similarly in both methods with additional distance mean. On the other hand, the kinetic energy is changed with additional distance for the Langevin dynamics method, but it is the same for the Monte Carlo simulation. With increased W in nanofilms, the total energy decreased at 300 K as shown in Figure 5 and Tables 1, 2, 3, 4, 5, and 6.

Conclusions

It is believed that nanofilm-incorporated sensors can bring dramatic changes to the biomedical industry. There are certain cases such as diabetes where regular tests by patients themselves are required to measure and control the sugar level in the body. The SnO_2 and WO_3 chemical sensors for liquids can be used for blood analysis (for example, detecting sodium or finding pH value). Sensors and biosensors have been widely used in the food industry to provide safety and quality control of food products as the contamination of foods caused by bacterial pathogens may result in numerous diseases.

The sensitivity could be measured when the detecting gas is mixed homogeneously with air. The sensitivity depends on the ethanol concentration at each work temperature. The sensitivity increases with increasing ethanol gas concentration. The response and sensitivity of the alcohol-treated sensor are found to be very high when compared to those of the air-treated sensor. The total energy depends linearly on the temperature in that it increased with additional temperature, so this interaction is endothermic.

Table 5 E_{total}, E_{pot}, and E_{kin} calculated for different percentages of SnO_2/WO_3 nanofilms (1.1.4) with chirality (3,3)

Number (C_2H_5OH)	80:20 (W/Sn)			60:40 (W/Sn)			40:60 (W/Sn)			20:80 (W/Sn)		
	E_{kin}	E_{pot}	E_{total}	E_{kin}	E_{pot}	E_{total}	E_{kin}	E_{pot}	E_{total}	E_{kin}	E_{pot}	E_{total}
Langevin dynamics												
2	2,210.62	15,257.10	17,467.72	2,610.16	17,432.68	20,042.84	3,065.21	19,358.78	22,423.99	3,412.01	21,421.35	24,833.36
4	2,219.86	15,292.77	17,512.63	2,663.95	17,169.90	19,833.85	3,165.18	19,574.31	22,739.49	3,459.65	21,415.72	24,875.37
8	2,164.23	16,591.41	18,755.64	2,678.74	18,129.12	20,807.86	3,084.77	20,868.51	23,953.27	3,474.81	22,874.86	26,349.67
14	2,402.37	17,790.82	20,193.19	2,895.48	19,780.01	22,675.48	3,369.85	22,050.74	25,420.59	3,619.75	23,950.49	27,570.24
20	3,695.35	23,113.83	26,809.18	4,700.47	26,356.08	31,056.55	5,416.69	28,721.93	34,138.62	5,786.23	29,874.15	36,251.63
25	3,298.99	22,089.16	25,388.15	3,651.98	24,348.83	28,000.81	4,281.28	26,668.50	30,949.77	4,719.27	28,339.91	33,059.18
30	5,049.80	25,379.65	30,429.45	5,600.27	28,120.61	33,720.89	5,987.73	30,976.19	36,963.92	6,396.02	32,997.49	39,393.52
35	5,393.05	31,294.44	36,687.49	5,965.39	33,824.96	39,790.34	6,544.17	36,430.97	42,975.14	7,085.10	38,844.76	45,929.85
40	5,028.14	27,665.13	32,693.26	5,549.61	30,147.64	35,697.25	5,380.29	32,404.76	37,785.05	6,038.96	34,147.74	40,186.70
45	12,069.53	32,404.33	44,473.86	12,727.84	34,802.41	47,530.25	13,177.77	36,924.81	50,102.58	13,543.40	39,169.52	52,712.92
Monte Carlo simulation												
2	59.61	17,476.86	17,536.47	59.61	20,066.75	20,126.36	59.61	22,438.33	22,497.94	59.61	24,863.32	24,922.93
4	75.87	17,508.01	17,583.88	75.87	19,847.76	19,923.63	75.87	22,757.18	22,833.04	75.87	24,888.85	24,964.72
8	108.38	18,719.21	18,827.59	108.38	20,780.24	20,888.63	108.38	23,932.30	24,040.68	108.38	26,364.94	26,473.32
14	157.15	20,113.38	20,270.53	157.15	22,613.16	22,770.32	157.15	25,379.32	25,536.47	157.15	27,531.85	27,689.00
20	205.92	26,685.96	26,891.89	205.92	30,959.45	31,165.37	205.92	34,047.59	34,253.52	205.92	30,252.63	36,985.23
25	238.44	25,209.70	25,448.14	238.44	27,869.02	28,107.46	238.44	30,811.64	31,050.08	238.44	32,930.97	33,169.41
30	287.21	30,123.06	30,410.27	287.21	33,415.40	33,702.61	287.21	35,942.61	36,761.09	287.21	39,118.61	39,405.82
35	327.85	36,439.57	36,767.42	327.85	39,539.66	39,867.52	327.85	44,457.25	44,785.11	327.85	45,686.76	46,014.61
40	352.24	32,410.93	32,763.17	352.24	35,403.28	35,755.52	352.24	37,542.95	37,895.19	352.24	39,930.83	40,283.07
45	409.14	43,147.25	43,556.39	409.14	46,068.35	46,477.49	409.14	48,902.73	49,311.88	409.14	51,386.80	51,795.94

The total energy in the two methods decreased initially, but it increased with additional number of ethanol molecules that have indicated interaction between SnO_2 nanofilms. The percentages of WO_3 were increased in the nanofilms, so the total energy decreased which indicated that the nanofilms stabled upon adsorption. The diffusion behavior can be clearly distinguished as either normal mode or single file mode in time scales of about 1,000 ps. It is possible to show that setting a threshold of 1 on the movement of the dihedrals of the backbone in a single Monte Carlo step, the mean quantities associated with the off-equilibrium dynamics are well reproduced, while the good description of higher moments requires smaller moves.

Methods

Langevin dynamics simulation

In this work, we use Langevin dynamics and Monte Carlo simulation methods for calculation of interaction systems. The Langevin dynamics method simulates the motion of molecules subjected to random collisions and frictional forces and can be used to model solvated systems without explicitly including solvent molecules [15]. It provides information on the time evolution of the molecular system. A Langevin dynamics simulation is set up in much the same way as a molecular dynamics simulation, with small differences due to the addition of the friction coefficient. Langevin dynamics is similar in principle to regular molecular dynamics. It includes frictional effects to simulate the presence of a solvent implicitly, where regular molecular dynamics requires inclusion of explicit solvent molecules. One term represents a frictional force, and the other, a *random* force $\vec{R}$. For example, the effects of solvent molecules not explicitly present in the system being simulated would be approximated in terms of a frictional drag on the solute as well as random kicks associated with the thermal motions of the solvent molecules. Since friction opposes motion, the first additional force is proportional to the particle's velocity and oppositely directed. Langevin's equation for the motion of atom i is [16]:

$$\vec{F_i} - \gamma_i \vec{\upsilon_i} + \vec{R_i}(t) = m_i \vec{a_i}, \tag{1}$$

where $\vec{F_i}$ is still the sum of all forces exerted on atom i by other atoms explicitly present in this system. This

Table 6 E_{total}, E_{pot}, and E_{kin} calculated for different percentages of SnO_2/WO_3 nanofilms (1.1.4) with chirality (6,6)

Number (C_2H_5OH)	80:20 (W/Sn)			60:40 (W/Sn)			40:60 (W/Sn)			20:80 (W/Sn)		
	E_{kin}	E_{pot}	E_{total}	E_{kin}	E_{pot}	E_{total}	E_{kin}	E_{pot}	E_{total}	E_{kin}	E_{pot}	E_{total}
Langevin dynamics												
2	3,817.97	31,158.54	34,976.51	4,041.47	32,267.54	36,309.00	5,109.32	40,236.10	45,345.42	5,758.93	44,949.41	50,708.34
4	4,246.10	35,553.67	39,799.77	4,246.10	35,553.67	39,799.77	5,142.00	40,002.49	45,144.49	5,820.00	44,520.45	50,340.46
8	3,802.52	31,103.23	34,905.74	4,431.18	35,065.37	39,496.55	5,170.54	40,022.17	45,192.71	5,799.21	44,854.30	50,653.50
14	3,817.97	31,158.54	34,976.51	4,444.51	35,444.35	39,888.86	5,109.32	40,236.10	45,345.42	5,758.93	44,949.41	50,708.34
20	3,706.21	33,321.15	37,027.36	4,466.97	37,843.44	42,310.40	5,173.90	41,796.93	46,970.82	5,878.20	46,798.62	52,676.82
25	4,359.98	36,145.42	40,505.41	5,012.30	40,023.27	45,035.57	6,279.97	45,541.76	51,821.74	6,582.70	50,952.08	57,534.77
30	4,864.99	38,000.46	42,865.45	5,994.86	42,804.52	48,799.38	7,263.04	47,117.77	54,380.81	8,065.98	52,145.61	60,211.59
35	5,050.85	39,953.08	45,003.93	5,903.17	44,866.43	50,769.60	6,776.62	49,169.95	55,946.57	7,496.28	53,577.81	61,074.10
40	7,846.96	42,015.53	49,862.49	9,853.55	51,521.68	61,375.23	9,183.96	50,993.38	60,177.34	9,926.74	54,968.60	64,895.34
45	5,399.89	41,925.54	47,325.43	6,224.97	45,811.98	52,036.95	7,047.15	50,277.66	57,324.81	7,623.16	54,855.18	62,478.35
Monte Carlo simulation												
2	180.64	31,882.55	32,063.19	180.64	36,303.64	36,484.28	180.64	40,340.47	40,521.11	180.64	45,439.63	45,620.26
4	119.22	35,430.47	35,549.69	119.22	39,878.80	39,998.02	119.22	45,254.13	45,373.35	119.22	50,473.20	50,592.42
8	151.73	39,522.59	39,674.32	151.73	39,522.59	39,674.32	151.73	45,250.92	45,402.65	151.73	50,762.30	50,914.04
14	200.51	50,786.41	50,986.91	200.51	50,786.41	50,986.91	200.51	50,786.41	50,986.91	200.51	50,786.41	50,986.91
20	249.28	36,960.39	37,209.67	249.28	42,270.60	42,519.88	249.28	46,945.01	47,194.29	249.28	52,688.51	52,937.79
25	289.92	40,409.42	40,699.34	289.92	44,944.90	45,234.82	289.92	51,745.62	52,035.54	289.92	57,505.35	57,795.27
30	330.56	42,700.64	43,031.20	330.56	48,649.13	48,979.70	330.56	54,252.67	54,583.23	330.56	60,074.48	60,405.04
35	371.21	44,836.86	45,208.06	371.21	50,607.44	50,978.64	371.21	55,802.74	56,173.95	371.21	60,956.07	61,327.28
40	411.85	49,316.07	49,727.92	411.85	60,917.38	61,329.23	411.85	59,687.30	60,099.15	411.85	64,377.98	64,789.83
45	452.49	47,089.70	47,542.20	452.49	51,817.49	52,269.99	452.49	57,089.73	57,542.22	452.49	62,273.49	62,725.98

equation is often expressed in terms of the collision frequency $\zeta = \gamma/m$. The friction coefficient is related to the fluctuations of the random force by the fluctuation-dissipation theorem:

$$\left\langle \vec{R_i}(t) \right\rangle = 0,$$

$$\int \left\langle \vec{R_i}(0).\vec{R_i}(t) \right\rangle dt = 6K_B T \gamma_i. \tag{2}$$

In simulations, it is often assumed that the random force is completely uncorrelated at different times, that is, the above equation takes the form:

$$\left\langle \vec{R_i}(t).\vec{R_i}(t') \right\rangle dt = 6K_B T \gamma_i \delta(t - t'). \tag{3}$$

The temperature of the system being simulated is maintained via this relationship between $\vec{R}(t)$ and γ. The jostling of a solute by a solvent can expedite barrier crossing, and hence, Langevin dynamics can search conformations better than Newtonian molecular dynamics ($\gamma = 0$).

The Monte Carlo method samples phase space by generating random configurations from a Boltzmann distribution at a given temperature. Averages computed from a properly equilibrated Monte Carlo simulation correspond to thermodynamic ensemble averages. Thus, the Monte Carlo method can be used to find average energies and equilibrium structural properties of complex interacting systems [17]. A sequence of successive configurations from a Monte Carlo simulation constitutes a trajectory in phase space [18,19]. The Monte Carlo method provides a more direct route to equilibrium structural and thermodynamic properties. However, these calculations can be quite long, depending upon the system studied [20]. Because of imperfections of the force field, the lowest energy base usually does not correspond to the native state in most cases, so the rank of native structure in those decoys produced by the force field itself is poor. Therefore, the rank of native structure can be relatively better when ranked by the second force field [21-23]. The acceptance ratio is a running average of the ratio of the number of accepted moves to attempted moves. Optimal values are close to 0.5; varying the step size can have a large effect on the acceptance ratio.

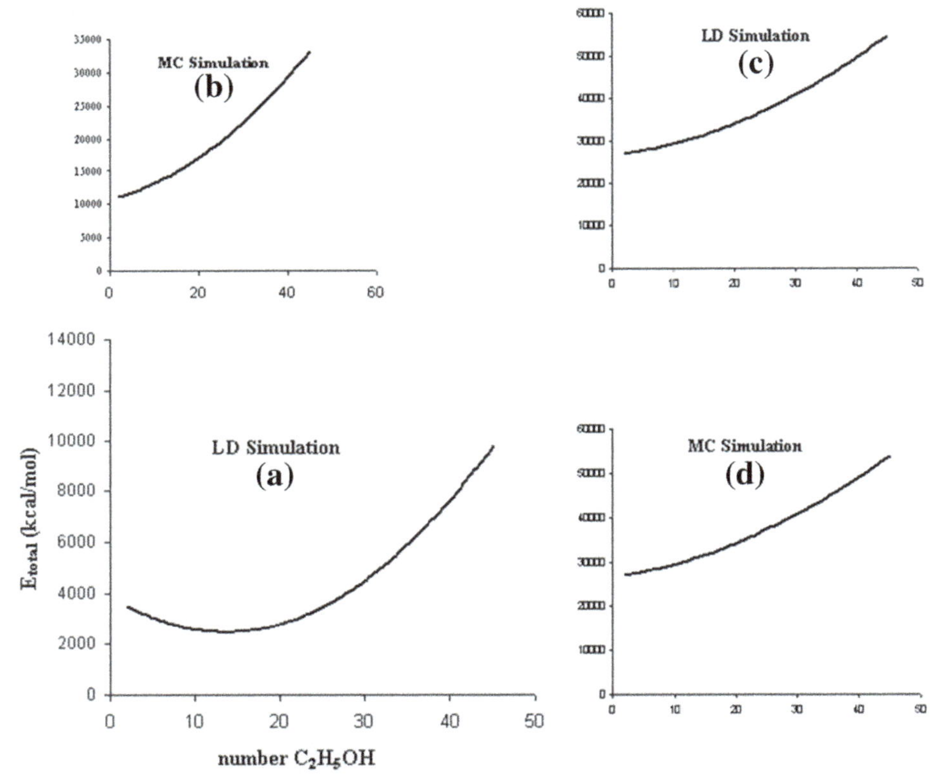

Figure 4 Total energy (kcal/mol) calculated for interaction of SnO_2 and ethanol molecules at 300 K. (a, b) Nanofilm (1.1.2) with chirality (3,3), **(c, d)** nanofilm (1.1.4) with chirality (3,3).

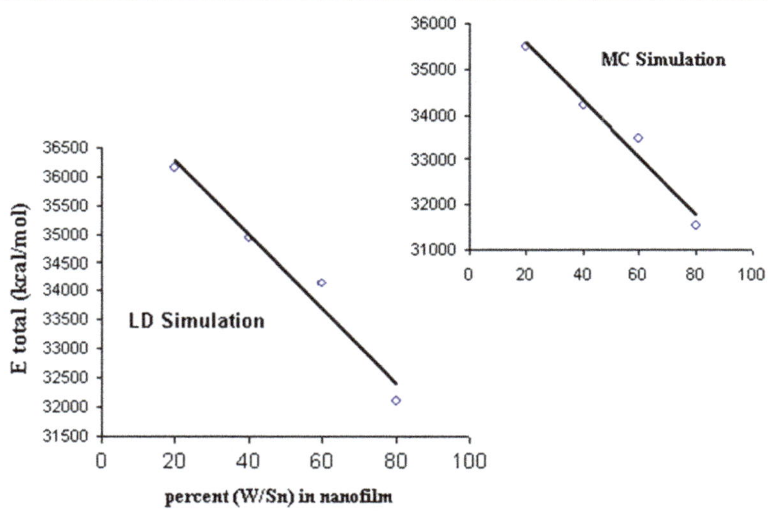

Figure 5 Total energy (kcal/mol) calculated for interaction of nanofilms and ethanol molecules at different percentages (W/Sn).

Competing interests

The author declares that he/she has no competing interests

Authors' contributions

The author did not provide this information.

References

1. Monajjemi, M, Mahdavian, L, Mollaamin, F, Honarparvar, B: Thermodynamic & electronic study of enol↔keto tautomerism for alcohol sensors based on carbon nanotube (CNT) as chemical sensors. Fuller. Nanotub. Car. N. **18**(1), 45–55 (2010)
2. Mahdavian, L, Monajjemi, M, Mangkorntong, N: Sensor response to alcohol and chemical mechanism of carbon nanotube gas sensors. Fuller. Nanotub. Car. N. **17**(5), 484–495 (2009)
3. Mahdavian, L: Thermodynamic study of alcohol on SnO_2 (100)-based gas nano-sensor. Physics and Chemistry of Liquids **49**(05), 626–638 (2011)
4. Paraguay-Delgado, F, Antunez-Flores, W, Miki-Yoshida, M, Aguilar-Elguezabal, A, Santiago, P, Diaz, R, Ascencio, JA: Microsc. Microanal. **10**, 340–351 (2004)
5. Pavesi, L, Dal Negro, L, Mazzoleni, C, Franzo, G, Priolo, F: Optical gain in silicon nanocrystals. Nature **408**, 440–448 (2000)
6. Comini, E, Guidi, V, Malagu, C, Martinelli, G, Pan, Z, Sberveglieri, G, Wang, ZL: Electrical properties of tin oxide two-dimensional nanostructures. J. Phys. Chem. B. **108**, 1882–1887 (2004)
7. Mahdavian, L: Highly sensitive SnO_2 (100) nano-crystal CH_3OH/C_2H_5OH gas sensor operating at different temperatures: Monte Carlo studies. J. Comput. Theor. Nanos. **8**(11), 2356–2361 (2011)
8. Batzill, M, Diebold, U: The surface and materials science of tin oxide. Prog. Surf. Sci. **79**, 47–154 (2005)
9. Hussain, M, Tajammul Hussain, S: Optical and gas sensing studies of transparent ZnO thin film deposited from a new precursor by ultrasonic aerosol assisted chemical vapor deposition. European Journal of Chemistry **1**(2), 96–101 (2010)
10. Banaś, J, Najbar, M, Tomašić, V: Kinetic investigation of the NO decomposition over V–O–W/Ti $(Sn)O_2$ catalyst. Catalysis Today **137**(2–4), 267–272 (2008)
11. Tsiakaras, PE: PtM/C (M = Sn, Ru, Pd, W) based anode direct ethanol–PEMFCs: structural characteristics and cell performance. J. Power Sources **171**(1), 107–112 (2007)
12. Wang, W, Skeel, RD: Analysis of a few numerical integration methods for the Langevin equation. Molecul. Phy. **101**(14), 2149–2156 (2003)
13. Batzill, M: Surface science studies of gas sensing materials: SnO_2. Sensors **6**, 1345–1366 (2006)
14. Jun, JM, Park, YH, Lee, CS: Characteristics of a metal-loaded SnO_2/WO_3 thick film gas sensor for detecting acetaldehyde gas. Bull. Korean Chem. Soc. **32**(6), 1865–1872 (2011)
15. Carotta, MC, Gherardi, S, Guidi, V, Malagu', C, Martinelli, G, Vendemiati, B, Sacerdoti, M, Ghiotti, G, Morandi, S, Bismuto, A, Maddalena, P, Setaro, A: $(Ti,Sn)O_2$ binary solid solutions for gas sensing: spectroscopic, optical and transport properties. Sensors and Actuators B: Chemical **130**(1), 38–45 (2008)
16. Skeel, SRD, Izaguirre, J: An impulse integrator for Langevin dynamics. Mol. Phys. **100**, 3885–3891 (2002)
17. YZh, S, Yi, L, Gao, YH: Thermodynamic and critical properties of dilute XY magnets: Monte Carlo study. Solid State Communications **149**(25–26), 1000–1003 (2009)
18. Jr, WAL, Mitas, L, Hammond, B: Quantum Monte Carlo for atoms, molecules and solids. Chemical Physics Letters **478**(1–3), 1–10 (2009)
19. Ripley, BD: Stochastic Simulation. Wiley, New York (1987)
20. Strecker, N, Moroz, V, Jaraiz, M: Introducing Monte Carlo diffusion simulation into TCAD tools. In: Technical Proceedings of the 2002 International Conference on Computational Nanoscience and Nanotechnology, vol. 2, p. 247. NSTI, Austin (2002)
21. Tian, B, Si, J, Zhao, Y, Wang, H, Hao, J: Approach of technical decision-making by element flow analysis and Monte-Carlo simulation of municipal solid waste stream. J. Environ. Sci. **19**(5), 633–640 (2007)
22. Mahdavian, L, Monajjemi, M: Alcohol sensors based on SWNT as chemical sensors: Monte Carlo and Langevin dynamics simulation. Microelectronics Journal **41**(2–3), 142–149 (2010)
23. Sun, S, Yin, G, Lee, Y, Wong, JTY, Zhang, TY: Effects of deformability and thermal motion of lipid membrane on electroporation: by molecular dynamics simulations. Biochem. Biophys. Res. Commun. **404**(2), 684–688 (2011)

2

Synthesis and nucleation-growth mechanism of almost catalyst-free carbon nanotubes grown from Fe-filled sphere-like graphene-shell surface

Muhammad Salleh Shamsudin[1,2,3*], Maryam Mohammad[1,2], Syazwan Afif Mohd Zobir[1,2], Noor Asnida Asli[1,2], Suriani Abu Bakar[5], Saifollah Abdullah[2], Syed Yusainee Syed Yahya[2] and Mohamad Rusop Mahmood[1,4]

Abstract

This finding focuses on the optimization of synthesis time for the transformation of Fe-filled spherical-like graphene shell (GS) to almost catalyst-free carbon nanotube (CNT) structure using two-stage catalytic chemical vapor deposition apparatus. The camphor oil and ferrocene were used as carbon precursor and catalyst respectively, following the variety growth of graphene-family nanomaterials for 2, 4, 6, 8, 10, 30, and 60 min at 800°C synthesis temperature. The graphene-family nanomaterial properties were characterized using field emission scanning electron microscope, high resolution transmission electron microscope, micro-Raman spectrometer, thermogravimetric, and carbon-hydrogen-nitrogen-sulfur/oxygen (CHNS/O) analyzer. The result of field emission scanning electron microscopy analysis reveals that the CNTs were observed with high aspect ratio at 60-min synthesis time. The dependence of integrated intensity ratio of D-band and G-band (I_D/I_G) presented that I_D/I_G ratio sharply decreases with longer synthesis time. At higher synthesis time, thermogravimetric and CHNS/O analysis of CNT can obviously improve with decreases of non-carbonaceous material and transition metal catalyst. The nucleation-growth model of Fe-filled spherical-like GS to almost catalyst-free CNT has been highlighted to explain the change in growth mode.

Keywords: Carbon nanotube, Graphene shell, Nucleation-growth mechanism, Catalyst, Camphor

Background

Carbon nanotubes (CNTs) play a large part in a new dimension into the knowledge of carbon science. Almost every day, new conceptual and experimental framework for the goals of CNT is identified [1-3]. Compared with more conventional materials, this carbon allotrope has extraordinary physical and chemical properties. In 1990s, single-walled CNT (SWCNT) was discovered by two different groups: (1) Bethune et al. from IBM, California, (2) and Iijima and Ichihashi who were affiliated with NEC Corporation, Japan, and made their own route of CNT synthesis [4,5]. However, the problem arises when Monthioux and Kuznetsov [6] raised this question few years ago, 'why did the scientific community seem to discover CNT in 1991 when they had actually been known for forty years?'. We believe this major issue is quite interesting, and one that we will never know. Oberlin et al. [7] unknowingly discovered SWCNT of a diameter less than 100 Å in 1976; however, they did not fully recognize the discovery and described their SWCNT as a carbon filament. At that time, journalists, industrialists, and the scientific community were not concerned with nano level structure. Anticipating a growing demand, large volumes of CNT seem to be in progress and ready for commercialization. For this reason, we could see that valuable outputs from CNT-based products will be available in the markets [8-15].

In recent years, there are many fundamental issues concerning CNT growth patterns which are still not clear. From a scientific point of view, there is an immediate challenge to produce CNT of the desired properties to suit

* Correspondence: M.S.Shamsudin@soton.ac.uk
[1]Centre of Nanoscience and Nanotechnology (NANO-SciTech Centre), Institute of Science, Universiti Teknologi MARA (UiTM), Shah Alam, Selangor Darul Ehsan 40450, Malaysia
[2]School of Physics and Material Studies, Faculty of Applied Sciences, Universiti Teknologi MARA (UiTM), Shah Alam, Selangor Darul Ehsan 40450, Malaysia
Full list of author information is available at the end of the article

individual applications. In this experiment, the effect of synthesis time (as short as 2 min) on the formation of graphene-family nanomaterials grown on transition metal catalyst without the use of a substrate was investigated. One possible explanation is the important role of synthesis time in the formation of CNT. Our modified two-stage catalytic chemical vapor deposition (CVD) apparatus is simple and economic equipment for high-quality formation of CNT. This synthesis method facilitates an improvement of the nucleation-growth mechanism of carbon-based material without using substrate. Detailed information will be discussed in the following sections.

Results and discussion

Surface morphology

Figure 1 illustrates the FESEM micrographs of the carbon black in powder form and shows the structural variation of low-dimensional graphene-family nanomaterial

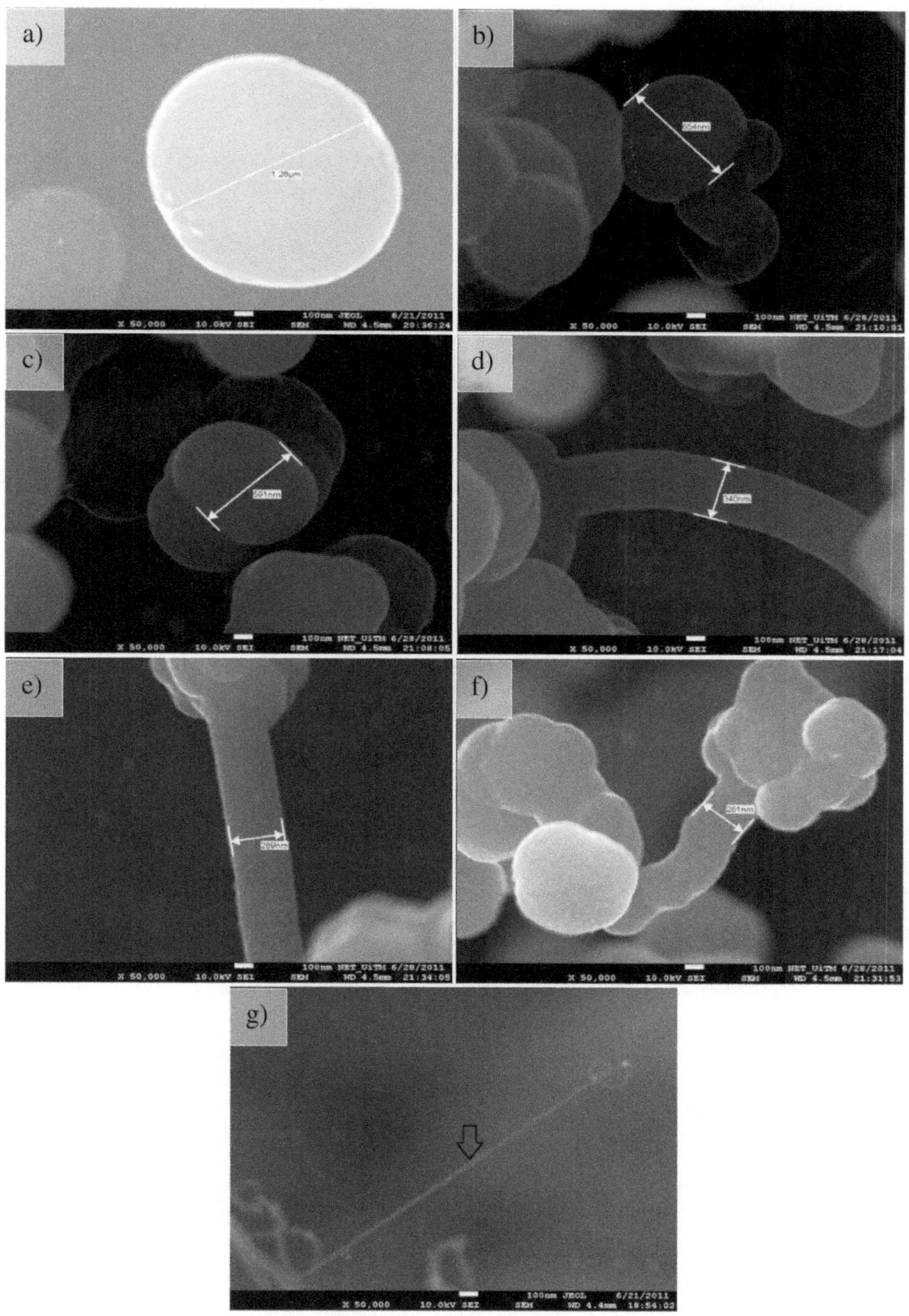

Figure 1 FESEM micrographs of graphene-family nanomaterial in powder form. (**a**) 2, (**b**) 4, (**c**) 6,(**d**) 8, (**e**) 10, (**f**) 30, and (**g**) 60 min.

influenced by the effect of synthesis time of 2, 4, 6, 8, 10, 30, and 60 min. In 2, 4, and 6 min (Figure 1a,b,c), it is shown that the appearance of the microsphere structure of the samples might be consisting of either partially amorphous carbon (a-C) or graphene shell [16,17]. The size of the microsphere structure is clearly seen decreasing with synthesis time. The diameter of the microsphere structure is approximately 1.26 μm, 654 nm, and 591 nm as shown in Figure 1a,b,c, respectively. However, after the synthesis time was further increased (Figure 1d,e,f,g), the microsphere structure vanished from the surface morphology of the sample. As synthesis time is increased from 8 to 60 min (as shown in Figure 1d,e,f,g), a tubular structure on the top surface of the carbon microsphere was formed. It is clearly observed that the diameter of the tubular structure decreases with the increase in synthesis time, resulting in narrower tubular structure. This result is in excellent agreement with earlier findings [18] and could be an additional information on the growth mechanism of carbon-based materials with respect to synthesis time. In a separate report, we have similar findings with those in the study by Kumar and Ando [19], where carbon microspheres consist of carbon and Fe elements as evident in Figures 1a and 2a. Based on the EDX spectrum, it is confirmed that the carbon was the dominant element, and small amounts of Fe were found in the sample. At longer synthesis times, the carbon microsphere diameter decreases which may be attributed to the catalytic chemical reaction. This suggests that up to the synthesis time of 8 min (as in Figure 1d), the nucleation energy is strong enough to break the outer surface of the carbon microsphere; thus, as shown diagrammatically in Figure 3, (1) the activated Fe catalyst leaves the core area of the carbon microsphere and (2) synthesizes a tubular structure on top of the carbon microsphere surface. Broadly speaking, substrate-catalyst interaction plays an important role in the formation of CNT. It can be either 'base- or tip-growth model' [20]. However, Kumar and Ando [19] failed to prove the nucleation-growth mechanism of the free-substrate CNT. It has been observed that the free-substrate CNT in this work demonstrates a new growth mechanism of CNT, supplementary to the previously studied tip- or base-growth model. In the next discussion, we try to promote extended nucleation-growth mechanism outline, which leads to the formation of free-substrate CNT, known as 'root-growth' mechanism for CNT. An optimal synthesis time for CNT formation in our experimental is 60 min due to (1) narrower diameter with higher aspect ratio of nanotubes and (2) less a-C produced. As presented in Figure 2b, the EDX spectrum of the as-grown CNT at 60-min synthesis time clearly indicates that 99.9 at.% of carbon element is present. From Figures 2b and 4, less a-C and Fe catalyst left after 60 min, producing 99.9 at.% of pure CNT. No characteristic peaks of impurities can be detected.

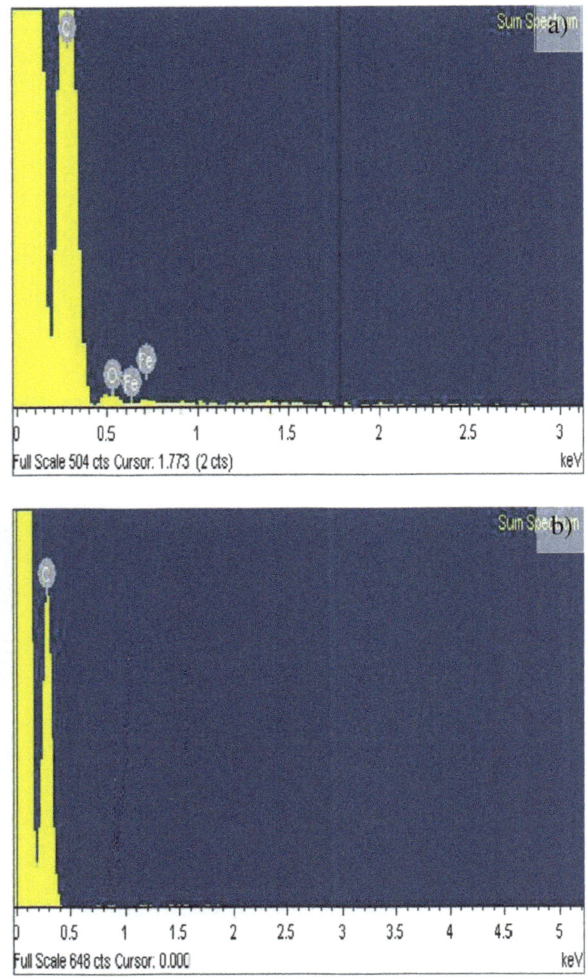

Figure 2 EDX spectrum of as-grown carbon powder. (**a**) Fe-filled spherical-like GS at 2-min synthesis time and (**b**) CNT at 60-min synthesis time.

TG analysis

Normally, TG analysis is used to estimate the purity of the sample in terms of carbonaceous material and metal catalyst present. A thermogram of the weight loss of the as-grown free-standing carbon black powder on the ferrocene synthesized at 800°C with different synthesis times is presented in Figure 4. In this experimental work, the TG analysis was performed in ambient oxygen, with a temperature increase of 20°C/min. For every case, TG analysis was done up to 1,000°C to evaluate the thermal stability and purity of the carbon black powder. We found that almost the same sharp decline in weight loss was consistent in the range of 500°C to 600°C (single-step decomposition of carbon black powder); this can be attributed to the oxidation behavior of carbonaceous material, including GS, SWCNT, and multi-walled CNT (MWCNT). It is surprising that the oxidation temperature and thermal stability of our best sample are much higher and lower, respectively,

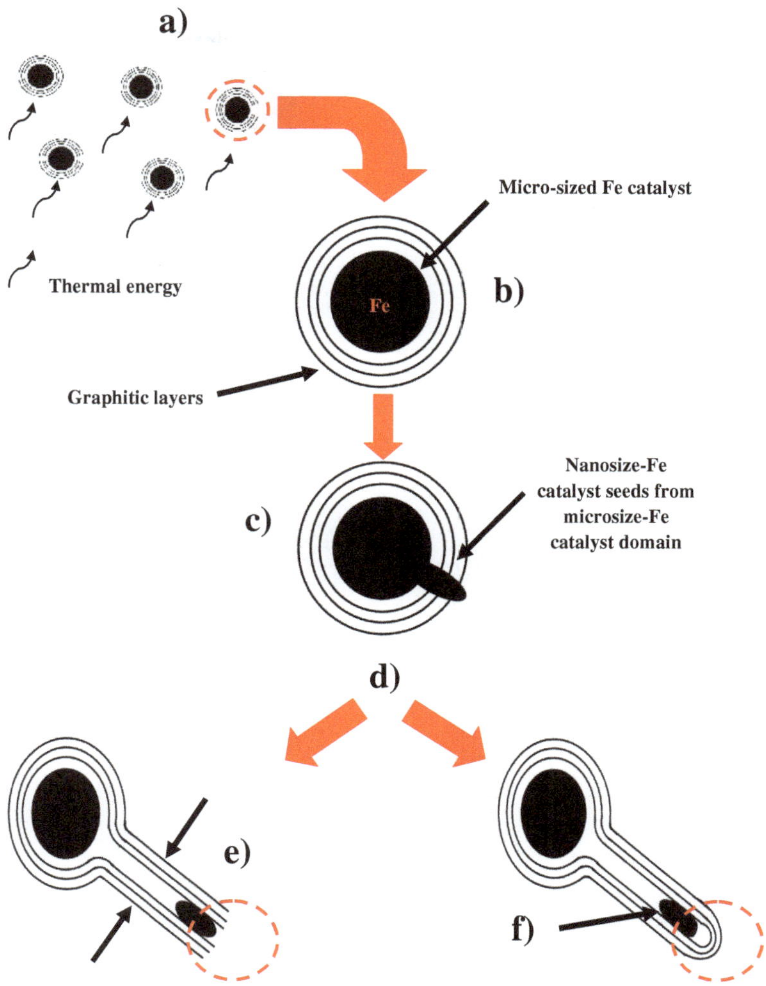

Figure 3 Possible growth mechanism for the formation of Fe-filled spherical GS to almost catalyst-free CNT structure. A two-stage catalytic CVD was used. (**a**) Formation of Fe-filled sphere-like GS, (**b**) a Fe-filled sphere-like GS consists of graphitic layers and reduced micro-size Fe catalyst, (**c**) activation of catalytic chemical reaction, (**d**) possibilities for re-ordering of carbon structure, (**e**) formation of nanotube structure and (**f**) formation of Fe nanoparticle catalyst.

compared to those of previous research works by Scheibe et al. [21] and Antunes et al. [22]. As in Figure 4, the purity and the residual weight of the crystalline carbon black powder and Fe catalyst respectively can be estimated by comparing the base line and the residual line. For 2, 4, 6, 8, 10, and 30-min synthesis time, the residual weights of the Fe catalyst were 7.81%, 5.65%, 5.10%, 4.73%, 4.64%, and 4.25% which imply that 92.19%, 94.35%, 94.9%, 95.27%, 95.36%, and 95.75% of the mass product, respectively, came from carbon. Under this condition, the highest CNT purity achieved is about 99.99% for 60-min synthesis time. To our knowledge, this is the first time that 99.9% purity CNT was treated using low temperature (approximately 800°C). Previous work by Huang et al. [23] reported 99.99% purity MWCNT with propylene on Fe-Mo-Al_2O_3 catalyst treated at higher temperature between 1,500°C and 2,150°C. It is observed that the oxidation temperature, purity, and thermal stability of CNT are dependent on the synthesis time of CNT sample preparation. In order to maintain the optimal performance of the CNT, high purity CNT is needed. This conclusion is consistent with the observations from the EDX analysis. The presence of impurities may significantly influence the behaviors of the CNT in existing and wide area of new application [24,25].

Micro-Raman analysis

The micro-Raman spectra obtained at different synthesis time are shown in Figure 5a,b, with radial breath mode (RBM) at low wave number regions between 100 and 500 cm^{-1} [26] and tangential mode at higher wave number regions between 1,000 and 2,000 cm^{-1} [27] observed at room temperature. Here, the authors concern

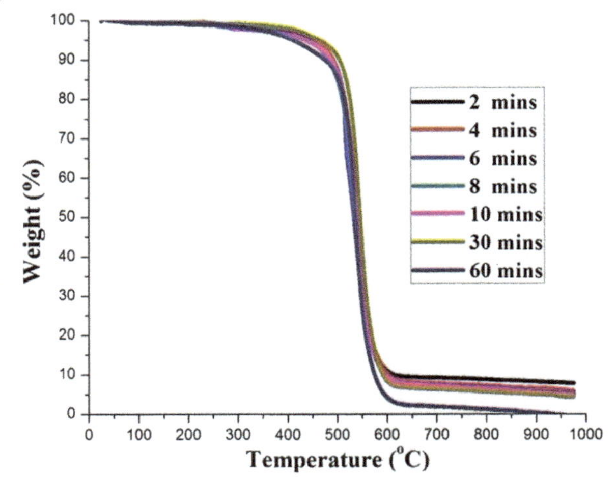

Figure 4 TGA patterns of as-grown free-standing carbon black powder at different synthesis times.

is on the quality of the graphene-family nanomaterials. From the micro-Raman analysis (results are shown in Figure 5a), obvious RBM peaks corresponding to the existence of SWCNT are present. According to Bandon's equation, the diameter (*d*, in nanometers) of the SWCNT can be estimated from the sharp RBM peak position, ω_{RBM} where d = 223.75 (cm^{-1})/ω_{RBM} (cm^{-1}). We notice that in the micro-Raman spectra (presented in Figure 5a), there were two ω_{RBM} peaks originating from 210.52 and 273.95 cm^{-1} (60-min synthesis time). This corresponds to the tube diameters of 1.063 and 0.897 nm for 60-min synthesis time. There is a narrow diameter distribution in SWCNT produced here. No growth of SWCNT was obtained at 2, 4, 6, 8, 10, and 30 min as there is insufficient nucleation energy of Fe to efficiently catalyze the hydrocarbon decomposition. According to the micro-Raman result as shown in Figure 5a, longer synthesis time was found to produce CNTs with narrower diameter which is associated with large specific surface area availability. This characteristic is very important to any application because their performance depends on the surface reaction process [28]. are two prevalent peaks in the micro-Raman spectrum of the graphitic mode (G-peak) and the disorder mode (D-peak) which are normally at 1,580 and 1,350 cm^{-1} in CNT, respectively. The G-peak is attributed to the ordered graphite structure, while the D-peak corresponds to non-crystal structure consisting of defects, dislocations, and lattice distortions in the carbon structure. As shown in Figure 5b, the G- and D-peak positions for 2, 4, 6, 8, 10, 30 and 60 min are always consistent at the same wave numbers (1,580 and 1,360 cm^{-1}, respectively). The graphitization of graphene-family nanomaterial can be obtained by comparing the relative integrated intensity of D-peak to G-peak (I_D/I_G). The lower the I_D/I_G value,

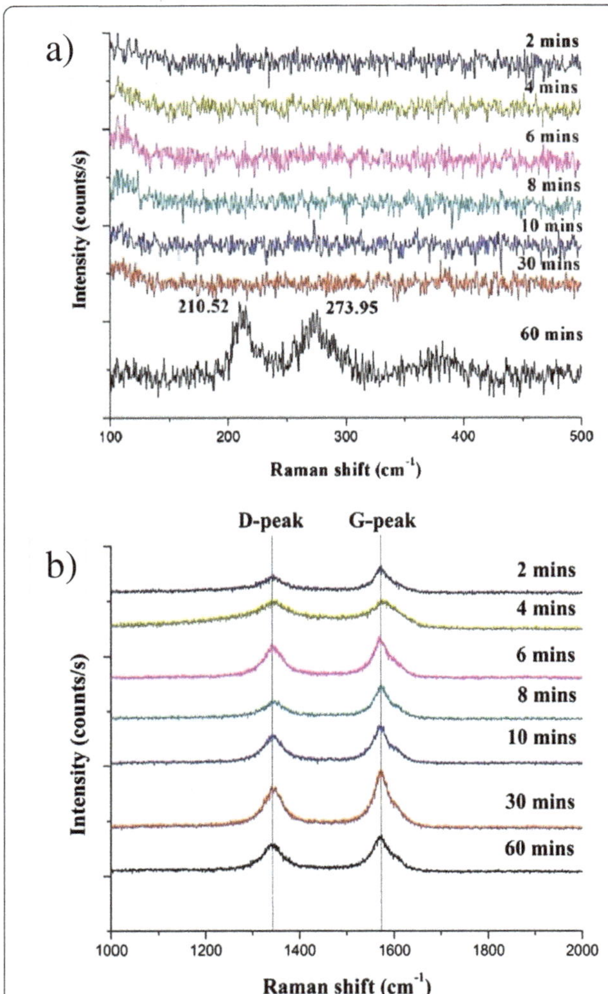

Figure 5 Micro-Raman patterns. Graphitic network at (**a**) lower and (**b**) higher frequencies (I_D/I_G ratio) for different synthesis times.

the higher is the degree of graphitization. The I_D/I_G is used also as a purity index to assess the quality of CNT with minor defect and with very low a-C content [29]. The lowest I_D/I_G ratio was found to be 0.791 for 60-min synthesis time, while the I_D/I_G values of the samples synthesized for 2, 4, 6, 8, 10, and 30 min (Figure 5b) were almost 1 (Table 1).

HRTEM analysis

The theoretical- and experimental-based understanding of the growth mechanisms of CNT would not have been possible without the use of TEM, officially invented by Siemens in 1939. TEM has enabled the confirmation of the crystallographic and morphologic structures of a variety of carbon-based materials. It is important to point out that in 1958, Radushkevich and Lukyanovich [30] highlighted the wealth of structural information that was possible from HRTEM, making this technique very

Table 1 Synthesis time, I_D peak position, I_G peak position, and I_D/I_G ratio

Synthesis time (minutes)	I_D peak position, ω_D (cm^{-1})	I_G peak position, ω_G (cm^{-1})	I_D/I_G ratio
2	1,360	1,580	0.998
4	1,360	1,580	0.972
6	1,360	1,580	0.968
8	1,360	1,580	0.944
10	1,360	1,580	0.914
30	1,360	1,580	0.830
60	1,360	1,580	0.791

interesting in the characterization of nanotubular-sized carbon growth, as it can be done *in situ*, allowing for direct observation in the characterized sample. In our experimental analysis, we hypothesize that the carbon microspheres and the nanotubes tested were of solid carbon structures. It was difficult to identify whether the cores of the carbon microsphere and CNT were hollow or solid. Based on a previous discussion on FESEM analysis, there are inherent limitations in FESEM analysis; this is only suitable for the analysis of the outer surface of the material, as its magnification is too low for the crystal structure to be observed in atomic level. Moreover, the analyzed sample could be either in a crystalline or amorphous phase. However, further investigation is needed to confirm the crystallographic structure of carbon microspheres and nanotubular structures.

- The important consideration is how to ensure (1) whether the nanofibers are nanotubes or not, and (2) whether the solid microspheres are core-shell microspheres or not, i.e., how to make them hollow-like structures.
- The other issue is how to ensure whether the crystal growth of CNT is in good agreement or not with the 'anisotropy of carbon deposition theory'.

Figure 6a displays a typical low magnification HRTEM image of a dispersed carbon microsphere and small tubular structure. The result indicates a clear spherical GS and CNT. As shown in Figure 6a,b, two types of materials can be clearly distinguished. The black spot represents the Fe nanoparticles (pointed with a black arrow). The electron beam cannot penetrate through the Fe element; therefore, it appears as a darker black color. The areas filled by the carbon element are semi-transparent; therefore, they appear gray. However, some studies suggest that those areas are Fe nanoparticles encapsulated by graphitic layers. This is supported by Figure 7b (inset of Figure 7a); this figure clearly shows that the nanotube is a multi-walled centrally hollow

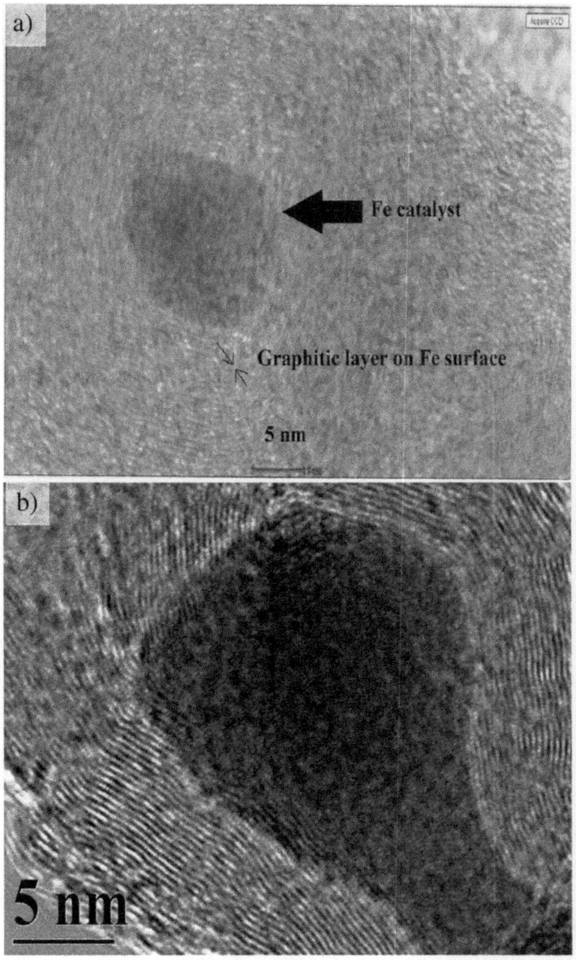

Figure 6 HRTEM images of Fe-filled sphere-like GS. At (**a**) low magnification and (**b**) high magnification.

tube, not a solid fiber. The lack of fringes (represent as individual cylindrical graphitic layers) inside the nanotube, as well as the lighter contrast as compared to the nanotube walls, is an evidence that the core of structure is hollow. The closing-cap CNT has been observed in the sample as shown in Figure 7c.

HS-SPME/GC-MS analysis

Previous studies [31] have used camphor oil as hydrocarbon precursor in CNT growth; however, such studies have not described the flavor volatile compounds. Therefore, HS-SPME/GC-MS analysis was undertaken as part of this study, as shown in Table 2. A total of 28 components were identified. This investigation has revealed further evidence of chromatogram structure for camphor oil. This consideration emphasizes the potential value of the nanostructured carbon material before it is submitted for the synthesis process.

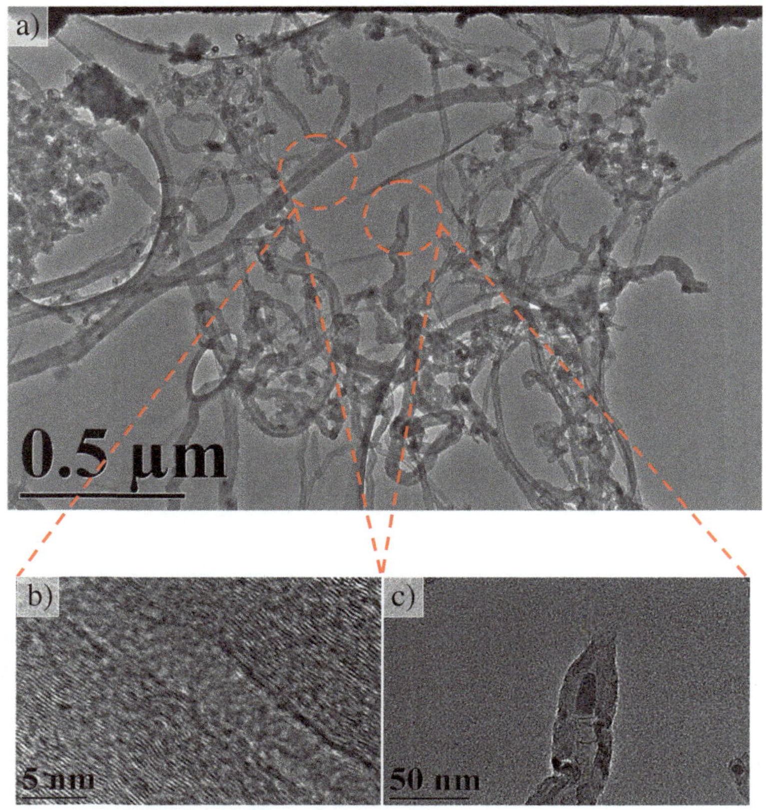

Figure 7 HRTEM images of the CNT. (**b**) and (**c**) are insets of (**a**).

Fe-filled sphere-like GS to almost catalyst-free CNT nucleation-growth mechanism: the possibilities are endless

Theoretically, the mystery of carbon growth at nanoscale is not completely solved [32]. A simple nucleation-growth mechanism of Fe-filled sphere-like GS to almost catalyst-free CNT was carried out to overcome some major experimental stumbling blocks. Our aim is to determine how the nanotube-like structure was grown from GS surfaces. The nature of camphor oil ($C_{10}H_{16}O$) itself is a monoterpene, a molecule that is built up biosynthetically from two isoprene units which obey the *isoprene rule*, $(C_5H_8)_n$, where n is the number of linked isoprene units. Interestingly, a previous study on camphor reported by Kareem [33] showed that the decomposition end-product sublimated from solid to volatile material at ordinary room temperature, and ambient air was calculated by fractional life method. Kumar and Ando [34,35] reported the mass production of CNT which was synthesized from botanical camphor oil rather than conventional hydrocarbon precursors. Unlike any other precursors, camphor oil is a non-toxic, reproducible precursor and is abundantly grown in Asian countries. Hence, the hydrocarbon precursor decomposition study is strongly dependent on environmental conditions [36] as evaluated using suitable analytical tools attached with a mass spectrometer [37] to extract derivative compounds from hydrocarbon precursor. This could explain why it is important to observe the preliminary study of kinetics of hydrocarbon precursor decomposition in selected gas ambient before proceeding to the selection of synthesis time. Bahrami et al. [38] reported and proposed the synthesis and nucleation-growth mechanism of CNT on Fe/MgO catalyst at 900°C to 1,000°C for 1 to 60 min. They found that Fe_xO_y was coated on MgO surface and reduced *in situ* in hydrocarbon precursor to Fe nanoparticles. Indirectly, tube-like structure was grown on Fe nanoparticles. This phenomenon can be either tip- or base-growth mechanism of CNT as supported by Charlier et al. [39].

Based on our understanding of the FESEM, EDX, micro-Raman, TG, and HRTEM analyses, the possibility of a nucleation-growth model of Fe-filled sphere-like GS to almost catalyst-free CNT is proposed as in Figure 3 as follows: A selected area in a large group of sphere-like GS structure was observed using HRTEM during exposure to heat under nitrogen ambient at 800°C under varying synthesis times (Figure 3a). The shape and diameter size of the Fe-graphene shell-CNT influenced by the synthesis time were investigated. Since the nature of the

Table 2 Chemical composition of camphor oil examined by HS-SPME/GC-MS analysis

Peak number	Retention time (min)	Compound	Amount (%)
1	4.66	α-Pinene	15.43
2	4.87	Camphene	0.89
3	5.38	Bicyclo [3.1.1] heptanes, 6, 6-dimethyl 1–2 methylene-	12.21
4	5.50	β-Pinene	2.54
5	5.81	α-Phellandrene	3.38
6	6.25	Eucalyptol	35.21
7	6.73	1, 4-Cyclohexadiene, 1-methyl-4-(1-methylethyl)-	2.36
8	6.89	Cyclopentasiloxane, decamethyl-	0.12
9	7.12	Cyclohexene, 1-methyl-4-(1-methylethylidene)-	0.67
10	7.31	1, 6-Octadien-3-ol, 3, 7-dimethyl-	0.94
11	7.71	2, 4, 6-Octatriene, 2, 6-dimethyl-	0.55
12	7.81	Cyclopentasiloxane, decamethyl-	0.25
13	8.12	Camphor	10.01
14	8.33	Borneol	0.35
15	8.49	3-Cyclohexen-1-ol, 4-methyl-1-(1-methylethyl)-	0.47
16	8.68	*p*-Menth-1-en-8-ol	0.42
17	8.77	Borane, diethyl (1-ethyl-1, 2-butadienyl)-	0.18
18	8.85	Naphthalene, 1, 2, 3, 4, 4a, 8a-hexahydro-	0.13
19	9.07	5-(4-Bromo-phenylcarbamoyl)-3-methyl-3H-imidazole-4-carboxylic acid	0.11
20	9.38	2-Butyn-1-ol	0.31
21	9.45	2-Cyclohexen-1-one, 2-methyl-5-(1-methylethenyl)-	0.1
22	9.79	2, 6-Octadienal, 3, 7-dimethyl-	0.23
23	10.48	Cyclohexasiloxane, dodecamethyl-	4.07
24	12.72	Cycloheptasiloxane, tetradecamethyl-	4.94
25	13.31	2(1H)-Pyrimidinone, 5-(4-methylphenoxy)-4-(4-nitrophenyl)-6-phenyl-	0.23
26	13.59	3, 7-Bis[(trimethylsilyl)oxy]-9-methoxy-1-methyl(6H)dibenzo[b, d]pyran-6-one	0.18
27	15.23	Phenethylamine, N-methyl-β,3,4-tris(trimethylsiloxy)-	2.28
28	19.17	Cyclohexasiloxane, octadecamethyl-	1.44

ferrocene compound consists of one iron atom, ten carbon atoms, and ten hydrogen atoms, in the first few minutes, cyclopentadienyl rings bound to the opposite sides of a central Fe atom break and decompose into free micro-sized Fe atomic cluster. At the same time, micro-sized Fe atomic cluster acts as a catalyst, promoting further growth on micro-sized Fe structure. Encapsulation of few-layer graphene consisting of free-carbon network due to the decomposition of camphor oil on micro-sized Fe structure surface was observed, as shown in Figure 3b. However, recently, Schaper and co-workers reported that it was a formation of intermediate iron carbide (Fe_3C) determined by the dissolution of carbon element in the quasi-liquid catalyst particle and its segregation in the form of graphitic layers in the case of super-saturation [40]. Recently, it was supported by Fedoseeva et al. [41] who also reported on the synthesis of defect-like carbon (also known as iron-containing rods covered by graphite-like shells and embedded into carbon matrix) by air oxidation and carbonization methods. As in Figures 6b and 3c, we could observe a portion of the nano-size Fe catalyst seeds from micro-size Fe catalyst domain. Thereby, the axial movements of nano-size Fe catalyst inside the sphere-like GS is due to the buildup competition between the interfacial force and the expulsive force between the graphitized tube walls and core material (micro-size Fe catalyst), respectively. On the other hand, during these structural transformations, the outer CNT diameter size remained almost constant due to compressive stress during the thickness growth of the graphitic tube walls. Ren et al. [42] proposed the thinner catalyst layer, producing thinner diameter size nanotubes. The length of a nanotube can be easily tailored by simply changing the synthesis time. Previous findings, such as those of Xiang et al. [43,44], have theoretically and experimentally confirmed

Figure 8 Schematic diagram and optical photograph. (**a**) The schematic diagram of two-stage catalytic CVD and (**b**) the optical photograph of the as-grown free-standing carbon black powder consisting of Fe-filled spherical-like GS or CNT.

that higher synthesis time is beneficial with regard to growing longer CNT. Reit et al. [45] reported the length of nanotube was slightly increased as synthesis time was increased. As a result, the electrochemical performance of the CNT super-capacitor is dependent on synthesis time. One possible explanation why the synthesis time could affect the length of nanotube is presented here. The CNT growth increases rapidly once a carbon atom becomes sufficiently saturated onto the catalyst surface as illustrated in Figure 1g. At current stage, it is found that our samples are free of camphor-feedstock diffusion resistance and Fe catalyst deactivation (i.e., causes the growth to decelerate), allowing CNT to grow on sphere-like GS surface. However, further investigation is needed to predict the critical length of CNT in different synthesis parameters, subjecting the CNT array to strong diffusion resistance. Moreover, further investigation is needed to demonstrate the possibility of cap-opening and cap-closing mechanisms as referred to in Figure 3e,f. Finally, this mechanism demonstrates how nanotube growth is being made and supports the validity of the concept of carbon-through-metal diffusion and of the dissolution of carbon element in metal catalyst.

Conclusion

In summary, we have presented a few explanations of camphor-grown sphere-like GS and CNT using a practical method (two-stage catalytic CVD) for various synthesis times. It is shown that the 60-min synthesis time is the ideal time in controlling the phase of structure graphitization, gives highest aspect ratio, renders the lowest integrated intensity (I_D/I_G) ratio and diameter size distribution, and provides excellent purity and thermal stability of CNT. It is reasonable to believe that growing high-quality CNT using two-stage catalytic CVD has great promises, which is more economical, have simple operating conditions, and reproducibility. However, to understand the actual growth mechanism inside the system, a future investigation optimizing the growth parameters is needed. Through this, it is hoped that these desirable properties of CNT will be recommended for existing and wide range of new applications.

Methods

The synthesis of CNT using two-stage catalytic CVD is very attractive because of its simplicity. The experimental details of the synthesis of CNT were described. Briefly, in this work, camphor oil has been explored (obtained from SAFC (Steinheim, Germany) with 99.8% purity) as a hydrocarbon source [46] to investigate the changes of structural and thermal properties due to synthesis time in two-stage catalytic CVD. The ferrocene (obtained from Sigma-Aldrich (MO, USA) with 98% purity) was used as a catalyst without pre-treatment. The optimized condition 10% *w*/*v* (weight-volume percentage) of ferrocene relative to camphor oil was placed in different alumina boats [47]. The alumina boats were positioned side by side along a quartz tube in the first furnace of the two-stage catalytic CVD apparatus. In this approach, we appraise the effect of short synthesis time (2, 4, 6, 8, 10, 30, and 60 min) which was systematically studied with fixed growth temperature around 800°C [48], followed by thermal annealing process for 30 min through a nitrogen gas ambient [49] in the second furnace. The experimental setup in Figure 8a is similar to that reported earlier from our group [50]. The process involved as-synthesized titanium dioxide (TiO_2) which was used as a catalyst prepared through a sol–gel method. However, in this process, the organometallic compound (ferrocene) was used as a catalyst. Under these conditions, sublimation and vaporization occurred on ferrocene and camphor oil, respectively, around 180°C in the first furnace. It has been observed that a 'mat-like' structure of carbon black powder grew inside the quartz tube of the second furnace; this was cooled down to room temperature. The as-grown free-standing carbon black powder was collected from the quartz tube as shown in Figure 8b and ready for characterization.

The characterization of the as-grown free-standing carbon black powder was carried out by field emission scanning electron microscopy (FESEM; JSM-7600-F, JEOL Ltd., Tokyo, Japan) operating at 10-kV electron high tension coupled to an energy dispersive X-ray (EDX). The protocol used for the sample preparation prior to FESEM characterization of each sample is as follows: The lint-free glove was used when touching the sample, sample holder, or stage to avoid contaminant and moisture. Small amounts of as-grown free-standing carbon black powder were mounted using carbon tape (carbon tape sticks on the top surface of the sample holder). Since our sample has good electrical properties, it was not coated with gold or platinum. An EDX analysis was based on dot mapping, which depends on the sample structure, to investigate the elemental analysis of the sample.

The diameter distribution, crystallinity, and impurities of the CNT were characterized by micro-Raman spectrometer (HR 800, Horiba Jobin Yvon, Edison, NJ, USA) at room temperature with Ar^+ ion laser line (λ = approximately 514.532 nm) wavelength source and 20-mW power. In our facility, micro-Raman spectroscopy requires minimal sample handling and preparation. We have examined small amounts of carbon black powder mounted on to glass slide and secured with slide holder. The focal point with ×100 magnification is the area where the specific part of the sample will be analyzed. Three measurements per sample were taken across it to check the homogeneity of the sample. The data have been computed with the help of the LabSpec® software (version 5.45.09; Hamilton, New Zealand).

Further information on the as-grown free-standing carbon black powder structure have been obtained using a thermogravimetric analyzer (TGA; PE Pyris 1 TGA Thermo Balance, Perkin Elmer Inc., Waltham, MA, USA) with a heating rate of 20°C/min under ambient oxygen and a CHNS/O analyzer (PE 2400 CHNS/O Elemental Analyzer) heated at 975°C with constant flow of helium stream and enriched with oxygen of 99.995% purity to evaluate significant synthesis time-dependent thermal change.

Micrographs were obtained using high-resolution transmission electron microscope (HRTEM; FEI, Technai G2 20S Twin) operating at an accelerating voltage of 200 kV. The sample was suspended in an ethanol-methanol mixture sonicated for 15 min. Using a pipette, a droplet of the suspension was taken, placed onto a holey carbon-coated 400 mesh copper grid, and dried overnight.

Finally, the qualitative and quantitative analyses of the camphor oil were examined by gas chromatography–mass spectrometry (GC-MS). Prior to this study, headspace solid-phase micro extraction (HS-SPME) technique was used in order to collect the flavor volatiles of the camphor oil. GC separation and identification were performed on a GC system coupled with a mass detector (6890 N Network GC System, Agilent Technologies Inc., Santa Clara, CA, USA), and the mass spectra of the extracted volatile compounds were computed using the normalization method through the ChemStation® Data Analysis software.

Competing interests
The authors declare that they have no competing interests.

Authors' contributions
MSS, MM, SAMZ, NAA, SAB, SA, SYSY, and MRM carried out the nanocarbon studies and participated in the experimental work on the growth and characterization of carbon nanotubes. All authors provided the same contributions in this article. All authors read and approved the final manuscript.

Authors' information
MSS is currently a graduate research assistant at the University of Southampton Malaysia Campus (USMC), where his studies focus on the growth of graphene. He is also one of the members in the Centre of Nanoscience and Nanotechnology, Universiti Teknologi MARA, where his research interest involves the synthesis and characterization of carbon nanotubes. He received a BSc degree in Physics from the Universiti Teknologi MARA, Malaysia in 2010. He

has authored and co-authored over 20 peer-reviewed papers in journals and conference proceedings in the field of nanostructured carbon materials including *Advances in Condensed Matter Physics*, *Journal of Spectroscopy*, *Journal of Nanomaterials*, and *Nano Hybrids*. He also serves as a reviewer for several national and international journals. SAMZ is currently a postgraduate student at the Universiti Teknologi MARA, where his research expertise is on the chemistry of carbon-carbon hybrids and oxide materials. He has published outstanding articles in several high reputation journals such as *Carbon*, *Materials Letters*, *Journal of Nanomaterials*, and *Materials Chemistry and Physics* related to the research area. MM joined the Centre of Nanoscience and Nanotechnology, Universiti Teknologi MARA in 2009 and currently is a postgraduate student under the supervision of Professor Mohamad Rusop. She holds a BSc in Physics from the Universiti Teknologi MARA in 2010. Her primary research interest focuses on the mechanistic study of carbon nanotubes from palm olein as a precursor. NAA is a postgraduate student in nanocarbon group at the Centre of Nanoscience and Nanotechnology, Universiti Teknologi MARA. Her research includes nanotube-porous silicon interface problems associated with the fabrication of field electron emission device. NAA has a BSc degree in Materials Technology from the same institution (2009). She is the author and co-author of more than 13 papers in refereed journals and conference proceedings including *Ionics* and *Microelectronic Engineering*. SAB is the head in the Nanotechnology Research Centre, Universiti Pendidikan Sultan Idris. Since 2012, she is also one of associate members in the Centre of Nanoscience and Nanotechnology, Universiti Teknologi MARA. She earned a PhD degree in nanomaterials in 2011, MSc degree in Materials Physics in 2005, and BSc degree in Industrial Physics from the same institution in 2002. Her main research interest is in the field of low dimensional carbon materials such as amorphous carbon, carbon nanotubes, and graphene from promising renewable hydrocarbon feedstock. She is also active in research publication, and until now, she has authored and co-authored more than 40 articles including conference proceedings, journals, patents, and book chapters such as *Materials Letters*, *Japanese Journal of Applied Physics*, *Journal of the Ceramic Society of Japan*, and *Carbon and Oxide Nanostructures* (*Advanced Structured Materials*). In 2011, she was awarded with a silver medal at the *39th Exhibition of Inventions Geneva* organized by WIPO, Geneva. SA is a professor of Physics at the Faculty of Applied Sciences, Universiti Teknologi MARA since 2009. In 2004, he holds a PhD degree in nanostructured materials at the University of Malaya, where his study focuses on the fabrication of porous silicon nanostructure. Currently, he extends his research interest in the fabrication of silicon quantum dot nanoparticles and one-dimensional materials (e.g., TiO_2, ZnO, and carbon nanotubes). He is also active in research publication such as *Surface Review and Letters*, *Optical Materials*, and *Microelectronic Engineering*. SYSY is an associate professor of Physics at the Faculty of Applied Sciences, Universiti Teknologi MARA. In 2005, he obtained a PhD degree at the School of Applied Physics, National University of Malaysia. He is also one of council member in American Society of Non Destructive Testing, Malaysian Association of Solid State Science, and Malaysian Institute of Physics. He has published several papers in conference proceedings such as *Advanced Materials Research* and *American Institute of Physics Proceedings* related to the low-dimensional carbon materials and superconductor physics. MRM is the head of the Centre of Nanoscience and Nanotechnology and a research fellow in the Centre of Nanoelectronic. He received a PhD degree in opto-electronic devices and nanotechnology in 2003 from the Nagoya Institute of Technology, Japan. He is a professor in exceptional class at Universiti Teknologi MARA. His research focuses on the opto-electronic devices and nanotechnology. He is the editor or editorial advisory board member of five international journals. He has published over 9 patents and 400 research papers in peer-reviewed journals in the field of nanomaterials with an H-index of 15 and a total citation over than 100.

Acknowledgments

The authors are so grateful to Prof. Dr. Norlida Kamarulzaman and Mrs. Norashikin Kamarudin (Centre of Nanomaterials, Institute of Science) for their valuable technical assistance and useful discussions on HRTEM analysis, and to Mr. M.Z. Borhan (School of Chemistry and Environmental Studies, Faculty of Applied Sciences) for his continuous support and helpful discussions on HS-SPME/GC-MS analysis. This project was supported by the Malaysian Ministry of Higher Education through the Fundamental Research Grant Scheme and MyMaster Scholarship program, the Research Management Institute through Excellence Fund under contact no. 600-RMI/ST/DANA 5/3/ Dst 435/2011, and the Institute of Graduate Studies through UiTM Postgraduate Teaching Assistance Scheme. MSS thanks the University of Southampton for the provision of facility. The authors gratefully acknowledge the anonymous reviewers for making recondite information available.

Author details

[1]Centre of Nanoscience and Nanotechnology (NANO-SciTech Centre), Institute of Science, Universiti Teknologi MARA (UiTM), Shah Alam, Selangor Darul Ehsan 40450, Malaysia. [2]School of Physics and Material Studies, Faculty of Applied Sciences, Universiti Teknologi MARA (UiTM), Shah Alam, Selangor Darul Ehsan 40450, Malaysia. [3]University of Southampton Malaysia Campus (USMC), No. 3, Persiaran Canselor 1, Kota Ilmu Educity@Iskandar, Nusajaya, Johor Darul Takzim 79200, Malaysia. [4]NANO-Electronic Centre, Faculty of Electrical Engineering, Universiti Teknologi MARA (UiTM), Shah Alam, Selangor Darul Ehsan 40450, Malaysia. [5]Department of Physics, Faculty of Science and Mathematics, Universiti Pendidikan Sultan Idris (UPSI), Tanjung Malim, Perak Darul Ridzuan 35900, Malaysia.

References

1. Zobir, SAM, Zainal, Z, Keng, CS, Sarijo, SH, Yusop, M: Synthesis of carbon nanohorn–carbon nanotube hybrids using palm olein as a precursor. Carbon **54**, 492–494 (2013)
2. Azmina, MZ, Suriani. AB, Salina, M, Azira, AA, Dalila, AR, Rosly, J, Roslan, MN, Rusop, M: Variety of bio-hydrocarbon precursors for the synthesis of carbon nanotubes. Nano Hybrids **2**, 43–63 (2012)
3. Yusop, MZM, Ghosh, P, Yaakob, Y, Kalita, G, Sasase, M, Hayashi, Y, Tanemura, M: In situ TEM observation of Fe-included carbon nanofiber: evolution of structural and electrical properties in field emission process. ACS Nano **11**, 9567–9573 (2012)
4. Bethune, DS, Kiang, CH, de Vries, MS, Gorman, G, Savoy, R, Vazquez, J, Beyers, R: Cobalt-catalysed growth of carbon nanotubes with single-atomic-layer walls. Nature **363**, 605–607 (1993)
5. Iijima, S, Ichihashi, T: Single-shell carbon nanotubes of 1-nm diameter. Nature **363**, 603–605 (1993)
6. Monthioux, M, Kuznetsov, VL: Who should be given the credit for the discovery of carbon nanotubes? Carbon **44**(9), 1621–1623 (2006)
7. Oberlin, A, Endo, M, Koyama, T: Filamentous growth of carbon through benzene decomposition. J. Cryst. Growth **32**(3), 335–349 (1976)
8. Oriňáková, R, Oriňák, A: Recent applications of carbon nanotubes in hydrogen production and storage. Fuel **90**(11), 3123–3140 (2011)
9. Ren, X, Chen, C, Nagatsu, M, Wang, X: Carbon nanotubes as adsorbents in environmental pollution management: a review. Chem. Eng. J. **170**(2–3), 395–410 (2011)
10. Upadhyayula, VKK, Deng, S, Mitchell, MC, Smith, GB: Application of carbon nanotube technology for removal of contaminants in drinking water: a review. Sci. Total. Environ **408**(1), 1–13 (2009)
11. Ferguson, AJ, Blackburn, JL, Kopidakis, N: Fullerenes and carbon nanotubes as acceptor materials in organic photovoltaics. Mater. Lett. **90**, 115–125 (2013)
12. Khodakovskaya, MV, Kim, B-S, Kim, JN, Alimohammadi, M, Dervishi, E, Mustafa, T, Cernigla, CE: Carbon nanotubes as plant growth regulators: effects on tomato growth, reproductive system, and soil microbial community. Small **9**(1), 115–123 (2013)
13. Grobert, N: Carbon nanotubes–becoming clean. Mater Today **10**(1–2), 28–35 (2007)
14. Gould, P: Carbon nanotubes help bones grow: biomaterials. Mater Today **9**(5), 19 (2006)
15. Sealy, C: Water on the nanoscale: carbon nanotubes. Mater Today **7**(12), 12 (2006)
16. Zobir, SAM, Abdullah, S, Zainal, Z, Sarijo, SH, Rusop, M: Synthesis of carbon nano- and microspheres using palm olein as the carbon source. Mater. Lett. **78**, 205–208 (2012)
17. Wu, H-C, Hong, C-T, Chiu, H-T, Li, Y-Y: Continuous synthesis of carbon spheres by a non-catalyst vertical chemical vapor deposition. Diamond Relat. Mater **18**(4), 601–605 (2009)
18. Suriani, AB, Azira, AA, Nik, SF, Roslan, MN, Rusop, M: Synthesis of vertically aligned carbon nanotubes using natural palm oil as carbon precursor. Mater. Lett **63**(30), 2704–2706 (2009)
19. Kumar, M, Ando, Y: Controlling the diameter distribution of carbon nanotubes grown from camphor on a zeolite support. Carbon **43**(3), 533–540 (2005)

20. Dijon, J, Szkutnik, PD, Fourier, A, de Monsabert, TG, Okuno, H, Quesnel, E, Muffato, V, de Vito, E, Bendiab, N, Bogner, A, Bernier, N: How to switch from a tip to base growth mechanism in carbon nanotube growth by catalytic chemical vapour deposition. Carbon **48**(13), 3953–3963 (2010)
21. Scheibe, B, Borowiak-Palen, E, Kalenczuk, RJ: Oxidation and reduction of multiwalled carbon nanotubes—preparation and characterization. Mater. Charact. **61**(2), 185–191 (2010)
22. Antunes, EF, de Resende, VG, Mengui, UA, Cunha, JBM, Corat, EJ, Massi, M: Analyses of residual iron in carbon nanotubes produced by camphor/ferrocene pyrolysis and purified by high temperature annealing. Appl. Surf. Sci. **257**(18), 8038–8043 (2011)
23. Huang, W, Wang, Y, Luo, G, Wei, F: 99.9% purity multi-walled carbon nanotubes by vacuum high-temperature annealing. Carbon **41**(13), 2585–2590 (2003)
24. Köhler, AR, Som, C, Helland, A, Gottschalk, F: Studying the potential release of carbon nanotubes throughout the application life cycle. J. Cleaner. Prod. **16**(8–9), 927–937 (2008)
25. Huang, B-R, Lin, T-C: A novel technique to fabricate horizontally aligned CNT nanostructure film for hydrogen gas sensing. Int. J. Hydrogen. Energy. **36**, 15919–15926 (2011)
26. Liu, P, Liu, L, Zhang, Y: Alignment characterization of single-wall carbon nanotubes by Raman scattering. Phys. Lett. A **313**(4), 302–306 (2003)
27. Ritter, U, Scharff, P, Dmytrenko, OP, Kulish, NP, Prylutskyy, YI, Belyi, NM, Gubanov, VA, Komarova, LA, Lizunova, SV, Shlapatskaya, VV, Bernas, H: Radiation damage and Raman vibrational modes of single-walled carbon nanotubes. Chem. Phys. Lett. **447**(4–6), 252–256 (2007)
28. Nijkamp, MG, Raaymakers, JEMJ, van Dillen, AJ, de Jong, KP: Hydrogen storage using physisorption—materials demands. Appl. Phys. A. **72**(5), 619–623 (2001)
29. Azira, AA, Suriani, AB, Rusop, M: Carbon nanotubes formation from Fe/Ni/Mg by camphor oil decomposition. J. Ceram. Soc. Jpn. **119**(1386), 125–128 (2011)
30. Radushkevich, LV, Lukyanovich, VM: About the structure of carbon formed by thermal decomposition of carbon monoxide on iron substrate. Soviet J. Phys. Chem. **26**, 88–95 (1952)
31. Kumar, M, Ando, Y: Carbon nanotubes from camphor: an environment-friendly nanotechnology. J. Phys. Conf. Ser. **61**, 643–646 (2007)
32. Yuan, Q, Xu, Z, Yakobson, BI, Ding, F: Efficient defect healing in catalytic carbon nanotube growth. Phys. Rev. Lett. **108**(24), 245505 (2012)
33. Kareem, SA: Kinetics of camphor decomposition in ambient condition. Res. Commun. Chem. **1**(1), 8–11 (2000)
34. Kumar, M, Ando, Y: Gigas growth of carbon nanotubes. Defense Sci. J. **58**(4), 496–503 (2008)
35. Kumar, M, Ando, Y: Camphor—a botanical precursor producing garden of carbon nanotubes. Diamond Relat. Mater. **12**(3–7), 998–1002 (2003)
36. Sharma, S, Kalitaa, G, Hirano, R, Hayashi, Y, Tanemura, M: Influence of gas composition on the formation of graphene domain synthesized from camphor. Mater. Lett. **93**, 258–262 (2013)
37. Suriani, AB, Roslan, MN, Rusop, M: Vertically aligned carbon nanotubes synthesized from waste cooking palm oil. J. Ceram. Soc. Jpn. **118**(10), 963–968 (2010)
38. Bahrami, B, Khodadadi, A, Mortazavi, Y, Esmaieli, M: Short time synthesis of high quality carbon nanotubes with high rates by CVD of methane on continuously emerged iron nanoparticles. Appl. Surf. Sci. **257**(23), 9710–9716 (2011)
39. Charlier, J-C, De Vita, A, Blasé, X, Car, R: Microscopic growth mechanisms for carbon nanotubes. Sci. **275**(5300), 647–649 (1997)
40. Schaper, AK, Hou, H, Greiner, A, Phillipp, F: The role of iron carbide in multiwalled carbon nanotube growth. J. Catal. **222**(1), 250–254 (2004)
41. Fedoseeva, YV, Bulusheva, LG, Okotrub, AV, Vyalikh, DV, Huo, J, Song, H, Zhou, J, Chen, X: Effect of oxidation and heat treatment on the morphology and electronic structure of carbon-encapsulated iron carbide nanoparticles. Mater. Chem. Phys **135**(1), 235–240 (2012)
42. Ren, ZF, Huang, ZP, Xu, JW, Wang, JH, Bush, P, Siegal, MP, Provencio, PN: Synthesis of large arrays of well-aligned carbon nanotubes on glass. Sci. **282**, 1105–1107 (1998)
43. Xiang, R, Yang, Z, Zhang, Q, Luo, G, Qian, W, Wei, F, Kadowaki, M, Einarsson, E, Maruyama, S: Growth deceleration of vertically aligned carbon nanotube arrays: catalyst deactivation or feedstock diffusion controlled? J. Phys. Chem. C. **112**(13), 4892–4896 (2008)
44. Xiang, R, Luo, G, Yang, Z, Zhang, Q, Qian, W, Wei, F: Temperature effect on the substrate selectivity of carbon nanotube growth in floating chemical vapor deposition. Nanotechnol. **18**(41), 415703–415706 (2007)
45. Reit, R, Nguyen, J, Ready, WJ: Growth time performance dependence of vertically aligned carbon nanotube supercapacitors grown on aluminum substrates. Electrochim. Acta. **91**, 96–100 (2013)
46. Asli, NA, Suriani, AB, Shamsudin, MS, Yusop, SFM, Rusop, M, Abdullah, S: Novel method: coral like structure of align carbon nanotubes (A-CNTs) on porous silicon template (PSiT) without catalyst; green approach. Adv. Mater. Res. **364**, 222–227 (2012)
47. Shamsudin, MS, Abdullah, S, Rusop, M: Structural and thermal behaviors of iron-filled align carbon nanotubes formulated by two-stage catalytic chemical vapor deposition. Adv. Mater. Res. **364**, 191–195 (2012)
48. Shamsudin, MS, Asli, NA, Abdullah, S, Yahya, SYS, Rusop, M: Effect of synthesis temperature on the growth iron-filled carbon nanotubes as evidenced by structural, micro-raman, and thermogravimetric analyses. Adv. Condens. Mat. Phys. **2012**, 420619 (2012)
49. Shamsudin, MS, Suriani, AB, Abdullah, S, Yahya, SYS, Rusop, M: Impact of thermal annealing under nitrogen ambient on structural, micro-Raman, and thermogravimetric analyses of camphoric-CNT. J. Spectro. **2013**, 167357 (2013)
50. Shamsudin, MS, Achoi, MF, Asiah, MN, Ismail, LN, Suriani, AB, Abdullah, S, Yahya, SYS, Rusop, M: An investigation on the formation of carbon nanotubes by two-stage chemical vapor deposition. J. Nanomater. **2012**, 972126 (2012)

3

Effect of catalyst on growth of diamond by plasma-enhanced chemical vapor deposition

Karim Zare[1,2,3*] and Morasa Amani Malkeshi[1,3]

Abstract

The nanodiamonds (NDs) were produced on glass substrates by plasma-enhanced chemical vapor deposition (PECVD) method. Acetylene (C_2H_2) diluted in H_2 were used as the reaction gasses as source of carbon and diluting gasses, respectively. The NDs have structures with high quality and density obtained by the selective Ni catalyst. The synthesized NDs have cubic structures with an average diameter of about 300 nm. This research focuses on the evolution of nickel catalyst under hydrogen/acetylene gasses at 320°C. The presence of the ND structure was confirmed by Raman spectroscopy, X-ray diffraction, and scanning electron microscopy analyses. It was found that the selective catalyst played an important role in producing ND structures by PECVD method on glass substrate in this research.

Keywords: Nanodiamonds, Nickel, Cubic structure, SEM, Raman, XRD, AFM

Background

The revolutionary discovery of nanodiamonds (NDs) in 1980s led to intense research activities and efforts in the domain of carbon science and nanostructures [1,2]. The fascinating properties of these unique materials have opened a great number of potential applications [3-5] of these nanostructures. Despite these stunning technical progresses, there is still much struggle in the development of a synthesis method suitable for experimental and commercial applications [6-8].

A leading candidate is the chemical vapor deposition (CVD) technique. CVD methods [9] are widely utilized techniques to fabricate nanostructures such as the ND structures in large quantities, and much progress has been made on the yield, the synthesis costs, or the purity of the products. CVD diamond growth includes steps of nucleation and formation of a continuous nanostructure, and surface morphology is closely related to these stages [6-8].

Nucleation and growth of NDs are induced by the decomposition of carbon-containing gasses (C_2H_2, CH_4, etc.) over a Ni catalyst at experimental conditions. A key reaction step for ND growth seems to be the diffusion of carbon through the Ni catalyst, and their catalytic action depends on the type of precursors, the type of substrate, and the reaction gasses used.

In the synthesis of cubic NDs, each carbon atom is coordinated tetrahedrally by four others, and the resulting structure is a cubic structure. This nanostructure is a consequence of the sp^3 hybridization of the bonding orbital. In fact, diamond is the hardest material known and is a wide-gap semiconductor [4,9-13].

One of the major issues in the growth of high-quality diamond by CVD is the choice of a suitable substrate. Many substrates can be used for the growth of diamond, such as Si, Mo, W, Ti, and Nb. Si is the most commonly used substrate in many researches. For example, the Fujishima group utilized Si wafers as the substrate in their research, which is needed for the efficient growth of diamond [14]. Single-crystal growth of diamond is being carried out by a growing number of research groups [15], in which diamond was used as the substrate material, but in this research, the synthesis of diamond was carried out using glass-coated Ni nanoparticle catalyst as the desired substrate. Most recently, the Carnegie groups reported high-quality synthetic single-crystal diamond in desired research [16]. The single-crystal CVD diamonds have been used by Element Six and Sumitomo in electronics applications and by Apollo Diamond as gems [17-20]. High growth rate of diamond fabricated by low-

* Correspondence: k-zare@sbu.ac.ir
[1]Department of Chemistry, Science and Research Campus, Islamic Azad University, Tehran, Iran
[2]Department of Chemistry, Shahid Beheshti University, Evin, Tehran, Iran
Full list of author information is available at the end of the article

pressure microwave plasma CVD is the most promising technology for fabricating low-cost and high-quality large diamond [21].

The feasibility of the plasma-enhanced chemical vapor deposition (PECVD) system to synthesize NDs using the Ni nanoparticle catalyst has been investigated with acetylene diluted in hydrogen. The NDs show cubic structures and result from the etching of carbon network on the selective catalyst and continuous precipitation on the glass substrate. In order to elucidate the nature of the Ni catalyst, the synthesis uses an atomic force microscopy (AFM) system, and the samples quenched after different pyrolysis times were investigated by X-ray diffraction (XRD), Raman spectroscopy, and scanning electron microscopy (SEM) which have been intensively employed for characterizing CVD diamond.

Results and discussion

Figure 1 shows the ND with body-centered cubic structure. The process of synthesis of NDs by PECVD method involves Ni catalyst deposition and growth of ND structures in specific conditions. Transition Ni nanoparticles are applied as catalyst for ND growth. Here, this metal was deposited on the glass substrate by a physical technique such as sputtering by PECVD method. Proper sputtering technique of the substrates was very important for the successful synthesis of NDs in our research. Ni with the capability of decomposing the hydrocarbon of C_2H_2 and ND formation is employed as the catalyst in the PECVD process. However, it has been found that only the C_2H_2 molecules are accounted for the decomposing ability of Ni, but not for ND formation. It was possible to render the substrate surfaces covered with grains of Ni catalyst of a certain size. The morphology of the Ni nanoparticle catalyst utilized in the synthesis of NDs is studied by the AFM technique.

Figure 2 shows a typical AFM image of the Ni nanoparticles deposited on the glass substrate. The deposited Ni catalyst is densely packed and of uniform diameter over the glass substrate. The Ni grain morphologies determine the quality of NDs grown on the substrate surface. Single-crystalline diamonds seem to be structured by the aggregation of nanoparticles.

The morphology of the structures was characterized by SEM technique. The SEM micrographs in Figure 3 show the cubic morphology of the single-crystal diamond, which grew on the deposited Ni nanoparticles on the glass substrate by PECVD method. In this figure, the SEM images show the ND structures synthesized with a high density and uniform diameter around 300 nm on the substrate surface. It is noticeable that Ni nanoparticles on the surface play an important role in the formation of nucleation sites; the density and distribution of Ni nanoparticles affect the synthesis and the type of NDs. In fact, Ni nanoparticles served as the well-defined nucleation sites to allow the growth of aligned NDs by PECVD system directly on the substrate surface, and the decomposition of the hydrocarbon of C_2H_2 and H_2 gasses is more sufficient that it led to less deposition of amorphous carbon. The Ni nanoparticles can act typically as catalyst under specific conditions for ND nucleation resulting in growth. The diffraction of carbon through the Ni metal particle is indicated by an increase of the metal cell parameters identified.

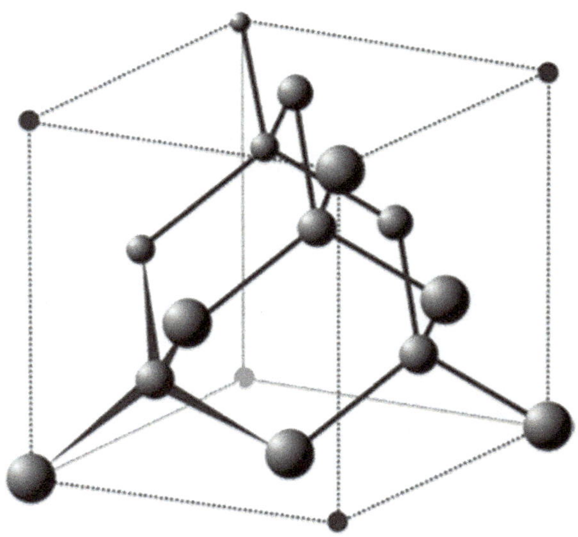

Figure 1 Structure of diamond.

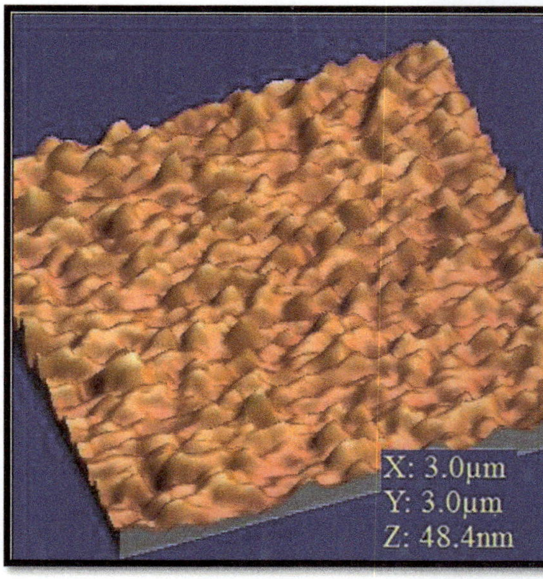

Figure 2 AFM image of glass substrates coated with Ni catalyst by PECVD method.

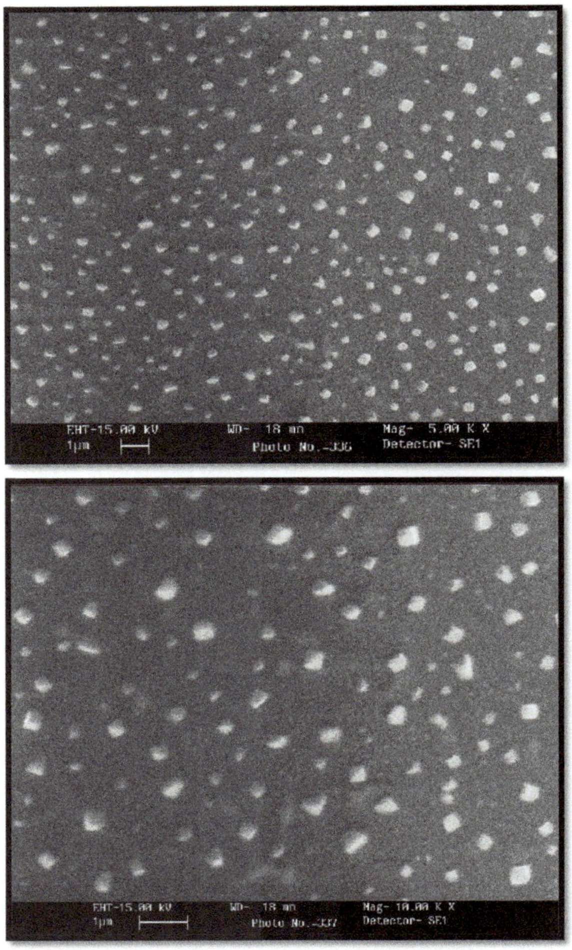

Figure 3 Scanning electron micrograph of nanodiamond synthesized using Ni catalyst on glass substrate.

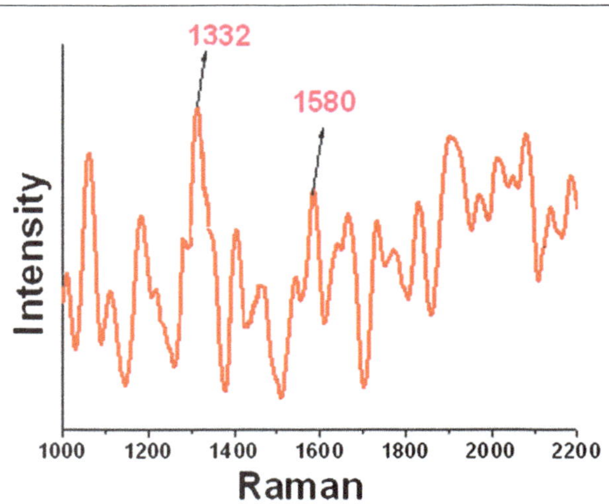

Figure 4 Raman spectra of diamond structure deposited on glass substrate.

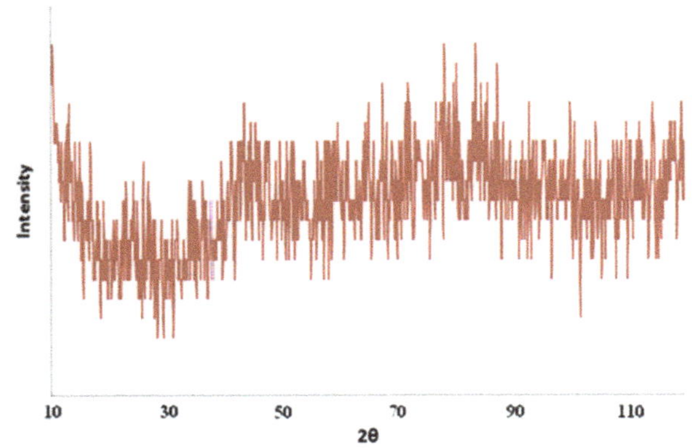

Figure 5 XRD patterns of diamond deposited on the substrate by PECVD method.

Raman spectroscopy is the most extensive and useful tool utilized to characterize CVD diamond to distinguish between different allotropes of carbon structures [22]. Figure 4 shows the Raman spectra of the ND structures grown using Ni catalyst on the glass substrate. In

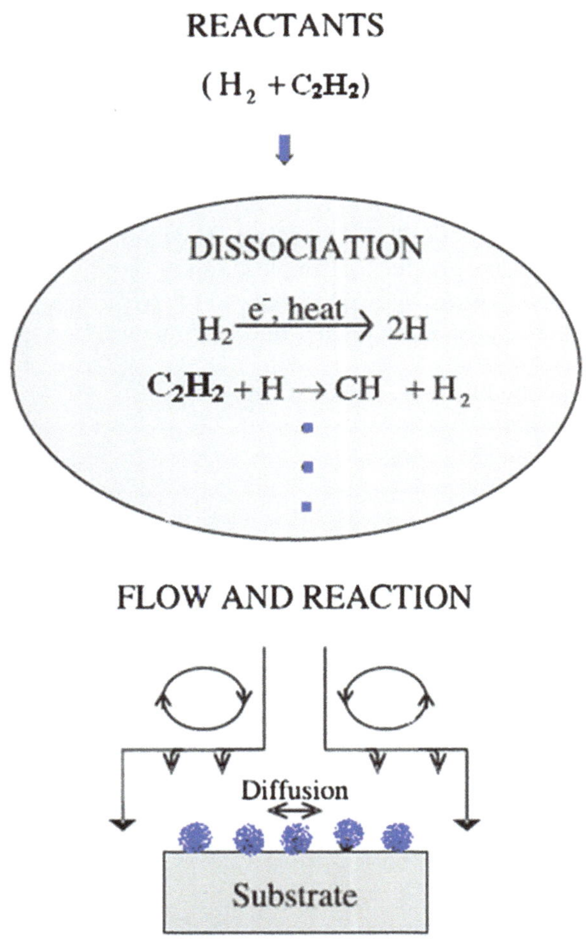

Figure 6 Schematic of the process occurring during diamond CVD by Ni catalyst.

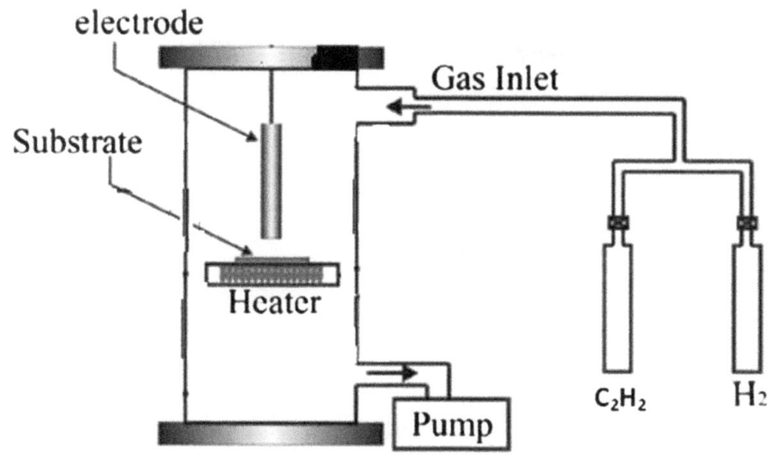

Figure 7 A schematic of plasma-enhanced CVD apparatus.

addition, a diamond peak around 1,332 cm^{-1} for the sp^3 bonded carbon and a peak around 1,580 cm^{-1} for the sp^2 bonded carbon in this structure were mainly observed. The spectra indicate the existence of good diamond phase with high quality.

XRD is a technique as powerful as Raman spectroscopy which provides a fingerprint of the presence of diamond phase. In addition to Raman spectroscopy, XRD can be utilized to confirm that diamond synthesis has been achieved by the Ni catalyst. Figure 5 demonstrates the single shape for cubic diamond. In many studies, it can be seen that the (111) plane is faster in a cubic structure for diamond, while the (100) plane is the fastest one in an octahedral structure of diamond [23]. The ND structures synthesized on glass substrate as-coated with Ni catalyst were determined by X-ray diffraction. Figure 5 illustrates the XRD spectra of the deposited ND structures produced at 320°C temperature and using the selective catalyst. The diffraction peaks show the presence of carbon material. From XRD spectra of the sample, one peak at around 43.70°C was ascribed to the (111) plane of the cubic diamond structures.

Table 1 Parameter range for diamond synthesis by plasma-enhanced CVD method

Parameter	Type/value
Substrate	Glass
Carbon source	Acetylene
Substrate temperature	320°C
Total pressure	15 Torr
Time	120 min

Conclusion

In summary, NDs were successfully synthesized with Ni catalyst by PECVD method on the glass substrate. In this research, the effect of the Ni catalyst and substrate on ND synthesis was investigated. It was found that the reaction of the catalyst has significant effect on the growth of NDs. Ni plays a crucial role in the PECVD synthesis of NDs, and therefore, improving the desired characteristics of catalyst will enhance the obtained ND quality as well as the process yield. The SEM images show that NDs can be grown using Ni as the selective catalyst with high density and surface morphology of the structures. The type of diamond synthesized in this research has cubic structures. From the Raman and XRD analyses of the NDs, it was proved that the Ni catalyst affected the quality and purity of the ND growth. The results were more promising than the experiments with Ni catalyst.

Methods

This process deals with many fundamental aspects of CVD, such as gas phase chemistry, complex heat and nucleation, surface chemistry, and diffusion as shown in Figure 6, that were utilized for all experiments in this field. Synthesis of NDs has been carried out by PECVD technique. Figure 7 describes a direct current (DC)-PECVD apparatus consisting of an electrode and a substrate with a heater. The electrode and substrate holder placed on the heater are made of Cu, while the substrate on the holder is made of glass. The substrate is set underneath the electrode. The Ni nanoparticle catalyst is deposited with the DC-PECVD system, and the sputtering of Ni nanoparticles is controlled by deposition time (120 min). The system is operated at a DC power of P_{DC} = 600 W. The etching step utilized only acetylene gas, and at the growth step, the gas mixture of acetylene

(C_2H_2) and hydrogen (H_2) is filled during the discharge, where the flow of gas mixture ratio is expressed by $P_{C2H2}/P_{H2} = 30\%$. The glass substrate is heated up to a temperature $T = 320°C$, which is measured by a thermocouple directly connected to the copper substrate holder. After which, the gas mixture of C_2H_2 and H_2 is fed and then the total pressure is kept at 15 Torr. The investigated desired optimum condition growth to control ND structures is shown in Table 1. Three most fundamental techniques, Raman spectroscopy, XRD, and SEM, are most intensively used to identify diamond structures. AFM is utilized for the morphological and structural investigation of NDs.

Competing interests
The authors declare that they have no competing interests.

Authors' contributions
KZ and MAM both contributed to this research work. Both authors read and approved the final manuscript.

Author details
[1]Department of Chemistry, Science and Research Campus, Islamic Azad University, Tehran, Iran. [2]Department of Chemistry, Shahid Beheshti University, Evin, Tehran, Iran. [3]Young Researchers and Elites Club of Science and Research Campus, Islamic Azad University, Tehran, Iran.

References
1. Gracio, J, Fan, QH, Madaleno, JC: Diamond growth by chemical vapour deposition. J. Phys. **43**, 374017 (2010)
2. Qi, X, Qin, C, Zhong, W, Au, C, Ye, X, Du, Y: Large-scale synthesis of carbon nanomaterials by catalytic chemical vapor deposition: a review of the effects of synthesis parameters and magnetic properties. Materials **3**, 4142–4174 (2010)
3. Luo, SY, Kuo, JK, Yeh, B, Sung, CY, Dai, CW, Tsai, TJ: The tribology of nanocrystalline diamond. Mater. Chem. Phys. **72**, 133–135 (2001)
4. Huang, LY, Xu, KW, Lu, J, Guelorget, B: Analysis of nano-scratch behavior of diamond-like carbon films. Surf. Coat. Technol. **154**, 232–236 (2002)
5. Ferreira, NG, Silva, LLG, Corat, EJ: Electrochemical activity of boron-doped diamond electrodes grown on carbon fiber cloth. Diam. Relat. Mater. **11**, 657–661 (2002)
6. Chayahara, A, Mokuno, Y, Tsubouchi, N, Yamada, H: Development of single-crystalline diamond wafers. Synthesiology **3**(4), 259–267 (2011)
7. Hemley, RJ, Chen, Y-C, Yan, C-S: Growing diamond crystals by chemical vapor deposition. Elements **1**, 105–108 (2005)
8. Zhao, N, Cui, Q, He, C, Shi, C, Li, J, Li, H, Du, X: Synthesis of carbon nanostructures with different morphologies by CVD of methane. Mater. Sci. Eng. A. **460–461**, 255–260 (2007)
9. Hoffman, A: Mechanism and properties of nanodiamond films deposited by the DC-GD-CVD process. J. Springer **192**, 125–142 (2005)
10. Chow, L, Zhou, D, Hussain, A, Kleckley, S, Zollinger, S, Wang, H: Chemical vapor deposition of novel carbon materials. J Thin Solid Films **368**, 193–197 (2000)
11. Tinchev, SY, Dyulgerska, YP, Nikolova, P, Grambole, D, Kreissig, U, Babeva, TZ: Optical properties of PECVD deposited DLC films prepared with air addition. J. Optoelectronics Adv. Mater. **8**(1), 308–311 (2006)
12. Cheung, CL, Kurtz, A, Park, H, Leiber, CM: Diameter-controlled synthesis of carbon nanotubes. J. Phys. Chem. B **106**, 2429–2433 (2002)
13. Schirru, F, Kisielewicz, K, Nowak, T, Marcczewska, B: Single crystal diamond detector for radiotherapy. J. Phys. D: Appl. Phys. **43**, 265101 (2010)
14. Fazaneh, F, Faal Hamedani, N: Preparation of carbon nanotubes by CVD process over nanoparticles of Ni-Ce-Zr mixed oxides. J. Sci. Islam. Repub. Iran **19**(2), 119–123 (2008)
15. Biener, J, Ho, DD, Wild, C, Woerner, E, Biener, MM, El-dasher, BS, Hicks, DG, Eggert, JH, Celliers, PM, Collins, GW, Teslich Jr, NE, Kozioziemski, BJ, Haan, SW, Hamza, AV: Diamond spheres for inertial confinement fusion. Nucl. Fusion **49**, 112001 (2009)
16. Martineau, PM, Lawson, SC, Taylor, AJ, Quinn, SJ, Evans, DJF, Crowder, MJ: Identification of synthetic diamond grown using chemical vapor deposition (CVD). Gems and Gemology **40**(1), 2–25 (2004)
17. Yan, CS, Mao, HK, Li, W, Qian, J, Zhao, Y, Hemley, RJ: Ultrahard diamond single crystals from chemical vapor deposition. Physica Status Solidi (A) **201** (4), 25–27 (2004)
18. Wang, W, Moses, T, Linares, RC, Shigley, JE, Hall, M, Bulter, JE: Gem-quality synthetic diamonds grown by a chemical vapor deposition (CVD) method. Gems and Gemology **39**(4), 268–283 (2003)
19. Okushi, H: High quality homoepitaxial CVD diamond for electronic devices. Diamond Relat. Mater. **10**(3–7), 281–288 (2001)
20. Isberg, J, Hammersberg, J, Johansson, E, Wikstrom, T, Twitchen, DJ, Whitehead, AJ, Scarsbrook, GA: High carrier mobility in single-crystal plasma-deposited diamond. Science **297**(5587), 1670–1672 (2002)
21. Charles, SJ, Butler, JE, Feygelson, BN, Newton, MF, Carroll, DL, Steeds, JW, Darwish, H, Mao, HK, Yan, CS, Hemley, RJ: Characterization of nitrogen doped chemical vapor deposited single crystal diamond before and after high pressure, high temperature annealing. Physica Status Solidi (A) **201**(11), 2473–2485 (2004)
22. Yan, C, Yogesh, K, Vohrat, H, Russell, J: Very high growth rate chemical vapor deposition of single-crystal diamond. Appl. Phys. Sci. **99**(20), 12523–12525 (2002)
23. Aislinn, H, Sirk, C, Donald, R: Electrochemical synthesis of diamond like carbon films. J. Electrochem. Soc. **155**, 45–55 (2008)

4

Fabrication of various electrical resistances producing Zn nanowires and subsequent oxidation fabricating ZnO nanowires in PAA template by periodic and pulsed electrochemical deposition

Mahmood Moradi[1,2*], Mina Zamanian[1] and Mohammad Noormohammadi[1]

Abstract

Highly ordered anodic alumina template was produced by hard anodization. Barrier layer thickness was then reduced by a thinning process, and in the following Zn nanowires, it was grown in pores by periodic (AC) and pulsed electrochemical deposition. The samples were placed in a furnace for oxidation and for the possible preparation of electrical resistors. By measuring the electrical resistance of nanowire arrays, we did the calculations to find the electrical resistivity of each ZnO nanowire.

Keywords: PAA template, Electrodeposition, ZnO nanowire, Electrical resistance

Background

A template of nanopores of porous anodic alumina (PAA) film is produce by aluminum anodization using acidic solutions and standard voltages. This structure can be engineered to form various shapes of pores [1].

In recent years, this structure has extensively developed in the industry, and some of its applications are magnetic memory [2] and manufactured electronic devices such as capacitors [3], biosensors [4], and template for synthesis of nanowires [5]. Filling the porous alumina template by various materials is one of the simple techniques that would produce a regular array of nanowires [5].

One method of filling the porous template is by electrochemical deposition [6]. This array of nanowires with almost a nonconducting alumina structure can be considered as a path for electron transition through a circuit, and freestanding wires can be used as electrical resistance. Measuring the electrical resistance of the template filled with nanowires makes it possible to use as a nanoresistor in microelectronic devices. Of course, in a very small scale, some physical laws cannot be established. For example, it is possible that electrical resistance would not follow the Ohm's law.

In general, one can classify the conductance of tiny wires into three regions according to their sizes: (1) region that follows material laws similar to bulk mode; (2) region that should be considered by quantum effects, with its dimensions less than the mean free path of electron in the material, and in this case, the electrical resistance values are discrete because of restrictions on electronic wave functions [7]; and (3) region between these two cases in which wires that have dimensions of the mean free path order and dimensions of grain boundaries can be reduced by decreasing the wire diameter, and thus, electrical resistance can be increased faster than the Ohm's law. Also, some equations are suggested to determine the electrical resistance in this region [7].

Of course, it can be said that for wires with diameters ≥100 nm, including all their materials, bulk laws are established. Here, we study using the mentioned range. First, we fabricate Zn nanowires in PAA template and then oxide it in a furnace. In the past decade, numerous studies have been made on both production and application of one-dimensional ZnO [8,9], e.g., the luminance

* Correspondence: moradi@susc.ac.ir
[1]Department of Physics, Science College, Shiraz University, Shiraz, Iran
[2]Nanotechnology Research Center, Shiraz University, Shiraz, Iran

Table 1 Resulting average of twice measured electrical resistance of PAA deposited and filled by Zn nanowires

Temperature	*R* (MΩ)	Error (MΩ)
200°C	2.27	0.03±
100°C	1.05	0.05±
18°C to 19°C	0.41	0.06±

properties of ZnO that increase by fabricating its nanowires in anodic aluminum oxide template [10].

Here, we measure the electrical resistance of porous anodic alumina template filled with Zn nanowires before and after oxidation and examine the provided various electrical resistances with oxidation at different temperatures while trying to introduce a method for measuring the electrical resistance of a single nanowire.

Results and discussion

As one can see in Table 1, there is a significant difference between the resulting values at different temperatures, and therefore, different electrical resistances are produced by this method.

We also approximated calculations to find the electrical resistance of a single nanowire for one of the cases in Table 1 before any oxidation (18°C to 19°C) and after oxidation of the sample at 200°C. In total, the electrical resistance of the sample after oxidation was produced from the initial resistance of aluminum, resistance of alumina, and resistance of some oxidized nanowires located at the electrical connection. By deducing the resistance of the sample before oxidation (initial resistance of alumina), an approximate value for the resistance of array of parallel nanowires can be obtained.

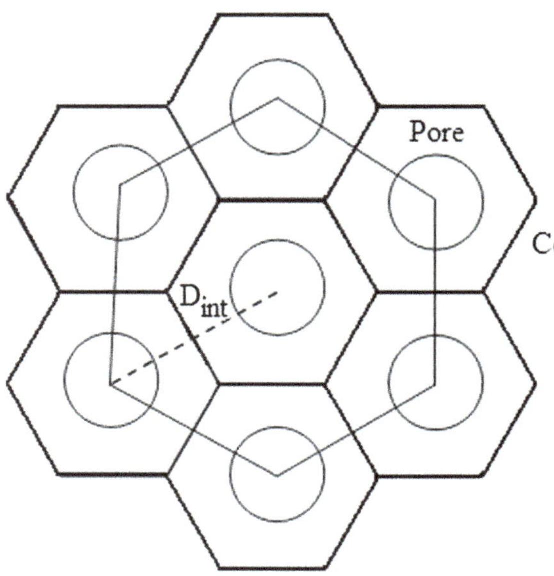

Figure 1 Two-dimensional honeycomb cell and its corresponding pore mouth of anodized alumina structure.

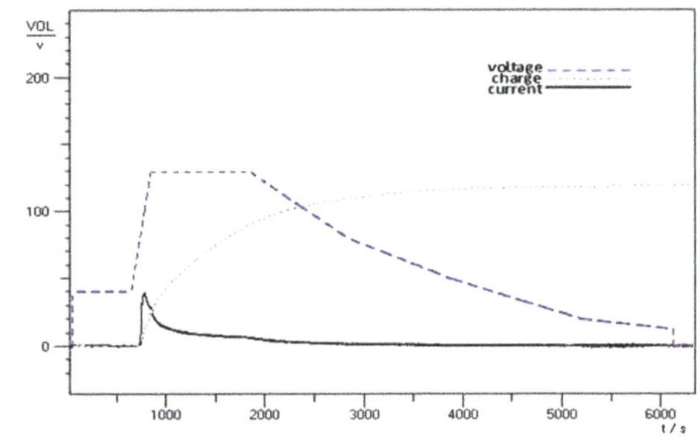

Figure 2 A typical curve of anodization at 130 V and thinning to 12 V.

The layer thickness under nanowires is about 15.6 nm. Thus, because the thickness is very thin, any change in the resistance of the alumina barrier layer due to annealing can be neglected. This assumption was also checked by another experiment, and in this work, we have measured the electrical resistance across the template filled with Ag nanowires produced by periodic electrodeposition. Thinning was continued until the voltage was reduced to 12 V, and electrodeposition was done at 18 V and 200 Hz, before and after oxidation at 300°C. The resulting values of resistances are 2.64 and 6.43 Ω m^{-2}, respectively. Because silver is a very good conductor (15.87 nΩ m in 20°C), we neglected its electrical resistance; also, its melting point is very high (961.78°C), but its oxidation after annealing (300°C for 17 h) is negligible. Thus, the resistance of the sample before oxidation is mainly due to the barrier layer resistance under the connected wires, and the resistance after oxidation is also due to the initial resistance of the

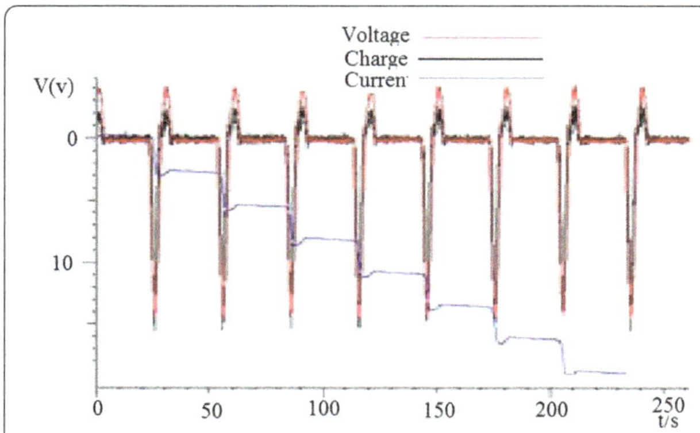

Figure 3 Voltage and current-time curves of pulsed electrochemical deposition. The range of voltage is −14 to +4 V with a relaxation time of 30 s, and the used solution is 0.4 M $ZnSo_4{\cdot}7H_2O$ and 0.3 H_3Bo_3.

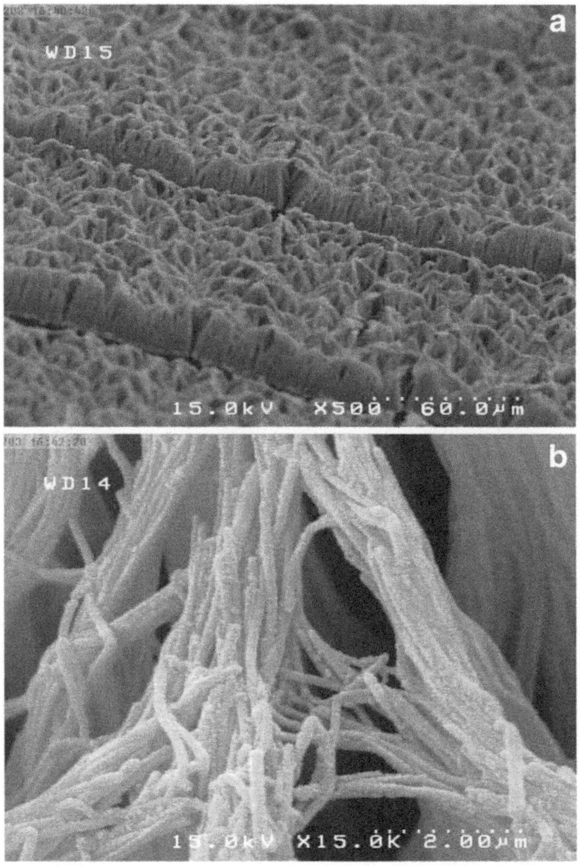

Figure 4 SEM images obtained from pulsed electrochemical deposition of Zn nanowires in PAA. (**a**) Arrays of nanowires and (**b**) some nanowires in bigger magnification.

barrier layer plus the added amount of resistance of this layer during oxidation and annealing.

By deducing the resistance values before and after oxidation, we can obtain the amount of increased resistance by annealing which is about 3.79 Ω m^{-2}. However, the resistance values in Table 1 are in megaohms, so this value is negligible for samples with such barrier layer thickness. Now, for ZnO nanowires with this approximation, we

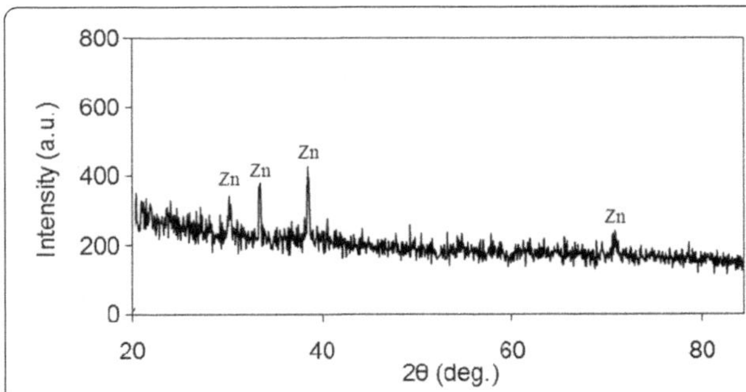

Figure 5 XRD image of Zn nanowires made by pulsed electrochemical deposition in PAA.

obtained the resistance of parallel array of the existing nanowires in the template to be 1.86 MΩ (difference between two values before and after oxidation).

Now, we calculated the number of pores per unit area using the hexagonal cells of the porous alumina structure (Figure 1). The hexagonal area is $\sqrt{3/2D_{int}^2}$, and the number density of pores per unit surface is $n = 1/\sqrt{3/2D_{int}^2}$. Also, a relation was established between the interpore distance (D_{int}) and the anodized voltage ($D_{\text{int}} = \lambda i \times U$) with the proportionality constant λ_i = 2.5 nm/V [11]. In our experiment, the anodized voltage is 130 V, so D_{int} = 325 nm and $n = 77.3 \times 10^7$ pore/cm^2.

The surface area of gold in the sample is 1.16 cm^2 (area of mask), so the number of pores in this area is 89.66×10^7. However, it is confirmed experimentally that about 5% to 10% of pores are completely filled up to the surface, and some wires can be connected to the gold layer because of the influence of gold particles into the pores. Thus, we could approximate the number of desired nanowires to 10% of the number of pores.

To find the electrical resistance of a single nanowire, we divided the amount of resistance of parallel nanowire arrays by 10% of the number of pores (89.66×10^6). Hence, the amount of electrical resistance of a single nanowire with an approximate length of 45 µm (the final length of pores after thinning) is 0.020 Ω.

As we have mentioned, this value is obtained by some approximate calculation which show the range of a single nanowire resistance. Unfortunately, we did not find similar results in other articles in order to compare with these values. Although the oxidation of the nanowire is low, the existence of such resistance is shown, so by increasing the time and oxidation temperature, we could obtain a complete oxide nanowire (ZnO) as an electrical resistance. In the next step, we can release the nanowire from the template as a nanoresistance for nanoelectronic purposes.

Conclusions

In this research, we initially produced nanoporous alumina templates by anodizing the aluminum foil and used them to deposit and make nanowires. We did this process by pulsed and periodic electrochemical deposition.

For the deposition process, the barrier layer at the bottom of the pores was made thin, and we then deposited Zn in pores and produced Zn nanowires. The STM micrograph and X-ray diffraction (XRD) pattern show the existence of the produced nanowires.

The electrical connection was created by depositing a gold layer using joule vapor deposition method. We measured the electrical resistance across the template before and after oxidation at some temperatures. Also, we had the opportunity to produce various resistances.

Table 2 Different parameters of anodization conditions and deposition in the periodic method in constant solution

Hard anodizing voltage (V)	Mild anodizing voltage (V)	Pore depth (μm)	Final voltage in thinning (V)	Deposition voltage (V)	Deposition frequency (Hz)	Observed color
86	68.8	0.5	12	18	100	Gray fill color
86	68.8	0.5	12	15	100	No color change
86	68.8	5	12	18	200	Medium gray
86	68.8	2	12	18	150	Medium gray
86	68.8	2	12	18	100	Medium gray
130	104	2.5	12	18 to ´2	100	Relatively dark
130	104	3	12	18 to ´2	100	Relatively dark

We then approximated calculations to measure the electrical resistance of a single nanowire. The results showed that we can produce nanoresistance for electrical circuits in microchips.

Methods

We chose aluminum foil with a purity of 99.999% to make a disc with a diameter of 12 mm and cleaned it by sinking in ultrasonic bath with acetone. After locating the sample in the anodizing reactor and polishing it with perchloric acid and ethanol aqueous solution with a volumetric ratio of 1:4, the voltage (20 V) and current (100 mA/cm^2) were applied for 3 min is to obtain a polished and smooth surface. The anodization process was done with the constant voltage regime and in 130 V at 0°C in 0.3-M oxalic acid electrolyte. This method is called hard anodization, and nanoporous anodic alumina was formed.

Anodization for 700 s was done until holes that reached as high as about 40 μm appeared and the thinning process of the barrier layer started. This layer is located at the bottom of the holes and acts as a resistor which prevents ion-electron exchange between the solution and the conductive back part of the template, though it could be penetrated as far as electrical conductance is possible by quantum phenomena and by the thinning barrier layer. This can be done by reducing the anodization voltage step by step. Thinning can be done manually or through programmable device.

The appropriate rates were experimentally obtained through the following process: initially, the voltage was 130 V and was decreased step by step, first, to 40 V by the rate of 3 V in 30 s, then to 20 V by the rate of 2 V in 30 s, to 10 V by 1 V in 30 s, and finally, to 5 V by the rate of 0.5 V in 30 s. A typical curve of anodization at 130 V and subsequent thinning by reducing the voltage to 12 V can be achieved during the process (Figure 2).

Temperature is also an important factor in the thinning process, and its increase prevents the high drop of the current. Hence, the temperature was gradually and manually increased, and when the voltage reached 40 V, the temperature experimentally increased to 15°C and remained in that value to the end.

Here, the deposition of metallic Zn nanowires, done to fill pores of the template, was done in two ways, by periodic and pulsed chemical electrodeposition. In both methods, the thinning procedure was required to provide electrical connection at the bottom of the holes, but after this stage, further experiment was done in two different methods.

In the pulsed method, appropriate solutions such as 0.4 M $ZnSo_4{\cdot}7H_2O$ and 0.3 H_3Bo_3 were used. After changing the anodizing acid with the deposition solution inside the reactor and setting up the system, we used the

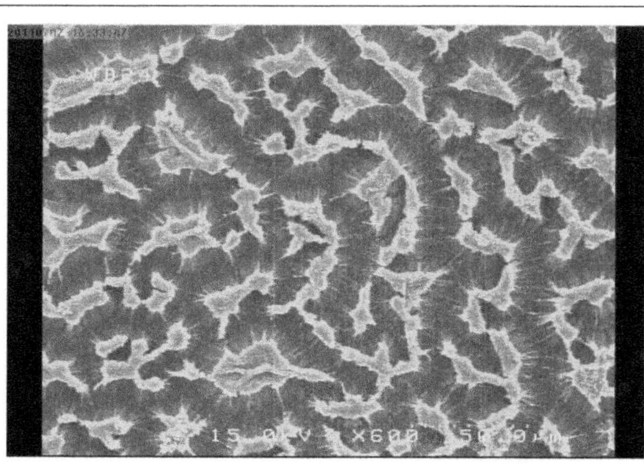

Figure 6 SEM images obtained from periodic electrochemical deposition of Zn nanowires in PAA. The images were obtained by removing some alumina from the template surface.

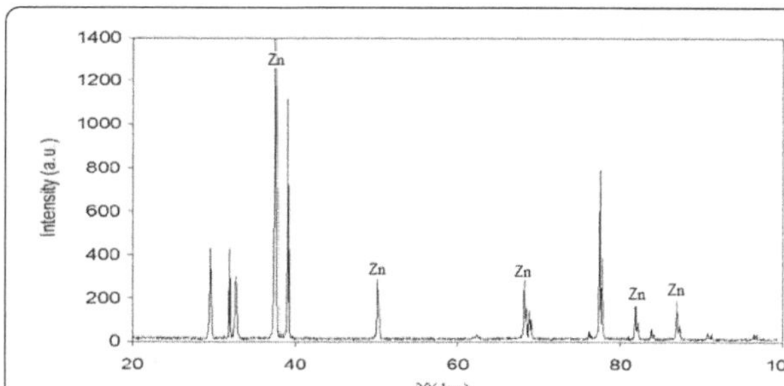

Figure 7 XRD image of Zn nanowires made by periodic electrochemical deposition in PAA.

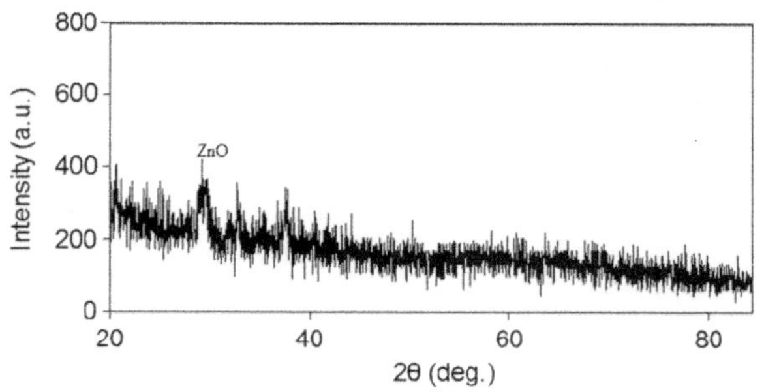

Figure 8 The XRD image of ZnO nanowires made by pulsed electrochemical deposition in PAA. The temperature increased in stages from 100°C to 500°C. For every stage, there was a 100°C increase in temperature until 500°C is reached and maintained; this condition was done for 15 h.

programmable AC/DC power supply, EC1000S model, to generate sine waves ranging from −14 to +4 V, with a relaxation time of 30 s for the deposition of Zn.

The voltage and current-time curves of this process are showed in Figure 3. In this figure, the slope of the charge curve shows the amount of deposition.

In this deposition, the graffiti cathode is placed inside the reactor against the anodized surface of the template. The anode located at the bottom of the holes connected through quantum tunneling with aluminum behind the barrier layer was formed on the template. Deposition was continued for about 1,000 s. The resulting scanning electron microscopy (SEM) image confirms the existence of the nanowires (Figure 4). The XRD pattern confirms it as well (Figure 5).

The surface of the sample was colored black after about 400 s, which signifies the filling of pores by Zn in this method. After 1,000 s, the color of sample began to turn silver, which shows the color change of the bulk material.

In the periodic (AC) method, we were able to generate nanowires in different conditions and with different parameters. As an example, for the solution that included 0.3 M $ZnSO_4{\cdot}7H_2O$ and 0.3 M H_3BO_3, nanowires were formed at different conditions, which resulted in different colors of the surface (Table 2). Better deposition showed a darker color.

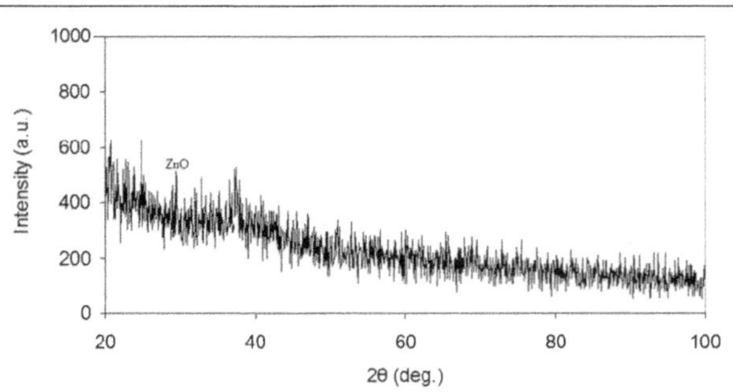

Figure 9 XRD image of ZnO nanowires made by periodic electrochemical deposition in PAA. The sample was placed in the oven for 40 h at 300°C.

Figure 10 Gold layer coating on PAA filled with Zn nanowires using a mask.

The constant solution included 0.3 M $ZnSO_4{\cdot}7H_2O$ and 0.3 M H_3BO_3. These parameters are needed in order to generate Zn nanowires and corresponding surface colors of the resulting sample.

The SEM image and the corresponding XRD of the resulting sample of this method show the existence of the nanowires (Figures 6 and 7).

Eventually, to obtain ZnO nanowires, we put the sample in the oven at different temperatures. We did this procedure without removing the template. The influence of oxygen to the nanowires takes place in the bare surface on top of the nanowires; also, it is due to the excess oxygen that existed in the porous body and around the wire, which had caused oxidation.

We placed the sample, obtained from pulsed deposition, in the furnace with a step-by-step increase in temperature from 100°C to 500°C. For each step, there is a 100°C increase in temperature until 500°C is reached and maintained; the process was done for 15 h. The XRD of the resulting sample shows the existence of ZnO nanowires (Figure 8).

The sample obtained from periodic deposition was placed in the oven for 40 h at 300°C, and the XRD result shows the existence of ZnO nanowires in this case (Figure 9).

In these two figures, some peaks of ZnO X-Ray pattern are shown as some of the Zn nanowires that are changed to ZnO nanowires. It should be noted that images were taken on glass and silicon substrates without the total dissolution of the template, but only the back if the aluminum was solved.

Our goal in this work is to fabricate electrical resistance using tiny nanostructure such as nanowires. For this purpose, we used the samples that have been made in the template using pulsed electrodeposition.

In order to make the pores wider and especially increase the pore diameter, during the thinning process and at a voltage of 50 V, we replaced the solution of anodization by 5% phosphoric acid for about 20 min in 30°C. This process helped us to have further deposition, and other existing reasons will be discussed later.

According to the SEM images of previous experiments of deposited samples by electrochemical deposition, we estimated that up to 5% to 10% of pores are filled to the top and covered the surface.

In this experiment, we slowly cleaned the layer that formed on the surface of the template with a suitable cloth and coated a gold layer on it using joule vapor deposition method. The mask is used to prevent the sample from creating a short circuit between the gold and aluminum substrates (Figure 10).

The plated gold is used in order to create an electrical connection for the measurement of electrical resistance across the sample. Also, the penetration of some gold particles into the pores caused to have more nanowires in the circuit that even reached to the sample surface. As mentioned before, the other reason for the widening of pores led to a wider pore mouth, therefore allowing the gold particles to penetrate better into the pores.

After coating the gold on the template, in order to measure the electrical resistance of the desired nanowires, we put a thin copper wire on the gold-covered surface, fixed it with silver glue, and covered all part of the surface with adhesive. The electrical resistance across the width of sample was then measured.

The creation of electric current was due to the transfer of electron flow because of quantum tunneling between the bottom layer Al and Zn nanowires, connecting the gold surface (the wires that are not completely filled did not contribute to the current transfer) and the copper wire.

Because the porous anodic alumina has high electrical resistance and is almost a nonconductive material while zinc is a good conductive material (59 nΩ m in 20°C), we can ignore the electrical resistance of the Zn nanowires. Therefore, it can be assumed that the resistance is only due to the electrical resistance of the barrier layer located under the wires that has thinned to about a few nanometers.

There is a direct relationship between the anodizing voltage and barrier layer thickness with the proportionality constant of 1.3 nm/V [12]; the thickness of this barrier layer is about 15.6 nm. Because it is very thin, there is electron tunneling conductance through it.

Now, we put the sample in a furnace for about 17 h in 100°C. The sample is then taken, and electrical resistance is measured again. We then put it in 200°C, repeat the experiment, and obtain the resulting average of the two measurements (Table 1).

Competing interests

The authors declare that they have no competing interests.

Authors' contributions

MM is the leader of the group and the main writer of this study's manuscript. MZ and MN both participated in writing the manuscript. All authors read and approved the final manuscript.

Acknowledgments

The authors kindly thank the team of Sherlyn C Machica for the language editing assistance provided for this article.

References

1. Moradi, M, Noormohammadi, M, Behzadi, F: Three-dimensional structural engineering of nanoporous alumina by controlled sprinkling of an electrolyte on a porous anodic alumina (PAA) template. J. Applied Phys. **44**, 045301 (2011)
2. Nielsch, K, Wehrspohn, PRB, Barthel, J, Kirschner, J, Gosele, U, Fischer, F, Kronmuller, H: Hexagonally ordered 100 nm period nickel nanowire arrays. Appl. Phys. Lett. **79**, 1360 (2001)
3. Sohn, JI, Kim, Y-S, Nam, C, Cho, BK, Seong, T-Y: Fabrication of high-density arrays of individually-isolated nanocapacitors using anodic aluminum oxide templates and carbon nanotubes. App. Phys. Lett. **87**, 123115 (2005)
4. Thompson, GE: Porous anodic alumina: fabrication, characterization and applications. Thin Solid Films **297**(1–2), 192–201 (1997)
5. Masuda, H, Fukada, K: Ordered metal nanohole arrays made by a two-step replication of honeycomb structures of anodic alumina. Science **268**, 1466 (1995)
6. Sun, M, Zangari, G, Shamsuzzoha, M, Metzger, RM: Electrodeposition of highly uniform magnetic nanoparticle arrays in ordered alumite. Appl. Phys. Let **78**, 2964–2966 (2001)
7. Durkan, C, Welland, ME: Size effects in the electrical resistivity of polycrystalline nanowires. The American Physical Society **61**, 20 (2000)
8. Liu, CH, Yiu, WC, Au, FCK, Ding, JX, Lee, CS, Lee, ST: Electrical properties of zinc oxide nanowires and intramolecular p–n junctions. Appl. Phys. Lett. **83**, 3186 (2003)
9. Lao, JY, Huang, JY, Wang, DZ, Ren, ZF: ZnO nanobridges and nanonails. Nano Lett. **3**, 235–238 (2003)
10. Wang, DL, Ruan, YF, Yang, HB, Zhang, LC, HE, L: Photonic band gaps and the luminance properties in anozic aluminum oxide/ZnO system [in Chinese]. J. Syn. Cryst. **38**, 83 (2009)
11. Wernick, S, Pinner, R, Sheasby, PG: In: Sheasby, PG (ed.) The Surface Treatment and Finishing of Aluminium and its Alloy, 5th edn. ASM International, Russell Township, OH (1987)
12. Nielsch, K, Choi, J, Schwirn, K, Wehrspohn, RB, Gösele, U: Self-ordering regimes of porous alumina: the 10% porosity rule. Nano Lett. **2**, 677–680 (2002)

5

Adsorption kinetics, thermodynamic studies, and high performance of CdO cauliflower-like nanostructure on the removal of Congo red from aqueous solution

Azadeh Tadjarodi[1*], Mina Imani[1] and Hamed Kerdari[2]

Abstract

In this work, CdO cauliflower-like nanostructure synthesized by mechanochemical method was employed to evaluate the adsorption ability of Congo red (CR) from the aqueous solution for the first time. UV-visible absorption spectroscopy was used to record the adsorption behavior. This special structure composed of nanorods and tubes with the high contact sites and surface area of 104 $m^2\ g^{-1}$ can be operated as a capable adsorbent to absorb the dye molecules via adsorption process. The adsorption capacity of this material (0.01 g) was studied in high concentrations of CR (50 to 300 mg L^{-1}) and represented an excellent efficiency to eliminate this toxic dye. Maximum adsorption capacity (q_{max}) calculated using Langmuir isotherm model, at room temperature and neutral pH, was found to be 588.24 mg g^{-1}. Electrostatic interactions were conceived as the main adsorption mechanism, and the calculated dimensionless separation factor (R_L), 0.023, indicated a favorable adsorption process. The kinetic and thermodynamic parameters for this proceeding were evaluated and confirmed the high performance of the synthesized adsorbent.

Keywords: Nanostructure; Adsorption; Kinetics; Congo red

Background

Dyes and pigments are the most common environmental pollutants widely used in various industries. Being toxic and harmful for environment and living organisms, these pollutants should be removed. A significant challenge that we deal with such as textile, food, paper, and dyeing industries has always been the wastewaters refinement of toxic contaminants before their discharge to the environment [1,2]. There are some techniques such as coagulation, flocculation, advanced oxidation processes, membrane filtration, and adsorption in order to eliminate the contaminants from wastewaters [3,4]. For this purpose, one of the most effective and economical methods is adsorption procedure which has attracted much attentions in recent years [5]. Many adsorbents are employed to decolorize wastewaters and remove pollutants such as magnetic materials [5], activated carbons [6], fly ash [7], and chitosan/montmorillonite nanocomposite [8]. In fact, using cadmium oxide as an adsorbent to purify the organic wastewaters like Congo red (CR) aqueous solution has not been reported yet. Congo red is a toxic and carcinogenic dye, which despite it being forbidden is still used in some developing countries in the textile industry and via their wastewaters is entered to the environment without treatment. Therefore, CR dye was selected as the model contamination to investigate the capability of the synthesized CdO nanostructure for decolorizing this dye from an aqueous solution.

Cadmium oxide, CdO, is a known n-type semiconductor with the direct band gap energy of 2.5 eV [9]. CdO is applied in solar cells [10], gas sensors [11], transparent electrodes [12], catalysts, photocatalysts [13,14], and photodiodes [15]. Numerous structures of cadmium oxide-like nanoparticles [16], nanowires [17], nanoneedles [18], and nanocrystals [19] have been reported in nanoscale. Among many structures of this compound, cauliflower structure indicates a quick growth in the materials science owing to novel and special morphology. This structure has a great importance due to its high

* Correspondence: tajarodi@iust.ac.ir
[1]Research Laboratory of Inorganic Materials Synthesis, Department of Chemistry, Iran University of Science and Technology, Tehran 16846-13114, Iran
Full list of author information is available at the end of the article

specific surface area and potential applications in various fields [20,21]. We easily synthesized this particular structure using a mechanochemical method, a cost-effective process, followed by heating treatment, which has been presented in our previous work in detail [21].

In this research, we employed the cadmium oxide cauliflower-like nanostructure to remove the CR dye from the solution for the first time. The results of Langmuir isotherm pattern indicate that this compound has the highest adsorption capacity among many reports published up to now. Meanwhile, the adsorption kinetics, thermodynamics parameters, and desorption process for this adsorbent were investigated.

Results and discussion

The structural description of the prepared adsorbent

The structural and morphological characterization of the prepared adsorbent, CdO cauliflower-like nanostructure, was carried out using the techniques clearly described in our previous work [21]. A series of experiments were carried out to investigate the synthesized compound adsorption performance after confirming the purposed special morphology (shown in Figure 1a,b,c,d). The SEM images indicated the cauliflower-like microstructure of this product. By recording the TEM images, it was denoted that this structure has been constructed from the nanorod bundles. In fact, it was found that nanorods are interwoven and formed the cauliflower-like structure.

Meanwhile, the XRD pattern from the prepared product clearly showed the formation of pure cubic CdO phase with lattice constant 4.695 Å and the space group *Fm3m* (Figure 2). The diffraction peaks at 2θ values of 32.90°, 38.20°, 55.20°, 65.80°, and 69.20° are in a close agreement with the 111, 200, 220, 311, and 222 planes (JCPDS-05-0640), respectively.

The surface area determination

The nitrogen adsorption and desorption experiments were performed to evaluate the surface area of this product. A distinct hysteresis loop recorded at the range of $0 < P/P_0 < 1$ indicated the isotherm curve of type V (Figure 3). The Brunauer-Emmett-Teller (BET) surface area of 104 $m^2\ g^{-1}$ proved the presence of high contact sites on the surface of this adsorbent. Likewise, the pore size distribution of 3.53 nm was obtained by the Barrett-Joyner-Halenda (BJH) plot using the desorption branch of the nitrogen isotherm (in the inset of Figure 3). This value introduces a mesoporous structure for the synthesized product. Therefore, there are the high surface contact sites, which can increase the adsorption percentage of dye molecules by this adsorbent in water.

Adsorption study

Effect of contact time and initial pH

The contact time effect of the adsorbent on the adsorption of CR in the range of 0.5 to 3 h was studied to determine the optimal demanded time for removing 100

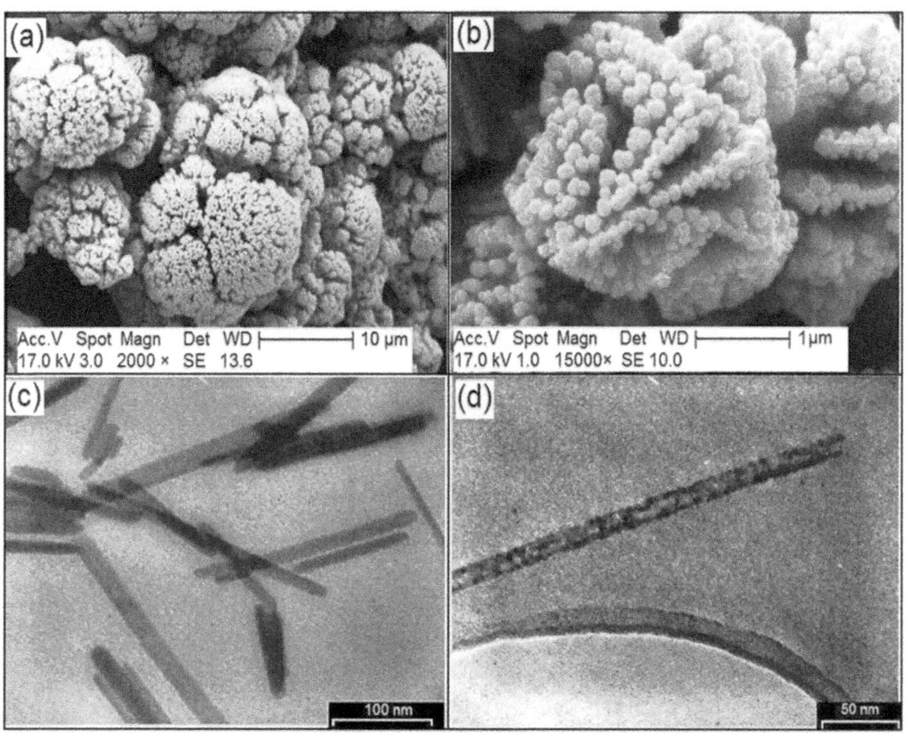

Figure 1 The CDO cauliflower-like nanostructure. (**a** and **b**) SEM images. (**c** and **d**) TEM images.

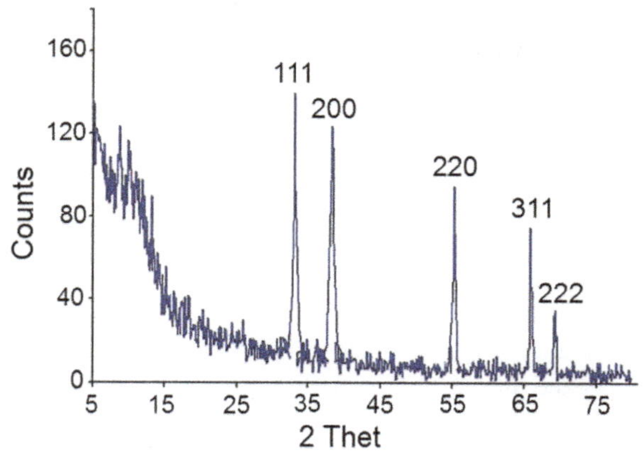

Figure 2 XRD pattern of the prepared product.

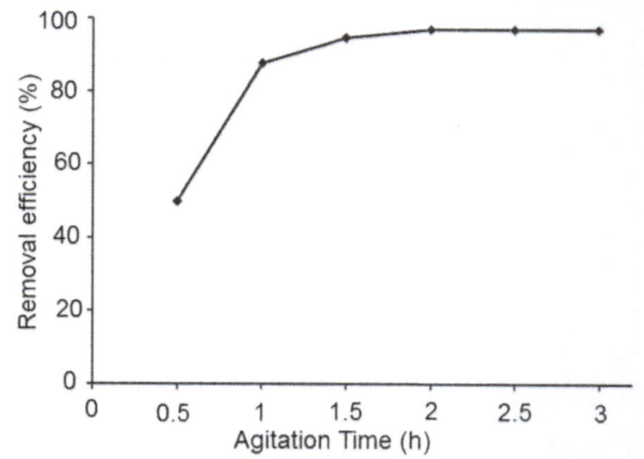

Figure 4 Removal of CR dye in various times. Conditions, 0.01 g CdO nanostructure, 25 mL of 100 mg L^{-1} CR, and neutral pH.

mg L^{-1} of CR solution at neutral pH. A certain amount of adsorbent, 0.01 g of CdO nanostructure, was added into 25 ml of CR solution. The adsorption of CR molecules on the adsorbent led to a decrease in the concentration of CR with time (Figure 4). According to this diagram, the maximum removal efficiency of CR dye (nearly 100%) occurs for about 3 h, and after this time, the adsorption efficiency of CR is constant. Therefore, agitation time of 3 h was selected for further studies. The pH is a significant controlling parameter that can intensely influence the adsorption efficiency of cationic or anionic dyes onto the adsorbent sites. The effect of solution pH on the adsorption percentage of CR by the cauliflower CdO nanostrucutre was studied via adjusting the pH of aqueous medium at the range of 4 to 9 using HCl and NaOH solutions (0.01 mol L^{-1}). The screening of these experiments indicated the highest removal efficiency of CR in a solution with pH of 6, which is close to the neutral pH (shown in Figure 5). As a result, the neutral pH was selected as the appropriate pH for per run.

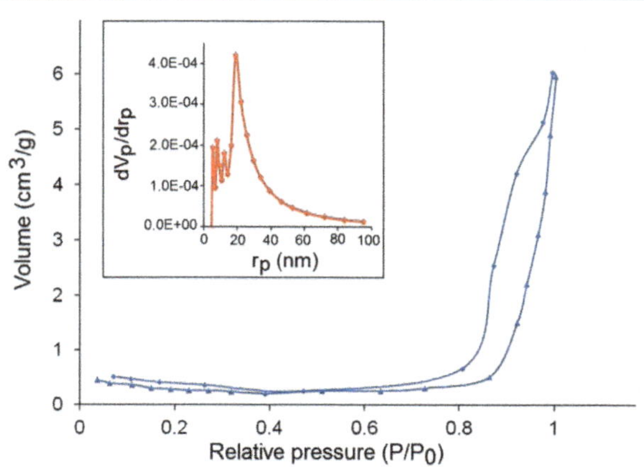

Figure 3 Nitrogen adsorption (diamond) and desorption (triangle) isotherm for the CdO cauliflower-like nanostructure. The inset shows the BJH plot of this product.

In fact, adsorption proceeding at neutral pH is a beneficial performance that few adsorbents show this feature. Most reports provided by researchers to remove the CR dye are at pH 5 and 6 [22,23].

Effect of the amount of CdO nanostructures

The effect of the amount of the adsorbent required for the maximum adsorption percentage of CR dye is shown in Figure 6. Different quantities of CdO nanostructures were investigated in the range of 0.005 to 0.015 g. Maximum percentage was specified when 0.01 g of CdO cauliflower-like nanostructure was used in 25-mL solution of CR dye with 100 mg L^{-1} concentration at neutral pH. Therefore, further studies were conducted on CR dye at an optimum amount of mentioned adsorbent.

Adsorption isotherm

Figure 7 illustrates the adsorption capacity of CdO cauliflower-like structure for the CR solution, which is studied by measuring the initial and final concentration of CR after 3 h of agitation in the dark. It was observed

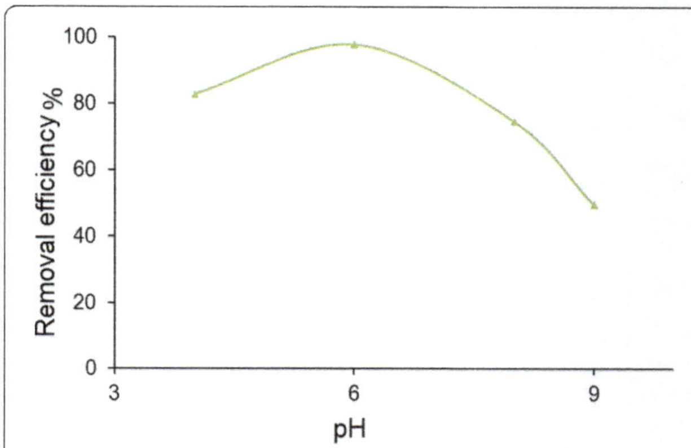

Figure 5 The adsorption amount of CR in various pHs. Conditions, 0.01 g CdO nanostructure, 25 mL of 100 mg L^{-1} CR.

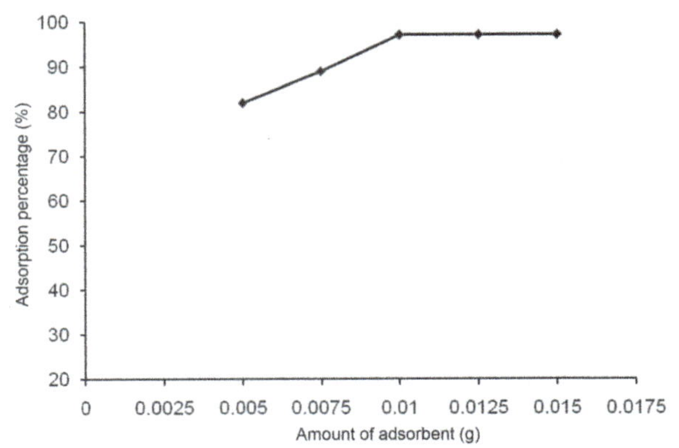

Figure 6 The effect of the amount of adsorbent on the adsorption of CR. Conditions, 25 mL of CR dye with 100 mg L^{-1} concentration and neutral pH.

that the adsorption of CR molecules increases with an enhancement in dye concentration and inclines to reach the saturation point at higher concentrations (250, 275, and 300 mg L^{-1}). Langmuir and Freundlich equations, the famous adsorption isotherm models, were applied to examine the relationship between the amount of CR adsorbed onto the CdO particles and its equilibrium concentration in solution. In fact, these adsorption isotherms were employed to interpret the interactions between CR molecules and adsorbent.

The linearized forms of Langmuir (Equation 1) and Freundlich (Equation 2) isotherm models are as follow:

$$\frac{C_e}{q_e} = C_e\left(\frac{a_L}{K_L}\right) + \left(\frac{1}{K_L}\right) \tag{1}$$

$$\text{Log}\, q_e = \text{Log}\, K_F + \frac{1}{n}\ \text{Log}\, C_e \tag{2}$$

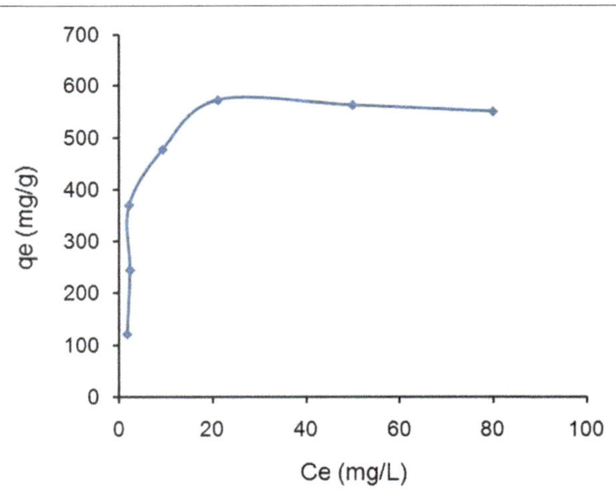

Figure 7 The adsorption isotherm for Congo red (CR) on the prepared CdO.

The parameters of these two isotherm models were calculated and given in Table 1. a_L (L mg^{-1}) and K_L (L g^{-1}) are the Langmuir constants. These constants were calculated from the slope and intercept of the plot of C_e/q_e vs. C_e shown in Figure 8a. K_F (mg$^{1-1/n}$ L$^{1/n}$ g^{-1}), and $n$is the Freundlich adsorption isotherm constant, which were obtained from the slope and intercept of linear plot of Log q_e vs. Log C_e (Figure 8b). In these equations, C_e is the equilibrium concentration of the CR in the solution (mg L^{-1}). Meanwhile, q_e, the amount of CR adsorbed (mg g^{-1}) per unit of adsorbent at equilibrium (mg g^{-1}), is calculated by the following Equation (3) [24]:

$$q_e = \frac{V(Ci-Cf)}{m} \tag{3}$$

where, C_i and C_f are the initial and final concentrations of CR in milligrams per liter, respectively; V is the volume of experimental solution in liters, and m is the weight of CdO cauliflower nanostructure in grams.

The results indicated that this adsorption process does not follow the Freundlich isotherm model but is in a good agreement with Langmuir model with reference to the obtained value of regression coefficients (R = 0.997). Although Langmuir's model does not consider the variation in adsorption energy, it obviously describes the adsorption method. It is based on the physical theory that the maximum adsorption capacity includes of a monolayer adsorption [22].

The maximum adsorption capacity (milligrams per gram) is indicated by [$q_m = K_L/a_L$], and it highly depends on the number and structure of adsorption sites. In our pervious study, we reported that cauliflower structure has been composed of tubes and rods with the average size of 68 nm [21]. As long as there are unoccupied sites, adsorption process will resume with increasing CR concentrations. However, as soon as all of the adsorbent sites are filled, an additional increase in the concentrations of CR solutions does not increase the amount of CR on adsorbent [22].

The maximum adsorption capacity ($q_{\max}$) for the adsorption of CR onto the CdO cauliflower structure was found to be 588.24 mg g^{-1}. This value is the highest value among the other adsorbents, which have been reported until now (see Table 2).

In fact, the high removal efficiency of CR from aqueous solution by cauliflower-like nanostructure can be referred to the presence of enormous adsorbing sites on the surface of this structure. An electrostatic interaction is established between these adsorbing sites and dye species, which lead to remove the CR molecules from the solution in a short time.

Table 1 Parameters of Langmuir and Freundlich isotherm equations

Langmuir model				Freundlich model		
a_L (L mg^{-1})	K_L (L g^{-1})	q_{max} (mg g^{-1})	R^2	K_F (mg$^{1-1/n}$ L$^{1/n}$ g^{-1})	n	R^2
0.36	212.76	588.24	0.997	121.92	2.63	0.919

Regression coefficients (R^2) for adsorption of CR onto CdO nanostructure at 25°C and neutral pH.

In order to express the essential characteristics of the presented Langmuir isotherm, a dimensionless constant separation factor (R_L) was used (Equation 4):

$$R_L = \frac{1}{(1 + K_L\, C_0)} \tag{4}$$

Generally, the nature of adsorption process can be described by several terms of R_L value; unfavorable ($R_L > 1$), linear ($R_L = 1$), favorable ($0 < R_L < 1$), and irreversible ($R_L = 0$). The obtained R_L value of 0.023 revealed that the adsorption of CR on the CdO cauliflower-like nanostructure is a favorable process.

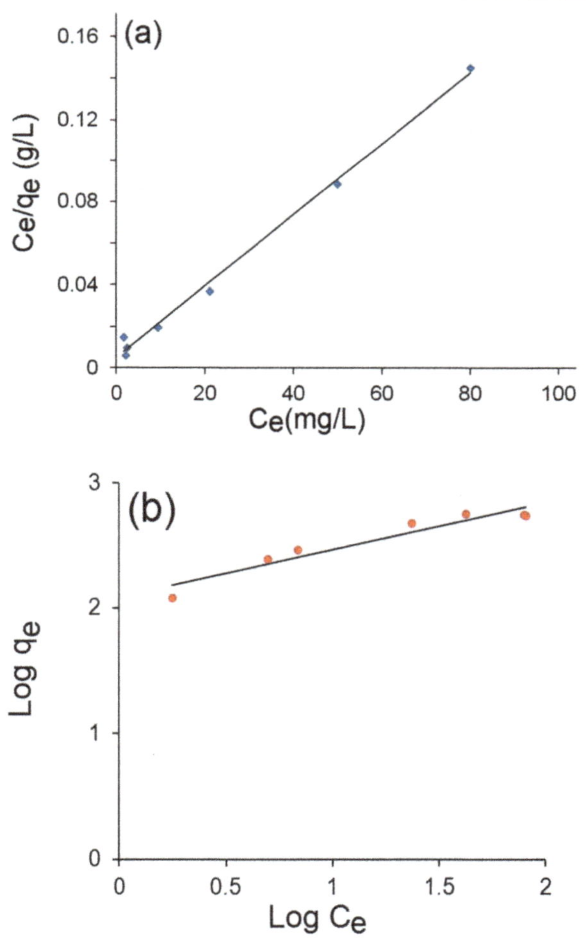

Figure 8 Langmuir (a) and Freundlich (b) adsorption isotherm models for the adsorption of CR. Conditions, ambient temperature, neutral pH, 25 mL of CR dye with known initial concentrations (in the range of 50 mg L^{-1}to 300 mg L^{-1}), and 0.01 g of adsorbent.

Adsorption kinetics

Figure 9 represents two kinetic models of pseudo-first-order (Figure 9a) and pseudo-second-order (Figure 9b) rate equations of adsorption process from CR dye onto CdO adsorbent as follow:

$$\text{Log}\,(q_e - q_t) = \text{Log}\, q_e - \frac{k_1 t}{2.303} \tag{5}$$

$$\frac{t}{q_t} = \frac{1}{k_2 q_e^2} + \frac{t}{q_e} \tag{6}$$

where, q_e and q_t are the amount of dye adsorbed (mg g^{-1}) at equilibrium and time, t (min), respectively. The parameters of k_1 and k_2 are the rate constants of pseudo-first-order (min^{-1}) and pseudo-second-order (g mg^{-1} min^{-1}) equations, which were calculated from the linear plots of log (q_e − q_t) vs. t and (t/q_t) vs. t, respectively. Table 3 shows the calculated kinetic parameters of both rate models. Comparing the correlation coefficients (R^2) of the mentioned models, the results revealed that the pseudo-second-order rate model is a better fit than the pseudo-first-order according to experimental adsorption data. In fact, this good agreement to the pseudo-second-order kinetic model presents the dependence of adsorption mechanism on the adsorbate and adsorbent [26].

Adsorption thermodynamic

The thermodynamic parameter of the performed adsorption process was evaluated through calculating of free

Table 2 Summary of reported CR adsorption capacities of various adsorbents

Type of adsorbent	q_{max} (mg g^{-1})	Reference
Bagasses fly ash	11.89	Acemioğlu [7]
Chitosan montmorillonite	96.62	Wang [8]
Maghemite nanoparticles	208.33	Afkhami and Moosavi [22]
Anilinepropylsilica xerogel	22.62	Pavan et al. [23]
CTAB-modified chitosan beads	352.5	Chatterjee et al. [24]
Fe_3O_4	149.7	Wang et al. 2012 [25]
$CoFe_2O_4$	244.5	Wang et al. 2012 [25]
CdO cauliflower structure	588.2	Present work

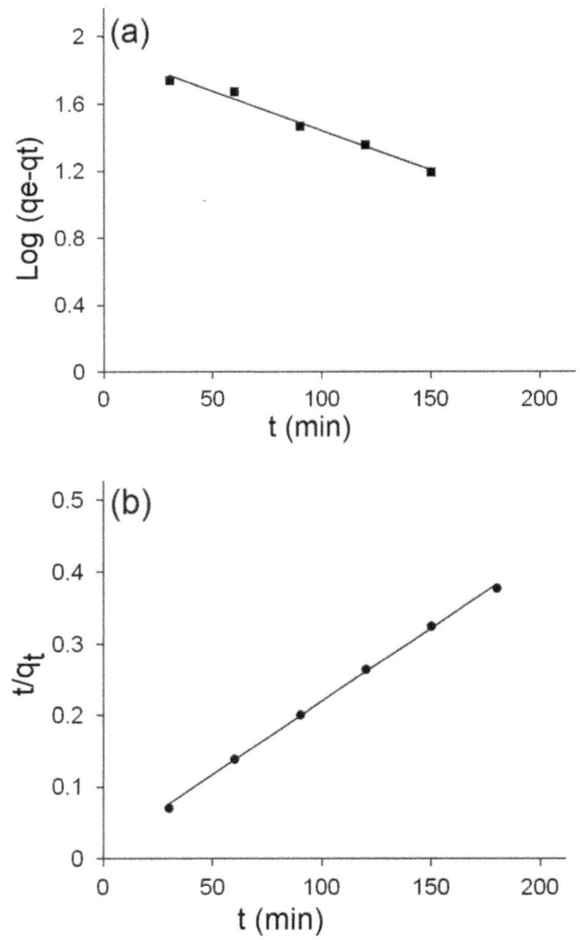

Figure 9 Adsorption kinetic for adsorption of CR on CdO cauliflower-like nanostructure. (a) Pseudo-first order and **(b)** Pseudo-second-order rate models (adsorption conditions, initial dye concentration 200 mg L^{-1}, natural pH, room temperature).

energy change (delta *G*, kilojoules per mole) and using the following equation (7):

$$\Delta G = -RT \ln K_c \quad (7)$$

Where, R is the gas constant (8.314 J mol^{-1} K^{-1}), T (K) is the temperature, and K_c (mL/g) is the standard thermodynamics equilibrium constant:

$$K_c = \left(\frac{q_e}{C_e}\right) \quad (8)$$

The obtained value of ΔG value is −11.48 kJ mol^{-1} (T = 298 K). This negative value of free energy change indicates the spontaneous nature of adsorption process of CR dye on this adsorbent [25]. The mentioned result confirms the high performance of CdO cauliflower-like nanostructure to remove the CR dye from aqueous solution by using adsorption process.

Table 3 The constant parameters of different rate models

Pseudo-first-order equation				Pseudo-second-order equation		
$q_{e(exp)}$ (mg g^{-1})	$q_{e(cal)}$ (mg g^{-1})	k_1 (min^{-1})	R^2	$q_{e(cal)}$ (mg g^{-1})	k_2 (g mg^{-1} min^{-1})	R^2
476.60	80.76	1.08×10^{-2}	0.983	476.19	3.02×10^{-4}	0.998

Desorption procedure

The adsorbed dye molecules were easily desorbed using an appropriate amount of acetone with magnetic stirring for 1 h. Desorption efficiency of 74% was calculated by relation of (9) as follows:

$$\text{Desorption efficiency } (\%) = \frac{\text{Amount of desorbed CR}}{\text{Amount of adsorbed CR}} \times 100 \quad (9)$$

It was found that such adsorbent not only possesses the high adsorption capability but also presents a good desorption efficiency for CR dye. Likewise, the implemented adsorbent was efficiently recovered by a heating treatment at 450°C for 2 h. The maximum adsorption capacity and high desorption ratio are the merits of this type of adsorbent to reduce the charges of purification treatments.

Conclusions

In the present work, the CdO cauliflower-like nanostructure was applied to remove the CR dye from aqueous solution in high concentrations (50 to 300 mg L^{-1}). The analysis of adsorption isotherm showed that this adsorption experiment is in a well accordance with the Langmuir model. The calculated maximum adsorption capacity (q_{max}) for this adsorbent was observed to be 588.24 mg g^{-1}, which is the highest value in comparison with the previous reports. It is inferred that the special structure of this compound having the numerous number of surface sites and strong interactions leads to this excellent adsorption performance and water treatment. Meanwhile, it was found that the adsorption process follows a pseudo-second-order rate model much better than a pseudo-first-order rate model. The kinetics and thermodynamics findings indicated the more efficient performance of this adsorbent to remove the CR dye from aqueous solution at a short time and without any additives. Therefore, this compound can

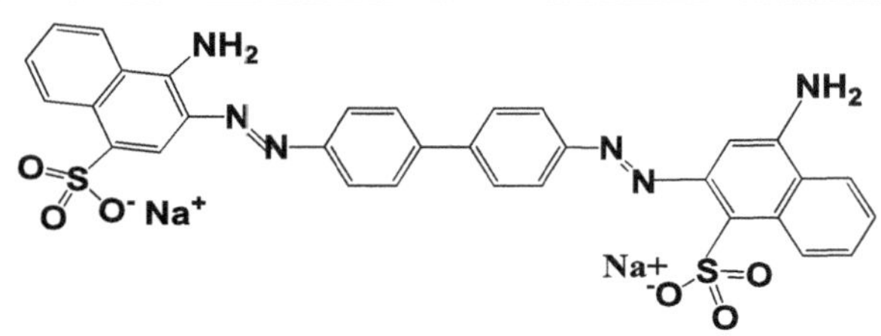

Figure 10 Structure of CR molecule.

be a promising candidate among other counterparts for water treatment.

Methods

Chemicals

Cadmium acetate dehydrate ($Cd(CH_3COO)_2.2H_2O$, pure, Merck, Whitehouse Station, NJ, USA) and acetamide (CH_3CONH_2, pure, Merck) were provided to synthesize the CdO adsorbent. Congo red (sodium sodium 3,3′-([1,1′-biphenyl]-4,4′-diyl)bis(4-aminonaphthalene-1-sulfonate) as a pollutant model was supplied from commercial source. All the chemicals were used without further purification.

Synthesis of adsorbent

The cauliflower-like nanostructure of cadmium oxide, CdO, was synthesized using mechanochemical method similar to the process reported in our previous work [21]. Briefly, $Cd(CH_3COO)_2$ $2H_2O$ and CH_3CONH_2 as the starting materials with a molar ratio of 3:4 were milled to each other at the milling rate of 1,800 rpm for 30 min. Then, the resulting precursor was calcined at 450°C for 2 h in air to prepare the product. The obtained product, CdO cauliflower-like nanostructure, was employed for dye adsorption proceeding.

Adsorption experiment

The Congo red dye was selected as an adsorbate to evaluate the performance of the resulting product at the adsorption proceeding of dye from the aqueous solution. The structure of this organic azo dye is illustrated in Figure 10. Adsorption operation was performed under the following conditions: 0.01 g of the prepared CdO cauliflower nanostructures as the adsorbent was added to 25 mL of CR dye aqueous solution with a known initial concentration (in the range of 50 to 300 mg L^{-1}) at neutral pH. The obtained suspension was stirred using a magnetic stirrer at room temperature for 3 h in the dark. Then, CR-loaded CdO powder was separated with centrifugation at 1,800 rpm for 5 min. In the specified time periods, the portions of suspension were taken away from the reaction vessel, and the concentration of the residual dye was measured using UV–vis spectrophotometer at an appropriate wavelength corresponding to the maximum absorption of CR (498 nm).

The percent removal of dye from the solution can be calculated by the following equation:

$$\%\ \text{Removal efficiency} = \frac{C_0 - C_i}{C_0} \times 100 \qquad (10)$$

where, C_0 is the initial concentration of dye, and C_i is the final concentration of dye after treatment with CdO nanostructure. The removal percentage of CR molecule increased with time due to its adsorption onto adsorbent sites. A double-beam UV spectrophotometer (Shimadzu UV-1700, Kyoto, Japan) was used to determine CR concentration in the supernatant solutions before and after adsorption.

Characterization

The prepared adsorbent, CdO cauliflower-like nanostructure, was characterized by utilizing the techniques clearly described in previous work reported in [21]. The surface area of the product was obtained by using the BET technique with Micromeritics (Gemini, Norcross, GA, USA) in the range of relative pressures from 0.0 to 1.0. Before employing, the sample was degassed at 200°C for 2 h. In addition, the pore size distribution was determined from the desorption branch of the isotherm curve using the BJH model.

Competing interests

The authors declare that they have no competing interests.

Authors' contributions

AT designed the related subject, managed all of the process, and conceived this study. MI carried out all of the experimental section and drafted the manuscript. HK participated in the manuscript preparation. All authors read and approved the final manuscript.

Acknowledgements

The financial support of this study by Iran University of Science and Technology and Iranian Nanotechnology Initiative is gratefully acknowledged.

Author details

[1]Research Laboratory of Inorganic Materials Synthesis, Department of Chemistry, Iran University of Science and Technology, Tehran 16846-13114, Iran. [2]Department of Chemistry, Saveh Branch, Islamic Azad University, Saveh 39187-366, Iran.

References

1. Guo, H, Lin, K, Zheng, Z, Xiao, F, Li, S: Sulfanilic acid-modified $P_{25}TiO_2$ nanoparticles with improved photocatalytic degradation on Congo red under visible light. Dye. Pigment. **9**, 1278–1284 (2012)
2. Ma, J, Jia, Y, Jing, Y, Yao, Y, Sun, J: Kinetics and thermodynamics of methylene blue adsorption by cobalt-hectorite composite. Dye. Pigment. **93**, 1441–1446 (2012)
3. Verma, AK, Dash, RR, Bhunia, P: A review on chemical coagulation/flocculation technologies for removal of colour from textile wastewaters. J. Environ. Manag. **93**, 154–168 (2012)
4. Cheng, S, Oatley, DL, Williams, PM, Wright, CJ: Characterisation and application of a novel positively charged nanofiltration membrane for the treatment of textile industry wastewaters. Water Res. **46**, 33–42 (2012)
5. Rahimi, R, Kerdari, H, Rabbani, M, Shafiee, M: Synthesis, characterization and adsorbing properties of hollow Zn-Fe_2O_4 nanospheres on removal of Congo red from aqueous solution. Desalination **280**, 412–418 (2011)
6. Purkait, MK, Maiti, A, DasGupta, S, De, S: Removal of Congo Red using activated carbon and its regeneration. J. Hazard. Mater. **145**, 287–295 (2007)
7. Acemioğlu, B: Adsorption of Congo red from aqueous solution onto calcium-rich fly ash. J. Colloid Interface Sci. **274**, 371–379 (2004)
8. Wang, L, Wang, A: Adsorption characteristics of Congo Red onto the chitosan/montmorillonite nanocomposite. J. Hazard. Mater. **147**, 979–985 (2007)
9. Grado-Caffaro, MA, Grado-Caffaro, M: A quantitative discussion on band-gap energy and carrier density of CdO in terms of temperature and oxygen partial pressure. Phys Lett A **372**, 4858–4860 (2008)
10. Yakuphanoglu, F: Nanocluster n-CdO thin film by sol-gel for solar cell applications. Appl. Surf. Sci. **257**, 1413–1419 (2010)
11. Kamble, AS, Pawar, RC, Patil, JY, Suryavanshi, SS, Patil, PS: From nanowires to cubes of CdO: ethanol gas response. J. Alloys Compd. **509**, 1035–1039 (2011)

12. Gupta, RK, Ghosh, K, Patel, R, Kahol, PK: Low temperature processed highly conducting, transparent, and wide bandgap Gd doped CdO thin films for transparent electronics. J. Alloys Compd. **509**, 4146–4149 (2011)
13. Singh, G, Kapoor, IPS, Dubey, R, Srivastava, P: Synthesis, characterization and catalytic activity of CdO nanocrystals. Mater Sci Eng: B **176**, 121–126 (2011)
14. Li, J, Ni, Y, Liu, J, Hong, J: Preparation, conversion, and comparison of the photocatalytic property of $Cd(OH)_2$, CdO. CdS and CdSe. J. Phys. Chem. Solid **70**, 1285–1289 (2009)
15. Yakuphanoglu, F, Caglar, M, Caglar, Y, Ilican, S: Electrical characterization of nanocluster n-CdO/p-Si heterojunction diode. J. Alloys Compd. **506**, 188–193 (2010)
16. Dong, W, Zhu, C: Optical properties of surface-modified CdO nanoparticles. Opt. Mater. **22**, 227–233 (2003)
17. Chang, Q, Chang, C, Zhang, X, Ye, H, Shi, G, Zhang, W, Wang, Y, Xin, X, Song, Y: Enhanced optical limiting properties in suspensions of CdO nanowires. Opt. Commun. **274**, 201–205 (2007)
18. Liu, X, Li, C, Han, S, Han, J, Zhou, C: Synthesis and electronic transport studies of CdO nanoneedles. Appl. Phys. Lett. **82**, 1950–1952 (2003)
19. Tadjarodi, A, Imani, M: Synthesis and characterization of CdO nanocrystalline structure by mechanochemical method. Mater. Lett. **65**, 1025–1027 (2011)
20. Ren, H-X, Huang, XJ, Yarimaga, O, Choi, YK, Gu, N: A cauliflower-like gold structure for superhydrophobicity. J. Colloid Interface Sci. **334**, 103–107 (2009)
21. Tadjarodi, A, Imani, M: A novel nanostructure of cadmium oxide synthesized by mechanochemical method. Mater. Res. Bull **46**, 1949–1954 (2011)
22. Afkhami, A, Moosavi, R: Adsorptive removal of Congo red, a carcinogenic textile dye, from aqueous solutions by maghemite nanoparticles. J. Hazard. Mater. **174**, 398–403 (2010)
23. Pavan, FA, Dias, SLP, Lima, EC, Benvenutti, EV: Removal of Congo red from aqueous solution by anilinepropylsilica xerogel. Dye. Pigment. **76**, 64–69 (2008)
24. Chatterjee, S, Lee, DS, Lee, MW, Woo, SH: Enhanced adsorption of congo red from aqueous solutions by chitosan hydrogel beads impregnated with cetyl trimethyl ammonium bromide. Bioresource Technol **100**, 2803–2809 (2009)
25. Wang, L, Li, J, Wang, Y, Zhao, L, Jiang, Q: Adsorption capability for Congo red on nanocrystalline MFe_2O_4 (M= Mn, Fe, Co, Ni) spinel ferrites. Chem. Eng. J. **181–182**, 72–79 (2012)
26. Ayad, M, Abu El-Nasr, A: Anionic dye (acid green 25) adsorption from water by using polyaniline nanotubes salt/silica composite. J Nanostruct Chem **3**, 3–11 (2012)

Catalytic effect of commercial nano-CuO and nano-Fe_2O_3 on thermal decomposition of ammonium perchlorate

Mostafa Mahinroosta

Abstract

Size reduction of the catalyst increases the surface area, and hence, the catalytic activity is also increased. The decrease in temperature of decomposition of ammonium perchlorate in the presence of nano-copper oxide and nano-ferric oxide is investigated within the scope of this study. Different mixes of ammonium perchlorate with nano-ferric oxide and nano-copper oxide were prepared. Differential scanning calorimetry test results show that addition of nanometer-sized ferric oxide and copper oxide leads to a significant decrease in higher decomposition temperature of ammonium perchlorate. The most significant decrease in the decomposition temperature was observed in the presence of 3% of nano-copper oxide.

Keywords: Ammonium perchlorate; Thermal decomposition; Catalyst; Nanostructure

Background

At nanoscale, the physicochemical characteristics of materials are considerably various from those found in larger scales and bulk materials [1]. Nanoparticles have many noteworthy properties due to their small size and very large specific surface area [2]. There is a growing interest in the synthesis and use of metal or metal oxide nanoparticles due to their large specific surface area and high activity in most catalytic processes [3,4]. It is noted that nanoparticles are not only effective in the catalytic processes such as gas-solid, gas–liquid–solid or liquid–solid, where the nanoparticles, as the catalyst, are the solid phase, but also effective in the solid-solid catalytic processes [5]. Transition metal oxide nanoparticles exhibit a broad class of materials that have been investigated extensively due to their interesting catalytic properties and wide scope of their potential applications [6]. It has been proven that nanometer-sized metals and metal oxides are effective to improve the decomposition of ammonium perchlorate (AP) [3-5,7-14]. Although the characteristics of AP thermal decomposition can be tailored to some extent by decreasing the particle size of AP, this method is rarely used due to the dangerousness of superfine AP [5,15]. Particle size distribution, morphology, and nanostructure of particles are very important characteristics and affect the kinetics of decomposition of AP [4]. Many researchers have studied the catalytic effects of nanostructures such as nanometer-sized metals and metal oxides on the thermal decomposition of AP and observed their tailoring effect on the thermal decomposition behavior of AP [3-5,7-14]. Table 1 presents a historical background about some important events in the field of the effect of nanostructures on thermal decomposition of AP. In this paper, the catalytic effect of commercial nano-sized ferric oxide and nano-sized copper oxide on the thermal decomposition of AP is investigated.

Correspondence: mahinroosta2010@gmail.com
No. 8, Negarestan 4, Nazarabad, Alborz 3331779638, Iran

Results and discussion

Transmission electron microscopy (TEM) images of nano-CuO and nano-Fe_2O_3 are shown in Figure 1. From this figure, it can be observed that the morphology of these nanoparticles is relatively spherical. Figure 2 shows the X-ray diffraction (XRD) patterns of commercial nano-CuO and nano-Fe_2O_3.

In Figure 2(a), the major peaks located at 2θ values of 20° to 75° correspond to the characteristic diffractions of monoclinic phase CuO [9]. Figure 2(b) can be readily indexed as the rhombohedral crystalline phase [11].

Table 1 Bibliographic history of the effect of nanostructure on thermal decomposition of AP

Author	Year	Significance
Weifan et al. [7]	2006	Nano-sized yttria
Meda et al. [8]	2007	Nano-aluminum
Duan et al. [16]	2007	Nano-sized MgO
Liu et al. [15]	2008	Nanometer-sized $CuFe_2O_4$
Duan et al. [5]	2008	Ni nanoparticles
Hongzhen et al. [3]	2008	Co nanoparticles
Satyawati et al. [4]	2008	Nano-ferric oxide
Wang et al. [9]	2009	CuO nanocrystals
Aijun et al. [10]	2011	Nano-$MnFe_2O_4$
Zhang et al. [11]	2011	Nano α-Fe_2O_3
Zhou et al. [12]	2011	ZnO/AP, Co_3O_4/AP, Fe_2O_3/AP nanocomposites
Chaturvedi and Dave [13]	2011	Nanometals
Alizadeh et al. [14]	2012	Nano-sized CuO, Co_3O_4, and $CuCo_2O_4$

Differential scanning calorimetry (DSC) curves for the thermal decomposition of AP in the presence of nano-sized CuO are illustrated in Figure 3. As can be seen in this figure, the thermal decomposition of pure AP occurs in two stages: the endothermic stage and the exothermic stage. By comparing these four curves, the endothermic stage happens at a temperature of about 240°C. This stage is related to crystal phase transition from orthorhombic to cubic. The other two peaks at 314°C and 455°C are the exothermic peaks that are associated with partial decomposition of AP and the formation of some NH_3 and $HClO_4$ via dissociation and sublimation and complete decomposition of AP and the formation of volatile products, respectively. It is obvious that the addition of nano-sized CuO to AP has no deep effect on the crystallographic phase transition temperature of AP, but it has considerable influence on the exothermic decomposition of AP. Two exothermic peaks are converted to a strong peak in the result of added nano-sized CuO, and this indicates the catalytic activity of nano-sized CuO in the thermal decomposition of AP. During the exothermic process, higher thermal decomposition temperatures of AP with nano-CuO contents of 1%, 2%, and 3% are 359°C, 348.9°C, and 347.12°C, respectively. At curves b, c, and d of Figure 3, it can be observed that the peaks become steep in the presence of nano-CuO which indicates that the rate of thermal decomposition of AP also increased.

Figure 4 presents the DSC curves for thermal decomposition of AP in the presence of nano-sized Fe_2O_3. From these curves, it can be found out that the thermal decomposition of AP is catalytically affected by nano-sized Fe_2O_3. The endothermic stage occurring at a temperature of about 240°C is not greatly influenced by nano-ferric oxide. It is clear that the addition of nano-sized Fe_2O_3 to AP has a significant effect on the exothermic temperatures of AP. During the exothermic process, higher thermal decomposition temperature of AP with nano-Fe_2O_3 contents of 1%, 2%, and 3% are 403.57°C, 374.18°C, and 347.87°C, respectively. Rate enhancement of thermal decomposition of AP in the presence of nano-Fe_2O_3 is drawn from steepness modes of the curves b, c, and d.

The results associated with decrease in higher thermal decomposition of AP in the presence of nano-sized CuO and nano-sized Fe_2O_3 are given in Table 2. As it is clear from this table, when nanometer-sized CuO and Fe_2O_3 are added to AP, the higher decomposition of AP is decreased considerably. From this table, it is obvious that the catalytic effect of adding 1% and 2% of nano-sized CuO with an average particle size of 40 nm is greater

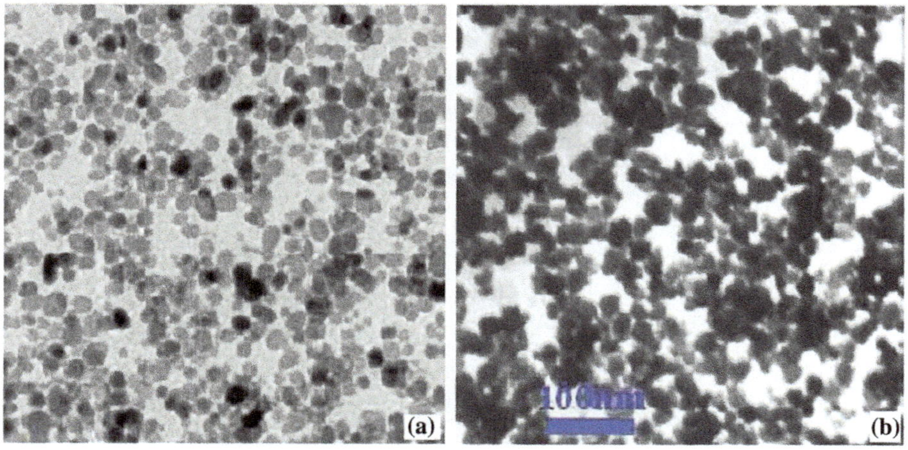

Figure 1 TEM images of (a) nano-CuO and (b) nano-Fe_2O_3.

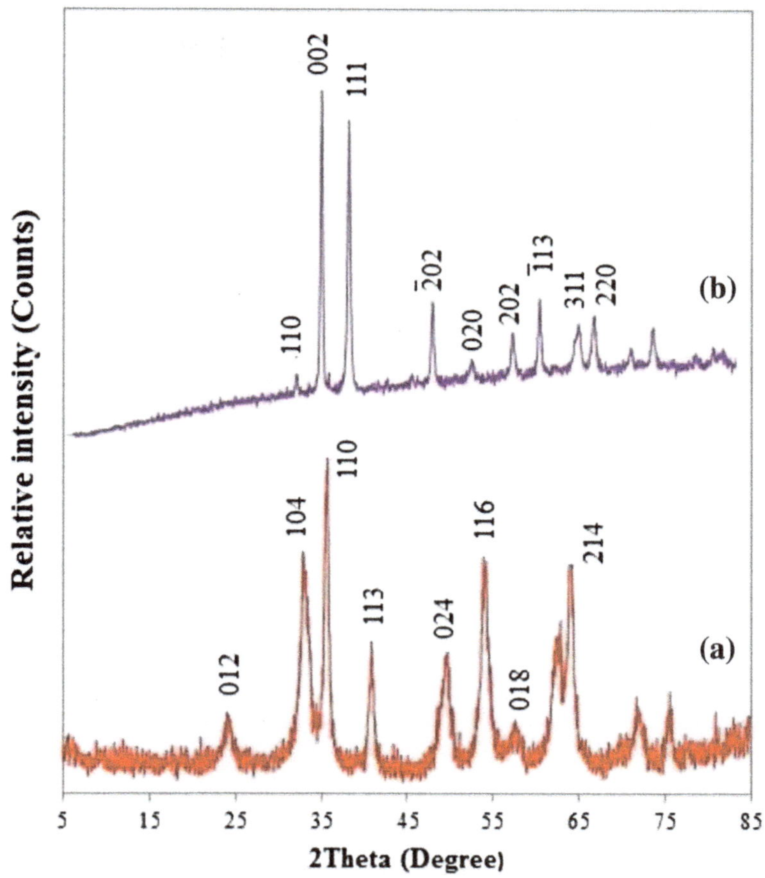

Figure 2 XRD patterns of (a) nano-Fe_2O_3 and (b) nano-CuO.

than that of the nano-sized Fe_2O_3 with an average particle size of 20 nm.

In Table 3, the results of the catalytic effect of as-synthesized nano-CuO and nano-Fe_2O_3 in some recent studies are given. As can be seen from this table, the catalytic effect of as-synthesized nano-CuO is higher than that of the as-synthesized Fe_2O_3. A similar comparison associated with commercial nano-sized CuO and nano-sized Fe_2O_3 is also concluded from this study.

Conclusions

The results of differential scanning calorimetry illustrate that nanometer-sized copper oxide and ferric oxide have a significant catalytic effect on the thermal decomposition of ammonium perchlorate. The presence of these nano-sized metal oxides reduces significantly the higher decomposition temperature of ammonium perchlorate. With increase of content of nanometer-sized metal oxide, the decrease in higher decomposition temperature of ammonium perchlorate becomes greater. Also, the catalytic effect of nano-sized copper oxide with larger particle size is more sizable than that of the nano-sized ferric oxide.

Methods

Materials

Ammonium perchlorate (monomodal 120 μm) was purchased from Merck (Darmstadt, Germany). Commercial nano-CuO and nano-Fe_2O_3 were purchased from Pishgaman Company located in Mashhad, Iran. Physical properties of nano-CuO and nano-Fe_2O_3 such as bulk density, actual density, specific surface area, and average particle size are given in Table 4. The Chemical compositions of these two nano-sized metal oxides are given in Tables 5 and 6.

Methods

X-ray diffraction analysis

XRD patterns of nano-CuO and nano-Fe_2O_3 was performed with a Philips (Amsterdam, The Netherlands) PW 1800 powder X-ray diffractometer using CuKα radiation at 40 kV and 30 mA.

Transmission electron microscopy

TEM images of nanoparticles were prepared on a Philips (Amsterdam, The Netherlands) transmission electron microscope operated at an accelerating voltage of 100 kV.

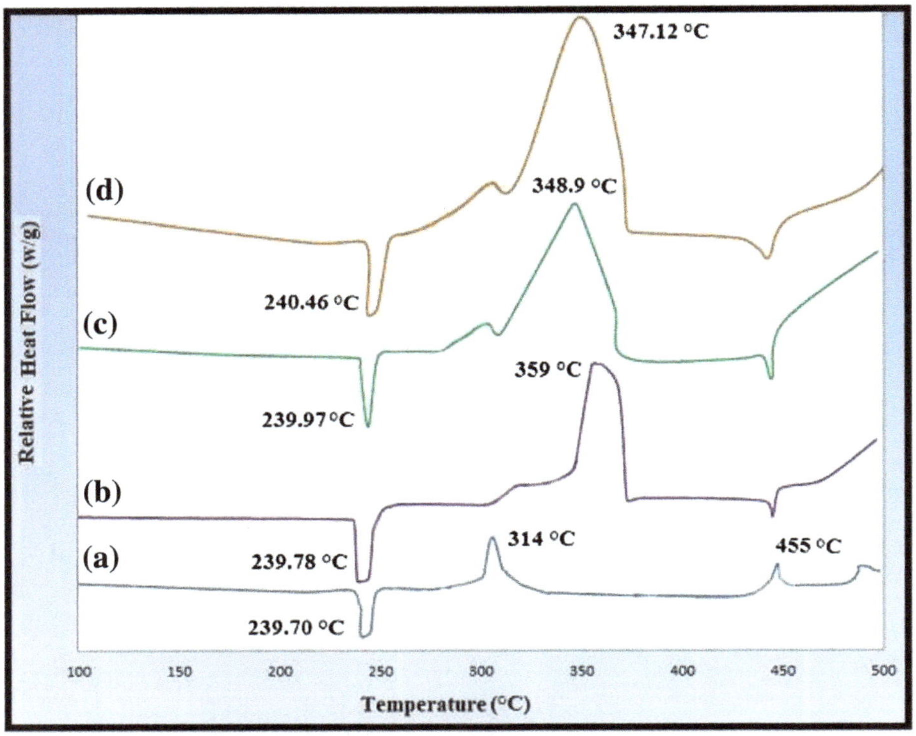

Figure 3 DSC curves related to (a) pure AP, (b) AP1C, (c) AP2C, and (d) AP3C.

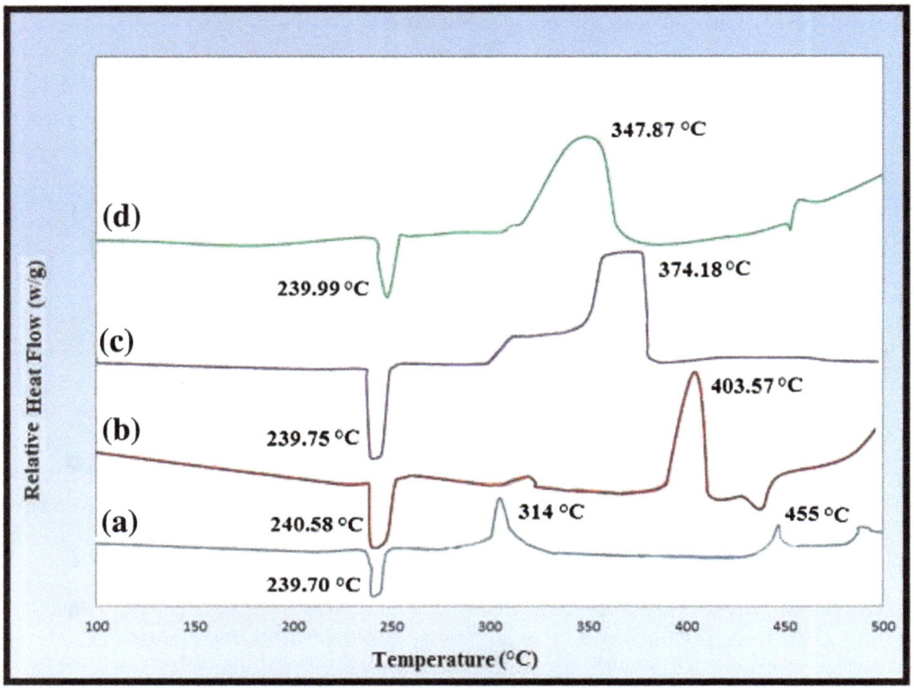

Figure 4 DSC curves related to (a) pure AP) (b) AP1F, (c) AP2F, and (d) AP3F.

Table 2 Decreased values of higher decomposition temperature of AP

	AP + nano-sized Fe_2O_3			AP + nano-sized CuO		
	3%	2%	1%	3%	2%	1%
Decrease in higher decomposition temperature (°C)	107.13	80.82	51.43	107.88	106.1	96

Table 3 Catalytic activity results of as-synthesized nano-sized CuO and nano-sized Fe_2O_3

Sample	Average particle size (nm)	Decrease in higher decomposition temperature of AP (°C)	Reference
AP + nano-CuO (chrysalis-like)	-	85	[9]
AP + nano-CuO	16.5	90.47	[14]
AP + nano-Fe_2O_3	30	57	[4]
AP + nano-Fe_2O_3 (sphere-like)	25	81	[11]
AP + nano-Fe_2O_3 (pod-like)	-	72	[11]

Table 4 Physical properties of nano-CuO and nano-Fe_2O_3

Nano-metal oxide	Bulk density (g/cm^3)	Actual density (g/cm^3)	Specific surface area (m^2/g)	Average particle size (nm)
Nano-CuO	0.79	6.40	20	40
Nano-Fe_2O_3	1.20	5.24	40 to 60	20

Table 5 Chemical composition of nano-CuO

Element	Mn	Pb	Fe	Mg	P	K	Ca	Sr	Zn	Co	Cd	Ba
Amount (ppm)	3.5	90	87	75	300	300	400	2.3	195	6.4	2.5	0.75

Table 6 Chemical composition of nano-Fe_2O_3

Element	Mn	Na	Al	S	SiO_2	P	Cr	Ca
Amount (ppm)	0.095	0.0005	0.0002	0.12	0.134	0.016	0.037	0.024

Figure 5 AP1C (1), AP2C (2), AP3C (3), and pure AP (4).

Differential scanning calorimetry

The thermal decomposition processes of the samples were characterized by DSC using Dupont (Wilmington, DE, USA) 2000 instrument at a heating rate of 10°C/min.

Sample preparation

The CuO nanoparticles were blent with AP in different contents of 1, 2, and 3 wt.% to prepare the samples for thermal decomposition experiments. These samples were labeled as AP1C (AP + 1% nano-CuO), AP2C (AP + 2% nano-CuO), and AP3C (AP + 3% nano-CuO). Also, in a similar way, the Fe_2O_3 nanoparticles were mixed with AP, and the samples were labeled as AP1F (AP + 1% nano-Fe_2O_3), AP2F (AP + 2% nano-Fe_2O_3), and AP3F (AP + 3% nano-Fe_2O_3). Before the thermal decomposition experiments using DSC technique, the samples were homogenized. Figures 5 and 6 show the samples used in this study.

Figure 6 AP1F (1), AP2F (2), AP3F (3), and pure AP (4).

Competing interests

The author has no competing interests.

Authors' information

MM is a student in M.Sc. of Chemical Engineering in Iran University of Science and Technology (IUST).

References

1. El-Sherbiny, IM, Salih, E, Reicha, FM: Green synthesis of densely dispersed and stable silver nanoparticles using myrrh extract and evaluation of their antibacterial activity. Journal of Nanostructure in Chemistry **3**, 8 (2013)
2. Maddah, H, Rezazadeh, M, Maghsoudi, M, NasiriKokhan, S: The effect of silver and aluminum oxide nanoparticles on thermophysical properties of nanofluids. Journal of Nanostructure in, Chemistry (2013)
3. Hongzhen, D, Xiangyang, L, Guanpeng, L, Lei, X: Synthesis of Co nanoparticles and their catalytic effect on the decomposition of ammonium perchlorate. Chin J Chem Eng **16**, 325–328 (2008)
4. Satyawati, SJ, Prajakta, RP, Krishnamurthy, NV: Thermal decomposition of ammonium perchlorate in the presence of nano-sized ferric oxide. Defense Science Journal **58**, 721–727 (2008)
5. Duan, H, Lin, X, Liu, G, Xu, L, Li, F: Synthesis of Ni nanoparticles and their catalytic effect on the decomposition of ammonium perchlorate. Materials Processing Technology **208**, 494–498 (2008)
6. Farhadi, S, Pourzare, K, Sadeghinejad, S: Simple preparation of ferromagnetic Co_3O_4 nanoparticles by thermal dissociation of the $[Co^{II}(NH_3)_6](NO_3)_2$ complex at low temperature. Journal of Nanostructure in Chemistry **3**, 16 (2013)
7. Weifan, C, Fengsheng, L, Leili, L, Yongxiu, L: Synthesis of nano-sized yttria via a sol–gel process based on hydrated yttrium nitrate and ethylene glycol and its catalytic performance for thermal decomposition of NH_4ClO_4. Journal of Rare Earths **24**, 543–548 (2006)
8. Meda, L, Marra, G, Galfetti, L, Severini, F, Luca, LD: Nano-aluminium as energetic material for rocket propellants. Mater Sci Eng **C27**, 1393–1396 (2007)
9. Wang, J, He, S, Li, Z, Jing, X, Zhang, M, Jiang, Z: Synthesis of chrysalis-like CuO nano-crystals and their catalytic activity in the thermal decomposition of ammonium perchlorate. J Chem Sci **121**, 1077–1081 (2009)
10. Aijun, H, Juanjuan, L, Mingquan, Y, Yan, L, Xinhua, P: Preparation of nano-$MnFe_2O_4$ and its catalytic performance of thermal decomposition of ammonium perchlorate. Chin J Chem Eng **19**, 1047–1051 (2011)
11. Zhang, Y, Liu, X, Nie, J, Yu, L, Zhong, Y, Huang, C: Improve the catalytic activity of α-Fe_2O_3 particles in decomposition of ammonium perchlorate by coating amorphous carbon on their surface. Journal of Solid State Chemistry **184**, 387–390 (2011)
12. Zhou, Z, Tian, S, Zeng, D, Tang, G, Xie, C: MOX (M = Zn, Co, Fe)/AP shell-core nanocomposites for self-catalytic decomposition of ammonium perchlorate. Alloys and Compounds **513**, 213–219 (2012)
13. Chaturvedi, S, Dave, PN: A review on the use of nanometals as catalysts for the thermal decomposition of ammonium perchlorate. Journal of Saudi Chemical Society **17**, 135–149 (2013)
14. Alizadeh-Gheshlaghi, E, Shaabani, B, Khodayari, A, Azizian-Kalandaragh, Y, Rahimi, R: Investigation of the catalytic activity of nano-sized CuO, Co_3O_4 and $CuCo_2O_4$ powders on thermal decomposition of ammonium perchlorate. Powder Technol **217**, 330–339 (2012)
15. Liu, T, Wang, L, Yang, P, Hu, B: Preparation of nanometer $CuFe_2O_4$ by auto-combustion and its catalytic activity on the thermal decomposition of ammonium perchlorate. Mater Lett **62**, 4056–4058 (2008)
16. Duan, G, Yang, X, Chen, J, Huang, G, Lu, L, Wang, X: The catalytic effect of nano-sized MgO On the decomposition of ammonium perchlorate. Powder Technol **172**, 27–29 (2007)

7

A potentiometric solid state copper electrode based on nanostructure polypyrrole conducting polymer film doped with 5-sulfosalicylic acid

Reza Ansari*, Zahra Mosayebzadeh, Majid Arvand and Ali Mohammad-khah

Abstract

A pencil graphite electrode (PGE) electrodeposited by nanostructure polypyrrole (PPy) conducting polymer film doped with 5-sulfosalicylic acid (SSA) as sensing material (PGE/PPy/SSA-Cu) was successfully developed for the detection of Cu (II) in aqueous solutions. For the preparation of the copper-sensing electrode, electrodeposition of a polypyrrole film was carried out using cyclic voltammetry in a solution containing 10.0 mM pyrrole and 5.0 mM 5-sulfosalicylic acid. The introduced electrode in the current paper can be fabricated simply and was found to possess high selectivity, showed wide working concentration range, fast response time (<20 s), long lifetime, and very good sensitivity to Cu (II) ion. The prepared electrode exhibited a linear Nernstian response within the copper ion concentration range of 1.0×10^{-5} to 1.0×10^{-1} M with a slope of 29.60 ± 0.30 mV per decade over the pH range of 4.0 to 6.0. Under the optimized conditions, a detection limit of 5.42 μM was obtained. Applications of this electrode for determination of copper ion in real samples and as an indicator electrode for potentiometric titration of Cu^{2+} ion using EDTA are reported.

Keywords: Pencil graphite electrode, Nanostructure polypyrrole film, Potentiometry, Copper sensor electrode

Background

The determination of copper assumes the importance in view of its widespread occurrence in environmental samples. As such, large concentrations of copper can be tolerated by human beings; however, excessive dosage and long-term exposure may cause irritation of the nose, mouth, and eyes, and it causes headache, stomachache, dizziness, vomiting, and diarrhea. Copper deficiency results in anemia while its accumulation resulting in Wilson disease [1-6].

A number of instrumental methods such as atomic absorption spectrometry (AAS), cold vapor AAS or flame atomic absorption spectrometry-electrothermal atomization, inductively coupled plasma optical emission spectroscopy, anodic stripping voltammetry, chromatography, gravimetric detection, or photometry have been proposed for its determination at low concentration level [7-11]. Most of the mentioned methods usually have sufficiently low detection limit and high selectivity but also in many cases, possess drawbacks such as high cost of equipment and expensive materials, time-consuming, and complicated operation. Thus, these methods are generally unsuitable for simple, low cost, and remote determination of copper.

Ion-selective electrodes (ISE) provide analytical procedures for such situations as they are fast, convenient, require minimum sample pretreatment, and may also be suitable for online analysis. In recent years, many substances have been used for electrode modification and determination of copper [12-20]. Conductive polymers (CPs) due to combination of tunable chemical and electronic properties of them are a promising group of compounds, which are widely applied in chemical sensors and a variety of other applications [21-23]. One type of sensors is of noble metal electrodes coated with a conductive polymer film. These electrodes, when used for potentiometric measurements, are called coated wire electrodes. This type of sensors, by eliminating the internal filling solution, provides new advantages such as good mechanical stability, simplicity, and possibility of miniaturization. The polymer layer is relatively easy to prepare and results in thick and homogenous coverage of the metal. Since this type of sensors is not selective to create a selective CP-based receptor,

* Correspondence: ransari@guilan.ac.ir
Chemistry Department, Faculty of Science, University of Guilan, PO Box: 41635–1914, Rasht, Iran

it requires a polymerizable and selective monomer that is also stable under the polymerization conditions. The CP's selectivity is controlled by doping the film from solution or by incorporating a recognition unit into the conjugated polymeric backbone. In terms of potentiometry, the film acts as a cation or anion exchanger and provides a cationic or anionic potentiometric response. Sensors based on conductive polymers have distinct advantages, such as simple preparation, fast response towards the analyte, and potential for miniaturization. Their lifetime is longer than classical PVC membrane electrodes because there is no receptor (ionophore) loss from the membrane due to the low solubility of the polymeric film in conventional organic and inorganic solvents [24-26]. Polypyrrole (PPy) has received the most attention owing to its convenience of preparation, high stability, and wide range of applications. The electroactive nature or the switching properties of PPy has been utilized as the basis of most proposed applications such as sensors, separation devices, rechargeable batteries, and controlled drug release. All other applications of this polymer are based on their electrical conductivity and unique electrochemical properties [27-29].

In this research, a new solid state electrode was prepared by the electrodeposition of a PPy nanolayer, which was doped with 5-sulfosalicylic acid (SSA) (Figure 1) onto the surface of pencil graphite electrode (PGE). We chose pencil graphite as the material of the electrode because it has a larger active electrode surface area and therefore is able to detect low concentrations and/or volumes of the analyte. The main original idea developed in the current research is drawn from the unique properties of the polypyrrole conducting polymer in which anions from the reaction solution can be incorporated into the polymer during electropolymerization as dopant or counterion. The incorporated anions termed as dopants can then induce its property into the polymer product. Choosing of 5-sulfosalicylic acid as dopant during electropolymerization of polypyrrole was based on its high affinity for the formation of a stable complex with Cu (II) ion [30]. Therefore, 5-sulfosalicylic acid was predicted to act as dopant during electropolymerization and as ionophore in the course of potentiometric determination of cupric ions. The thickness of the coated polymer film onto working electrode can be easily controlled by measuring the charge that passes through the cell during electropolymerization. The polymeric nanofilm (PPy/SSA) used for the preparation of the copper-sensing electrode is an electronically and ionically conducting material based on polypyrrole which allows better defined ion-to-electron transduction compared to the classical coated wire electrodes.

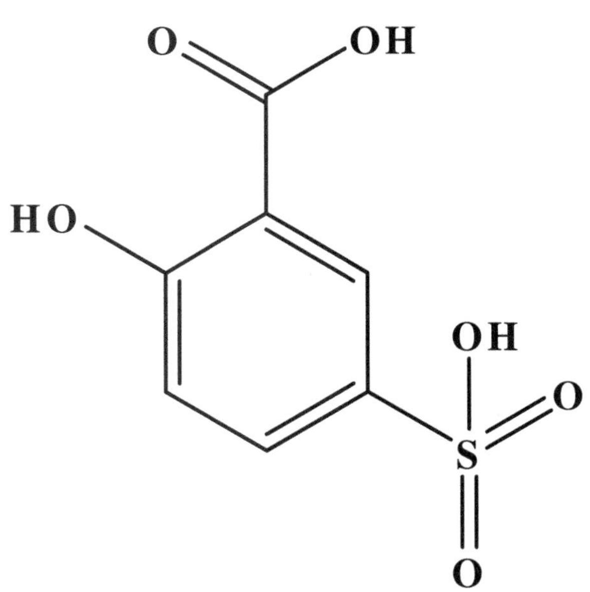

Figure 1 Chemical structure of 5-sulfosalicylic acid (SSA).

Results and discussion

The electrode composition effect on the potential response of the PGE/PPY/SSA-C

It is well known that the compositions of membranes have great effect on the characteristics of ISEs such as sensitivity, working concentration range, selectivity, and potential stability. Therefore, electrodes with different compositions have been prepared, and their potentiometric response characteristics were evaluated in order to find out the optimum electropolymerization conditions. The result of some investigated parameters is shown in Table 1. It was found that the best electrodeposition conditions for observing Nernstian response resulted when a cell solution containing 10-mM monomer (pyrrole), and 5.0-mM electrolyte (5-sulfosalicylic acid) was employed by cyclic voltammetry method, potential range of −1 to +1 (V), scan rate of 50 (mV/s), and cycle number of 20 (at optimal set pH = 5), (no. 13). The fabricated electrode was soaked for 1 h (optimized condition time) in 1.0×10^{-3} M Cu (II) solution before its use for potentiometric analysis. The electromotive force (EMF) values were measured using an accurate digital voltmeter.

Characterization

Scanning electron microscope (SEM) is a useful instrument for characterizing the surface morphology of the polymers. The micrographs obtained for polypyrrole used in this investigation is shown in Figure 2: (a) SEM of bare PGE, (b) PGE/PPY/Cl, (c) PGE/PPY/SSA-Cu, and (d) cross section of PGE/PPY/SSA-Cu. As the SEM pictures show, the PPy at the presence of SSA as a dopant is electrodeposited onto the pencil graphite electrode as nanoparticles which the average diameter of those is about 84 nm, and the layer thickness is about 200 nm

Table 1 Optimization of the membrane ingredients and technique parameters for fabrication of PGE/PPy/SSA-Cu electrode

Number	Composition (mM)		Cycle number	Slope (mV/decade)	Linear range (M)	R^2
	Py	SSA				
1	100	3	10	12.41	1.0×10^{-4} to 1.0×10^{-2}	0.970
2	100	5	10	16.10	1.0×10^{-4} to 1.0×10^{-2}	0.965
3	100	8	10	10.22	1.0×10^{-4} to 1.0×10^{-2}	0.982
4	100	5	20	25.60	1.0×10^{-4} to 1.0×10^{-2}	0.985
5	100	5	30	20.31	1.0×10^{-4} to 1.0×10^{-2}	0.990
6	50	5	10	15.43	1.0×10^{-5} to 1.0×10^{-2}	0.974
7	50	5	20	17.30	1.0×10^{-5} to 5.0×10^{-2}	0.980
8	50	5	30	18.25	1.0×10^{-5} to 5.0×10^{-2}	0.971
9	50	25	10	19.45	1.0×10^{-5} to 1.0×10^{-1}	0.973
10	50	25	20	24.82	1.0×10^{-5} to 1.0×10^{-1}	0.986
11	50	25	30	25.61	1.0×10^{-5} to 1.0×10^{-1}	0.963
12	10	5	10	22.10	1.0×10^{-5} to 1.0×10^{-1}	0.985
13[a]	10	5	20	29.60	1.0×10^{-5} to 1.0×10^{-1}	0.999
14	10	5	30	31.42	1.0×10^{-5} to 1.0×10^{-1}	0.988

[a]It was found that the best electrodeposition conditions for observing Nernstian response were achieved with electrode 13.

(Figure 2c,d). These images also demonstrate that a bulk quantity of nanoparticle size on the PPy exists.

Fourier transform infrared spectroscopy (FTIR) technique was used to ascertain the presence of SSA in polymer matrix. Additional file 1: Figure S1a shows the FTIR spectrum of pure PPy. The band at 1,570.54 and 1,384.60 cm^{-1} may be corresponded to typical polypyrrole ring vibration and =C–H band in plane vibration, respectively. Additional file 1: Figure S1b shows that the FTIR spectrum of PPy/SSA, which exhibits absorption peaks at 1,085.49, is assigned as S=O stretching band and at 1,387.76 and 1,635.47 may be corresponded to C=C stretching bands. Also, a broad and sharp band observed at 3,443.32 may be attributed to the presence of O–H. These results confirm the incorporation of SSA within the polymer structure which has the most important role of sensing the electrode toward Cu (II) ions.

Energy-dispersive X-ray spectroscopy (EDS) is another analytical technique which is used for chemical

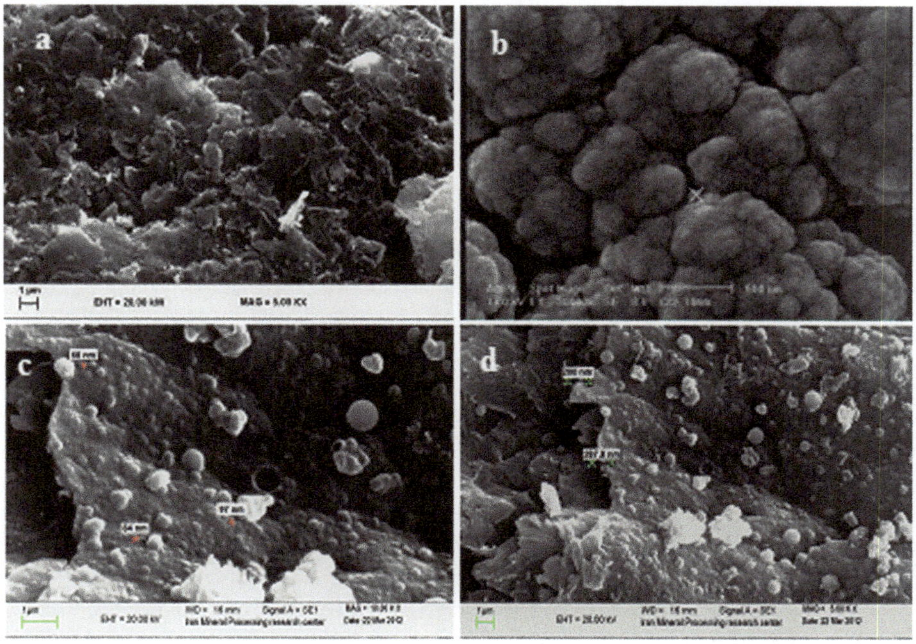

Figure 2 SEM image. Of (**a**) bare PGE, (**b**) PGE/PPY/Cl, (**c**) PGE/PPY/SSA-Cu, and (**d**) cross section of PGE/PPY/SSA-Cu.

characterization of a sample. Its characterization capabilities are due to the fundamental principle that each element has a unique atomic structure allowing unique set of peaks on its X-ray spectrum. The comparison of the EDS spectra of the polymer before and after doping with SSA is illustrated in Additional file 2: Figure S2a,b, respectively. The presence of a sharp and intense peak of sulfur in EDS spectrum appeared at 2.2 to 2.4 keV (Additional file 2: Figure S2b) which is a clear indication of incorporation of SSA into the polypyrrole film during electropolymerization.

Calibration curve and statistical data

Direct potentiometry using a working curve (a plot of EMF vs. log C) is usually used in potentiometric analysis with ion-selective electrodes. The calibration curve for analysis of Cu (II) ions was prepared using standard solutions of Cu (II) ion (pH of solution was adjusted about 5.0), and the potential readings were carried out using the cell setup with solid state Cu (II) electrode (PGE/PPy/SSA-Cu) and was compared with the result of PGE/PPy/SA (Figure 3, curves A and B, respectively).

As the results show (Figure 3, curve A), the fabricated electrode exhibits linear response over a working concentration range of 1.0 × 10^{-5} to 1.0 × 10^{-1} M with a Nernstian slope of 29.60 ± 0.30 mV per decade (for one electrode with n = 4) of copper ion concentrations. By extrapolating the linear parts of the ion-selective calibration curve, the detection limit of an ion-selective electrode can be calculated. In this work, the detection limit of the prepared sensor was 5.42 μM which was calculated by the extrapolating of the two segments of the calibration curve in Figure 3. The reproducibility of the Cu^{2+} electrode was also investigated. The standard deviations of the slope of four replicate constructed electrodes were ±0.50. The response characteristics of the prepared copper-sensing electrode are also summarized in Table 2.

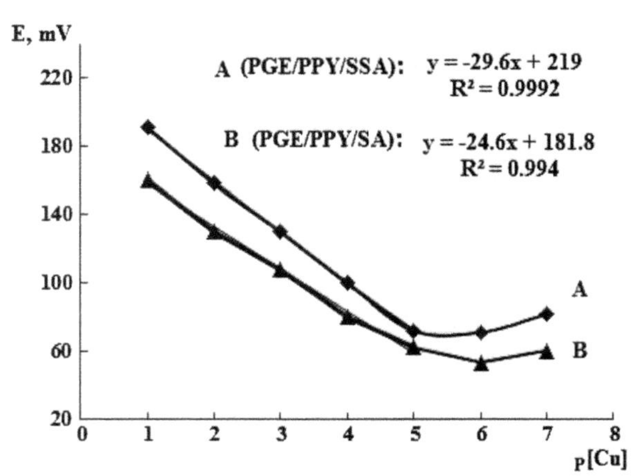

Figure 3 The calibration curve obtained for Cu (II) ion using PGE/PPy/SSA-Cu and PGE/PPy/SA electrodes.

Table 2 Properties of PGE/PPY/SSA-Cu electrode

Properties	Values/range
Preparation technique	Cyclic voltammetry (v = 50 mV s^{-1})
Preparation conditions	PPy (10.0 mM), SSA (5.0 mM)
Cycle numbers	20
Electrode type	Pencil graphite electrode (PGE)
Best pH range for response	4 to 6
Conditioning time	1 h (in 10^{-3} M Cu^{2+})
Linear range (Cu^{2+}, M)	1.0 × 10^{-5} to 1.0 × 10^{-1}
Electrode slope	29.60 (mV $decade^{-1}$)
Detection limit (M)	5.42 × 10^{-6}
Static response time (s)	20 s
Useful shelf lifetime	At least for 60 days

Effect of pH

The influence of the pH of the test solution on the potential response of the electrode for a solution containing 1.0 × 10^{-4} M Cu (II) was considered in the pH range of 2.0 to 7.0, and the results are shown in Additional file 3: Figure S3. As can be seen, the potential remains constant over a pH range of 4.0 to 6.0. At alkali media, the potential sharply decreased due to the formation of some hydroxyl complexes of Cu (II) ions in solution. On the other hand, at pH lower than 3.5, the interference of H^+ ions is more which is due to the high rate of diffusion of H^+ ions from the sample solution to electrode surface, and its resulting increase in potential that indicates the protonated ionophore possesses a poor response to the Cu (II) ions and strong response to H_3O^+ ions in solution.

Response time

Response time is the average time required for the electrodes to reach a potential response within ±1 mV of the final equilibrium value, after successive immersions in a series of testing solutions, each having a tenfold concentration difference. In this study, the practical response time of the sensor was recorded by changing the Cu^{2+} concentration in solution, over a concentration range of 1.0 × 10^{-5} to 1.0 × 10^{-1} M. The potentials vs. time traces are shown in Figure 4a. It is noticed that stable potentials were attained within 20 s. This short response time is most probably due to the fast exchange kinetics of complexation-decomplexation of Cu^{2+} ion with the sensing material on the tested solution-polymeric surface of electrode interface. In order to examine the stability of the potential reading of the electrode, it was also put in a 1.0 × 10^{-3} M of copper ion. As the results show (Figure 4b), the electrode reaches its equilibrium response in a very short time (approximately 20 s) and remained constant for approximately 4 min.

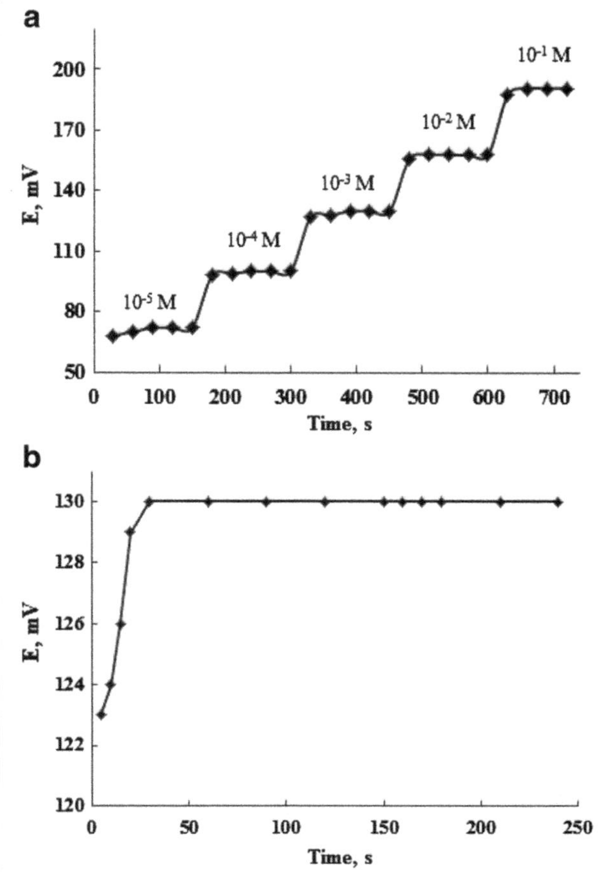

Figure 4 The potentials vs. time traces. (**a**) Dynamic response time and (**b**) static response time of electrode.

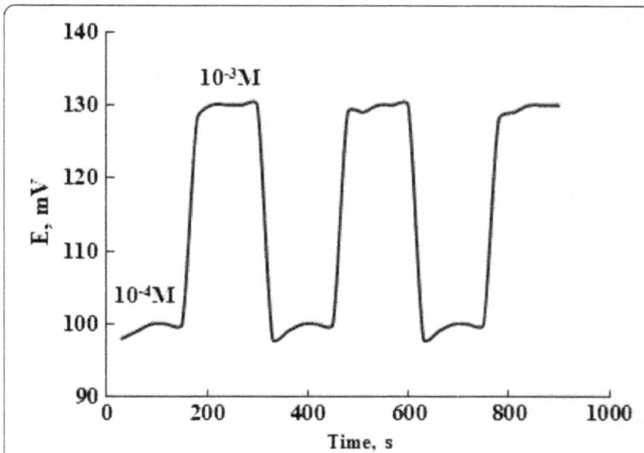

Figure 5 Reversibility of PGE/PPy/SSA-Cu electrode.

Reversibility and lifetime

Reversibility factor is one of the most important parameters in assessing a sensor performance for its practical application. In order to find out the reversibility of the response, the designed electrode was exposed repetitively in the Cu (II) solutions with two different concentrations ($1.0 \times 10^{-4}$and 1.0×10^{-3} M). The reported potential is the mean values of at least three measurements for each concentration. The consequences observed are shown in Figure 5 as the data fascinatingly indicate that the signal of the fabricated Cu (II) electrode is interestingly highly reversible.

The lifetime of the PGE/PPy/SSA-Cu electrode was carried out by performing periodic calibration with standard solutions and calculating the response and slope over the range 1.0×10^{-5} to 1.0×10^{-1} M Cu^{2+} solution (Table 3). It was found that the electrode worked well over the period of 2 months without showing any significant divergence in concentration range, slope, and response time. During non-usage, the electrode was stored in the air and before use; they were re-equilibrated by dipping in 1.0×10^{-3} M Cu^{2+} solution for 1 h.

Potentiometric selectivity

The potentiometric selectivity coefficients, which reflect the relative response of the membrane sensor towards the primary ions over other ions present in solution, are perhaps the most essential characteristic of an ISE. To investigate the membrane electrode selectivity, its potential response was monitored in the presence of various interfering foreign cations with the help of two different procedures, namely the so-called matched potential method (MPM) and the separated solution method (SSM). The resulting selectivity coefficients are presented in Table 4. As the results present, the electrode is very selective to Cu (II) ions over all the tested interfering ions.

Effect of temperature

The standard cell potentials (E°_{cell}) were determined at different temperatures from the respective calibration plots as the intercepts of these plots at p[Cu] = 0 and were used to determine the isothermal temperature coefficient (dE°/dt) of the cell [31]. By plotting of E°_{cell} vs. (t − 25), a straight line is produced (Equation 1). The slope of this line was taken as the isothermal temperature coefficient of the cell.

$$E^{o}_{cell} = E^{o}_{cell(25^{o}c)} + (dE^{o}/dt)_{cell}(t-25) \quad (1)$$

The standard potentials of the reference (Calomel) and the prepared electrodes were calculated using Equations 2 and 3, respectively.

$$E^{o}_{Hg_2Cl_2} = 0.241 - 0.00066\,(t-25) \quad (2)$$

$$E^{\circ}_{reference} + E^{\circ}_{cell} = E^{\circ}_{electrode} \quad (3)$$

A plot of $E^{\circ}_{electrode}$ vs. (t − 25) gives a straight line. The slope of this line was taken as the isothermal temperature coefficient of the prepared copper-sensing electrode. Trend of changes of PGE/PPy/SSA-Cu performance at various temperatures (20°C to 60°C) is depicted in Additional file 4: Figure S4. Results showed

Table 3 Lifetime behavior of PGE/PPy/SSA-Cu electrode

Time	Linear Range (M)	Slope (mV/decade)	Detection limit (M)
1 day	1×10^{-5} to 1×10^{-1}	29.60	5.42×10^{-6}
7 days	1×10^{-5} to 1×10^{-1}	29.60	5.42×10^{-6}
3 weeks	1×10^{-5} to 1×10^{-1}	29.30	5.63×10^{-6}
1 month	1×10^{-5} to 1×10^{-1}	29.06	5.85×10^{-6}
2 months	1×10^{-5} to 1×10^{-1}	28.40	6.14×10^{-6}

that the electrode exhibited good Nernstian behavior at temperature range between 20°C and 60°C. The amount of the isothermal temperature coefficient of the cell and the PGE/PPy/SSA-Cu electrode are obtained 1.26 and 1.27 (mV/°C), respectively. The small values of $(dE^{o}/dt)_{cell}$ and $(dE^{o}/dt)_{electrode}$ reveal the high thermal stability of the electrode within the investigated temperature range.

Effect of partially non-aqueous medium on the working of PGE/PPy/SSA-Cu electrode

The prepared electrode was investigated in partially non-aqueous media using methanol and ethanol mixture with water. Table 5 indicates that the slope remains unaltered with the addition of non-aqueous solvents. Hence, the new electrode can be successfully used in partially non-aqueous medium.

Some practical applications of the prepared copper-sensing electrode

The Cu^{2+} sensor was successfully used as an indicator electrode in the titration of 5.0 mL of a 1.0×10^{-2} M Cu (II) solution with 1.0×10^{-2} M of ethylenediaminetetraacetic acid (EDTA) (at pH = 5.0 acetate buffer) and vice versa. The resulting sigmoid titration curve is given in Figure 6. As it is seen, the amount of Cu (II) ions in solution can be determined with the fabricated electrode. Therefore, the PGE/PPy/SSA-Cu electrode can be used as indicator electrode in potentiometric titration of Cu (II) ions successfully.

The electrode was also successfully applied to the direct potentiometric measurement of copper (II) in a known copper solution prepared from electrical copper wire (purity >98%) using standard addition technique [28]. The experiments were run three times, and the results obtained are presented in Table 6. Statistical treatment of the data (t test) has shown that the potentiometric measurements of copper ion performed at several samples using the prepared copper-sensing electrode have high accuracy ($t_{exp} < t_{cri}$).

Table 4 Potentiometric selectivity coefficient values of PGE/PPy/SSA-Cu electrode

Interfering ion	K^{MPM}	K^{SSM}
Na^{+}	4.73×10^{-4}	3.65×10^{-4}
K^{+}	7.50×10^{-3}	4.58×10^{-3}
Mg^{2+}	8.10×10^{-4}	2.15×10^{-4}
Ca^{2+}	6.66×10^{-3}	1.78×10^{-4}
Ni^{2+}	5.45×10^{-3}	6.25×10^{-3}
Pb^{2+}	4.44×10^{-3}	2.65×10^{-3}
Mn^{2+}	5.23×10^{-3}	4.78×10^{-3}
Zn^{2+}	7.78×10^{-4}	4.60×10^{-4}
Co^{2+}	7.61×10^{-3}	3.16×10^{-3}
Fe^{2+}	6.66×10^{-3}	6.41×10^{-3}
Hg^{2+}	6.35×10^{-3}	2.21×10^{-3}
Fe^{3+}	6.87×10^{-4}	5.46×10^{-4}
Cd^{2+}	3.20×10^{-4}	4.16×10^{-4}

Table 5 Performance of PGE/PPy/SSA-Cu electrode in non-aqueous media

Solvent composition [% (v/v)]	Slope (mV per decade)	Working concentration range (M)
H_2O	29.60 ± 0.2	1×10^{-5} to 1×10^{-1}
CH_3OH		
10	29.42 ± 0.2	1×10^{-5} to 1×10^{-1}
20	29.05 ± 0.3	1×10^{-5} to 1×10^{-1}
30	27.32 ± 0.3	1×10^{-5} to 1×10^{-1}
C_2H_5OH		
10	29.60 ± 0.2	1×10^{-5} to 1×10^{-1}
20	29.25 ± 0.3	1×10^{-5} to 1×10^{-1}
30	28.45 ± 0.3	1×10^{-5} to 1×10^{-1}

Conclusions

The introduced indicator electrode based on polypyrrole conducting polymer can be easily prepared via simple electrochemical method (cyclic voltammetry). The copper sensor electrode fabricated in the current study showed response characteristics with Nernstian behavior over the concentration range 1.0×10^{-5} to 1.0×10^{-1} M

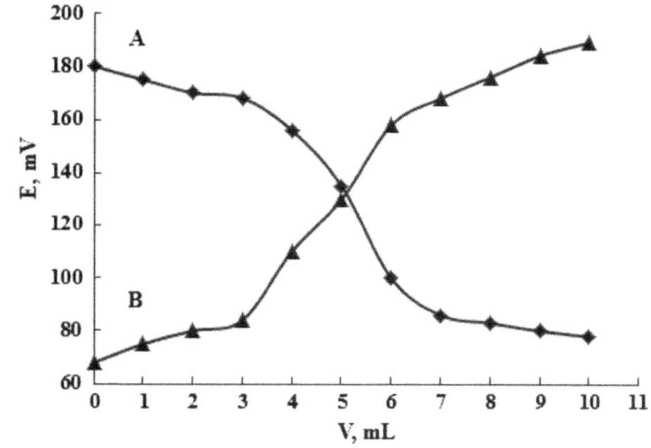

Figure 6 Potentiometric titration curves of 5.0 mL 1.0×10^{-2} M of Cu^{2+} with 1.0×10^{-2} M EDTA (curve A), and vice versa (curve B).

Table 6 Application of PGE/PPy/SSA-Cu electrode for potentiometric measurement of copper in known solutions prepared from copper electrical wire

No.	Copper known solution (M)	Measured (M)[a]	$t_{exp.}$	$t^{95\%}_{crit.}$
1	5.0×10^{-4}	$4.80\ (\pm 0.17) \times 10^{-4}$	2.04	4.30
2	1.0×10^{-3}	$0.88\ (\pm 0.15) \times 10^{-3}$	1.38	4.30
3	5.0×10^{-3}	$4.74\ (\pm 0.22) \times 10^{-3}$	2.05	4.30

[a]Average of three determinations.

of Cu (II) ion. The advantages such as fast response time (less than 20 s), low detection limit (5.42 μM), simplicity of preparation, low cost, high shelf lifetime (at least 60 days), high selectivity, sensitivity, accuracy, and precision as well as good dynamic linear range, high repeatability and reproducibility make this electrode potentially useful for measurement of free Cu (II) ions in real samples. The copper sensor electrode can also be used for potentiometric titration of copper ion with sharp and distinguished end point. The electrode was also used for determination of copper ion in real samples very successfully. Due to the high toxicity of copper ions to aquatic life, the finding in this study is very important from both analytical and environmental point of views.

Methods

Chemicals and apparatus

All chemicals used were of analytical grade (AR) and were prepared in distilled water. Pyrrole was obtained from Merck (Darmstadt, Germany, http://www.merckmillipore.com/chemicals) and was distilled before use. 5-sulfosalicylic acid and the nitrate or chloride salts of all cations were purchased from Merck with highest purity available and used without any further purification. Cu^{2+} solution (1.0 M) was prepared by dissolving AR grade $CuSO_4.5H_2O$ in distilled water and was used as stock solution. Standard solutions for calibration graph were then prepared by the successive diluting of the accurately prepared stock solution, and potential readings started from the least concentrated standard. For the investigation of selectivity, interfering metal ions were prepared as 0.010 M from their pure related salts. Electrochemical synthesis of polymer was carried out with an electrochemical analyzer (SAMA 500, SAMA research center, Esfahan, Iran). A Metrohm pH meter (model 827) (Metrohm, Filderstadt, Germany, http://www.metrohmusa.com) with a combined double junction glass electrode, calibrated against two standard buffer solutions at pH 4.0 and 7.0, was used for pH measurement. The pH of the solutions was adjusted using 0.10 M HCl and NaOH solutions. All potentiometric measurements were carried out at 25°C with a cell of the following type: SCE | Cu^{2+} (a_A) | PGE/PPy/SSA-Cu. EMF measurements were carried out using an accurate digital electronic voltmeter with high internal impedance to the nearest 0.10 mV. The readings were recorded when the potentials reached a stable and constant value (within 1 mV). The sensing material of the fabricated Cu (II) electrode obtained was characterized using FTIR, energy-dispersive X-ray spectroscopy, and SEM techniques. The surface morphology was examined using Philips XL30 scanning electron microscope (Amsterdam, The Netherlands, https://www.mems-exchange.org/equipment/E1175/ at an accelerating voltage of 20.0 kV with 10,000× magnification. The FTIR spectra of the electrode were recorded in Bruker, Madison, WI, USA (model Vertex 70) at room temperature in the region 4,000 to 500 cm^{-1}.

Preparation of electrode and procedure

A three-electrode system was employed for electropolymerization of pyrrole in aqueous solution containing monomer. Electropolymerization was carried out by cyclic voltammetry method. A PGE as working electrode, a Pt wire as auxiliary electrode, and a double junction saturated calomel electrode were used as the reference electrodes. PGE (with 0.70 mm diameter) was cleaned with concentrated HNO_3 (6.0 M) for removing any possible metal impurities. A PGE was also used for the electrodeposition of polypyrrole from a solution of pyrrole and KCl as electrolyte under the same conditions used for the PGE/PPy/SSA-Cu copper-sensing electrode termed as PGE/PPy/Cl. In order to have a good comparison, this electrode (PGE/PPy/Cl) was then used for surface morphological analysis and determination of copper ions. Since any potential stability, reversibility, and linear response for Cu^{2+} using the PGE/PPy/Cl electrode were not observed, all the potentiometric experiments were shown only for PGE/PPy/SSA-Cu electrode. Since salicylic acid (SA) is the metal-chelating ligand, for comparison and also further confirmation that 5-sulfosalicylic acid is indeed working as an ionophore, we also tested salicylic acid as an electrolyte for electrodeposition of polypyrrole termed as PGE/PPy/SA. The EMF (*E*) of the sensor electrode system dipped in the solution of Cu (II) ion was investigated given by the following equation (Equation 4):

$$E = k + S \log\left[Cu^{2+}\right] \qquad (4)$$

where *E* (mV) is the total potential developed between the sensing and reference electrodes and*k* (mV) is a constant which is characteristic of the particular ISE/reference pair. It can be calculated from the linear plot of *E* vs. log *C*, and *S* is the electrode slope. ISEs are not actually specific but are more selective toward target or analyte ion. So, it is important to determine the selectivity of a fabricated sensor electrode toward target ion or the possible interferences from other ions. Different methods are available for determination of selectivity of an ISE. In the current study, selectivity coefficient ($K^{pot}_{Cu^{2+},X^{n+}}$) values for a range of common interfering ions were calculated by both the MPM and the SSM, which are recommended by IUPAC [32,33]. According to the MPM, selectivity coefficient ($K_{A,B}^{MPM}$) is determined as follows:

$$K^{MPM} = \frac{\Delta A}{a_B} \quad (5)$$

where $\Delta A = a'_A - a_A$, a_A and a'_A are the initial and final primary ion activities (or concentrations), respectively, and a_B the interfering ion activity. In practice, a specified amount of primary ions (A) is added to a reference solution (primary ion with a fixed activity), and the membrane potential is measured. In a separate experiment, interfering ions (B) are successively added to an identical reference solution until the membrane potential matches the one obtained before with the primary ion. It should be noted that the values of a_A and a'_A for Cu^{2+} ion were taken as 1.0×10^{-4} and 1.0×10^{-3} M, respectively, at pH 5.0 and a_B was determined experimentally. In SSM, the EMF of a cell is measured for each of two separate solutions, one containing the ion A of the activity a_A and charge Z_A (but not B), the other containing the ion B with charge Z_B at the same activity $a_B = a_A$ (but not A). If the measured values are E_A and E_B, respectively, then the value of log $K_{A,B}^{Pot}$ may be calculated from Equation 6:

$$\log K_{A,B}^{Pot} = \frac{E_A - E_B}{2.303RT/Z_A F} - \left(1 - \frac{Z_A}{Z_B}\right)\log a_A \quad (6)$$

For the preparation of the calibration curve, standard solutions of Cu (II) ion were prepared with different concentrations (from 1.0×10^{-6} to 1.0×10^{-1} M), and the potentials were recorded after 20 s or less until the readings were stable. The maximum electrode potential fluctuation was less than 1.0 mV. The effect of temperature on the performance of the potentiometric electrodes was evaluated in a thermostat at different temperatures ranged from 20°C to 60°C.

Additional files

Additional file 1: Figure S1. FTIR spectra of (a) PGE/PPy/Cl and (b) PGE/PPy/SSA-Cu.

Additional file 2: Figure S2. EDS spectrum of (a) bare PGE and (b) PGE/PPy/SSA-Cu.

Additional file 3: Figure S3. Effect of pH on the potential response of PGE/PPy/SSA-Cu electrode.

Additional file 4: Figure S4. Calibration graphs of PGE/PPy/SSA-Cu at different temperatures.

Abbreviations

CP: conductive polymer; EDS: energy-dispersive X-ray spectroscopy; ISE: ion-selective electrode; MPM: matched potential method; PGE: pencil graphite electrode; PGE/PPy/Cl: polypyrrole conducting polymer doped with chloride ion; PGE/PPy/SSA-Cu: polypyrrole conducting polymer doped with 5-sulfosalicylic acid; PPy: polypyrrole; SA: sulfosalicylic acid; SSA: 5-sulfosalicylic acid; SSM: separated solution method.

Competing interests

The authors declare that they have no competing interests.

Authors' contributions

RA participated in the idea and the design of the study and interpretation of the results. ZM carried out the experiments and wrote the manuscript. MA and AM conceived of the study and participated in its design and coordination. All authors read and approved the final manuscript.

Acknowledgments

The authors appreciate the postgraduate students of the University of Guilan for the financial support of this work.

References

1. Greenwood, NN, Earnshow, A: Chemistry of the Elements, pp. 628–630. Pergamon, New York (1984)
2. Mazloum Ardakania, M, Mirhoseinib, SH, Salavati-Niasari, M: Copper selective electrode based on 1,8-Bis(2-hydroxynaphthaldiminato)3,6-dioxaoctane. Acta Chim. Slov. **53**, 197–203 (2006)
3. Ansari, R, Fallah Delavar, A, Aliakbar, A, Mohammad-khah, A: Solid-state Cu (II) ion-selective electrode based on polyanilin-conducting polymer film doped with copper carmosin dye complex. J. Solid State Electrochem **16**, 869–875 (2012)
4. Jeong, DC, Lee, HK, Jeon, S: Highly copper(II)-selective PVC membrane based on a Schiff base complex of N, N'-Bis-pyridin-2-ylmethylene-naphthalene-1,8-diamine as an ionophore. Bull. Korean Chem. Soc **27**(10), 1593–1596 (2006)
5. Underwood, EJ: Trace elements in human and animal nutrition, pp. 197–204. Academic Press, New York (1962)
6. Kumar Mahajan, R, Sood, P: Novel copper (II)-selective electrode based on 2,2':5',2''-terthiophenein PVC matrix. Int. J. Electrochem. Sci. **2**, 832–847 (2007)
7. Neudachina, LK, Oshintseva, EV, Skorik, YA, Vshivkova, AA: N-Aryl-3-Aminopropionic acids as selective reagents for the determination of copper in metallurgical products. J. Anal. Chem. **60**(3), 240–246 (2005)
8. Mashhadizadeh, MH, Pesteh, M, Talakesh, M, Sheikhshoaie, I, Ardakani, MM, Karimi, MA: Solid phase extraction of copper (II) by sorption on octadecyl silica membrane disk modified with a new Schiff base and determination with atomic absorption spectrometry. Spectrochim. Acta. B. **63**, 885–888 (2008)
9. Campos, RC, Santos, HR, Grinberg, P: Determination of Copper, Iron, Lead, Nickle in gasoline by electrothermal atomic absorption spectrometryusing three component solutions. Spectrochim. Acta. B **57**, 15–28 (2002)
10. Asan, A, Andac, M, Isildak, I: Flow-injection spectrophotometric determination of copper(II) ion at low mu g L^{-1} levels using 4 benzylpiperidinedithiocarbamate. Anal. Sci. **17**, 1125–1127 (2001)
11. Mohadesi, A, Salmanipour, A, Mohammadi, Z, Pourhatami, A, Taher, MA: Stripping voltammetric determination of copper (II) on an overoxidized polypyrrole functionalized with Nitroso-R. J. Braz. Chem. Soc. **19**(5), 956–962 (2008)
12. Ganjali, MR, Ghafarloo, A, Faridbod, F, Norouzi, P: Copper-selective PVC membrane sensor. Int. J. Electrochem. Sci. **7**, 3706–3716 (2012)
13. Stankovic, D, Roglic, G, Jovic, M, Andjelkovic, I, Mutic, J, Manojlovic, D: New potentiometric sensor for copper determination in waste waters based on a new type of modified GC electrode. Anal. Bioanal. Electrochem. **4**(3), 238–245 (2012)
14. Brin, S, Buzuk, M, Generali, E, Brali, M: Improving the response of copper (II) selective PVC membrane electrode by modification of N_2S_2donor ligand. Acta Chim. Slov. **57**, 318–324 (2010)
15. Essi, M: Chalcogenide sensors for copper (II) detection in polluted waters. Chalcogenide Letters **8**(2), 103–110 (2011)
16. Shokrollahi, A, Abbaspour, A, Ghaedi, M, Haghighi, AN, Kianfar, AH, Ranjbar, M: Construction of a new Cu^{2+} coated wire ion selective electrode based on 2-((2-(2-(2-(2 hydroxy-5-methoxybenzylidene amino) phenyl) disufanyl) phenylimino) methyl)-4 methoxy phenol Schiff base. Talanta **84**(1), 34–41 (2011)
17. Maksymiuk, K: Chemical reactivity of polypyrrole and its relevance to polypyrrole based electrochemical sensors. Electroanalysis **18**(16), 1537–1551 (2006)
18. Mosayebzadeh, Z, Ansari, R, Mohammad-khah, A, Arvand, M: Electrochemical preparation of a copper ion selective electrode based on polypyrrole conducting polymer doped with Ponceau 4R Azo dye. Anal. Bioanal. Electrochem. **5**(1), 109–129 (2013)

19. Buzuk, M, Brinic, S, Generalic, E, Bralic, M: Copper (II) ion selective PVC membrane electrode based on S, S'-bis(2-aminophenyl)ethanebis(thioate). Croat Chem Acta **82**(4), 801–806 (2009)
20. Jeong, E, Lee, HK, Ahmed, MS, Seo, H, Jeon, S: Copper (II) selective PVC membrane electrodes based on Schiff base 1,2-Bis (E-2-hydroxy benzylidene amino) anthracene-9,10-dione complex as an Ionophore. Bull Korean Chem Soc **31**(2), 401–405 (2010)
21. Trojanowicz, M: Application of conducting polymers in chemical analysis. Microchim Acta **143**, 75–91 (2003)
22. Aminur Rahman, MD, Kumar, P, Park, DS, Shim, YB: Electrochemical sensors based on organic conjugated polymers. Sensors **8**, 118–141 (2008)
23. Ansari, R, Mosayebzadeh, Z: Removal of anionic dye Eosin Y from aqueous solutions using conducting electroactive polymers. Iranian Polym J **19**(7), 541–551 (2010)
24. Pankaj, K, Kumar, SH, Sukhjeet, K: Conducting polymer based potentiometric sensors. Res. J. Chem. Environ. **16**(3), 125–133 (2012)
25. Gupta, N, Sharma, S, Ahmad Mir, I, Kumar, D: Advances in sensors based on conducting polymers. J. Sci. Ind. Res. **65**, 549–557 (2006)
26. Michalska, A: All-solid-state ion selective and all-solid-state- reference electrodes. Electroanalysis **24**(6), 1253–1265 (2012)
27. Migdalski, J, Blaz, T, Lewenstam, A: Conducting polymer-based ion-selective electrodes. Anal. Chim. Acta **322**(3), 141–149 (1996)
28. Ansari, R, Fallah Delavar, A, Mohammad-khah, A: A solid state Cr(VI) ion-selective electrode based on polypyrrole. Microchimica Acta **178**(1–2), 71–79 (2012)
29. Ansari, R, Fallah Delavar, A, Mohammad-khah, A: Solid-State ion selective electrode based on polypyrrole conducting polymer nanofilm as a new potentiometric sensor For Zn^{2+} ion. J Solid State Electrochem. **16**(10), 3315–3322 (2012)
30. Charles, V, Rajendra, S: Composition and stability of 5-sulfosalicylate complexes of beryllium and copper. J. Am. Chem. Soc. **81**(23), 6159–6163 (1959)
31. Khalil, S: Ion-selective electrode for the determination of trazodone in tablets. Analyst **124**, 139–142 (1999)
32. Bakker, E, Buhlmann, P, Pretsch, E: Carrier-based ion-selective electrodes and bulk optodes. 1. General characteristics. Chem. Rev. **97**, 3083–3132 (1997)
33. Umezawa, Y, Umezawa, K, Sato, H: Selectivity coefficients for ion-selective electrodes: Recommended methods for reporting K_{AB}^{pot} values. Pure Appl Chem **67**, 507–518 (1995)

Superior photocatalytic degradation of azo dyes in aqueous solutions using $TiO_2/SrTiO_3$ nanocomposite

Loghman Karimi[1*] and Salar Zohoori[2]

Abstract

Research on photocatalytic degradation rate of azo dyes using $TiO_2/SrTiO_3$ nanocomposite in the photocatalysis process was the main goal of the present study. The $TiO_2/SrTiO_3$ nanocomposite was first synthesized under ultraviolet (UV) irradiation and sonication, and then, its photocatalytic activity has been examined under UV irradiation. The absorbance of samples was measured using a UV-vis spectrophotometer. The structure and morphology of the nanocomposite were investigated by scanning electron microscopy and the crystalline structure by X-ray diffraction spectroscopy. The results reveal that adding nano-$SrTiO_3$ to nano-TiO_2 showed the most promising photocatalytic activity toward dye degradation.

Keywords: $TiO_2/SrTiO_3$ nanocomposite, Photocatalytic degradation, Azo dyes, Irradiation, UV

Background

Environmental problem of toxic wastewater and infected waters is one of the main subjects that researchers work on. Due to this, organic dyes are one of the main industrial wastewater pollutions. More than 50% of textile dyes are azoic dyes which are recognized by their nitrogen π-bonds [1,2]. Textile and industrial dyes contain large groups of organic compounds that are produced more than 700,000 tons per year. About 1% to 20% of the world's dye products are included in the textile wastewater during the dyeing process [3-5].

The use of semiconductors such as TiO_2, ZnO, Fe_2O_3, and CdS as photocatalysts is interesting for the degradation of organic pollution. Due to the optical and electrical properties, low cost, high photocatalytic activity, chemical stability, and non-toxicity of nano-titanium dioxide, it is used as a common photocatalyst [6,7].

Although there are numerous advantages in utilizing the TiO_2, there are some disadvantages for the pure one, some of which are listed here: first, the bandgap larger than 3.2 eV causes low efficiency of nano-TiO_2. Furthermore, the electron-hole recombination rate is too high, resulting in low photocatalytic efficiency [8-11]. One of the research fields for performance improvement of TiO_2 photocatalysts is the retard of a charge recombination between excited electrons in the conduction band and holes in the valence band, which are formed by UV irradiation. Supporting the nanoparticle of a semiconductor or a metal, into which excited electrons in the conduction band of TiO_2 can be injected, is one of the methods to increase the charge separation efficiency of TiO_2 [12-15].

The perovskite oxides recognized as photocatalysts have ABO_3 formulas, where A is a rare earth metal with a large ionic radius or alkaline earth metal, and B is a transition metal with a small ionic radius. The alkali metal acts as the ionic balance (place in A) and the titanate framework plays the main role in the structure and properties with Ti in the B sites [16-19]. The presence of Sr in $SrTiO_3$ gives more ionic properties in comparison with SrO, and Ti is more covalent in $SrTiO_3$ than in TiO_2; this causes the reduction of the acidity of Ti ions and the increase of covalent property. Thus, bound formation properties might be different with titania and can lead to different photocatalitical reaction [20]. In comparison with other oxidants, the multi-cation oxide of $SrTiO_3$ is more capable of tuning the chemical and physical properties by altering the compositions and also has a larger number of photocatalytic sites [21]. By adding $SrTiO_3$ on the

* Correspondence: loghman.karimi@gmail.com
[1]Young Researchers and Elites Club, Yazd Branch, Islamic Azad University, Yazd, Iran
Full list of author information is available at the end of the article

TiO_2, its demerits can be reduced. Zhang et al. produced TiO_2-$SrTiO_3$ composite heterostructures by hydrothermal treatment and proved the significant improvement in photoelectrochemical performance of TiO_2 nanotube [22].

In this study, we dedicated our efforts to synthesize photocatalysts with a better quality. The photocatalytic degradation of dark green 6 and reactive orange 72 with $TiO_2/SrTiO_3$ nanocomposite under 20- and 400-W UV irradiations is investigated. Finally, for the investigation of the $TiO_2/SrTiO_3$ nanocomposite photocatalytic efficiency, its photocatalytic activity will be compared with nano-titanium dioxide and nano-strontium titanate.

Results and discussion

Photocatalyst performance

The results acquired indicate that adding nano-$SrTiO_3$ to nano-TiO_2 particles had a tangible effect on its photocatalytic activity (Figure 1C), and its photocatalytic degradation efficiency was higher than either $SrTiO_3$ or TiO_2. When the photocatalyst is illuminated by a light with energy higher than its bandgap energy, electron-hole pairs diffuse out to the surface of the photocatalyst. At the surface, the electrons and holes can either take part in the chemical reaction with electron donors and acceptors or recombine. In the absence of suitable electron and hole scavengers, the recombination of electron-hole pairs will occur within a very short time. Therefore, observed enhancement on the photocatalytic activity of TiO_2 by supporting the $SrTiO_3$ is thought to be explained on the basis of the increase in charge separation efficiency and the excited electrons in the conduction band of TiO_2 by UV irradiation being injected to the conduction band of $SrTiO_3$ to suppress the recombination between the electrons and holes in TiO_2. In order to realize an effective suppression of the recombination, it is reasonably essential to have a contact of $SrTiO_3$ particle with TiO_2.

Particle size and surface area are the effective parameters of photocatalytic activity of nanomaterials. In our previous work, we revealed that the nano-TiO_2 has smaller particle size and higher surface area than the nano-$SrTiO_3$ [23]. Therefore, photocatalytic degradation efficiency of the nano-TiO_2 is more than the nano-$SrTiO_3$ (Figure 1A,B).

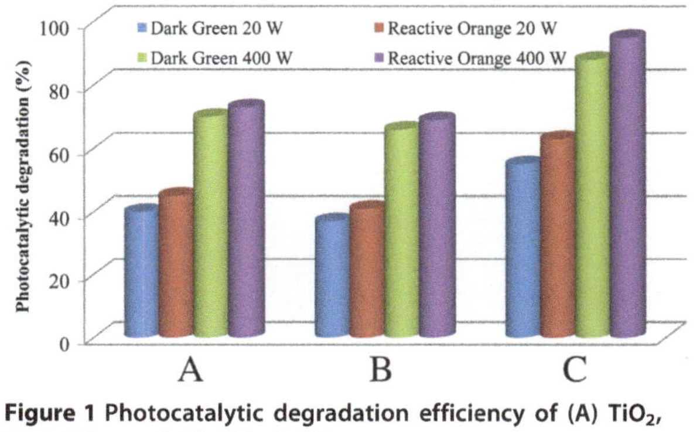

Figure 1 Photocatalytic degradation efficiency of (A) TiO_2, (B) $SrTiO_3$, and (C) $TiO_2/SrTiO_3$.

For both dark and reactive dyes, the percentage of photocatalyst degradation with a 400-W UV lamp was higher than that with 20 W. This occurs for two reasons: first of all, there is more radiation intensity, and second, there is a lower wavelength that resulted in the electron excitation of nanomaterials.

The higher degree of degradation of the reactive dye compared with the direct dye is due to the chromophore structures of the dyes, which are monoazo and trisazo, respectively [24].

SEM-EDS measurement

A scanning electron microscopy (SEM) image (Figure 2A) provides morphological information of the $TiO_2/SrTiO_3$ nanocomposite. It is thoroughly possible to recognize the nanoparticles on the composite. It may be clearly seen that particle distribution is uniform in the $TiO_2/SrTiO_3$ nanocomposite.

Energy-dispersive X-ray spectrometry (EDS) is an analytical technique used for the elemental analysis or chemical

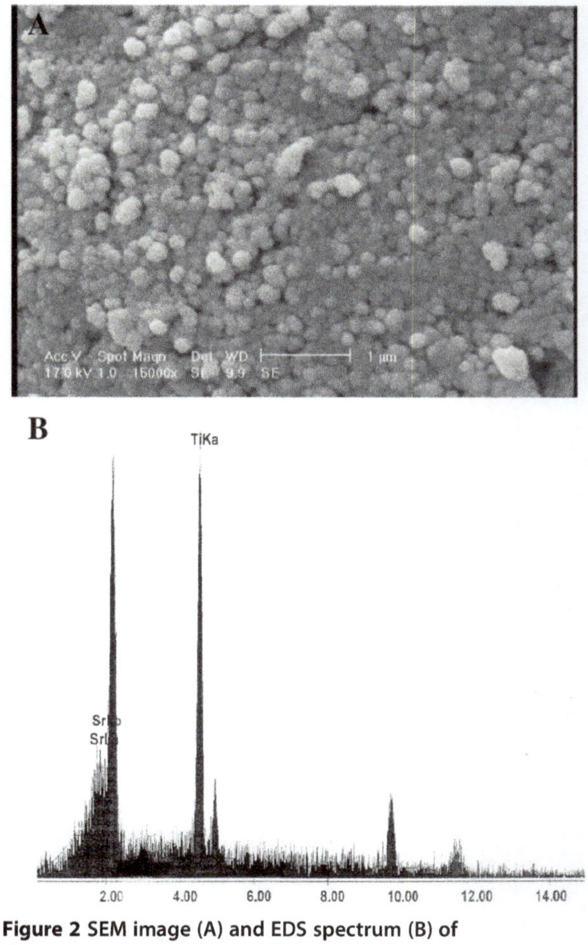

Figure 2 SEM image (A) and EDS spectrum (B) of the nanocomposite.

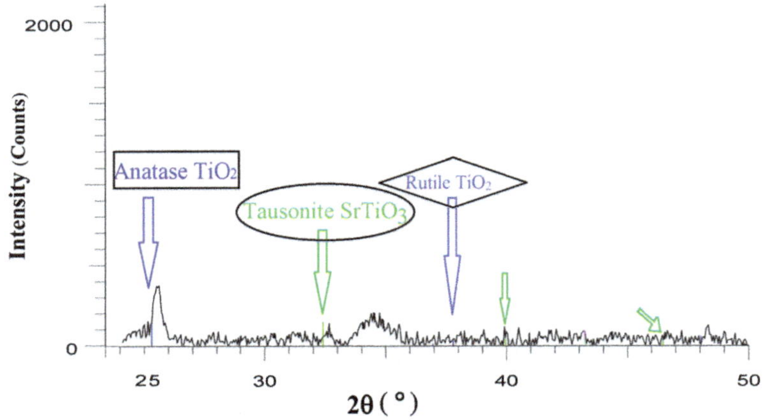

Figure 3 XRD pattern of the nanocomposite.

characterization of a sample and detects the elements using X-rays that are emitted from the specimen. Energy peaks correspond to the various elements in the sample. The presence of strontium and titanium particles on the $TiO_2/SrTiO_3$ nanocomposite was proved using EDS (Figure 2B).

XRD spectroscopy

The crystalline status of nanoparticles was studied by X-ray powder diffraction (XRD). The XRD pattern of the $TiO_2/SrTiO_3$ nanocomposite is shown in Figure 3. In the XRD pattern, the peak is around $2\Theta = 25.3°$ related to the anatase structure of nano-TiO_2, and in $2\Theta = 37.9°$, the peaks of the rutile structure of TiO_2 exist. One of the components of the fertilizing bath was $SrTiO_3/TiO_2$, as a result when the peaks of the tausonite structure of $SrTiO_3$ appeared around $2\Theta = 32.3°$. Also, the crystal size was calculated, and for the $TiO_2/SrTiO_3$, it was 201.77 Å.

Conclusion

In the present work, effective enhancement in photocatalytic activity of TiO_2 by supporting $SrTiO_3$ is experimentally confirmed. Through the XRD pattern, SEM image, and EDS spectrum, the presence of nanomaterials on the nanocomposite sample was verified. The produced nanocomposite provides a higher photocatalytic degradation efficiency in comparison with nano-strontium titanate or nano-titanium dioxide. On the composite system, strontium titanate plays the role of electron acceptor in order to accelerate photodecomposition.

Methods

Materials and equipment

Nano-strontium titanate powder (P.N.517011) with an average particle size of less than 100 nm and nano-titania (P-25) with an average particle size of about 21 nm were provided by Sigma-Aldrich (St. Louis, MO, USA) and Degussa Companies (Dusseldorf, Germany), respectively. Commercially available dark green 6 (CI 30295) and reactive orange 72 (CI 17754) were obtained from Alvan Sabet Company, Hamedan, Iran, and their structures are shown in Figure 4.

The dye solution was mixed with a magnetic stirrer during reaction (MR Hei, Heidolf, Germany). Samples after photocatalytic treatment were filtered through a Millipore filter (0.45 μm) membrane (Millipore Filter Corporation, Bedford, Massachusetts). The pH of the

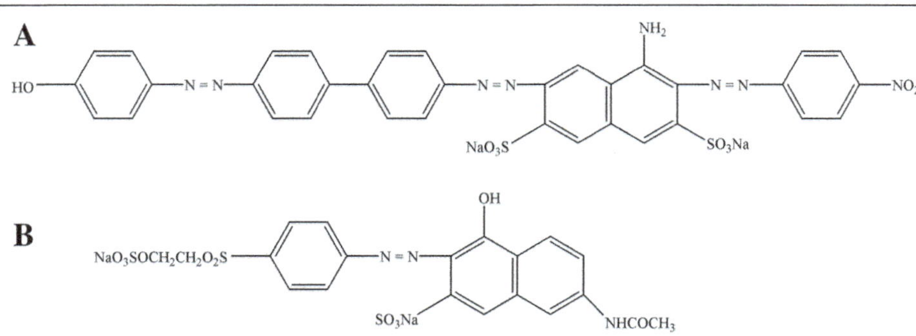

Figure 4 Structure of (A) dark green 6 (B) reactive orange 7.

solution was measured using a LI 120 pH meter (ELICO, Hyderabad, India), and concentrations of dyes were determined using a Carry 100 UV-vis spectrophotometer (Varian, Australia).

A UV-A 20W lamp (Sylvania, Belgium), with 365-nm wavelength radiation and light intensity of 0.2 to 0.4 $\mu W \cdot cm^{-2}$, and a UV400 W lamp (Philips, HPA 400 s, Guildford, England, UK), with higher radiation of wavelength between 300 and 400 nm along with the light spectrum having an intensity of 800 $\mu W \cdot cm^{-2}$ for UV-A, 115 $\mu W \cdot cm^{-2}$ for UV-B, and 25 $\mu W \cdot cm^{-2}$ for UV-C were used as light sources. An ultrasonic bath model EURONDA SPA 4D, 40 kHz (Vicenza, Italy) was used to mix the components.

SEM image and EDS spectra were obtained using a Philips XL30 scanning electron microscope (Amsterdam, Netherlands). The crystalline structure of the $TiO_2/SrTiO_3$ nanocomposite was characterized by XRD (Bruker D8 Discover X-ray diffractometer, Karlsruhe, Germany).

$TiO_2/SrTiO_3$ nanocomposite preparation

Six milligrams of nano-TiO_2 and 4 mg of nano-$SrTiO_3$ were added into 100 mL of distilled water. After 1 h of sonication and irradiation (UV-A light) of the suspension, the mixture was centrifuged and then dried at 30°C. Finally, the obtained product was calcined at 300°C for 2 h.

Procedure

The concentration of the dye in the solution was calculated by a computer program using a calibration curve. The program determines the absorbance of the dye solution at a maximum wavelength of dyes: dark green 6, 623 nm; reactive orange 72, 433 nm. The first step was the preparation of dye solution by distilled water. Then, nanomaterials were added. Firstly, the solution mixture was stirred for 15 min without irradiation in order to get equilibrium of dye adsorption. Then, the solution was irradiated with two lamps (20 and 400 W) for 1 h, and during irradiation, it was continuously stirred with a rate of 200 rpm and a temperature of 25°C, and the real dye solution pH for the direct and reactive dyes were 6.6 and 6.4, respectively. After irradiation, the samples were purified with the Millipore filter. The decolorization

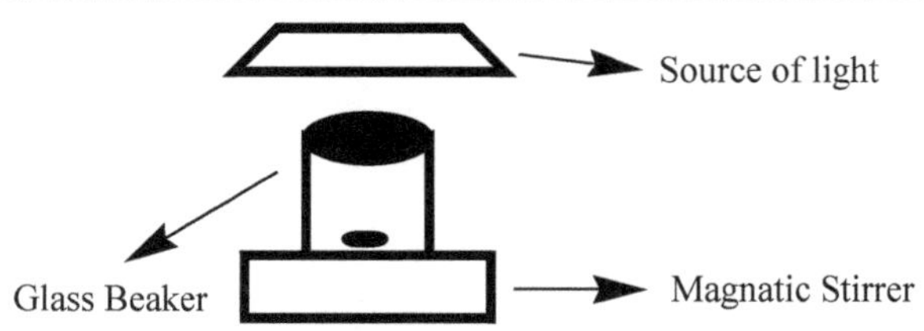

Figure 5 Schematic of photocatalytic dye degradation.

and photocatalytic degradation efficiency have been calculated as follows:

$$\text{Efficiency } (\%) = \frac{C_0 - C_e}{C_0} \times 100,$$

where C_0 and C_e correspond to the initial and final concentrations of the dyes before and after photo irradiation. In this equation, Efficiency % shows the dye photocatalyst degradation percent [25]. For the investigation of the $TiO_2/SrTiO_3$ nanocomposite photocatalytic efficiency, its degradation percent was compared with nano-TiO_2 and nano-$SrTiO_3$. Basic parameters were photocatalyst concentration of 0.1% (on weight of bath), 20 mg/L dye, 25°C temperature, 1 h irradiation and real pH. Figure 5 shows the schematic of dye photocatalytic degradation.

Competing interests

The authors declare that they have no competing interests.

Authors' contributions

LK participated in the idea of the study, the design of the study, interpretation of the results, and writing the manuscript for publication. SZ carried out the experiment and participated in its design, coordination, and sequence of alignments. Both authors read and approved the final manuscript.

Acknowledgments

This work was supported by the Young Researchers and Elites Club, Yazd Branch, Islamic Azad University, Yazd, Iran under Grant no. 88363.

Author details

[1]Young Researchers and Elites Club, Yazd Branch, Islamic Azad University, Yazd, Iran. [2]Department of Textile Engineering, Yazd Branch, Islamic Azad University, Yazd, Iran.

References

1. Song, S, Xu, L, He, Z, Ying, H, Chen, J, Xiao, X, Yan, B: Photocatalytic degradation of C.I. Direct Red 23 in aqueous solutions under UV irradiation using $SrTiO_3/CeO_2$ composite as the catalyst. J. Hazard. Mater. **152**, 1301–1308 (2008)
2. Lee, J, Choi, SP, Thiruvenkatachari, R, Shim, WG, Moon, H: Evaluation of the performance of adsorption and coagulation processes for the maximum removal of reactive dyes. Dyes Pigm **69**, 196–203 (2006)
3. Madhavan, J, Grieser, F, Ashokkumar, M: Degradation of orange-G by advanced oxidation processes. Ultras. Sonochem. **17**, 338–343 (2010)
4. Isaev, AB, Aliev, ZM, Adamadzieva, NK, Alieva, NA, Magomedova, GA: The photocatalytic oxidation of azo dyes on Fe_2O_3 nanoparticles under oxygen pressure. Nanotech. Rus. **4**, 475–479 (2009)
5. Mahvi, AH, Ghanbarian, M, Nasseri, S, Khairi, A: Mineralization and discoloration of textile wastewater by TiO_2 nanoparticles. Desalination **239**, 309–316 (2009)
6. Xua, J, Aoa, Y, Fua, D, Yuana, C: Low-temperature preparation of anatase titania-coated magnetite. J. Phys. Chem. Sol. **69**, 1980–1984 (2008)
7. Hegde, M, Nagaveni, K, Roy, S: Synthesis, structure and photocatalytic activity of nano TiO_2 and nano $Ti_{1-x}M_xO_{2-\delta}$ (M = Cu, Fe, Pt, Pd, V, W, Ce, Zr). PRAMANA J. Phys. **65**, 641–645 (2005)
8. Linsebigler, A, Lu, G, Yates, JT: Photocatalysis on TiO_2 surfaces: principles, mechanisms, and selected results. Chem. Rev. **95**, 735–758 (1995)
9. Wang, J, Wen, FY, Zhang, ZH, Zhang, XD, Pan, ZJ, Zhang, P, Kang, PL, Tong, J, Wang, L, Xu, L: Investigation on degradation of dyestuff wastewater using visible light in the presence of a novel nano TiO_2 catalyst doped with upconversion luminescence agent. J. Photochem. Photobiol. A **180**, 189–195 (2006)

10. Zyoud, AH, Zaatar, N, Saadeddin, I, Ali, C, Park, D, Campet, G, Hilal, HS: CdS-sensitized TiO_2 in phenazopyridine photo-degradation: Catalyst efficiency, stability and feasibility assessment. J. Hazard. Mater. **173**, 318–325 (2010)
11. Ueda, M, Matsuo, SOY: Preparation of tabular TiO_2–$SrTiO_3$-δ composite for photocatalytic electrode. Sci. Tech. Adv. Mater. **5**, 187–193 (2004)
12. Montazer, M, Behzadnia, A, Moghadam, MB: Superior self-Cleaning features on wool fabric using TiO_2/Ag nanocomposite optimized by response surface methodology. J. Appl. Polym. Sci. **125**, E356–E363 (2012)
13. Montazer, M, Behzadnia, A, Pakdel, E, Rahimi, MK, Moghadam, MB: Photo induced silver on nano titanium dioxide as an enhanced antimicrobial agent for wool. J. Photochem. Photobiol. B **103**, 207–214 (2011)
14. Sung-Suh, HM, Choi, JR, Hah, HJ, Koo, SM, Bae, YC: Comparison of Ag deposition effects on the photocatalytic activity of nanoparticulate TiO_2 under visible and UV light irradiation. J. Photochem. Photobiol. A **163**, 37–44 (2004)
15. Subramanian, V, Wolf, E, Kamat, P: Semiconductor–metal composite nanostructures. To what extent do metal nanoparticles improve the photocatalytic activity of TiO_2 films? J. Phys. Chem. B **105**, 11439–11446 (2001)
16. Boudali, A, Khodja, MD, Amrani, B, Bourbie, D, Amara, K, Abada, A: First-principles study of structural, elastic, electronic, and thermal properties of $SrTiO_3$ perovskite cubic. Phys. Lett. A **373**, 879–884 (2009)
17. Subramanian, V, Roeder, RK, Wolf, EE: Synthesis and UV-Visible-light photoactivity of noble-metal-$SrTiO_3$ composites. Ind. Eng. Chem. Res. **45**, 2187–2193 (2006)
18. Wang, J, Yin, S, Zhang, Q, Saito, F, Sato, T: Mechanochemical synthesis of $SrTiO_{3-x}F_x$ with high visible light photocatalytic activities for nitrogen monoxide destruction. J. Mater. Chem. **13**, 2348–2352 (2003)
19. Niishiro, R, Kato, H, Kudo, A: Nickel and either tantalum or niobium-codoped TiO_2 and $SrTiO_3$ photocatalysts with visible-light response for H_2 or O_2 evolution from aqueous solutions. Phys. Chem. Chem. Phys. **7**, 2241–2245 (2005)
20. Chang, C, Ray, B, Paul, DK, Demydov, D, Klabunde, KJ: Photocatalytic reaction of acetaldehyde over $SrTiO_3$ nanoparticles. J. Molec. Catal. A: Chem. **281**, 99–106 (2008)
21. Wei, X, Xu, G, Ren, Z, Xu, C, Shen, G, Han, G: PVA-assisted hydrothermal synthesis of $SrTiO_3$ nanoparticles with enhanced photocatalytic activity for degradation of RhB. J. Am. Ceram. Soc. **91**, 3795–3799 (2008)
22. Zhang, J, Bang, JH, Tang, C, Kamat, PV: Tailored TiO_2 – $SrTiO_3$ heterostructure nanotube arrays for improved photoelectrochemical performance. ACS Nano **4**, 387–395 (2010)
23. Karimi, L, Zohoori, S, Yazdanshenas, ME: Photocatalytic degradation of azo dyes in aqueous solutions under UV irradiation using nano-strontium titanate as the nanophotocatalyst. J. Saudi Chem. Soc (2011). in press
24. Karimi, L, Mirjalili, M, Yazdanshenas, ME, Nazari, A: Effect of nano TiO_2 on self-cleaning property of cross-linking cotton fabric with succinic acid under UV irradiation. Photochem. Photobiol. **86**, 1030–1037 (2010)
25. Chen, CY: Photocatalytic degradation of azo dye reactive orange 16 by TiO_2. Water Air Soil Pollut. **202**, 335–342 (2009)

9

Preparation of transparent, conductive ZnO:Co and ZnO:In thin films by ultrasonic spray method

Said Benramache[1*], Boubaker Benhaoua[2] and Hamza Bentrah[3]

Abstract

This paper examines the growth of undoped and doped thin films with (Co and In) on glass substrate at 350°C using ultrasonic spray technique. We have investigated the influence of doping concentrations ranging from 0 to 4 wt.% on structural, optical, and electrical properties of ZnO thin films. Zinc acetate dehydrate, $CoCl_3$ $4H_2O$ or $InCl_3$, ethanol, and monoethanolamine were used as a starting materials, dopant source, solvent, and stabilizer, respectively. The X-ray diffraction analysis indicated that the undoped and doped ZnO thin films have polycrystalline nature and hexagonal wurtzite structure with (002) preferential orientation. The maximum average crystallite sizes of ZnO:Co and ZnO:In were 55.46 and 45.78 nm at concentrations of 2 wt.% Co and 3 wt.% In, respectively, indicating that the crystallinity of doped films improved after doping. The optical absorption spectra showed that all undoped and doped ZnO films are transparent within the visible wavelength region. The band gap energy of ZnO:Co thin films increased after doping from 3.25 to 3.36 eV; however, the optical gap of ZnO:In decreases after doping from 3.25 to 3.18 eV, indicating the increase and decrease, respectively, in the transition tail width. The electrical conductivity of doped films is stabilized after doping. Transparent, conductive Co-doped ZnO thin films deposited by ultrasonic spray technique are of good quality.

Keywords: ZnO; Thin film; Semiconductor doping; Transparent conducting oxides; Ultrasonic spray

Background

Zinc oxide (ZnO), which is one of the most important binary II-VI semiconductor compounds, has a hexagonal wurtzite structure and a natural n-type electrical conductivity with a direct energy wide band gap of 3.37 eV at room temperature and a large exciton binding energy (approximately 60 meV) [1]. The resistivity values of ZnO films may be adjusted between 10^{-4} and 10^{-1} Ω cm by changing the annealing conditions and doping [2]. Transparent conducting oxides are widely used in microelectronic devices, light emitting diodes, thin films, antireflection coatings for transparent electrodes in solar cells [3,4], and gas sensors in surface acoustic wave devices [5], varistors, spintronic devices, and lasers [6].

ZnO thin films which can be produced by several techniques such as reactive evaporation, molecular beam epitaxy [7], magnetron sputtering technique, pulsed laser deposition [8], sol-gel technique, chemical vapor deposition, electrochemical deposition [9], and spray pyrolysis [10] have been reported. Among these, we will focus more particularly in this paper the spray ultrasonic technique that is a low-cost method suitable for large-scale production. It has several advantages in producing nanocrystalline thin films, such as relatively homogeneous composition, simple deposition on glass substrate because of the low substrate temperatures involved, easy control of film thickness, and fine and porous microstructure. It is possible to alter the mechanical, electrical, optical, and magnetic properties of ZnO nanostructures.

The doped ZnO thin films have various applications such as transparent conductors, in ferromagnetism, semiconductors, and in piezoelectric and solar cells; moreover, the films have low resistivity and good optical gap energy at low temperature and are transparent in the visible region. There are several reports on ZnO nanostructures doped with different elements, such as Al, Ga, Mg, Li, P, N, Ni, In, and Co [11-15]. ZnO:Co and ZnO:In films have been extensively studied because they exhibit high mobility, good optical transparency, and good electrical conductivity and have lower material cost.

In this paper, the CZO and IZO thin films were deposited on glass substrate by ultrasonic spray technique; at

* Correspondence: benramache.said@gmail.com
[1]Material Sciences Department, Faculty of Science, University of Biskra, 07000, Algeria
Full list of author information is available at the end of the article

a substrate temperature of 350°C, we have studied the effect of the doping concentration on structural, optical, and electrical properties of ZnO thin films. The main goal for this research is to find the optimum doping concentration which gives highly semiconducting properties of ZnO:Co and ZnO:In thin films.

Experimental procedure

Preparation of spray solution

ZnO solution was prepared by dissolving 0.1 M Zn $(CH_3COO)_2$ and $2H_2O$ in the solvent containing equal volumes of ethanol solution (99.995%) of absolute purity; then we added drops of monoethanolamine solution as a stabilizer; and the mixed solution was stirred at 50°C for 120 min to yield a clear and transparent solution.

The doped solution was prepared by adding to the precedent solution $CoCl_3{\cdot}4H_2O$ or $InCl_3$ as dopant source. The weight percentages of (Co/Zn) and (In/Zn) were 1, 2, and 3 for ZnO:Co and 2, 3, and 4 for ZnO:In. The solution became clear and homogeneous after stirring for 120 min at 50 to 70°C.

The substrate was R217102 glass in a size of 1 cm × 1 cm × 0.1 cm; prior to pumping, the substrate (R217102 glass) was cleaned with alcohol in an ultrasonic bath and blow-dried with dry nitrogen gas.

Deposition of thin films

The resulting solutions were sprayed on the heated glass substrates by ultrasonic nebulizer system (Sonics) which transforms the liquid to a stream formed with uniform and fine droplets of 35 μm average diameter (given by the manufacturer). The deposition was performed at a substrate temperature of 350°C with 2 min deposition time [15-17].

Characterization

Crystallographic and phase structures of the thin films were determined by X-ray diffraction (Bruker AXS-8D; Bruker Corporation, Billerica, MA, USA) with CuKα radiation (λ = 0.1541 nm) in the scanning range was between 2θ = 25° and 2θ = 55°. The optical properties of the deposited films was measured in the range of 300 to 800 nm using by an ultraviolet-visible spectrophotometer (UV, Lambda 35; PerkinElmer Inc., Waltham, MA, USA), and the electrical conductivity of the films was measured in a coplanar structure obtained with evaporation of four golden stripes on film surface. All spectra were measured at room temperature in air.

Results and discussion

Crystalline structure

The X-ray diffraction patterns of Co-doped ZnO and In-doped ZnO thin films with doping levels are presented in Figures 1 and 2, respectively. Here the films were deposited on glass substrate at a substrate temperature of 350°C [18]. One can be seen, the diffraction peaks at 2θ = 31.74°, 34.52°, and 36.41 corresponding to the (100), (002), and (101) planes, respectively, of undoped ZnO film (0 wt.%). As discussed in our previous paper [15], this result showed that the thin films were polycrystalline and had a hexagonal wurtzite structure. Moreover, as can be seen in Figures 1 and 2, the ZnO:Co and ZnO:In thin films deposited at different concentrations have a (002) diffraction peak which is the highest one in our publication [15]. The films doped at 2 wt.% Co and 3 wt.% In have higher and sharper diffraction peaks indicating an improvement in (002) peak intensity as compared to other films. The films exhibit polycrystalline structure, which is a hexagonal wurtzite structure, in these spectra [19,20]. This indicates that all films have preferential *c*-axis orientation along the (002) plane [21,22], because of the high-intensity direction and different degrees of substitution of Zn^{2+} ions by Co^{2+} and In^{3+} ions. This result indicates that the doping level of the films improved the structural properties.

In order to attain the detailed structure information (Table 1), the crystallite size *G* (002) along the *c*-axis

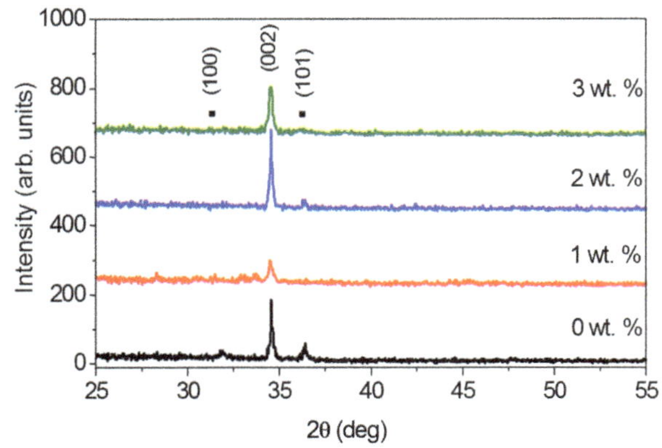

Figure 1 X-ray diffraction spectra of ZnO thin films doped with cobalt at different concentrations.

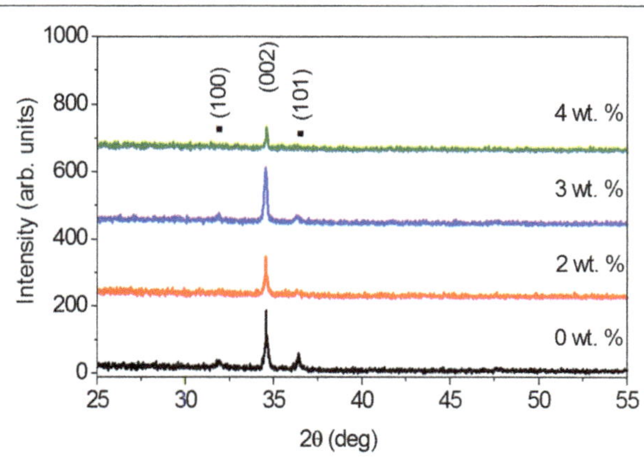

Figure 2 X-ray diffraction spectra of ZnO thin films doped with indium at different concentrations.

Table 1 Structure information of undoped and doped ZnO thin films

	ZnO films	ZnO:Co films			ZnO:In films		
	0%	1%	2%	3%	2%	3%	4%
$2\theta_{(002)}$ (deg)	34.52	34.49	34.56	34.58	34.54	34.52	34.58
$d_{(002)}$ (A°)	2.6005	2.5983	2.5932	2.5918	2.5947	2.6005	2.5918
c (A°)	5.1923	5.1967	5.1865	5.1836	5.1894	5.1923	5.1836
FWHM (deg)	0.245	0.236	0.150	0.207	0.20	0.182	0.26
G (nm)	33.28	35.25	55.46	40.19	41.60	45.78	32.00
Strain ε (%)	−0.263	−0.179	−0.375	−0.431	−0.319	−0.263	−0.431

The Bragg angle 2θ, the interplanar spacing d_{hkl}, the lattice parameter c, the full width at half-maximum FWHM β, the crystallite size G, and strain ε measured for undoped and doped ZnO thin films.

was calculated according to the Scherrer equation [23]:

$$G = \frac{0.9\lambda}{\beta\cos\theta}, \quad (1)$$

where G is the crystallite size, λ is the X-ray wavelength (λ = 1.5406 A°), β is the full width at half-maximum (FWHM), and θ is Bragg angle of (002) peak.

According to the hexagonal symmetry, the lattice constant can be calculated by the following formula [24]:

$$d_{hkl} = \left(\frac{4}{3}\frac{h^2 + hk + k^2}{a^2} + \frac{l^2}{c^2}\right)^{-\frac{1}{2}}, \quad (2)$$

where a, c are the lattice parameters, h, k, l are the Miller indices of the planes, and d_{hkl} is the interplanar spacing.

The lattice mismatch between ZnO film and substrate can result in varying degrees of stress during the deposition process of ZnO thin films. Therefore, the strain also affected significantly the structures and properties of ZnO films to some extent. For ZnO films with wurtzite structure, the strain can be obtained by following formula [10]:

$$\varepsilon = \frac{c - c_0{}^{0}}{c_0} \times 100\%, \quad (3)$$

where ε is the mean strain in ZnO thin films (Table 1), c is the lattice constant of ZnO thin films, and c_0 is the lattice constant of bulk (standard c_0 = 0.5206 nm).

In Figure 3 we have reported the variation of the crystallite size as a function of doping level. As can be seen, the crystallite size increased for undoped ZnO and doped films with increasing doping level up to 2 wt.% Co and 3 wt.% In. In this region the values of lattice parameters c decreased (Table 1), and the strain of the films is also decreased, indicating the strain along the c-axis. The increase of the crystallite size has been indicated by the enhancement of the crystallinity and c-axis orientation of ZnO thin films by Zhu et al. [25]. The decrease in the crystallite size with the doping level indicates the deterioration in crystallinity of doped films, as found by Bao et al. [26].

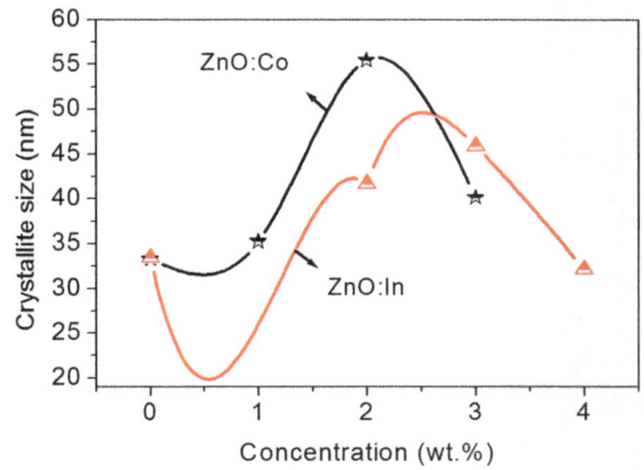

Figure 3 Variation of crystallite size of undoped ZnO and doped films as a function of doping level.

Optical characteristics

Figures 4 and 5 show the absorbance spectra of ZnO:Co and ZnO:In samples. The thin films were deposited at different concentrations at a temperature of 350°C for 2 min. It is clear that the absorbance at higher than 400 nm wavelength is low. This confirms that the absorbance value and the tail height of doped films were obtained at different concentrations and reach a minimum concentration of 0 wt.%, followed by an increase and then decrease, with the doping level further increasing. As seen from this figure, the absorption edge shifts in ZnO:Co thin films are better than in ZnO:In. We found that the values of absorption edge shifts are decreased and increased with increasing doping level, which shows that the optical band gap of the CZO films is broadened after doping with cobalt [15] and the optical band gap of the IZO films narrowed after doping with indium. The films obtained at concentrations of 2 wt.% Co

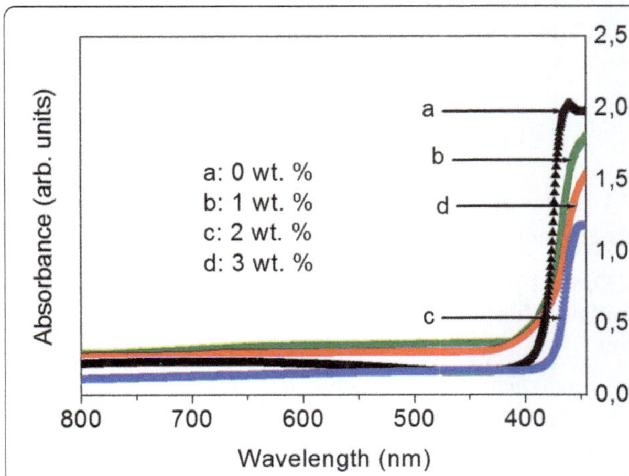

Figure 4 The absorbance spectra of ZnO:Co samples as a function of doping level.

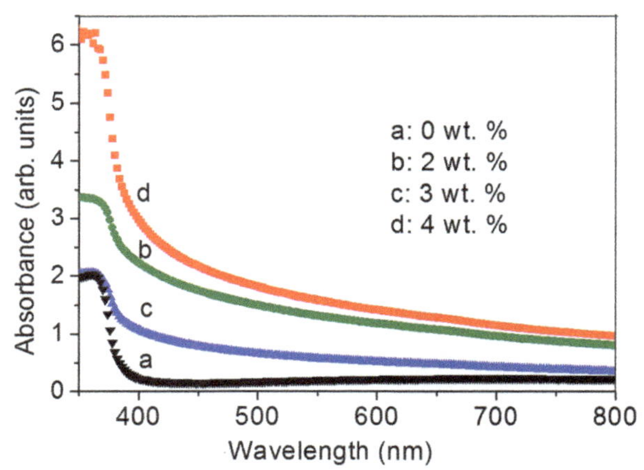

Figure 5 The absorbance spectra of ZnO:In samples as a function of doping level.

and 3 wt.%. In show lower absorption after doping at lower wavelength, which indicated the enhancement of the crystallinity [18,27]. In addition, the difference of the absorbance between undoped and doped films can be observed clearly at wavelength smaller than 400 nm. These results show that the produced ZnO:Co and ZnO:In thin films could be used in photovoltaic applications due to the sharp increase of absorbance in the strong absorption region.

In order to further investigate the effect of doping level on doped films, the optical band gap energy E_g was measured from the absorbance spectra using the following relations [28-30]:

$$A = \alpha\, d = -\ln T \tag{4}$$

$$(Ah\nu)^2 = B(h\nu - E_g), \tag{5}$$

where A is the absorbance, and d is the film thickness; T the transmittance spectra of thin films; α is the absorption coefficient values; B is a constant; hν is the photon energy and E_g is the band gap energy of the semiconductor. Besides, we have used the Urbach energy (Table 2), which is related to the disorder in the film network, expressed as follows [31]:

$$A = A_0 \exp\left(\frac{h\upsilon}{E_u}\right), \tag{6}$$

where A is the absorbance, A_0 is a constant, and E_u is the Urbach energy.

Table 2 Optical properties and electrical conductivity measured as a function of doping level for doped ZnO thin films

	ZnO films	ZnO:Co films			ZnO:In films		
	0%	1%	2%	3%	2%	3%	4%
E_g (eV)	3.250	3.295	3.362	3.30	3158	3.185	3.066
E_u (meV)	209	183	108	210	423	328	376
σ (Ω cm^{-1})	7.55	6.56	7.63	5.78	6.19	7.54	7.82

Figure 6 shows the typical variation of $(Ah\nu)^2$ as a function of photon energy (hν) used for optical band gap, which is determined by extrapolation of the linear region to $(Ah\nu)^2 = 0$ [15,32].

Figure 6 shows a plot of $(Ah\nu)^2$ vs. photon energy hν for undoped ZnO thin films. A linear dependence of $(Ah\nu)^2$ on hν at higher photon energies indicates that the films are essentially direct transition-type semiconductors. The inset shows the drawn ln A as a function of photon energy (hν) to deduce the Urbach energy.

As clearly seen in the Figure 7, the optical gap energy increased with increasing Co doping from 0 to 2 wt.% and the value is found to be 3.25 to 3.36 eV, which results in the narrowing of the conduction band E_C and the valence band E_V and causes the motion of E_C upwards and E_V downwards; hence, Co doping causes the band gap broadening [15].

In the four elaborated films (Figure 7), it could be noticed that the band gap energy of IZO films decreases after doping from 3.25 to 3.18 eV with In doping increase from 0 to 3 wt.%. This decrease in the band gap with the increasing indium concentration might be explained in terms of electron concentration dependence of the band gap shift in In-doped ZnO films. The optical band gap is related to the transition of the electrons from valance band to Fermi energy levels in the conduction band for degenerate semiconductors. According to Moss-Burstein effect, the band values should increase with In concentration, but the E_g value decreased with doping. This can be explained by the increase of the defects (such as an increase of the Urbach energy in Figure 8), impurities, or incomplete substitution of dopant with the host atoms [33-36]. The same phenomena are carried out by Tubtimtae and Lee, where they have observed a red shift of the absorption edges

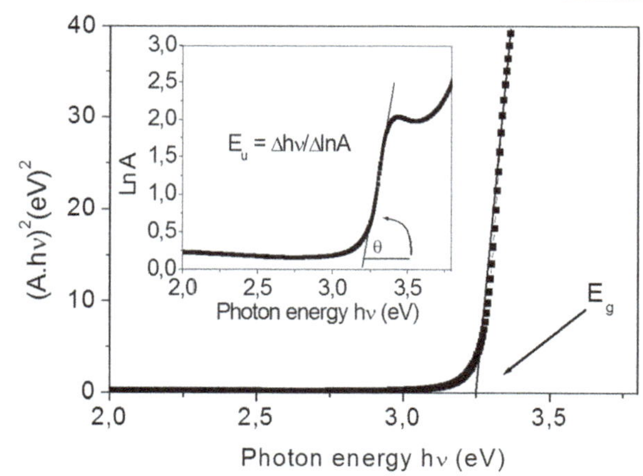

Figure 6 The typical variation of $(Ah\upsilon)^2$ as a function of photon energy of undoped ZnO film. The inset shows the drawn of ln A as a function of photon energy (hν) for deduce the Urbach energy.

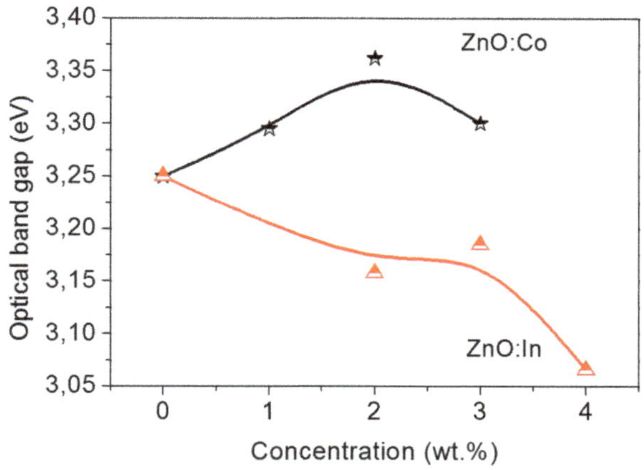

Figure 7 Variation of band gap (E_g) as a function of doping level for doped ZnO thin films.

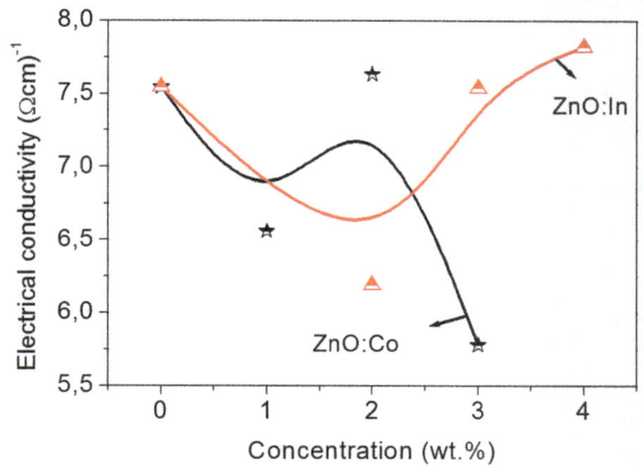

Figure 9 Variation of electrical conductivity (σ) as a function of doping level for doped ZnO thin films.

from 3.37 eV (undoped ZnO) to 3.34 eV (In-doped ZnO) [37].

From Figure 8 one can see that the Urbach energy decreases after doping with cobalt and increases after doping with indium reaching a minimum of about 0.108 and 0.328 eV for 2 wt.% Co and 3 wt.% In, respectively, as expressed in the literature [15,18,38].

Electrical conductivity

Figure 9 shows the variation of the electrical conductivity σ of undoped ZnO and doped films with Co and In as a function of doping level. As can be seen, the electrical conductivity decreases first and then increases, reaching a maximum of about 7.63 and 7.82 Ω cm^{-1}for 2 wt.% Co and 4 wt.% In, respectively. The increase in the conductivity of the films with doping has been explained by the displacement of the electrons [39]; the latter are coming from the ions In^{3+} and Co^{2+} donors in the substitutional sites of Zn^{2+}, which resulted in an increased carrier density. The decrease of the electrical conductivity with the increase of the doping level are explained by increasing of the disorder in the films; hence, the potential barriers are increased [40,41] because the introduced atoms are segregated into the grain boundaries.

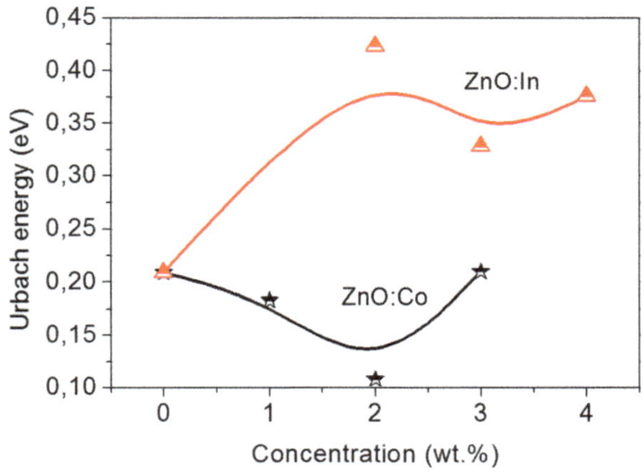

Figure 8 Variation of Urbach energy (E_u) as a function of doping level for doped ZnO thin films.

Conclusions

In conclusion, highly transparent, conductive ZnO:Co and ZnO:In thin films have been deposited on glass substrate by ultrasonic spray at a substrate temperature of 350°C. The crystalline structure, conductivity, and optical properties were investigated. All the films are polycrystalline structure wurtzite with (002) preferential orientation and, thus, are favorable to the diffusion of atoms absorbed on the substrate. The maximum values of crystallite size G = 33.28, 55.46, and 45.78 nm are attained in ZnO, ZnO:Co, and ZnO:In films at concentrations of 0, 2, and 3 wt.%, respectively. The absorption edge shifts in ZnO:Co thin films are better than in ZnO: In. We found that the values of absorption edge shifts are decreased and increased with the increasing doping level, which shows that the optical band gap of the ZnO: Co films is broadened after doping with cobalt from 3.25 to 3.36 eV and the optical band gap of the ZnO:In films is narrowed after doping with indium from 3.25 to 3.18 eV. The electrical conductivity of doped films is stabilized after doping with cobalt and indium. Transparent, conductive C-doped ZnO thin films deposited by ultrasonic spray technique are of good quality. The amount of doping concentration is 2 wt.% Co and 3 wt.% In, in ZnO thin films.

Methods

Authors did not provide this information.

Competing interests

Authors did not provide this information.

Authors' contributions

Authors did not provide this information.

Acknowledgments

The authors would like to thank Prof. A. Si Ahmed, Prof. A. Chari, Dr. S. Rahmane, Mr. B. Gasmi, Mr. H. Bentrah, Dr. O. Belahssen, Dr. D. Bensahal, and Mr. C. Mahboub for their helpful counseling.

Author details

[1]Material Sciences Department, Faculty of Science, University of Biskra, 07000, Algeria. [2]VTRS Laboratory, Institute of Technology, University of El-Oued, 39000, Algeria. [3]Mechanical Department, Faculty of Technology, University of Biskra, 07000, Algeria.

References

1. Ma, L, Ai, X, Huang, X, Ma, S: Effects of the substrate and oxygen partial pressure on the microstructures and optical properties of Ti-doped ZnO thin films. Superlattice. Microst. **50**, 703–712 (2011)
2. Ko, YD, Kim, KC, Kim, YS: Effects of substrate temperature on the Ga-doped ZnO films as an anode material of organic light emitting diodes. Superlattice. Microst. **51**, 933–941 (2012)
3. Benramache, S, Benhaoua, B, Chabane, F, Lemadi, FZ: Influence of growth time on crystalline structure. Conductivity and optical properties of ZnO thin films. J. Semicond **34**, 023001-1 (2013)
4. Khomchenko, VS, Kryshtab, TG, Savin, AK, Zavyalova, LV, Roshchina, NN, Rodionov, VE, Lytvyn, OS, Kushnirenko, VI, Khachatryan, VB, Adame, JAA: Fabrication and properties of ZnO:Cu and ZnO:Ag thin films. Superlattice. Microst. **42**, 94–98 (2007)
5. Venkatachalam, S, Iida, Y, Kanno, Y: Preparation and characterization of Al doped ZnO thin films by PLD. Superlattice. Microst. **44**, 127–135 (2008)
6. Rahmane, S, Djouadi, MA, Aida, MS, Barreau, N, Abdallah, B, Zoubir, NH: Power and pressure effects upon magnetron sputtered aluminum doped ZnO films properties. Thin Sol. Film. **519**, 5–10 (2010)
7. Wang, ZY, Hu, LZ, Jie, Z, Jie, S, Wang, ZJ: Effect of the variation of temperature on the structural and optical properties of ZnO thin films prepared on Si (1 1 1) substrates using PLD. Vacuum **78**, 53–57 (2005)
8. Mosbah, A, Moustaghfir, A, Abed, S, Bouhssira, N, Aida, MS, Tomasella, E, Jacquet, M: Comparison of the structural and optical properties of zinc oxide thin films deposited by d.c. and r.f. sputtering and spray pyrolysis. Surf. Coat. Technol **200**, 293–296 (2005)
9. Mosbah, A, Abed, S, Bouhssira, N, Aida, MS, Tomasella, E: Preparation of highly textured surface ZnO thin films. Mater. Sci. Eng. B **129**, 144–149 (2006)
10. Bahsi, ZB, Oral, AY: Effects of Mn and Cu doping on the microstructures and optical properties of sol-gel derived ZnO thin films. Opt. Mater. **29**, 672–678 (2007)
11. Benramache, S, Benhaoua, B, Chabane, F: Effect of substrate temperature on the stability of transparent conducting cobalt doped ZnO thin films. J. Semicond. **33**, 093001-1 (2012)
12. Yamada, T, Nebiki, T, Kishimoto, S, Makino, H, Awai, K, Narusawa, T, Yamamoto, T: Dependences of structural and electrical properties on thickness of polycrystalline Ga-doped ZnO thin films prepared by reactive plasma deposition. Superlattice. Microst. **42**, 68–73 (2007)
13. Duclère, JR, Novotny, M, Meaney, A, O'Haire, R, McGlynn, E, Henry, MO, Mosnier, PJ: Properties of Li-, P- and N-doped ZnO thin films prepared by pulsed laser deposition. Superlattice. Microst **38**, 397–405 (2005)
14. Abed, S, Aida, MS, Bouchouit, K, Arbaoui, A, Iliopoulos, K, Sahraoui, B: Non-linear optical and electrical properties of ZnO doped Ni thin films obtained using spray ultrasonic technique. Opt. Mater. **33**, 968–972 (2011)
15. Benramache, S, Benhaoua, B: Influence of substrate temperature and cobalt concentration on structural and optical properties of ZnO thin films prepared by ultrasonic spray technique. Superlattice. Microst. **52**, 807–815 (2012)
16. Djelloul, A, Aida, MS, Bougdira, J: Photoluminescence, FTIR and X-ray diffraction studies on undoped and Al-doped ZnO thin films grown on polycrystalline α-alumina substrates by ultrasonic spray pyrolysis. J. Lumin **130**, 2113–2117 (2010)
17. Zebbar, N, Kheireddine, Y, Mokeddem, K, Hafdallah, A, Kechouane, M, Aida, MS: Structural, optical and electrical properties of n-ZnO/p-Si heterojunction prepared by ultrasonic spray. Mater. Sci. Semicond. Process. **14**, 229–234 (2011)
18. Benramache, S, Benhaoua, B: Influence of annealing temperature on structural and optical properties of ZnO:In thin films prepared by ultrasonic spray technique. Superlattice. Microst. **52**, 1062–1070 (2012)
19. Kavak, H, Tuzemen, ES, Ozbayraktar, LN, Esen, R: Optical and photoconductivity properties of ZnO thin films grown by pulsed filtered cathodic vacuum arc deposition. Vacuum **83**, 540–544 (2009)
20. Wang, Y, Chu, B: Structural and optical properties of ZnO thin films on (111) CaF_2 substrates grown by magnetron sputtering. Superlattice. Microst. **44**, 54–61 (2008)
21. Rolo, AG, de Campos Ayres, J, Viseu, T, de Lacerda-Arôso, T, Cerqueira, MF: The annealing effect on structural and optical properties of ZnO thin films produced by r.f. sputtering. Superlattice. Microst **42**, 265–269 (2007)
22. Zhang, Z, Bao, C, Yao, W, Ma, S, Zhang, L, Hou, S: Influence of deposition temperature on the crystallinity of Al-doped ZnO thin films at glass substrates prepared by RF magnetron sputtering method. Superlattice. Microst. **49**, 644–653 (2011)
23. Rani, S, Suri, P, Shishodia, PK, Mehra, RM: Synthesis of nanocrystalline ZnO powder via sol-gel route for dye-sensitized solar cells. Sol. Energ. Mat. Sol. C **92**, 1639–1645 (2008)
24. Benramache, S, Benhaoua, B, Khechai, N, Chabane, F: Elaboration and characterisation of ZnO thin films. Matériaux & Techniques **100**, 573–580 (2012)
25. Zhu, BL, Zhao, XZ, Su, FH, Li, GH, Wu, XG, Wu, J, Wu, R: Low temperature annealing effects on the structure and optical properties of ZnO films grown by pulsed laser deposition. Vacuum **84**, 128–130 (2010)
26. Bao, D, Gu, H, Kuang, A: Sol-gel-derived *c*-axis oriented ZnO thin films. Thin Sol. Film. **312**, 37–41 (1998)
27. Nian, H, Hahn, SH, Koo, KK, Kim, JS, Kim, S, Shin, EW, Kim, EJ: Preparation and characterization of sol-gel Li and Al codoped ZnO thin films. Mater. Lett. **64**, 157–160 (2010)
28. Verma, A, Khan, F, Kumar, D, Kar, M, Chakravarty, BC, Singh, SN, Husain, M: Sol-gel derived aluminum doped zinc oxide for application as anti-reflection coating in terrestrial silicon solar cells. Thin Solid Films **518**, 2649–2653 (2010)
29. Wu, C, Shen, J, Ma, J, Wang, S, Zhang, Z, Yang, X: Electrical and optical properties of molybdenum-doped ZnO transparent conductive thin films prepared by dc reactive magnetron sputtering. Semicond. Sci. Technol. **24**, 125012–125014 (2009)
30. Bouraiou, A, Aida, MS, Meglali, O, Attaf, N: Potential effect on the properties of CuInSe2 thin films deposited using two-electrode system. Curr. Appl. Phys. **11**, 1173–1178 (2011)
31. Benouis, CE, Benhaliliba, M, Juarez, AS, Aida, MS, Chami, F, Yakuphanoglu, F: The effect of indium doping on structural, electrical conductivity, photoconductivity and density of states properties of ZnO films. J. Alloys Compd. **490**, 62–67 (2010)
32. Liu, J, Ma, SY, Huang, XL, Ma, LG, Li, FM, Yang, FC, Zhao, Q, Zhang, XL: Effects of Ti-doped concentration on the microstructures and optical properties of ZnO thin films. Superlattice. Microst. **52**, 765–773 (2012)
33. Chen, KJ, Hung, FY, Chang, SJ, Hu, ZS: Microstructures, optical and electrical properties of In-doped ZnO thin films prepared by sol-gel method. Appl. Surf. Sci. **255**, 6308–6312 (2009)
34. Keskenler, EF, Turgut, G, Dogan, S: Investigation of structural and optical properties of ZnO films co-doped with fluorine and indium. Superlattice. Microst. **52**, 107–115 (2012)
35. Jung, MN, Lee, ES, Jeon, TI, Gil, KS, Kim, JJ, Murakami, Y, Lee, SH, Park, SH, Lee, HJ, Yao, T, Makino, H, Chang, JH: Synthesis and investigation on the extrinsic carrier concentration of indium doped ZnO tetrapods. J. Alloys Compd. **481**, 649–653 (2009)
36. Yousefi, R, Sheini, FJ, Zak, AK, Mahmoudian, MR: Effect of indium concentration on morphology and optical properties of In-doped ZnO nanostructures. Ceram. Int. **38**, 6295–6301 (2012)
37. Tubtimtae, A, Lee, MW: ZnO nanorods on undoped and indium-doped ZnO thin films as a TCO layer on nonconductive glass for dye-sensitized solar cells. Superlattice. Microst. **52**, 987–996 (2012)
38. Daranfed, W, Aida, MS, Hafdallah, A, Lekiket, H: Substrate temperature influence on ZnS thin films prepared by ultrasonic spray. Thin Solid Films **518**, 1082–1084 (2009)

39. Hafdallah, A, Yanineb, F, Aida, MS, Attaf, N: In doped ZnO thin films. J. Alloys Compd. **509**, 7267–7270 (2011)
40. Mosbah, A, Aida, MS: Influence of deposition temperature on structural, optical and electrical properties of sputtered Al doped ZnO thin films. J. Alloys Compd. **515**, 149–153 (2012)
41. Lucio-Lopez, MA, Luna-Arias, MA, Maldonado, A, Olvera, ML, Acost, DR: Preparation of conducting and transparent indium-doped ZnO thin films by chemical spray. Sol. Energ. Mat. Sol. C. **90**, 733–741 (2006)

10

Decoration of multi-walled carbon nanotubes (MWCNTs) with different ferrite nanoparticles and its use as an adsorbent

Ahmed A Farghali[1], Mohamed Bahgat[2], Waleed M A ElRouby[1*] and Mohamed H Khedr[1]

Abstract

A simple and inexpensive synthesis route to produce multi-walled carbon nanotubes (MWCNTs) decorated with $CoFe_2O_4$ and $Co_{0.5}Ni_{0.5}Fe_2O_4$ nanoparticles using a simplified hydrothermal precipitation is reported. Transmission electron microscopy and X-ray diffraction confirm the formation of 14.9 nm $CoFe_2O_4$ and 26 nm $Co_{0.5}Ni_{0.5}Fe_2O_4$ nanoparticles on MWCNTs surface. The prepared composite was used for methyl green adsorption. The effects of various parameters, such as temperature, initial dye concentration, and composite dosage, were investigated. Experimental results have shown that the amount of adsorbed dye increased with increasing initial dye concentration, composite dosage, and temperature. The adsorption kinetic data were analyzed using pseudo-first -order, pseudo-second-order, and intraparticle diffusion models. It was found that the pseudo-second-order kinetic model was the most appropriate model in describing the adsorption kinetics. The adsorption isotherms of methyl green onto MWCNTs decorated with $CoFe_2O_4$ and $Co_{0.5}Ni_{0.5}Fe_2O_4$ nanoparticles were determined at 298, 313, and 323 K. Equilibrium data were fitted to the Langmuir and the Freundlich isotherm models, then the isotherm constants were determined. The equilibrium data were best represented by the Langmuir isotherm model. Thermodynamic parameters such as changes in the free energy of adsorption, enthalpy, entropy, and activation energy were calculated.

Keywords: MWCNTs; Decoration; $CoFe_2O_4$; $Co_{0.5}Ni_{0.5}Fe_2O_4$; Methyl green; Adsorption

Background

Many industries, such as textile, paper, plastic, and dyestuffs, consume substantial amount of water and also use chemicals and dyes to color their products during the manufacturing process [1,2]. Color is the first contaminant recognized in wastewater, and the presence of very small amounts of dyes in water is highly visible and undesirable [3]. Most of these dyes contain aromatic rings, which make them carcinogenic and mutagenic [4,5]. Therefore, the removal of dyes is currently of high importance for environmental remediation. Adsorption technology is one of the most effective methods for dye and toxic removal due to its low cost, high efficiency, simplicity, and insensitivity to toxic substances [6,7].

* Correspondence: waleedmohamedali@yahoo.com
[1]Nanotechnology Department, Faculty of Postgraduate Studies for Advanced Sciences, Beni-Suef University, Beni-Suef 62111, Egypt
Full list of author information is available at the end of the article

Due to their large specific surface area and small, hollow, and layered structures, nanomaterials have recently drawn much attention for dye removal applications. For instance, carbon nanotubes (CNTs) are attracting increasing research interest as promising adsorbents for harmful cations, anions, and other organic and inorganic impurities present in natural sources of water [8-10]. However, it is difficult to separate CNTs from aqueous solutions because of their small size. There are serious concerns over the health and environmental risks of CNTs once they have been released into the environment [11]. It should be noted that CNTs can enter cells, causing damage to plants, animals, and human beings [12]. Thus, there is potential for CNTs to become another source of environmental contaminant if the use of CNTs is not responsibly managed. Compared with traditional centrifugation and filtration methods, the magnetic separation method is considered as a rapid and effective technique for separating adsorbents from environmental applications [13].

Magnetic separation technology is a fast and easy method for separating magnetic adsorbents from an aqueous solution. In recent years, magnetic separation technology, combined with the adsorption process, has been widely used for dye removal from wastewaters [14-16]. To date, only a few studies have been conducted on the adsorption of dyes by magnetic CNTs. Qu et al. have prepared multi-walled carbon nanotubes (MWCNTs) filled with Fe_2O_3 for removal of methylene blue and neutral red from aqueous solution [17]. Gong et al. have synthesized magnetic MWCNT nanocomposites as adsorbent for the removal of cationic dyes: methylene blue, neutral red, and brilliant cresyl from aqueous solution [18]. Madrakian et al. used magnetic-modified MWCNTs for the removal of cationic dye crystal violet, thionine, janus green B, and methylene blue from water samples [19].

In the present work, the main purpose is to demonstrate a simple and general procedure for the decoration of MWCNTs with magnetic ferrite ($CoFe_2O_4$ or $Co_{0.5}Ni_{0.5}Fe_2O_4$). The decorated MWCNTs were used and evaluated as possible sorbents for the removal of methyl green from aqueous solution. The effect of methyl green concentration, temperature, and contact time on the adsorption process was investigated. Kinetics and thermodynamics studies have been performed, and the results have been analyzed. Thermodynamic parameters, such as ΔG^0, ΔH^0, and ΔS^0, were calculated.

Results and discussion

Characterizations of synthesized materials

Figure 1 shows TEM images of the pure and decorated MWCNTs. Figure 1a shows the morphological structure of MWCNTs where the crystalline tubular structure of nanotubes are observed. It is observed that the nanotubes have clear inner channels with lengths of some microns. Figure 1b displays typical TEM images of MWCNTs decorated with $CoFe_2O_4$ nanoparticles. It shows the morphology and the size distribution of $CoFe_2O_4$ nanoparticles. It can be seen that the size of nanoparticles is distributed from 8.3 to 18.9 nm, and the mean particle size is about 14.9 nm. $CoFe_2O_4$ nanoparticles are seen as dense aggregates. The well-distributed nanoparticles deposited onto the carbon nanotubes demonstrate that the MWCNTs pretreatment processing was effective, which resulted in many active sites on the carbon nanotubes. Figure 1c shows the typical TEM image of MWCNTs decorated with $Co_{0.5}Ni_{0.5}Fe_2O_4$ nanoparticles. $Co_{0.5}Ni_{0.5}Fe_2O_4$ nanoparticles dispersed on the sidewalls of the carbon nanotubes and the inner cavity of the tube still clear.

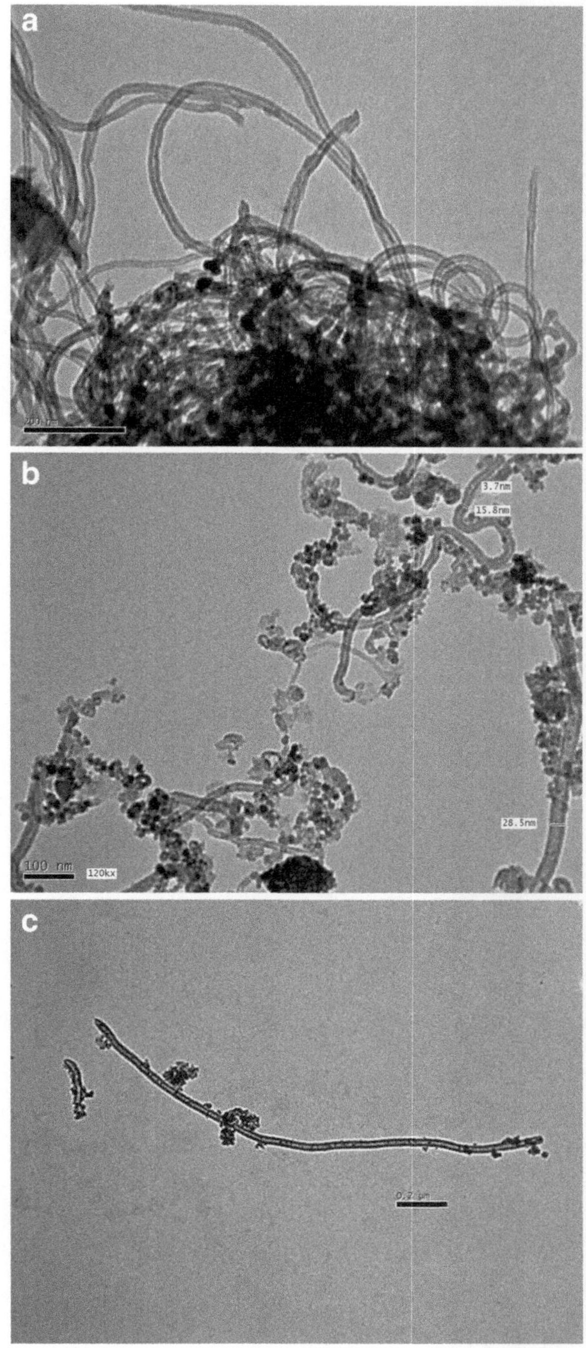

Figure 1 TEM images of pure MWCNT and MWCNT decorated with $CoFe_2O_4$ nanoparticles and with $CO_{0.5}Ni_{0.5}Fe_2O_4$ nanoparticles. Pure MWCNT **(a)**, MWCNTs decorated with $CoFe_2O_4$ nanoparticles **(b)**, and MWCNTs decorated with $Co_{0.5}Ni_{0.5}Fe_2O_4$ nanoparticles **(c)**.

In X-ray diffraction (XRD) patterns of $CoFe_2O_4$ and $CoFe_2O_4$/MWCNTs the diffraction peak at 18.4°, 30.09°, 35.55°, 43018°, 57.08°, and 62.66° are reflections of $CoFe_2O_4$. Nevertheless, in XRD patterns of $CoFe_2O_4$/MWCNTs, the additional peaks could be well seen at 26.066° which is corresponding to graphite (Figure 2).

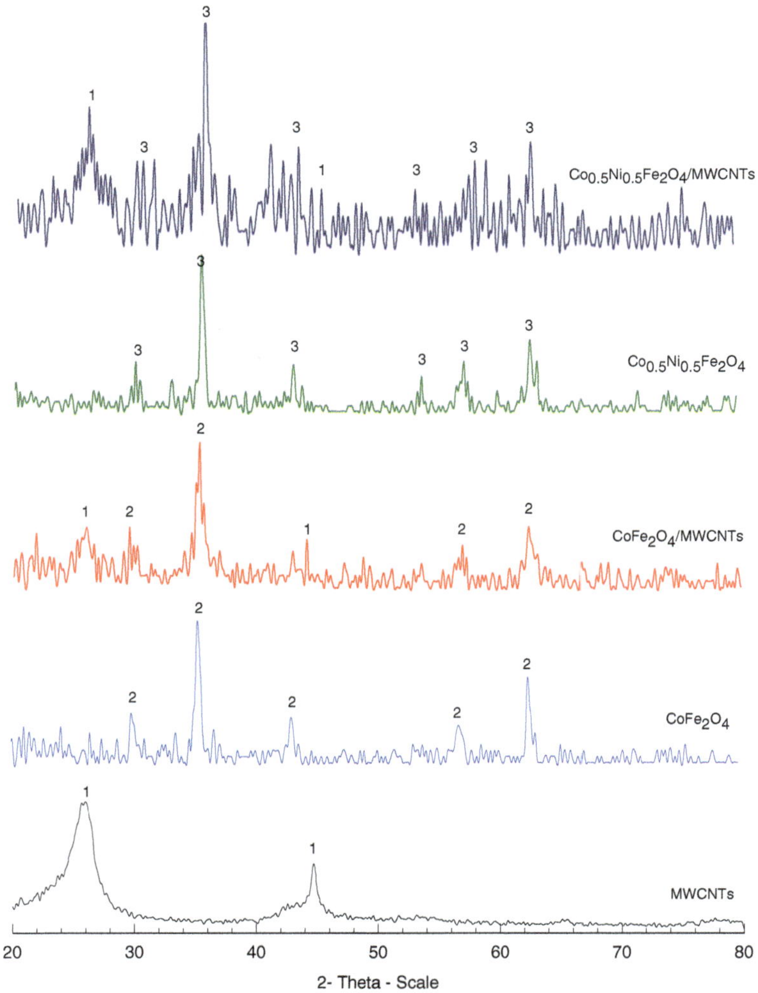

Figure 2 XRD diffraction patterns of MWCNTs, $CoFe_2O_4$, $CoFe_2O_4$/MWCNTs, $Co_{0.5}Ni_{0.5}Fe_2O_4$ and $Co_{0.5}Ni_{0.5}Fe_2O_4$/MWCNTs. 1. Graphite, 2. $CoFe_2O_4$, 3. $Co_{0.5}Ni_{0.5}Fe_2O_4$.

The crystallite size of $CoFe_2O_4$ on MWCNTs surface was calculated using the Scherer equation (Equation 1) [20].

$$D = 0.89\lambda / B \cos\theta, \tag{1}$$

where $\lambda = 0.154056$ nm, B is the full peak width at half maximum value, and θ is the diffraction angle.

It was found that the average crystallite size is about 14.9 nm. XRD patterns of $Co_{0.5}Ni_{0.5}Fe_2O_4$ and $Co_{0.5}Ni_{0.5}Fe_2O_4$/MWCNTs were also shown in Figure 2. The sample exhibits characteristic diffraction peaks of $Co_{0.5}Ni_{0.5}Fe_2O_4$ at 2θ of 30.22°, 35.6°, 43.18°, 57.28°, and 62.84°, but in XRD of $Co_{0.5}Ni_{0.5}Fe_2O_4$/MWCNTs, there is an additional peak at 26.066° which is corresponding to graphite. The average crystallite size of the $Co_{0.5}Ni_{0.5}Fe_2O_4$ particles is determined using the Scherer equation and found it to be about 26.2 nm.

Fourier transform infrared spectroscopic analysis of oxidized MWCNTs has been shown in Figure 3. The Fourier transform infrared spectroscopy (FTIR) study of oxidized MWCNTs confirms the defective sites at the surface of MWCNTs, and the presence of function groups > C = C (1,642/cm), > C = O (1,025/cm), = CH_2 (2,857/cm, 2,925/cm), and -OH (3,439/cm). This leads to the hydrophilic nature of MWCNTs. These functional

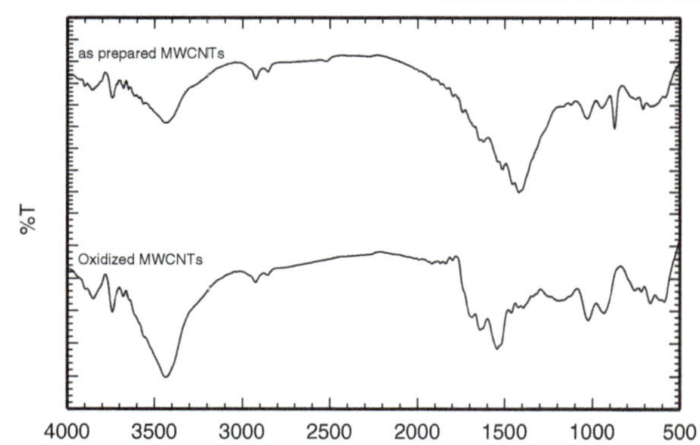

Figure 3 FTIR spectra of the as-prepared and oxidized MWCNTs.

groups may also act as anchoring sites for ferrite particles in the decoration process and also in the adsorption of dye molecules.

N_2 adsorption-desorption isotherm of MWCNTs decorated with $CoFe_2O_4$ and MWCNTs decorated with $Co_{0.5}Ni_{0.5}Fe_2O_4$ nanoparticles are shown in Figure 4. There are hysteresis loops clearly visible in the isotherm, which is associated with capillary condensation in mesopores. These mesopores include mesopore-sized inner cavities and aggregated pores resulted from aggregation of decorated MWCNTs. Surface area by the Brunauer-Emmett-Teller (BET) method, micropore volume by the t-plot method, and pore size distributions by the BJH equation are obtained for MWCNTs decorated with $CoFe_2O_4$ and $Co_{0.5}Ni_{0.5}Fe_2O_4$ as listed in Table 1. This adsorption isotherm exhibited a type-II shape. It was observed that there was a small closed adsorption-desorption hysteresis loop with a relative pressure above 0.4, which is suggested to be due to the mesopores with a capillary condensation [21].

Adsorption of methyl green dye

Effect of adsorbent dosage on adsorption capacity of methyl green

The mechanism of MWCNT adsorption for methyl green dyes may be derived from two reasons: the first reason might be based on the van der Waals interactions occurring between the hexagonally arrayed carbon atoms in the graphite sheet of MWCNTs and the aromatic backbones of the dyes [22]; the second reason might be due to the electrostatic attraction between the positive cationic dyes and the negative charged MWCNTs adsorbent surface.

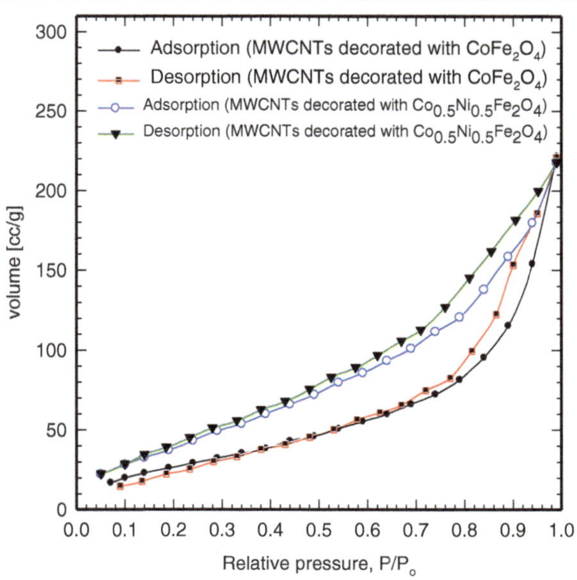

Figure 4 Adsorption-desorption hysteresis loop of MWCNTs decorated with $CoFe_2O_4$ and MWCNTs decorated with $Co_{0.5}Ni_{0.5}Fe_2O_4$.

Table 1 Surface area measurements for MWCNTs decorated with $CoFe_2O_4$ and MWCNTs decorated with $Co_{0.5}Ni_{0.5}Fe_2O_4$ nanoparticles

Measurement	MWCNT decorated with $CoFe_2O_4$	MWCNT decorated with $Co_{0.5}Ni_{0.5}Fe_2O_4$
Surface area (m^2/g)	109.54	176
Total pore volume (cm^3/g)	0.0505	0.076
Average pore diameter (nm)	18.96	17.33
Micro pore volume (cm^3/g)	0.109	0.16

It was observed that the percentages of adsorbed dye increased as the decorated MWCNT dosages were increased over the range 0.6 to 1.4 g/L as shown in Figure 5. The increase in the percentage of the removed dye with an adsorbent dosage can be attributed to an increase in the adsorbent surface, which increased the availability of the adsorption sites.

Effects of dye concentration

The amount of adsorbed dye per unit of decorated MWCNT mass increased as initial dye concentration increased due to the increase in the driving force of the concentration gradient as shown in Figure 6. The adsorption process at different concentrations was rapid in the initial 30 min then gradually decreased as adsorption proceeded until equilibrium was reached. This progression is expected to be based on the large number of vacant surface sites available for adsorption during the initial stage. After a certain time, the remaining vacant surface sites are difficult to be occupied due to the repulsive forces between dye molecules on the decorated MWCNTs and bulk phases [23].

Effect of temperature

To study the effect of temperature on the adsorption of methyl green dye onto MWCNTs decorated with $CoFe_2O_4$ or $Co_{0.5}Ni_{0.5}Fe_2O_4$ nanoparticles, the experiments were performed at temperatures of 298, 313, and 323 K as shown in Figure 7. It was observed that the equilibrium adsorption capacity of methyl green onto both decorated MWCNTs increased with increasing the temperature due to increasing the mobility of dye molecules.

Equilibrium uptake increased with the increasing of methylene blue concentrations at the present experimental range. This is a result of the increase in the driving force from the concentration gradient. In the same conditions, if the concentration of methyl green in the

Table 2 Isotherm parameters for removal of methyl green by the decorated MWCNTs at different temperatures

Isotherms	Parameters	Temperatures (K)					
		MWCNT decorated with $CoFe_2O_4$			MWCNT decorated with $Co_{0.5}Ni_{0.5}Fe_2O_4$		
		298 K	313 K	323 K	298 K	313 K	323 K
Langmuir	q_0 (mg/g)	91.74	96.16	103.3	172.98	244.25	257.7
	K_L (L/mg)	0.0556	0.0698	0.0822	0.072	0.088	0.102
	R_L	0.043	0.0345	0.095	0.0335	0.0276	0.024
	R^2	0.997	0.997	0.995	0.992	0.991	0.992
Freundlich	K_F (mg/g (L/mg)$^{1/n}$)	19.98	22.264	22.624	42.52	51.94	65.75
	n	3.57	3.62	3.67	3.89	3.3	3.79
	R^2	0.926	0.902	0.900	0.98	0.95	0.96

solution was bigger, the active sites of MWCNTs decorated with $CoFe_2O_4$ or $Co_{0.5}Ni_{0.5}Fe_2O_4$ nanoparticles were surrounded by much more methyl green ions, and the process of adsorption would be carried out sufficiently. Therefore, the amount absorbed at equilibrium (q_e) increased with the increase in equilibrium methyl green concentrations. From Figure 7, the adsorption capacity of methyl green onto MWCNTs decorated with $CoFe_2O_4$ are 85.07, 89.41, and 96.24 mg/g at 298, 313, and 323 K, respectively, while the adsorption capacity for MWCNTs decorated with $Co_{0.5}Ni_{0.5}Fe_2O_4$ are 167.95, 288.40, and 243.89 mg/g at 298, 313, and 323 K, respectively. The increase of the equilibrium adsorption with the increase in temperature indicated that the adsorption of methyl green ions onto $CoFe_2O_4$/MWCNT composite was endothermic in nature. The higher adsorption capacity of MWCNTs decorated with $Co_{0.5}Ni_{0.5}Fe_2O_4$ than MWCNTs decorated with $CoFe_2O_4$ may be attributed to the high surface area of MWCNTs decorated with $Co_{0.5}Ni_{0.5}Fe_2O_4$ or may be due to the small crystallite size of $CoFe_2O_4$ nanoparticles on the MWCNTs surface which can act as a blocker for the active sites on the tube surfaces (Figure 1).

Adsorption isotherms

The quantity of the dye that could be adsorbed over MWCNTs decorated with $CoFe_2O_4$ or $Co_{0.5}Ni_{0.5}Fe_2O_4$ nanoparticles surface is a function of concentration, which could be explained by the adsorption isotherms. In the present study, the Langmuir [24] and Freundlich [25] isotherms are tested for methyl green dye adsorption.

The Langmuir adsorption isotherm assumes that adsorption takes place at specific homogeneous sites

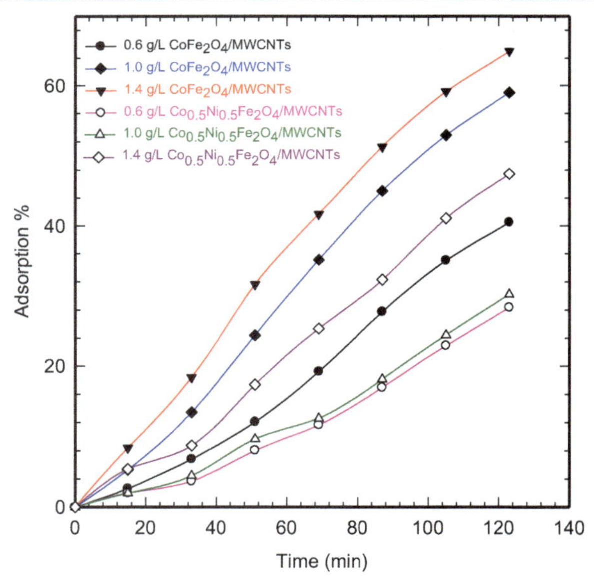

Figure 5 Effects of decorated MWCNT dosages on the adsorption of methyl green dye. Dye concentration = 100 mg/L and T = 298 K.

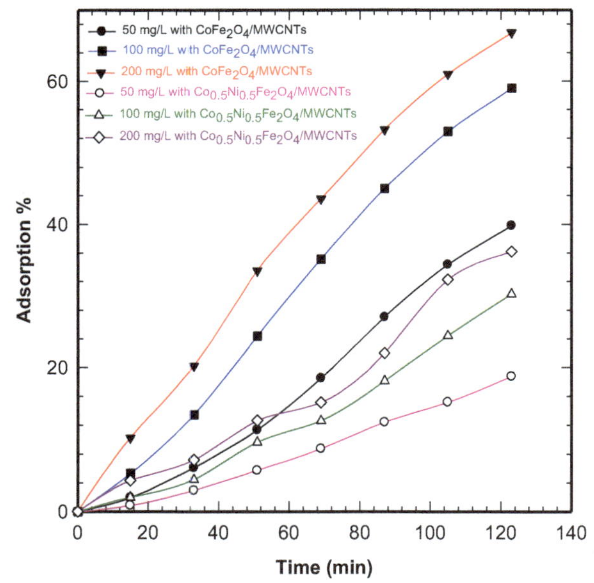

Figure 6 Effects of dye concentration on the adsorption of methyl green dye. $CoFe_2O_4$/MWCNTs and $Co_{0.5}Ni_{0.5}Fe_2O_4$/MWCNTs = 1 g/L and T = 298 K).

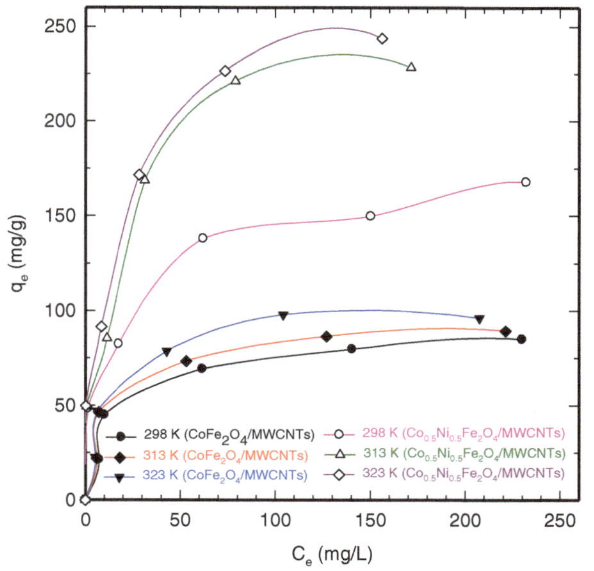

Figure 7 Adsorption isotherms of methyl green onto MWCNTs decorated with $CoFe_2O_4$ and $Co_{0.5}Ni_{0.5}Fe_2O_4$ at different temperatures.

within the adsorbent. The Langmuir isotherm can be written in this form:

$$C_e/q_e = 1/(q0K_L) + (1/q0)C_e, \tag{2}$$

where C_e (mg/L) is the equilibrium concentration, q_e (mg/g) is the amount of adsorbate adsorbed per unit mass of adsorbate, and q_o and K_L are the Langmuir constants related to the adsorption capacity and the rate of adsorption, respectively. When C_e/q_e was plotted against C_e, a straight line with slope $1/q_0$ was obtained (Figure 8a,c), indicating that the adsorption of methyl green onto MWCNTs decorated with $CoFe_2O_4$ or $Co_{0.5}Ni_{0.5}Fe_2O_4$ nanoparticles follows the Langmuir isotherm. The Langmuir constants K_L and q_o were calculated from this isotherm, and their values are listed in Table 2. Another important parameter, R_L, is called the separation factor or the equilibrium parameter which is determined from the relation [26]:

$$R_L = 1/\ [1 + K_L C_o] \tag{3}$$

where K_L is the Langmuir constant (l/mg) and C_o (mg/L) is the highest dye concentration. The value of R_L indicates the type of the isotherm to be either unfavorable ($R_L > 1$), linear ($R_L = 1$), favorable ($0 < R_L < 1$), or irreversible ($R_L = 0$). R_L values for methyl green adsorption onto MWCNTs decorated with $CoFe_2O_4$ or $Co_{0.5}Ni_{0.5}Fe_2O_4$ nanoparticles were calculated and found to be less than 1 and greater than zero indicating the favorable adsorption (Table 2).

The Freundlich isotherm is an empirical equation employed to describe the heterogeneous systems [27]. The Freundlich equation is as follows:

$$ln\ q_e = ln\ K_F + (1/n) ln\ C_e, \tag{4}$$

where q_e is the amount adsorbed at equilibrium (mg/g) and C_e is the equilibrium concentration of methyl green. n and K_F are Freundlich constants, n giving an indication of how favorable the adsorption process and K_F (mg/g (L/mg)$^{1/n}$) is the adsorption capacity of the adsorbent. The slope $1/n$ ranging between 0 and 1 is a measure of the adsorption intensity or the surface heterogeneity, becoming more heterogeneous as its value gets closer to 0 [28]. The plot of $ln\ q_e$ versus $ln\ C_e$ (Figure 8b,c) gives straight lines with slope $1/n$ reflecting that the adsorption of methyl green also follows the Freundlich isotherm. Accordingly, Freundlich constants (K_F and n) were calculated and listed in Table 2.

Kinetics analyses

The kinetic analysis of temperature effect was evaluated. The adsorption capacity increased with the temperature, indicating that the mobility of dye molecules increased with temperature, and the adsorption was endothermic. Additionally, increasing the temperature reduces the viscosity of the solution and increases the diffusion rate of dye molecules.

The pseudo-first-order, pseudo-second-order, and intraparticle diffusion models were adopted to test the experimental data and thereby elucidated the kinetic adsorption process. The pseudo first-order model can be expressed as follows:

$$ln(q_e - q) = ln(q_e) - k_1 t, \tag{5}$$

where q_e and q are the amounts of methyl green adsorbed on MWCNTs decorated with $CoFe_2O_4$ or $Co_{0.5}Ni_{0.5}Fe_2O_4$ nanoparticles at equilibrium and at various times t (mg/g) and k_1 is the rate constant of the pseudo-first-order model for the adsorption (per minute) [29]. The values of q_e and k_1 can be determined from the intercept and the slope of the linear plot of ln ($q_e - q$) versus t (Figure 9a,b).

The pseudo-second-order model is as follows:

$$t/q = 1/k_2 q_e^2 + t/q_e, \tag{6}$$

where q_e and q are the amounts of dye adsorbed onto MWCNTs decorated with $CoFe_2O_4$ or $Co_{0.5}Ni_{0.5}Fe_2O_4$ nanoparticles at equilibrium and at various times t (mg/g), and k_2 is the rate constant of the pseudo-second-order model for adsorption (g/mg/min) [29]. The slope and intercept of the linear plot of t/q as a function of t (Figure 9c,d) yielded the values of q_e and k_2. Additionally,

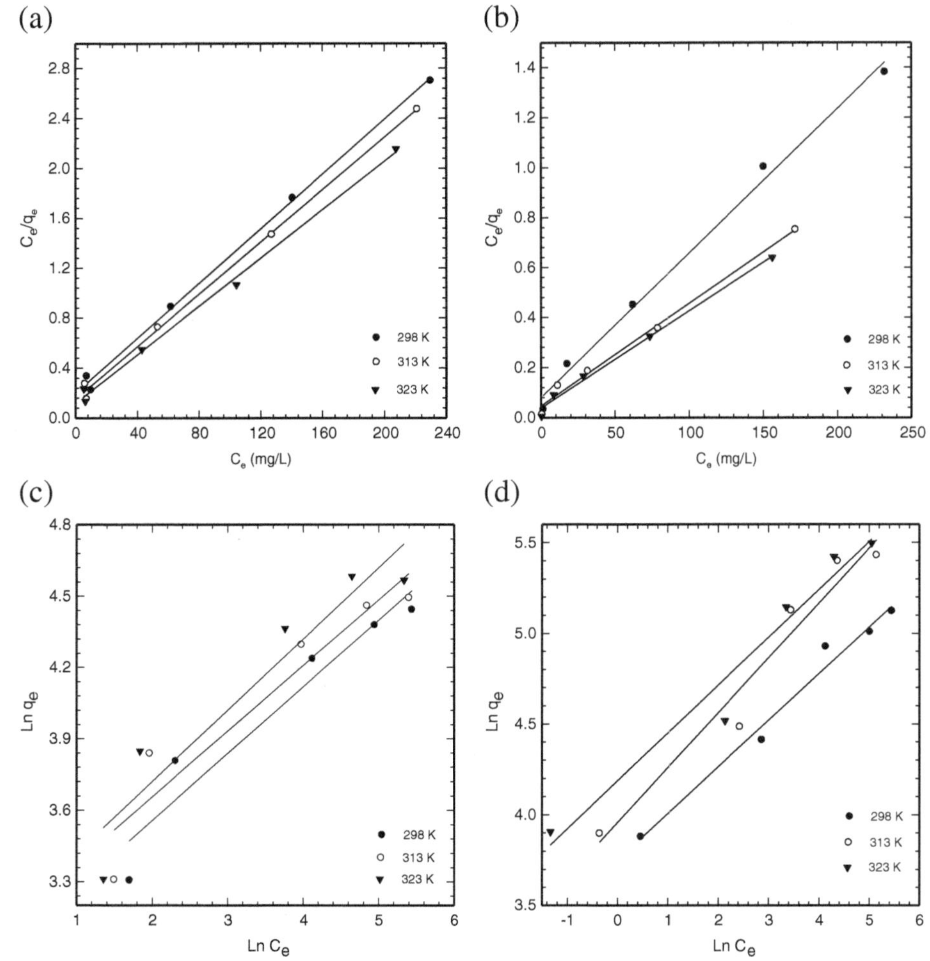

Figure 8 Langmuir isotherms and Freundlich isotherms. Langmuir isotherms **(a,c)** for MWCNTs decorated with $CoFe_2O_4$ and $Co_{0.5}Ni_{0.5}Fe_2O_4$ nanoparticles, respectively, and Freundlich isotherms **(b,d)** for MWCNTs decorated with $CoFe_2O_4$ and $Co_{0.5}Ni_{0.5}Fe_2O_4$ nanoparticles, respectively, for methyl green dye adsorption at different temperatures.

Table 3 Coefficients of pseudo-first and second-order adsorption kinetic models and intraparticle diffusion model (methyl green = 100 mg/L, MWCNTs decorated with $CoFe_2O_4$ or $Co_{0.5}Ni_{0.5}Fe_2O_4$ = 1 g/L)

Orders model	Parameters	Temperatures (K)					
		MWCNT decorated with $Co_{0.5}Ni_{0.5}Fe_2O_4$			MWCNT decorated with $CoFe_2O_4$		
		298 K	323 K	313 K	298 K	323 K	313 K
Pseudo-first-order model	$q_{e\ Cal.}$ (mg/g)	51.92	87.452	83.12	80.125	19.22	30.96
	$q_{e\ Exp.}$ (mg/g)	44.96	91.59	85.40	82.39	47.58	46.43
	K_1 (min^{-1})	9.747	5.29×10^{-3}	4.26×10^{-3}	3.28×10^{-3}	0.013	0.014
	R^2	0.99	0.977	0.99	0.985	0.90	0.977
Pseudo-second-order model	$q_{e\ Cal.}$ (mg/g)	45.15	87.452	83.12	80.125	74.6	46.55
	$q_{e\ Exp.}$ (mg/g)	44.96	91.59	85.40	82.39	47.58	46.43
	K_2 (g/mg min)	0.00046	1.08×10^{-4}	7.66×10^{-5}	6.15×10^{-5}	0.0012	0.00092
	R^2	0.995	0.992	0.998	0.991	0.998	0.998
Intraparticle diffusion model	k_i (mg/g $min^{0.5}$)	3.214	4.98	4.061	3.4	2.05	2.958
	C (mg/g)	0.591	−7.685	−8.66	−12.056	22.44	9.22
	R^2	0.981	0.991	0.999	0.997	0.90	0.958

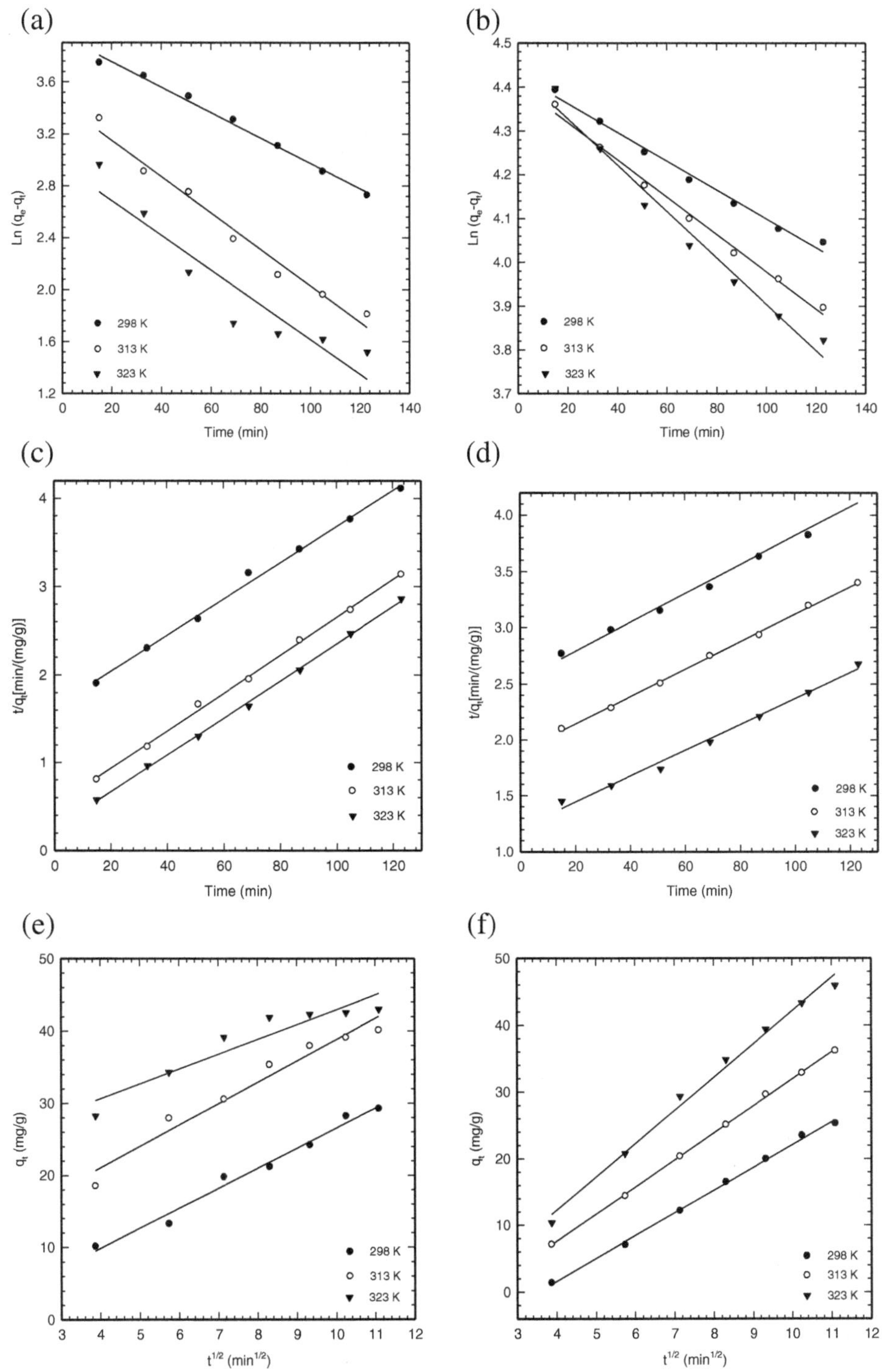

Figure 9 Regressions of kinetic plots at different temperatures. Pseudo-first-order model **(a)** $CoFe_2O_4$/MWCNTs, **(b)** $Co_{0.5}Ni_{0.5}Fe_2O_4$/MWCNTs; pseudo-second-order model **(c)** $CoFe_2O_4$/MWCNTs, **(d)** $Co_{0.5}Ni_{0.5}Fe_2O_4$/MWCNTs; intraparticle diffusion model **(e)** $CoFe_2O_4$/MWCNTs, **(f)** $Co_{0.5}Ni_{0.5}Fe_2O_4$/MWCNTs.

the initial adsorption rate h (mg/g/min) can be determined using the equation $h = k_2 q_e^2$. The adsorption process on porous adsorbents generally has a four-stage bulk diffusion, film diffusion, intraparticle diffusion, and finally, adsorption of the solute onto the surface. Typically, bulk diffusion and adsorption are assumed to be rapid and therefore not rate determining. Since the pseudo-second-order model cannot identify the diffusion mechanism, kinetic results were analyzed using the intraparticle diffusion model to elucidate the diffusion mechanism. Film diffusion was negligible, and intraparticle diffusion was the only rate-controlling step. The intraparticle diffusion model is expressed as follows:

$$q = k_i t^{1/2} + C, \tag{7}$$

where C is the intercept and k_i is the intraparticle diffusion rate constant (mg/g/min$^{0.5}$), which can be determined from the slope of the linear plot of q versus $t^{1/2}$ [30] (Figure 9e,f). Table 3 presents the kinetic parameters for the removal of methyl green by MWCNTs decorated with $CoFe_2O_4$ or $Co_{0.5}Ni_{0.5}Fe_2O_4$ nanoparticles. The R^2 value of the pseudo-second-order model exceeded 0.99; moreover, the q value ($q_{e,cal}$) derived from the pseudo-second-order model was consistent with the experimental q values ($q_{e,exp}$). Hence, this study showed that the pseudo-second-order model best represents adsorption kinetics. A similar phenomenon has been observed in the adsorption of Acid Blue 93 by natural sepiolite [31], and Acid Red 57 by surfactant-modified sepiolite [32]. If the regression of q versus $t^{1/2}$ is linear and passes through the origin, intraparticle diffusion is then the sole rate-limiting step [32]. Although the regression was linear, the plot did not pass through the origin (Table 3), indicating that adsorption involved intraparticle diffusion that was not the only rate-controlling step. Other kinetic mechanisms may control the adsorption rate, which is a similar finding to that obtained from other studies of adsorption [31]. The values of C were helpful in determining the boundary thickness: a larger C value corresponds to a greater boundary layer diffusion effect [33]. The C values increased with the temperature (298 to 323 K), and so, increasing the temperature promoted the boundary-layer diffusion effect.

Adsorption thermodynamics

The thermodynamic parameters, namely free energy ($\Delta G°$), enthalpy ($\Delta H°$), and entropy ($\Delta S°$) have an important role to determine spontaneity and heat change for the adsorption process. Thermodynamic parameters were calculated using the following relations [34]:

$$\Delta G° = -RT\ ln(K_L), \tag{8}$$

$$ln(K_L) = (\Delta S°/R) - (\Delta H°/RT), \tag{9}$$

where K_L is the Langmuir equilibrium constant (L/mol), R is the gas constant (8.314 J/mol/K), and T is the temperature (K). $\Delta H°$ and $\Delta S°$ parameters can be calculated from the slope and intercept of the plot $ln\ K_L$ versus 1/T. From Equation 10, $\Delta G°$ was calculated using $ln\ K_L$ values for different temperatures. Results are summarized in Table 4. It can be seen that $\Delta G°$ values at temperatures 298, 313, and 323 K are negative. Hence, the adsorption process was a spontaneous process. The decrease in $\Delta G°$ value with the increase of temperature indicates the efficient adsorption at higher temperature. The positive $\Delta H°$ value reflected that the adsorption process is endothermic, and there is a strong interaction between MWCNTs decorated with $CoFe_2O_4$ or $Co_{0.5}Ni_{0.5}Fe_2O_4$ nanoparticles and methyl green. Since methyl green ions travel through solution and reach the adsorption sites, it is necessary for them to be stripped out (at least partially) of their hydration shell, that requires energy input. Thus, the positive value of $\Delta H°$ indicates that the adsorption is increasing with temperature. Moreover, the positive value of $\Delta S°$ indicates that the degrees of freedom increased at the solid–liquid interface during the adsorption of methyl green onto MWCNTs decorated with $CoFe_2O_4$ or $Co_{0.5}Ni_{0.5}Fe_2O_4$ nanoparticles and reflected the affinity of the composite toward methyl green ions in aqueous solutions [35].

The pseudo-second-order model was identified as the best kinetic model for the adsorption of methyl green onto MWCNTs decorated with $CoFe_2O_4$ or $Co_{0.5}Ni_{0.5}Fe_2O_4$ nanoparticles. Accordingly, the rate constants (k_2) of the pseudo-second-order model were adopted to calculate the activation energy of the adsorption process using the Arrhenius equation [36]:

$$ln(k_2) = ln(A) - (E_a/RT), \tag{10}$$

where k_2, A, E_a, R, and T are the rate constant of the

Table 4 Thermodynamic parameters of methyl green dye adsorption onto decorated MWCNTs at various temperatures

Temperature (K)	MWCNT decorated with $CoFe_2O_4$				MWCNT decorated with $Co_{0.5}Ni_{0.5}Fe_2O_4$			
	$\Delta G°$ (kJ/mol)	$\Delta H°$ (kJ/mol)	$\Delta S°$ (J/mol/K)	E_a (kJ/mol)	$\Delta G°$ (kJ/mol)	$\Delta H°$ (kJ/mol)	$\Delta S°$ (J/mol/K)	E_a (kJ/mol)
298	−8.023	31.189	68.64	12.44	−21.7	17.38	35.9	11.075
313	−9.019				−22.27			
323	−9.746				−22.59			

pseudo-second-order model (g/mg/min), the Arrhenius factor, the activation energy (kJ/mol), the gas constant (8.314 J/mol/K), and the temperature (K), respectively. The activation energy could be determined from the slope of the plot of $ln(k_2)$ versus $1/T$ (Table 4). Nollet et al. [37] mentioned that the physisorption process normally had activation energy of 5 to 40 kJ/mol, while chemisorption had higher activation energy (40 to 800 kJ/mol). Therefore, $\Delta H°$, $\Delta G°$, and E_a values confirm that the adsorption of methyl green onto MWCNTs decorated with $CoFe_2O_4$ or $Co_{0.5}Ni_{0.5}Fe_2O_4$ nanoparticles was a physisorption process. Lazaridis and Asouhidou [38] stated that in a diffusion-controlled process, the activation energy of adsorption was less than 25 to 30 kJ/mol. Based on the results of activation energy and the intraparticle diffusion model (Table 4 and Figure 9e,f), this study proposes that the adsorption involved intraparticle diffusion, that was not the only rate-controlling step, and the other kinetic models might control the adsorption rate.

Conclusions

MWCNTs decorated with $CoFe_2O_4$ and $Co_{0.5}Ni_{0.5}Fe_2O_4$ nanoparticles have been successfully prepared by a hydrothermal precipitation method. The prepared composite exhibits a homogeneous dispersion of MWCNTs in the matrix and coating of $CoFe_2O_4$ and $Co_{0.5}Ni_{0.5}Fe_2O_4$ nanoparticles on MWCNTs. This study investigated the removal of methyl green from aqueous solution by the prepared composites. It was found that $Co_{0.5}Ni_{0.5}Fe_2O_4$/MWCNTs composite has a high adsorptive capacity than $CoFe_2O_4$/MWCNTs for methyl green adsorption. The equilibrium adsorption capacity of methyl green increased with the increase in temperature, methyl green concentration, and adsorbent materials. The adsorption kinetics could be quite successfully fitted by a pseudo-second -order kinetic equation. The Langmuir and Freundlich adsorption isotherm models were used to express the adsorption phenomenon of the methyl green. The equilibrium data were well described by the Langmuir model. Thermodynamic analyses indicated that the adsorption of methyl green onto MWCNTs decorated with $CoFe_2O_4$ and $Co_{0.5}Ni_{0.5}Fe_2O_4$ nanoparticles was endothermic, spontaneous, and physisorption process.

Methods

Material and methods

All the reagents were of analytical grade and used as received without further purification. Cobalt nitrate [$Co(NO_3)_2 \cdot 6H_2O$], nickel nitrate [$Ni(NO_3)_2 \cdot 6H_2O$], and ferric nitrate [$Fe(NO_3)_3 \cdot 9H_2O$] were obtained from WINLAB, Baths, UK. Sodium hydroxide (NaOH) was obtained from Bio Chem Laboratories, Inc., Grand Rapids, MI, USA and methyl green (Table 5) was obtained from Gurr microscopy materials, BDH chemicals Ltd., Poole, England. MWCNTs were produced by chemical vapor deposition using acetylene cracking over Fe-Co/$CaCO_3$ catalyst/support. The method of synthesis had been described before [39]. The raw product contains the support, catalyst particles, and a few amorphous carbons as impurities. The as-grown MWCNTs were purified using a two-step purification procedure involving treatment with diluted HCl and then with mixture of concentrated nitric acid/sulfuric acid (3:1 by volume, respectively).

For the decoration of MWCNTs, a specific amount of oxidized MWCNTs was first ultrasonicated in 100 ml of distilled water for 30 min. Afterward, this suspension was mixed with a solution of analytical grade $Co(NO_3)_2 \cdot 6H_2O$ and $Fe(NO_3)_3 \cdot 9H_2O$ in which the Co/Fe molar ratio was maintained at 1:2, in case of decoration of MWCNTs with $CoFe_2O_4$ nanoparticles. The mixture of $Ni(NO_3)_2 \cdot 6H_2O$, $Co(NO_3)_2 6H_2O$ and $Fe(NO_3)_3 \cdot 9H_2O$, in which the Ni/Co/Fe molar ratio was maintained at 1:1:4, was used in case of decoration of MWCNTs with $Co_{0.5}Ni_{0.5}Fe_2O_4$ nanoparticles. Then, a sodium hydroxide solution (6 molar) was added dropwise into the above mixture with vigorous stirring until the pH value reaches 10. The mixture was stirred for another 30 min for complete reaction then the solution was neutralized. The produced mixture was then placed in a Teflon-lined autoclave and maintained at 220°C for 10 h. The obtained precipitates were rinsed repeatedly with water and ethanol, and then, dried at 100°C for 12 h in a vacuum oven. The prepared powder was denoted as $CoFe_2O_4$/MWCNTs and $Co_{0.5}Ni_{0.5}Fe_2O_4$/MWCNTs.

Table 5 Structure and characteristics of Methyl green dye

Name	C.I. number	Formula	Molecular weight	λ_{max}	Chemical structure
Methyl green	42585	$C_{26}H_{33}N_3Cl_2$	458.5	630 to 634, 420 nm	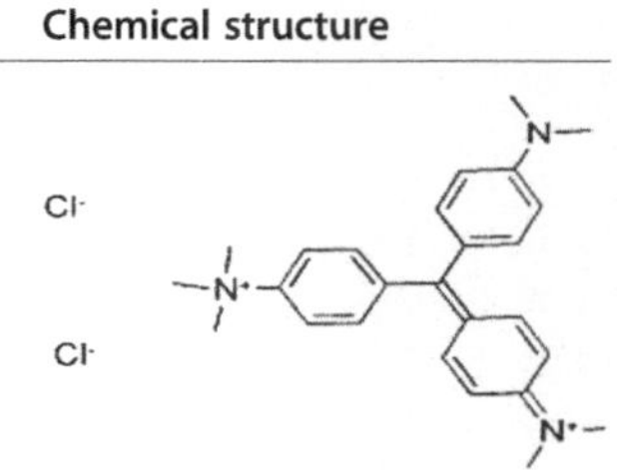

The synthesized and decorated MWCNTs were characterized by powder XRD analysis (D8 Advance, Bruker AXS, Inc, Madison, WI, USA), transmission electron microscopy (TEM, Jeol JEM-1230, Akishima-shi, Japan), and FTIR (JASCO 410, Mary's Court Easton, MD, USA). The BET surface area was determined from adsorption isotherms using a Quantachrome NOVA automated gas sorption system report.

Dye adsorption experiments

For studying the effect of dye concentration on the adsorption process equal volumes, 50 ml of various methyl green dye concentrations, were taken in number of 250-ml beakers. Definite weight of the adsorbent materials ($CoFe_2O_4$/MWCNTs or $Co_{0.5}Ni_{0.5}Fe_2O_4$/MWCNTs) was added in each beaker and shacked well. At different interval times, 5 ml of the sample solution was withdrawn and the change in characteristic absorption at the specific beaks was measured using an ultraviolet–visible (UV–vis) spectrophotometer (Jasco 530), from which the concentration of dye was inferred.

For evaluation, the effect of adsorbent materials concentration on the adsorption process equal volumes, 50 ml of definite methyl green dye concentrations, were taken in a number of 250-ml beakers. Different weights of the adsorbent materials were added in each beaker and shacked well. At different interval times, 5 ml of the sample solution was withdrawn and the change in characteristic absorption at the specific peaks was measured using an UV–vis spectrophotometer (Jasco 530), from which the concentration of dyes was inferred.

For identifying the adsorption isotherm and kinetics adsorption experiments were performed using 250-ml glass bottles, specific amount of the adsorbent materials were inserted into the bottles along with 50 ml of the methyl green dye solutions. The initial concentration of the dye (C_0) was varied from 50 to 400 mg/L. The glass bottles were sealed and placed within a temperature control box to maintain water temperature. Temperatures that were studied include 298, 313, and 333 K. At the end of the equilibrium period, the suspensions were separated for later analysis of the dye concentration. The amount of dye adsorption at equilibrium q_e (mg/g) was calculated from the following equation:

$$q_e = V(C_0 - C_e)/W, \tag{11}$$

where C_0 and C_e (mg/L) are the liquid-phase concentrations of methyl green dye at initial and equilibrium, respectively. V (L) is the volume of the solution, and W (g) is the mass of the used adsorbent. The concentration of the dye before and after adsorption was determined using a spectrophotometer (Jasco 530).

Kinetic experimental procedures were identical to the equilibrium tests. The effect of contact time on the amount of adsorbed methyl green dye was investigated at specific initial dye concentration and at varying temperatures (298, 313, and 333 K). The amounts of adsorbed dye on the adsorbent materials at any time, t, were calculated from the concentrations in solutions before and after adsorption. At any time, the amount of adsorbed dye (mg/g) onto the adsorbent materials was calculated from the mass balance equation as follows:

$$q_t = V(C_0 - C_t)/W, \tag{12}$$

where q_t is the amount of adsorbed dye on the adsorbent materials at any time (mg/g); C_0 and C_t are the initial and liquid-phase concentrations of dye at any time (mg/L), respectively; V (L) is the volume of dye solution, and W (g) is the mass of the adsorbent materials.

Competing interests

The authors declare that they have no competing interests.

Authors' contributions

WR carried out the experimental part and wrote the manuscript for publication. AF participated in the idea and the design of the study. MB, MK, and AF as supervisors, participated in this work. All authors read and approved the final manuscript.

Acknowledgements

The authors gratefully acknowledge the financial support of this work by Beni-Suef University nanotechnology research team.

Author details

[1]Nanotechnology Department, Faculty of Postgraduate Studies for Advanced Sciences, Beni-Suef University, Beni-Suef 62111, Egypt. [2]Minerals Technology Department Central Metallurgical R & D Institute (CMRDI), Helwan 11421, Egypt.

References

1. Tandon, S, Nandini, N: Adsorption efficiency of carbon prepared from agricultural waste materials in removing basic dye from aqueous solution. Asian J. Chem. **23**, 636–640 (2011)
2. Baskaran, P, Venkatraman, B, Arivoli, S: Kinetics of adsorption of ferrous ion onto acid activated carbon from *Zea mays* dust. J. Chem. **8**, 185–195 (2011)
3. Mohammadi, N, Khani, H, Gupta, VK, Amereh, E, Agarwal, S: Adsorption process of methyl orange dye onto mesoporous carbon material - kinetic and thermodynamic studies. J. Colloid Interf. Sci. **362**, 457–462 (2011)
4. Loghman, K, Salar, Z: Superior photocatalytic degradation of azo dyes in aqueous solutions using TiO2/SrTiO3 nanocomposite. J. Nanostruct. Chem. **3**, 32 (2013)
5. Manoj Kumar Reddy, P, Rama Raju, B, Karuppiah, J, Linga Reddy, E, Subrahmanyam, C: Degradation and mineralization of methylene blue by dielectric barrier discharge non-thermal plasma reactor. Chem. Eng. J. **217**, 41–47 (2013)
6. Panic, VV, Madzarevic, ZP, Volkov-Husovic, T, Velickovic, SJ: Poly (methacrylic acid) Based Hydrogels as Sorbents for Removal of Cationic Dye Basic Yellow 28: Kinetics, Equilibrium Study and Image Analysis. Chem. Eng. J. **217**, 192–204 (2013)
7. Shi, Y, Kong, X, Zhang, C, Chen, Y, Hua, Y: Adsorption of soy isoflavones by activated carbon: kinetics, thermodynamics and influence of soy oligosaccharides. Chem. Eng. J. **215–216**, 113–121 (2013)
8. Machado, FM, Bergmann, CP, Fernandes, THM, Lima, EC, Royer, B, Calvete, T, Fagan, SB: Adsorption of Reactive Red M-2BE dye from water solutions by

multi-walled carbon nanotubes and activated carbon. J. Hazard. Mater. **192**, 1122–1131 (2011)
9. Wang, JP, Yang, HC, Hsieh, CT: Adsorption of phenol and basic dye on carbon nanotubes/carbon fabric composites from aqueous solution. Sep. Sci. Technol. **46**, 340–348 (2011)
10. Tang, WW, Zeng, GM, Gong, JL, Liu, Y, Wang, XY, Liu, YY, Liu, ZF, Chen, L, Zhang, XR, Tu, DZ: Simultaneous adsorption of atrazine and Cu (II) from wastewater by magnetic multi-walled carbon nanotube. Chem. Eng. J. **211–212**, 470–478 (2012)
11. Plata, D, Gschwend, P, Reddy, C: Industrially synthesized single-walled carbon nanotubes: compositional data for users, environmental risk assessments, and source apportionment. Nanotechnol. **19**, 185706 (2008)
12. Fraczek-Szczypta, A, Menaszek, E, Blazewicz, S: Some observations on carbon nanotubes susceptibility to cell phagocytosis. J. Nanomat. **2011**, 473516 (2011)
13. Ngomsik, AF, Bee, A, Draye, M, Cote, G, Cabuil, V: Magnetic nano- and microparticles for metal removal and environmental applications: a review. Comptes. Rendus. Chimie **8**, 963–970 (2005)
14. Yao, Y, Liu, S, Ma, LP, Sun, H, Wang, S: Synthesis, characterization, and adsorption properties of magnetic Fe_3O_4@ graphene nanocomposite. Chem. Eng. J. **184**, 326–332 (2012)
15. Yao, Y, Miao, S, Ping Ma, L, Sun, H., Wang, S.: Fabrication of Fe3O4/SiO2 core/shell nanoparticles attached to graphene oxide and its use as an adsorbent. J. Colloid Interf. Sci. **379**, 20–26 (2012)
16. Konicki, W, Pełech, I, Mijowska, E, Jasińska, I: Adsorption of anionic dye Direct Red 23 onto magnetic multi-walled carbon nanotubes-Fe_3C nanocomposite: Kinetics, equilibrium and thermodynamics. Chem. Eng. J. **210**, 87–95 (2012)
17. Qu, S, Huang, F, Yu, S, Chen, G, Kong, J: Magnetic removal of dyes from aqueous solution using multi-walled carbon nanotubes filled with Fe_2O_3 particles. J. Hazard Mater. **160**, 643–647 (2008)
18. Gong, JL, Wang, B, Zeng, GM, Yang, CP, Niu, CG, Niu, QY, Zhou, WJ, Liang, Y: Removal of cationic dyes from aqueous solution using magnetic multi-wall carbon nanotube nanocomposite as adsorbent. J. Hazard Mater. **164**, 1517–1522 (2009)
19. Madrakian, T, Afkhami, A, Ahmadi, M, Bagheri, H: Removal of some cationic dyes from aqueous solutions using magnetic-modified multi-walled carbon nanotubes. J. Hazard Mater. **196**, 109–114 (2011)
20. Bansal, RC: Activated Carbon Adsorption. Taylor and Francis Group. CRC Press, Boca Raton (2005)
21. Bansal, RC, Goyal, M: Activated Carbon Adsorption. CRC Press, Boca Raton, Taylor and Francis Group (2005)
22. Fugetsu, B, Satoh, S, Shiba, T, Mizutani, T, Lin, YB, Terui, N, Nodasaka, Y, Sasa, K, Kiyoko Shimizu, O, Akasaka, T: Caged multiwalled carbon nanotubes as the adsorbents for affinity-based elimination of ionic dyes. Environ. Sci. Technol. **38**, 6890–6896 (2004)
23. Mall, ID, Srivastava, VC, Agarwal, NK: Removal of Orange-G and Methyl Violet dyes by adsorption onto bagasse fly ash–kinetic study and equilibrium isotherm analyses. Dyes pigm. **69**, 210–223 (2006)
24. Langmuir, I: The adsorption of gases on plane surfaces of glass, mica and platinum. J. Am. Chem. Soc. **40**, 1361–1403 (1918)
25. Freundlich, H: Over the adsorption in solution. J. Phys. Chem. **57**, 385–470 (1906)
26. Hall, K, Eagleton, L, Acrivos, A, Vermeulen, T: Pore-and solid-diffusion kinetics in fixed-bed adsorption under constant-pattern conditions. Ind. Eng. Chem. Fund. **5**, 212–223 (1966)
27. Wang, S, Boyjoo, Y, Choueib, A: A comparative study of dye removal using fly ash treated by different methods. Chemosphere **60**, 1401–1407 (2005)
28. Haghseresht, F, Lu, G: Adsorption characteristics of phenolic compounds onto coal-reject-derived adsorbents. Energ. Fuel. **12**, 1100–1107 (1998)
29. Demirbas, E, Dizge, N, Sulak, M, Kobya, M: Adsorption kinetics and equilibrium of copper from aqueous solutions using hazelnut shell activated carbon. Chem. Eng. J. **148**, 480–487 (2009)
30. Weber, W, Morris, J: Kinetics of adsorption on carbon from solution. J Sanit Eng Div, J. Sanit. Eng. Div. Am. Soc. Civ. Eng **89**, 31–60 (1963)
31. Ozcan, AS, Erdem, B, Ozcan, A: Adsorption of Acid Blue 193 from aqueous solutions onto Na-bentonite and DTMA-bentonite. J. Colloid Interf. Sci. **280**, 44–54 (2004)
32. Zcan, A, Zcan, AS: Adsorption of Acid Red 57 from aqueous solutions onto surfactant-modified sepiolite. J. Hazard Mater. **125**, 252–259 (2005)
33. Kannan, N, Sundaram, MM: Kinetics and mechanism of removal of methylene blue by adsorption on various carbons–a comparative study. Dyes. pigm. **51**, 25–40 (2001)
34. Karagz, S, Tay, T, Ucar, S, Erdem, M: Activated carbons from waste biomass by sulfuric acid activation and their use on methylene blue adsorption. Bioresour. Technol. **99**, 6214–6222 (2008)
35. Chen, C, Li, X, Zhao, D, Tan, X, Wang, X: Adsorption kinetic, thermodynamic and desorption studies of Th (IV) on oxidized multi-wall carbon nanotubes. Colloid. Surface A. **302**, 449–454 (2007)
36. Doan, M, Alkan, M: Adsorption kinetics of methyl violet onto perlite. Chemosphere **50**, 517–528 (2003)
37. Nollet, H, Roels, M, Lutgen, P, Van der Meeren, P, Verstraete, W: Removal of PCBs from wastewater using fly ash. Chemosphere **53**, 655–665 (2003)
38. Lazaridis, N, Asouhidou, D: Kinetics of sorptive removal of chromium (VI) from aqueous solutions by calcined Mg-Al-CO_3 hydrotalcite. Water Res. **37**, 2875–2882 (2003)
39. Bahgat, M, Farghali, A, El Rouby, WMA, Khedr, M: Synthesis and modification of multi-walled carbon nano-tubes (MWCNTs) for water treatment applications. J. Anal. Appl. Pyrol **92**, 307–313 (2011)

Synthesis of clay-CNTs nanocomposite

Sajjad Sedaghat

Abstract

Nanocomposite of clay-carbon nanotubes (CNTs) is prepared by using carboxyl-modified multiwalled carbon nanotubes and amino-functionalized organophilic montmorillonite (clay). The nanocomposite is characterized by scanning electron microscopy to evaluate the morphology. The findings show uniform dispersion of CNTs.

Keywords: Carbon nanotubes, Montmorillonite, Carboxyl-modified multiwalled carbon nanotubes, Organo clay

Background

Carbon nanotubes (CNTs) were first discovered by Iijimain1991 [1]. Since then, CNTs have been the focus of considerable research because of their physicochemical properties. The first observed CNTs were multiwalled carbon nanotubes (MWCNTs) (see Figure 1a), consisting of up to several tens of graphitic shells with adjacent shell separation of 0.34 nm, diameters of 1 nm, and large length/diameter ratio. MWCNTs can be considered as elongated fullerene [2]. A few months after Iijima's discovery, Ebbesen and Ajayan published their work on the bulk synthesis of MWCNTs by arcing graphite electrodes in inert atmospheres at optimum current and pressure conditions. However, it is necessary to note that the first images of CNTs were obtained by Oberlin et al. by pyrolysis of benzene and ferrocene at 1,000°C in the mid-1970s. In 1993, smaller-diameter, single-walled carbon nanotubes (SWCNTs) (see Figure 1b) were independently discovered by Iijima and Ichihashi and Bethune et al. using arc-discharge method and metal (iron and cobalt) as catalysts. CNTs have been termed 'materials of the twenty-first century' due to their unique properties such as functional mechanical, thermal, electrical, and optoelectronic properties which depend on the atomic arrangement (how the sheets of graphite are 'rolled'), the diameter and length of the tubes, the morphology, or nanostructure [3,4].

The adsorption-related applications of CNTs to solve environmental pollution problems have received considerable attention in recent years. CNTs are relatively new adsorbents and hold interesting positions in carbon-based adsorptive materials for many reasons [5].

Correspondence: sajjadsedaghat@yahoo.com
Department of Chemistry, Islamic Azad University, Shahr-eQods Branch, Tehran, Iran

The surface functional group density, rather than total surface area, becomes the primary determinant of inorganic pollutant adsorption capacity. The activation of CNTs plays an important role in enhancing the maximum adsorption capacity because of the modification in the surface morphology and surface functional groups. Activation of CNTs under oxidizing conditions with chemicals such as HNO_3, $KMnO_4$, H_2O_2, NaClO, H_2SO_4, KOH, and NaOH have been widely reported [6]. After activation, the metallic impurities, amorphous carbon, and catalyst support materials are removed, and the surface characteristics are altered due to the introduction of new functional groups as shown in Figure 2.

Nanotechnology is being used or considered in many applications targeted to provide cleaner and more efficient energy supplies and uses. While many of these applications may not affect energy transmission directly, each has the potential to reduce the need for the electricity, petroleum distillate fuel, or natural gas [7].

Clay minerals may be divided into four major groups, mainly in terms of the variation in the layered structure. The kaolinite group has three members, including kaolinite, dickite, and nacrite, each with a formula of $Al_2Si_2O_5(OH)_4$. The same formula indicates that the members of this group are polymorphs, meaning that they have same formula but different structures. Each member is composed of silicate sheets (Si_2O_5) bonded to aluminum oxide/hydroxide layers $Al_2(OH)_4$ the two types of layers are tightly bonded. These clays are used as fillers in ceramics, paint, rubber, paper, plastics and as plasticizer in molding sands. Montmorillonite, talc, pyrophyllite, saponite, and nontronite are a few members of the larger smectite clay group. The general formula for the chemical structure of this group is (Ca,Na,H)(Al,Mg,

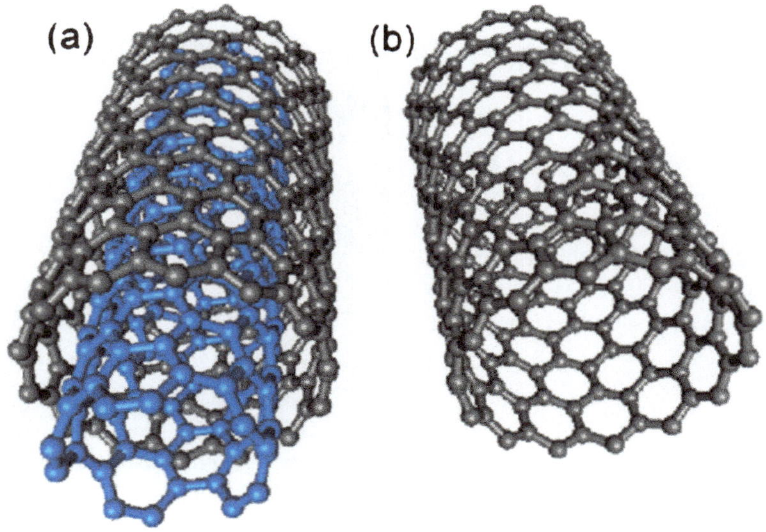

Figure 1 Structure presentations of (a) MWCNT and (b) SWCNT.

Fe,Zn)$_2$(Si,Al)$_4$O10(OH)$_2$XH$_2$O. The important difference among the members of this group is seen in the chemical characteristics. The layer structure contains silicate layers sandwiching an aluminum oxide/hydroxide layer ($Al_2(OH)_4$) [8].

One of the members, talc, has traditional presence in facial powder. 'Nanoclay' is the term generally used when referring to a clay mineral with a phyllosilicate or sheet structure with a thickness of the order of 1 nm and surfaces of perhaps 50 150 nm in one dimension. The mineral base can be natural or synthetic and is hydrophilic. The clay surfaces can be modified with specific methods to render them organophilic and therefore compatible with organic polymers adsorption of clay minerals can occur in two forms, selective and non-selective [9]. Selective sorption is also named as specific or non-exchangeable sorption, while non-selective sorption is named as exchangeable sorption. In the process of non-selective adsorption, clay minerals adsorb ions through exchangeable cations and electrostatically bound counter ions in the diffusion layer which is formed because of the surface charge. Such ions are electrostatically bound to the surface and the surface charge is a factor controlling the sorption density. Selectively adsorbed ions are assumed to form inner sphere complexes with surface hydroxide sites that is, they are primarily bound to the surface of clay minerals by chemical bond. This bond is highly dependent upon the structure of the surface and its interaction with a specific ion and, hence, exhibits substantially different behavior from electrostatic bonding. In recent decades, clays such as montmorillonite, vermiculite, kaolinite, and mica were reported as inexpensive products in a large diversity of industrial processes, pharmaceutical industry, cosmetics, organic synthesis, and environmental remediation [10]. In this paper we have used

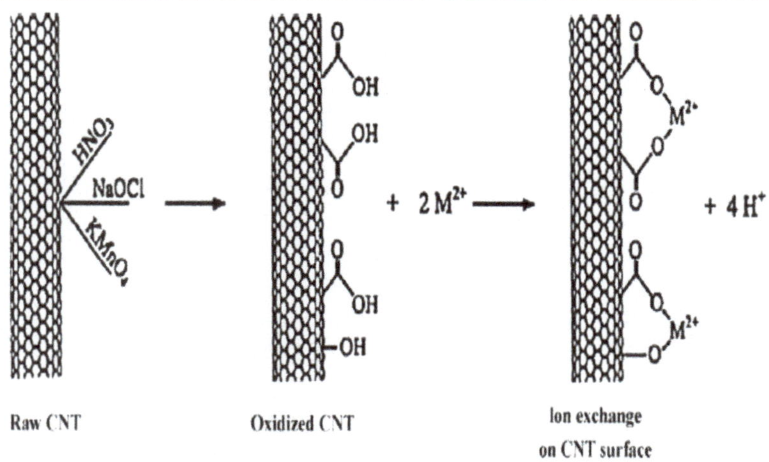

Figure 2 Schematic of the major mechanism for sorption of divalent metal ions onto CNT surfaces.

amino-functionalized clay and carboxyl-functionalized MWCNTs to synthesize the nanocomposite.

Results and discussion

The SEM analysis of nanocomposite

The failure surface of the MMT, OMMT, and clay-CNTs nanocomposite e film was observed via SEM. The SEM photographs were taken with a magnification 625,1,250, and ×2,500 at room temperature. The study of the SEM images shows the differences between the MMT, OMMT, and clay-CNTs nanocomposite the reference sample (Figures 3, 4, and 5).

Conclusion

Because of a hydrophilic nature of the MMT, the organo-modification of the MMT is an important step in the preparation of MMT/HMDA composite. The organo-modification could change the hydrophilicity of Na-MMT to be more hydrophobic to generate a microchemical environment appropriate to the intercalation of MMT. Furthermore, the increase in the layer-to-layer spacing in the MMT is also desired. Primary aliphatic amines such as hexamethylenediamine and quaternary ammonium salts have been the most effective modifying reagents in the preparation of MMT/HMDA composite. Findings from the SEM images show that OMMT and clay-CNTs nanocomposite have uniform disperse and unique morphology. The clay-CNTs nanocomposite is favorable for many applications such as nanosponges for oil split remediation.

Methods

Materials

Montmorillonite (clay) was purchased from Fluka (Fluka Chemical Corporation, St. Gallen, Switzerland) and carboxyl-functionalized MWCNTs were purchased from China. In all experiment, double distilled water was used.

Figure 3 SEM image of montmorillonite.

Figure 4 SEM image of modified montmorillonite.

Synthesis of organic montmorillonite

A 1.0 g of MMT was added in 40 mL of distilled water then mixed and heated to 80°C for 15.0 min. In a solution of 2.5 mM HCL, 0.04 g of hexamethylenediamine was added and mixed, and then the two abovementioned solutions were mixed together overnight at 80°C. The precipitate was removed by filtration and washed with distilled water, then dried in an oven at 60°C overnight.

Synthesis of clay-CNTs nanocomposite

A 0.1 g of the as-synthesized organoclay is mixed in 30 mL of DMF and sonicated for 40 min. Then 0.1 g of carboxyl-functionalized MWCNTs is added and heated to 90°C and refluxed for 24 h. The precipitate is washed with distilled water and dried in an oven.

Characterization

The as-synthesized nanocomposite has been characterized by scanning electron microscopy (SEM) using a XL-30 Philips model scanning electron microscope (Royal Philips Electronics, Amsterdam, The Netherlands).

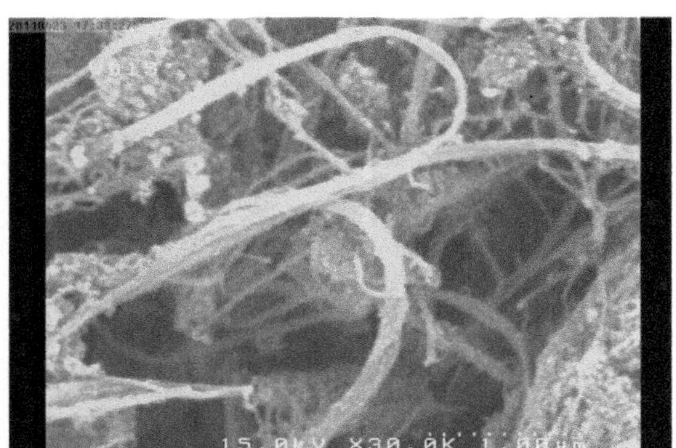

Figure 5 SEM image of clay-CNTs nanocomposite.

Competing interest
The author declares no competing interest.

Acknowledgment
The author would be like to thank Islamic Azad University, Shahr-e Qods Branch for the kindly support.

References

1. Iijima, S: Helical microtubules of graphitic carbon. Nature **354**, 56–58 (1991)
2. Endo, M, Kroto, HW: Formation of carbon nanofibers. J. Phys. Chem. **96**, 6941–6944 (1992)
3. Ebbesen, TW, Ajayan, PM: Large scale synthesis of carbon nanotubes. Nature **358**, 220–222 (1992)
4. Oberlin, A, Endo, M, Koyama, T: Filamentous growth of carbon through benzene decomposition. J. Cryst. Growth **32**, 335–349 (1976)
5. Iijima, S, Ichihashi, T: Single-shell carbon nanotubes of 1-nm diameter. Nature **363**, 603–605 (1993)
6. Bethune, DS, Kiang, CH, deVries, MS, Gorman, G, Savoy, R, Vazquez, J, Beyers, R: Cobalt catalyzed growth of carbon nanotubes with single atomic-layer walls. Nature **363**, 605–607 (1993)
7. Britz, DA, Khlobystov, AN: Noncovalent interactions of molecules with single walled carbon nanotubes. Chem. Soc. Rev. **35**, 637–659 (2006)
8. Avouris, P: Molecular electronics with carbon nanotubes. Acc. Chem. Res. **35**, 1026–1034 (2002)
9. Thostenson, ET, Ren, ZF, Chou, TW: Advances in the science and technology of carbon nanotubes and their composites: a review. Compos. Sci. Technol. **61**, 1899–1912 (2001)
10. Zhao, YL, Stoddart, JF: Noncovalent functionalization of single-walled carbon nanotubes. Acc. Chem. Res. **421**, 1161–1171 (2009)

12

One-step electrosynthesis of multi-walled carbon nanotube/poly-ortho-aminophenol composite film and investigation of its electrocatalytic properties

Narges Ajami* and Niloufar Bahrami Panah

Abstract

In this study, one-step electrochemical polymerization method was developec for the fabrication of multi-walled carbon nanotubes/poly-ortho-aminophenol (MWCNTs/PoAP) composite film to detect oxytetracycline. The fabricated nanocomposite shows good stability and enhanced electrocatalytic ability for the detection of oxytetracycline. The resulting composite can be used as electrode material in an electrochemical capacitor. Scanning electron microscopy and electrochemical methods have been used to characterize the MWCNTs/PoAP nanocomposite film.

Keywords: Linear sweep voltammetry; Multi-walled carbon nanotube; Oxytetracycline; Poly ortho aminophenol

Background

Conducting polymers are important materials of considerable interest in many fields including corrosion inhibitors, batteries, organic electronics, electrochromic devices and sensors [1-6]. Among Conducting polymers, poly ortho aminophenol, has been research focus [7,8], which is due to its good redox reversibility, unique doping characteristics, good stability, and its high conductivity. Multi-walled carbon nanotubes (MWCNTs) have chemical stability, highly accessible surface area, unique structure, narrow distribution size, and low resistivity [9-13]. It has been shown experimentally that the introduction of carbon nanotubes (CNTs) into a polymer matrix improves the electric conductivity as well as the mechanical property of the original polymer matrix [14-16]. Oxytetracycline is one of the most common tetracyclines that are widely used for the treatment of infectious diseases [17]. Modified electrodes are being used continuously in the voltammetric determination of organic compounds because of their efficiency and the selectivity that can be obtained by varying the modifier. Herein, we describe a simple strategy for the aligning of disordered CNTs within the conducting polymer matrix by *in situ* electropolymerization using an ionic surfactant as the supporting electrolyte. In this study, the modified electrode was fabricated successively by electropolymerization of ortho-aminophenol (oAP) in the presence of dodecyl benzene sulfonic acid sodium (DBSA) onto a platinum electrode (PtE) surface. The prepared electrode was characterized by linear sweep voltammetry, cyclic voltammetry, and scanning electron microscopy (SEM) techniques. It exhibited excellent electrochemical response to oxytetracycline with good stability and sensitivity and featured with a low detection limit and wide linear range.

* Correspondence: n_ajami@pnu.ac.ir
Department of Chemistry, Payame Noor University, PO BOX, 19395–3697, Tehran, IRAN

Results and discussion

Characterization of the modified electrode

Figure 1 shows the cyclic voltammograms of the electropolymerization of ortho-aminophenol monomers in the presence of MWCNs. The effect of DBSA, as an anionic dopant, on the electropolymerization of the polymer in aqueous solutions has been investigated. DBSA increases the conductivity of the polymer [7]. The broad peak on the first positive sweep (Inset of Figure 1) between 0.61 and 0.75 V is due to the oxidation of the hydroxyl group of oAP and amino group, which leads to the polymerization of the monomer [18]. In the second cycle, a pair of redox peaks appears at 0.34 and 0.31 V, which related to the formation of a cyclic dimmer of the polymer, the 3-aminophenoxazone (3APZ). As it can be seen in Figure 1, during continuous cycling, the current response of this pair of redox peaks diminishes slowly, and at the same time, another redox peak appears between 0.1 and 0.23 V. This pair of redox peaks is due to the oxidation and reduction of PoAP, indicating the formation of PoAP [19]. Therefore, 3APZ plays the role of a monomer in the formation of PoAP as evident from the disappearance of the system at 0.34 V and corresponding growth of the polymer film between 0.1 and 0.23 V.

The cyclic voltammograms of the PoAP film and MWCNTs/PoAP composite have a couple of strong and broad oxidation and reduction waves in a solution of 0.5 M $NaClO_4/HClO_4$ (Figure 2). Figure 2a shows a pair of redox peaks at around 0.12 V attributed to the redox process of the phenoxazine units. The evanescence of a pair of redox peaks at around 0.34 V due to the absence of the oAP monomer, indicating that the redox peaks have been generated during the oxidation process of the oAP. To elucidate the effect of MWCNTs on the property of PoAP films, electrochemical performance of composite films was evaluated by carrying out cyclic voltammogram measurements in 0.5 M $NaClO_4/HClO_4$ (Figure 2b). The cyclic voltammograms of the MWCNTs/PoAP films show a higher background current in the potential sweep. The peak current of a MWCNTs/PoAP composite is stronger than that of pure a PoAP film. The voltammetric behavior of both films is similar and shows capacitive-like responses. A comparison between the voltammograms of MWCNTs/PoAP and pure PoAP electrodes demonstrates that the composite not only shows a higher background current in the potential sweep but also there exists faradic currents, which are due to the participation of the loaded MWCNTs. Owing to the higher currents in the cyclic voltammogram of the MWCNTs/PoAP, a larger capacitance for capacitors equipped with MWCNTs/PoAP is predictable and the MWCNTs/PoAP composite acts as a supercapacitor with a specific capacitance of 395 $F{\cdot}g^{-1}$. The specific capacitance was determined by the cyclic voltammetry technique. The specific capacitance (*SC*) was calculated from the cyclic voltammogram, according to the following equation [20]:

$$SC = \frac{2i}{sm},$$

where i is the average cathodic current, s denotes the applied potential sweep rate, and m is the mass of each active material on the electrode.

Figure 3 shows the SEM micrographs of the pure PoAP and MWCNTs/PoAP composite. As it can be seen from Figure 3a, PoAP/Pt has an incompact and rough structure. After the synthesis of MWCNTs/PoAP onto the surface of Pt (Figure 3b), a uniform and compact layer was obtained, indicating a strong interaction between the polymer and carbon nanotube [21]. It was clearly observed that the MWCNTs/PoAP composite (Figure 3b) has a larger surface area in comparison with PoAP/Pt films. The nanocomposite film formed a homogenous and three-dimensional film structure.

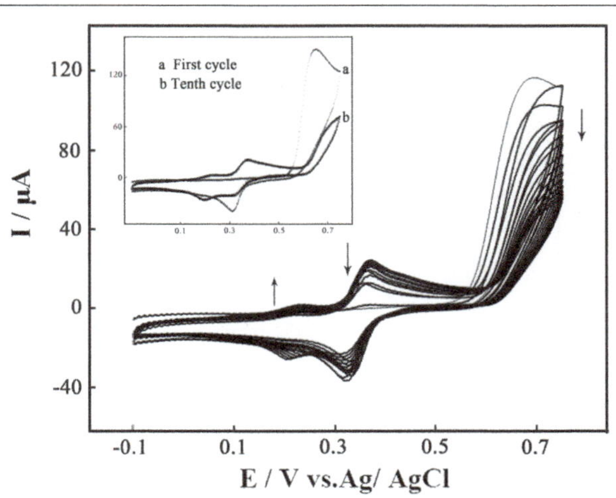

Figure 1 Cyclic voltammograms of oAP electropolymerization in 2 mM oAP and 1% MWCNTs in 0.05 M $NaClO_4/HClO_4$ with 0.01 M DBSA. Scan rate: $0.1V{\cdot}s^{-1}$. Inset: The first and tenth cycle of cyclic voltammograms of oAP electropolymerization.

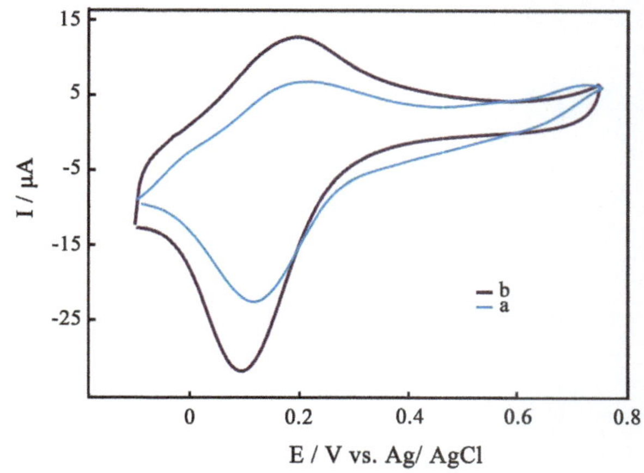

Figure 2 Cyclic voltammograms of (a) PoAP/PtE and (b) MWCNTs/PoAP/PtE obtained in 0.5 M $NaClO_4/HClO_4$ solution. Scan rate: 100 $mV{\cdot}s^{-1}$.

Electrocatalytic oxidation of oxytetracycline

The linear sweep voltammogram of different electrodes in a solution of oxytetracycline are shown in Figure 4. It can be seen that the oxytetracycline peak at the bare platinum electrode is relatively weak (Figure 4a) which is due to the slow electron transfer. Figure 4b shows the oxidation of oxytetracycline on PoAP/Pt electrode. It is obvious that the presence of the PoAP film leads to the great reinforcement

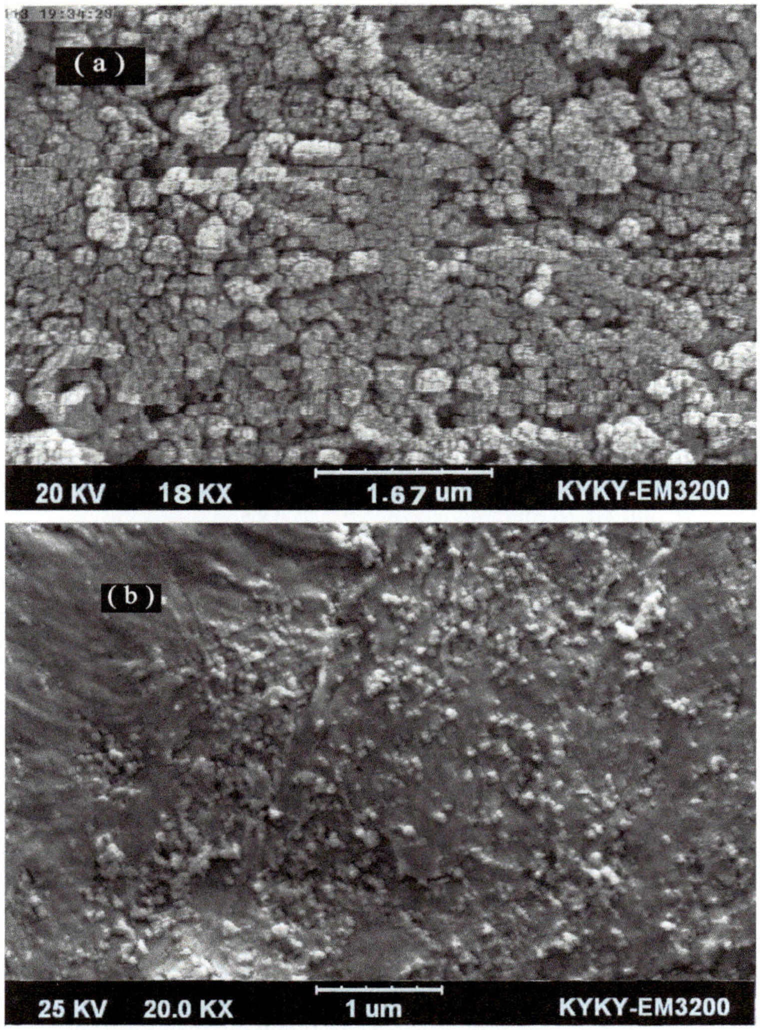

Figure 3 SEM micrographs of pure PoAP (a) and MWCNTs/PoAP nanocomposite (b).

of the anodic peak. The presence of MWCNTs on the bare electrode improves the electrocatalytic oxidation of oxytetracycline (Figure 4c). In comparison with other electrodes, MWCNTs/PoAP/Pt electrode exhibits better electrochemical response (Figure 4d). Therefore, the combination of PoAP and MWCNTs results in the increase of the oxytetracycline current response.

To some extent, the current of oxytetracycline is dependent on the thickness of MWCNTs/PoAP composite, which can easily be adjusted by controlling the number of potential cycles during the electropolymerization. The maximum anodic peak current of oxytetracycline is observed when 50 potential cycles were used for the fabrication of modified electrode (Figure 5) and then the electrochemical response decreased. So, the number of cycles was found to affect the sensitivity and linearity of the sensor. This indicates that increasing the MWCNTs/PoAP composite thickness after a specified extent leads to the reduced transfer

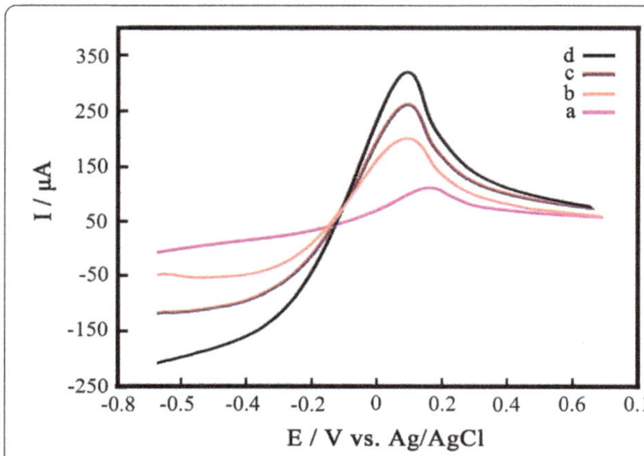

Figure 4 Linear sweep voltammograms of (a) bare platinum, (b) PoAP/PtE, (c) MWCNTs/PtE, and (d) MWCNTs/PoAP/PtE. Buffer solution of pH = 2.0 containing 0.3 μM oxytetracycline. Scan rate: 100 mV·s^{-1}.

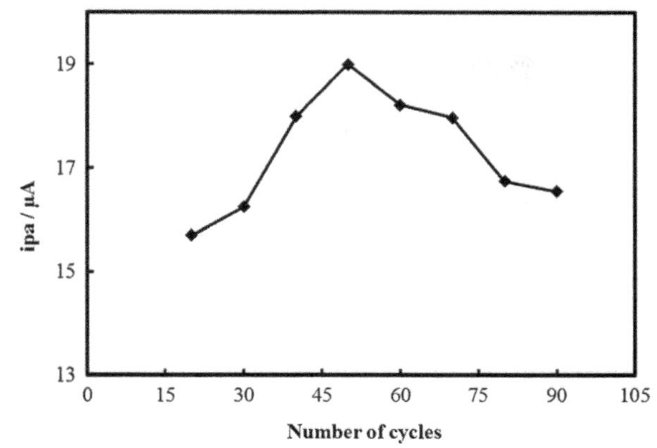

Figure 5 Dependence of anodic peak current (i_{pa}) of MWCNTs/PoAP/PtE on number of potential cycles during electropolymerization of composite.

rate of electron in the film. Thus, 50 cycles of electropolymerization for fabrication the MWCNTs/PoAP composite was selected for this study and it exhibited favorable analytical performance.

For further investigation, the dependence of the anodic peak current (i_{pa}) of oxytetracycline with scan rate (v) was also studied for MWCNTs/PoAP/PtE. As shown in Figure 6, the response of the MWCNTs/PoAP/PtE to oxytetracycline increases as the scan rate increases up to 70 $mV{\cdot}s^{-1}$ and then decreases. The MWCNTs/ PoAP produced at a faster scan rate were found to form a loose film with a low recognition capacity. Therefore, the optimum scan rate has been found to be 70 $mV{\cdot}s^{-1}$.

Conclusion

MWCNTs/PoAP composite were successfully fabricated through the electrochemical polymerization on the surface of platinum electrode. The SEM micrograph of the composite shows a porous morphology that wrapped on the surface of MWCNTs. The composite have good response to oxytetracycline at low potential. MWCNTs and the cross-linked structure of PoAP exhibit an electrocatalytic activity for the oxidation of oxytetracycline, which greatly facilitated the electron-exchange between oxytetracycline and the electrode and decreased the overpotentials for the oxidation of oxytetracycline. The sensor produced several advantages over the conventional, including low cost, sustainable electrocatalytic ability. It shows a distinguished stability. The sensor opens a new doorway for the application of MWCNTs/PoAP as analytical devices. It might find application as an amperometric sensor with good stability and reproducibility for the determination of oxytetracycline which is reported in our other research.

Methods

All materials (ortho-aminophenol, dodecyl benzene sulfonic acid sodium, perchloric acid, sodium perchlorate, and oxytetracycline) were with analytical grade (Merck, Whitehouse Station, NJ, USA). Typical MWCNTs with outer diameter in the range of 5 to 20 nm and lengths up to several hundreds of nanometers were used. Solutions were prepared using doubly distilled water. Electrochemical studies were carried out in a one-compartment three-electrode cell powered by a Metrohm-Autolab potentiostat/galvanostat (model 12/30/302, Utrecht, The Netherlands). An Ag/AgCl electrode and a platinum foil with a large area were used as reference and auxiliary electrode, respectively. A MWCNTs/PoAP composite film was used as a working electrode. The morphology of the obtained films was characterized using a KYKY-EM3200 field emission

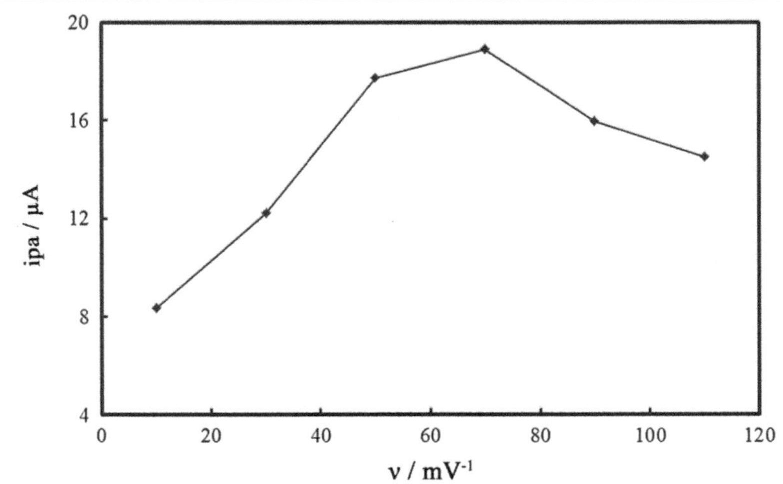

Figure 6 Effect of scan rate (v) on anodic peak current (i_{pa}) of MWCNTs/ PoAP/PtE. In a solution of pH = 2.0 containing 0.3 μM oxytetracycline.

scanning electron microanalyzer (Beijing, China). The pH values of the solutions were determined using a pH meter (827, Metrohm, Riverview, FL, USA). Numerous amounts of purified MWCNTs were dispersed ultrasonically (Sigma-Aldrich-Bandelin Sonopuls ultrasonic homogenizer, Munich, Germany) into a dodecyl benzene sulfonic acid sodium solution. All the experiments were carried out at 22 ± 1°C.

The preparation of the modified electrode (MWCNTs/ PoAP/ PtE)

Bare platinum electrode was polished by 0.05 μm alumina powder and then washed ultrasonically in water and ethanol, respectively. For the preparation of MWCNTs/poly-ortho-aminophenol (PoAP) composite film on platinum electrode, the aqueous micellar medium was prepared by dissolving 2.0 mM oAP in an aqueous solution containing 0.2 mg mL^{-1} purified MWCNTs, 0.01 M DBSA, and 0.05 M $NaClO_4/HClO_4$ ultrasonically for 10 min. The MWCNTs/PoAP/PtE was formed by dipping PtE into the prepared mixture solution and then scanning in the potential range from −0.1 to 0.75 V at 0.1 $V{\cdot}s^{-1}$ for 50 cycles. The PoAP/PtE were prepared in the same way but without the addition of MWCNTs.

Competing interests

The authors declare that they have no competing interests.

Authors' contributions

NA carried out all of the experiments and analysis and drafted and rechecked the whole manuscript, and NBP rechecked the manuscript. Both authors read and approved the final manuscript.

Acknowledgements

The authors gratefully acknowledge the instrumental support from Payame Noor University.

References

1. Firoz Babu, K, Subramanian, SP, Kulandainathan, M: Functionalisation of fabrics with conducting polymer for tuning capacitance and fabrication of supercapacitor. Carbohydr. Polym. **94**, 487–495 (2013)
2. Hangarter, C, Chartuprayoon, N, Hernández, S, Choa, Y, Myung, N: Hybridized conducting polymer chemiresistive nano-sensors. Nano. Today. **8**, 39–55 (2013)
3. Gaihre, B, Ashraf, S, Spinks, G, Innis, P, Wallace, G: Comparative displacement study of bilayer actuators comprising of conducting polymers, fabricated from polypyrrole, poly(3,4-ethylenedioxythiophene) or poly(3,4-propylenedioxythiophene. Sens. Actuators. A: Physical. **193**, 48–53 (2013)
4. Ehsani, A, Mahjani, MG, Jafarian, M, Naeemy, A: An electrochemical study of the synthesis and properties of multi-walled carbon nanotube/poly ortho aminophenol composites. Synth. Met. **161**, 1760–1765 (2011)
5. Ayad, M, Abu El-Nasr, A: Anionic dye (acid green 25) adsorption from water by using polyaniline nanotubes salt/silica composite. J. Nanostr. Chem. (JNSC) **3**(1), 3 (2012)
6. Shokrieh, MM, Saeedi, A, Chitsazzadeh, M: Mechanical properties of multi-walled carbon nanotube/ polyester nanocomposites. J. Nanostru. Chem. (JNSC) **3**, 20 (2013)
7. Ojani, R, Raoof, J-B, Fathi, S: Poly (*o*-aminophenol) film prepared in the presence of sodium dodecyl sulfate: application for nickel ion dispersion and the electrocatalytic oxidation of methanol and ethylene glycol. Electrochim. Acta. **54**, 2190–2196 (2009)
8. Hu, D, Peng, C, Chen, GZ: Electrodeposition of nonconducting polymers: roles of carbon nanotubes in the process and products. ACS. Nano. **4**, 4274–4282 (20`0)
9. Kan, K, Zhou, H, Li, C, Zhu, A, Xing, Z, Zhao, Z: Imprinted electrochemical sensor for dopamine recognition and determination based on a carbon nanotube/polypyrrole film. Electrochim. Acta. **63**, 69–75 (2012)
10. Lota, G, Fic, K, Frackowiak, E: Carbon nanotubes and their composites in electrochemical applications. Energ. Environ. Sci. **4**, 1592–1605 (2011)
11. Taleshi, F, Hosseini, AA: Synthesis of uniform MgO/CNT nanorods by precipitation method. J. Nanostruc. Chem. (JNSC) **3**, 4 (2012)
12. Hesabi, M, Hesabi, M: The interaction between carbon nanotube and skin anti-cancer drugs: a DFT and NBO approach. J. Nanostr. Chem. (JNSC) **3**(1), 22 (2013)
13. Farghali, AA, Bahgat, M, Elrouby WM, A, Khedr, MH: Decoration of multi-walled carbon nanotubes (MWCNTs) with different ferrite nanoparticles and its use as an adsorbent. J. Nanostru. Chem. (JNSC) **3**, 50 (2013)
14. Bi, YH, Huang, ZL, Zhao, YD: The interface behavior and biocatalytic activity of superoxide dismutase at carbon nanotube. Biosens. Bioelectro. **21**, 1350–1354 (2006)
15. Zhou, H, Cheng, X, Rao, L, Duan, Y: Poly (3, 4-ethylenedioxythiophene)/multiwall carbon nanotube composite coatings for improving the stability of microelectrodes in neural prostheses applications. Acta. Biomater. **9**, 6439–6449 (2013)
16. Branzoi, F, Branzoi. V, Musina, A: Coatings based on conducting polymers and functionalized carbon nanotubes obtained by electropolymerization. Prog. Org. Coat. **76**, 632–638 (2013)
17. Chopra, I, Roberts, M: Tetracycline antibiotics: mode of action, applications, molecular biology, and epidemiology of bacterial resistance. Microbiol. Mol. Biol. R. **65**, 232–260 (2001)
18. Mu, S: Electrochemical copolymerization of aniline and *o*-aminophenol. Synth. Met. **143**, 259 (2004)
19. Zhang, L, Lian, J: Electrochemical synthesis of copolymer of aniline and *o*-aminophenol and its use to the electrocatalytic oxidation of ascorbic acid. J. Anal. Chem. **611**, 51–59 (2007)
20. Reddy, ALM, Ramaprabhu, S: Nanocrystalline metal oxides dispersed multiwalled carbon nanotubes as supercapacitor electrodes. J. Phys. Chem. C. **111**, 7727–7773 (2007)
21. Gajendran, P, Saraswathi, R: Enhanced electrochemical growth and redox characteristics of poly(*o*-phenylenediamine) on a carbon nanotube modified glassy carbon electrode and its application in the electrocatalytic reduction of oxygen. J. Phys. Chem. C. **111**, 11320–11328 (2011)

13

A facile one-step synthesis of PTSA-doped tetraaniline nanostructure/magnetite nanoparticles via self-assembly method

Keloth Basavaiah*, Yarramsetti Pavankumar and Alamanda Vara Prasada Rao

Abstract

Tetraaniline nanostructures/magnetite nanoparticles (Fe_3O_4 NPs) have been prepared via *in situ* self-assembly method using ammonium persulphate as an oxidant in the presence of *p*-toluene sulphonic acid as a dopant as well a soft template. The effect of the concentration and molar ratio of *p*-toluene sulphonic acid to aniline on the morphology and size of the nanostructures, and the crystallinity, thermal stability and magnetic properties of the nanocomposites have been studied by UV-visible spectroscopy, Fourier transform infrared spectroscopy, X-ray diffraction, scanning electron microscopy, transmission electron microscopy, thermogravimetric analysis and vibrating sample magnetometry. The spectroscopic results indicated the interaction between the tetraaniline nanostructures and the Fe_3O_4 NPs. The microscopic results show that the Fe_3O_4 NPs were coated on tetraaniline nanostructures. The saturation magnetization values and the thermal stability of the nanocomposites were found to depend on the molar ratio of *p*-toluene sulphonic acid to aniline.

Keywords: Intrinsic conducting polymer; Tetraaniline; Magnetite nanoparticles; *In situ* self-assembly; Superparamagnetism

Background

Since their discovery, intrinsic conducting polymers (ICPs) have emerged as potential materials for many technological applications due to their unique electronic properties covering the whole insulator-semiconductor-metal range, mechanical flexibility and relative ease of processing. Among the most commonly studied ICPs is polyaniline (PANI) due to its ease of synthesis, unique non-redox doping/dedoping chemistry, relatively high conductivity and environmental stability [1], and many technological applications in molecular electronics, light emitting diodes, sensors, actuators, solar cells, light weight batteries and antistatic and anticorrosion coatings [2].

Recently, PANI nanostructures such as nanotubes, nanorods, nanofibers and nanowires have attracted attention over its bulk counterpart because of their low dimensionality and high surface-to-volume ratio. PANI nanostructures have been prepared by various methods such as interfacial polymerization, plasma polymerization, enzymatic polymerization, electrospinning and rapid mixing polymerization in the presence of specific structure-directing materials such as zeolite channels, porous membranes, opal, surfactants, polymers and bulky organic acid dopants [3-9]. In hard-template method, it is very difficult to retain ordered PANI nanostructures because harsh experimental conditions are required for the removal of hard templates [10,11]. However, the soft-template method has the advantage in preparing large quantities of PANI nanostructures by being simple and cost effective. In this method, large quantities of PANI nanostructures are produced using structure director in the form of surfactants and bulky organic acids. The external dopant acid forms complexes with monomer via the acid-base chemistry and, as a formed complex, acts as a soft template for the growth of the PANI nanostructures.

Even though PANI nanostructures have been studied extensively, a very little emphasis is given on the synthesis of the nanostructures of PANI oligomers such as

* Correspondence: klbasu@gmail.com
Department of Inorganic and Analytical Chemistry, Andhra University, Visakhapatnam, Andhra Pradesh 530003, India

tetraaniline, octaaniline and hexadecaaniline, which have been studied experimentally and theoretically as model compounds for PANI. Among the PANI oligomers, tetraaniline (TANI) has received great attention because it is soluble in common solvents, exhibits similar electroactivity as PANI and has well-defined structure, designed end group and monodispersed molecular weight. The most useful form of PANI is emeraldine (EBPANI), which has ca. 50% amine and ca. 50% imine nitrogen, in which one of its four rings is quinoid, the remaining three being benzenoid. TANI is a perfect model compound for PANI since its chemical structure resembles that of EBPANI.

ICP nanocomposites, in particular, with magnetic nanoparticles (MNPs), have also received a great deal of attention because they are multifunctional materials with many technological applications like magnetic electrocatalysis, non-linear optic processes, electromagnetic interference shielding, antibiofouling, separation and purification of biomolecules, as carriers for targeted drug delivery and biosensors [12-15]. Among all MNPs, magnetite nanoparticles (Fe_3O_4 NPs) have attracted attention because they have many potential applications due to their low toxicity, high saturation magnetization biocompatibility. In most of these applications, magnetic nanoparticles are found to be easily dispersed in aqueous solution, which have particle sizes less than the critical size, without forming any aggregation. However, Fe_3O_4 NPs easily aggregate because of dipole-dipole interaction and van der Waals attractive forces between the nanoparticles, which results in poor magnetic properties and low dispersibility. One of the important approaches is, therefore, to overcome these limitations by protecting the Fe_3O_4 NPs with macrocyclic surfactants, polymers, inorganic metals, or oxides. The advantage of using a protective layer around each nanoparticle is that it not only avoids aggregation but also provides a platform for further functionalization. Among all the protective layer materials, PANI has received more attention. However, like other ICPs, the solution processing of PANI is difficult because of the presence of intrachain and interchain H-bonding interactions arising from the N-quinonoid and benzenoid units.

Here, we report a novel, scalable and simple one-pot synthesis of *p*-toluene sulphonic acid (PTSA)-doped TANI/Fe_3O_4 NP nanocomposites via micelle-assisted *in situ* chemical oxidative polymerization using ammonium peroxydisulphate (APS) as an oxidant. In this technique, PTSA acts as the dopant as well as the soft template for the synthesis of TANI/Fe_3O_4 nanocomposites. The effect of PTSA concentration and molar ratio of PTSA to monomer on the morphology, crystallinity, magnetic properties and thermal properties of the nanocomposites have been investigated.

Results and discussion

PTSA-doped TANI/Fe_3O_4 nanocomposites have been prepared by oxidative chemical polymerization using APS as an oxidant. PTSA has been found to play a dual role in the formation of TANI nanostructures as a dopant and as a surfactant. Due to the presence of methyl group and hydrophilic '-SO_3H' group, PTSA can form micelles in water. PTSA can also form micelles by forming a complex with NPPD via the acid-base interaction between the -SO $_3$H group of the PTSA and the "-NH_2" group of the NPPD. Thus, the reaction medium consists of PTSA micelles, PTSA-NPPD micelles and free NPPD. Moreover, the concentration of the PTSA micelles, PTSA-NPPD micelles and anilinium cations present in the reaction solution is dependent on the molar ratio of PTSA to NPPD which results in the variation of morphology and size of the TANI/Fe_3O_4 nanocomposites. As formed, the PTSA micelles or the PTSA-NPPD micelles serve as "soft template" for the growth of the TANI nanostructures. The presence of excess free PTSA can effectively prevent the formation of larger particles due to steric hindrance of bulky polymeric acids surrounding each nanoparticle and, thereby, producing nanometre-size polymer. Moreover, the hydrophilic "-SO_3^-" group of PTSA limits the size of the Fe_3O_4 NPs by forming a coating layer around the particle, while the hydrophobic methyl group acts as a barrier to prevent the Fe_3O_4 NPs from aggregating, which results from the repulsive forces among the surroundings nanoparticle.

The molecular structure of the PTSA-doped TANI/Fe_3O_4 nanocomposite was investigated by UV-visible spectroscopy, FTIR spectroscopy and powder X-ray diffraction (XRD). Figure 1a,b,c,d,e represents the powder XRD patterns of the PTSA-doped TANI/Fe_3O_4 nanocomposites prepared with 1:1, 2:1, 3:1, 4:1 and 5:1 molar ratio of PTSA

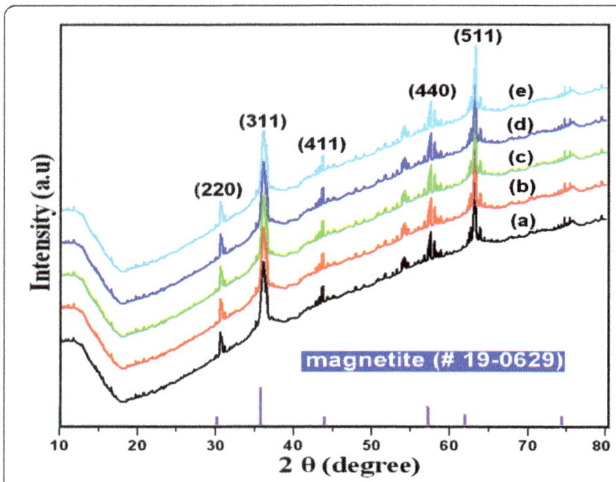

Figure 1 XRD patterns of PTSA-doped TANI/Fe_3O_4 nanocomposites prepared with different molar ratios of PTSA to NPPD.

to NPPD. The XRD patterns show both the broad amorphous peaks and sharp crystalline peaks of TANI. The broad peaks were observed at 2θ = 19.8° and 25°, while the sharp crystalline peaks, at 2θ = 11°, 13° and 22°. The sharp crystalline peaks originated from the increase in the interplanar distance due to the inclusion of PTSA in the tunnels between the TANI chains. The XRD patterns centred at 2θ = 31°, 36°, 44°, 58° and 63° can be indexed as (220), (311), (400), (511) and (440) planes of the magnetite, respectively, which are in good agreement with the standard magnetite JCPDS card (card no. 19-0629).

The UV-vis spectra of the PTSA-doped TANI/Fe_3O_4 nanocomposites prepared with 1:1, 2:1, 3:1, 4:1 and 5:1 molar ratios of PTSA to NPPD are presented in Figure 2. The UV-vis spectra of the nanocomposites dispersed in dimethyl sulphonic oxide (DMSO) show three prominent absorption bands at 285, 335 and 595 nm. The absorption band at 285 is ascribed to the $\pi \rightarrow \pi^*$ transition in the benzenoid rings of PTSA, indicating the doping of PTSA to TANI in the nanocomposites. The bands at 335 and 595 nm are due to the $\pi \rightarrow \pi^*$ transition of the benzenoid rings in the TANI chain and the transition caused by the interchain charge transfer from the two adjacent benzenoid rings to the quinoid ring of the TANI chain, respectively [16]. Both these absorption bands were shifted to the lower wavelength region from those of pure TANI, indicating that there is some interaction between Fe_3O_4 NP and TANI.

Figure 3 depicts the FTIR spectra of the PTSA-doped TANI/Fe_3O_4 nanocomposite synthesized using the different molar ratios of PTSA to NPPD (1:1, 2:1, 3:1, 4:1 and 5:1). The bands at 3,180, 1,596 and 1,493 cm^{-1} are ascribed to the N-H stretching vibration, C=C stretching mode of quinoid phenyl and benzenoid phenyl rings, respectively. The band at 1,408 cm^{-1} is due to the stretching frequency of the B-N=Q moiety (where B refers to benzenoid phenyl rings, and Q refers to quinoid phenyl ring). The C-N stretching and the C-H aromatic in-plane and out-of-plane deformations of the 1,4-disubstituted benzene rings were confirmed by the presence of band 1,304, 1,136 and 836 cm^{-1} [17]. The PTSA doping to TANI/Fe_3O_4 is confirmed by the presence of characteristic bands due to the symmetric and asymmetric O=S=O stretching vibrations of the -SO_3H group of PTSA (1,038 and 1,008 cm^{-1}). The bands at 580 and 454 cm^{-1} were ascribed to the intrinsic Fe-O stretching vibrations of tetrahedral and octahedral Fe^{3+} of the magnetite nanoparticle in the nanocomposites [18].

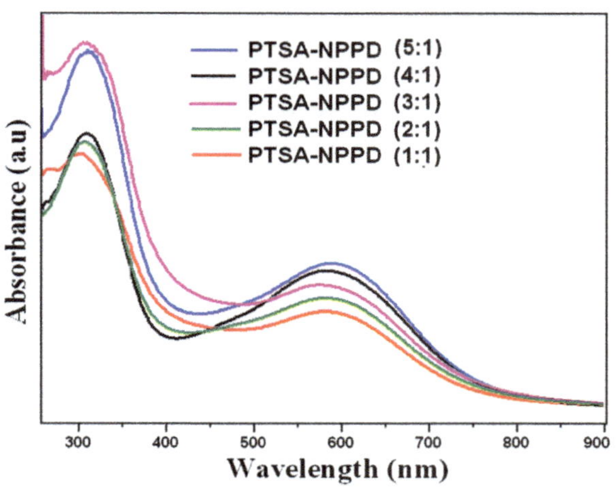

Figure 2 UV-visible spectra of the PTSA-doped TANI/Fe_3O_4 nanocomposites.

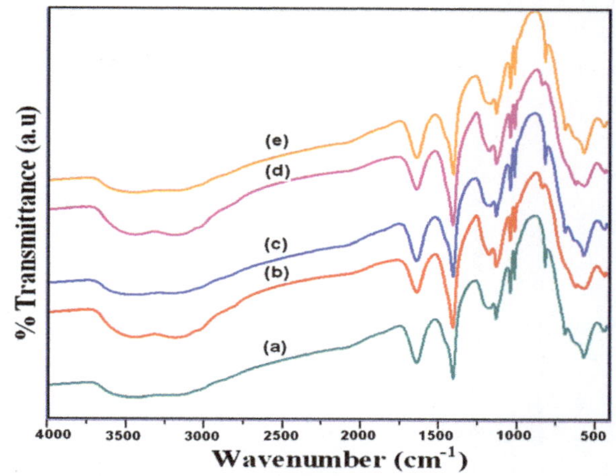

Figure 3 FTIR of PTSA-doped TANI/Fe_3O_4 prepared with different molar ratios of PTSA to NPPD. (a) 1:1, **(b)** 2:1, **(c)** 3:1, **(d)** 4:1 and **(e)** 5:1.

The morphology of the TANI/Fe_3O_4 NP nanocomposites was investigated by transmission electron microscopy (TEM) and SEM. The SEM images (Figure 4) show the rod-like morphology of the PTSA-doped TANI/Fe_3O_4 nanocomposite prepared at 5:1 molar ratio of PTSA to NPPD. It can be seen from the SEM image (Figure 5) that no rod-like morphology was obtained for the PTSA-doped TANI/Fe_3O_4 nanocomposite prepared with 4:1, 3:1, 2:1 and 1:1 molar ratios of PTSA to NPPD. The presence of carbon, nitrogen, oxygen, sulphur and iron in the TANI/Fe_3O_4 nanocomposites is confirmed by energy dispersive X-ray analysis (EDAX) (Figure 5f).

Figure 6 shows the TEM images of the as-prepared PTSA-doped TANI/Fe_3O_4 nanocomposites in direct reaction dispersion and dispersion obtained from final dried powder dispersed in DMSO. The TEM image (Figure 6a) clearly reveals the rod-like morphology of the TANI/Fe_3O_4 nanocomposites prepared with the 5:1 molar ratio of PTSA/NPPD. A closer look at the nanorods reveals that the average diameter is 40 to 95 nm, and the length of the nanorods is several hundred nanometres. Figure 6b represents the TEM image of the PTSA-doped TANI/Fe_3O_4 nanocomposite after being dispersed in DMSO, which

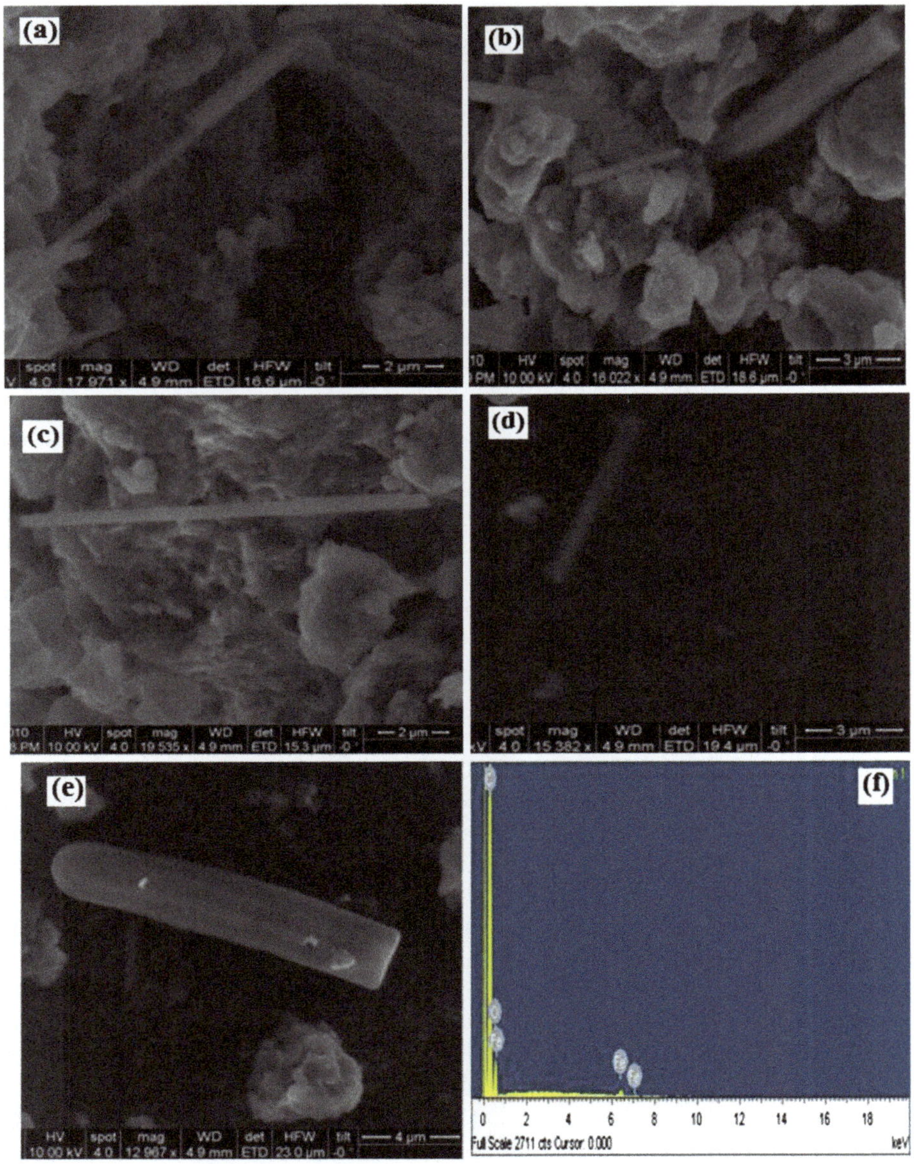

Figure 4 Images of the PTSA-doped TANI/Fe_3O_4 with the 5:1 molar ratio of PTSA/NPPD. (a, b, c, d and **e)** SEM and **(f)** EDX.

shows spherical Fe_3O_4 NPs in the composite with average size of 7 to 10 nm. The TEM images of the PTSA-doped TPANI/Fe_3O_4 nanocomposite samples prepared with 4:1, 3;1, 2:1 and 1:1 molar ratios of PTSA to NPPD are presented in Additional file 1: Figure S1, which show spherical nanoparticles without any aggregation. The SAED patterns (see in Additional file 1: Figure S2) show sharp, well-defined diffraction rings, which reveal the crystallinity of the Fe_3O_4 NPs. The corresponding interplanar spacing calculated from the diffraction patterns matched with that of the standard bulk Fe_3O_4 (JCPDS, card no. 19-0629).

The room temperature field-dependent magnetization curves of the PTSA-doped TANI/Fe_3O_4 nanocomposites prepared with 5:1,4:1, 3:1, 2:1 and 1:1 molar ratios of PTSA to NPPD are shown in Figure 7. The nanocomposite samples exhibited superparamagnetic behaviour as evidenced by zero remanence, zero coercivity and the absence of hysteresis loops. The saturation magnetization (*M*s) values of the PTSA-doped TANI/Fe_3O_4 are 46, 34, 33, 32 and 10 emu g^{-1}, corresponding to 1:1, 2:1, 3:1, 4:1 and 5:1 PTSA-to-NPPD molar ratios. The *M*s values are much lower than the theoretical *M*s values of bulk magnetite of 92 emu g^{-1} [19]. The decrease of the *M*s values is due to the finite size effect and surface effect. Below the critical particle size, the total magnetization decreases due to increased dispersion in the exchange integral and finally approaches the

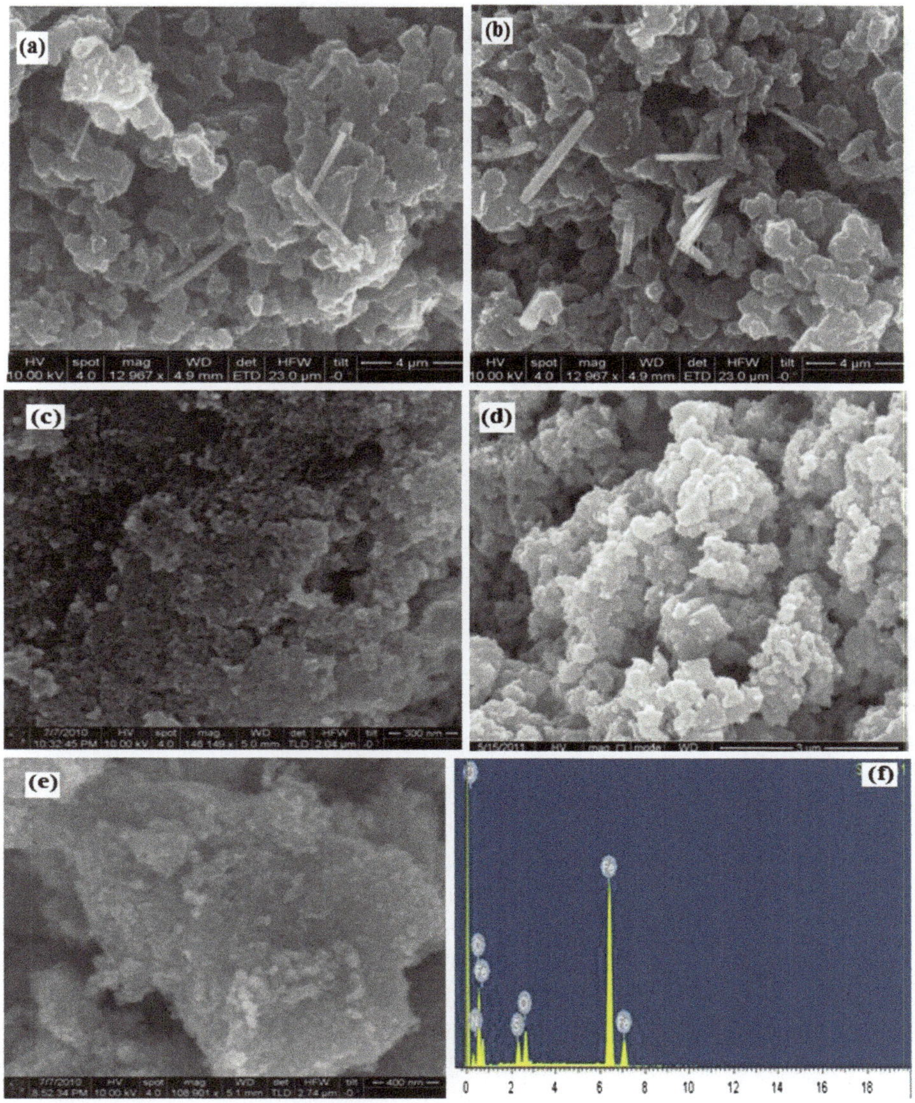

Figure 5 Images of the PTSA-doped TANI/Fe_3O_4 nanocomposites prepared with different molar ratios of PTSA to NPPD. The nanocomposites in (**a** and **b**) 4:1, (**c**) 3:1, (**d**) 2:1, (**e**) 1:1 molar ratios. (**f**) EDX of the PTSA-doped TPANI/Fe_3O_4.

superparamagnetic state. The decrease of the *M*s values may also be due to the increase of the weight ratio of the non-magnetic PTSA and TANI in the nanocomposites.

Figure 8 shows the TGA curve of the pure TANI and PTSA-doped TANI/Fe_3O_4 nanocomposites prepared with various molar feed ratios of PTSA to NPPD under N_2 atmosphere at a heating rate 20°C/minute in the temperature range of 50°C to 840°C. A close comparison of the thermal transitions between them revealed that the TANI/Fe_3O_4 nanocomposites are more thermally stable than pure TANI. Three characteristic weight loss regions were observed for both pure TANI and all PTSA-doped TANI/Fe_3O_4 nanocomposites. The first weight loss before 130°C corresponds to the removal of water, excess of unbound PTSA and other volatile impurities in the composites. The second stage of weight loss below 350°C is associated to the loss of bound PTSA and the decomposition of TANI initiated by the dopant acid. The final continuous weight loss occurring between 350°C and 650°C is attributed to the breakdown of the main chain of TANI. The TGA curves of the composites revealed higher final residue and better thermal stability due to the higher packing of TANI and the retarding effect of Fe_3O_4 NPs, respectively.

Conclusions

PTSA-doped TANI/Fe_3O_4 nanocomposites were prepared by *in situ* oxidative chemical polymerization technique using APS, an oxidant, via micelle-assisted self-assembly method. The SEM and TEM images clearly show that the

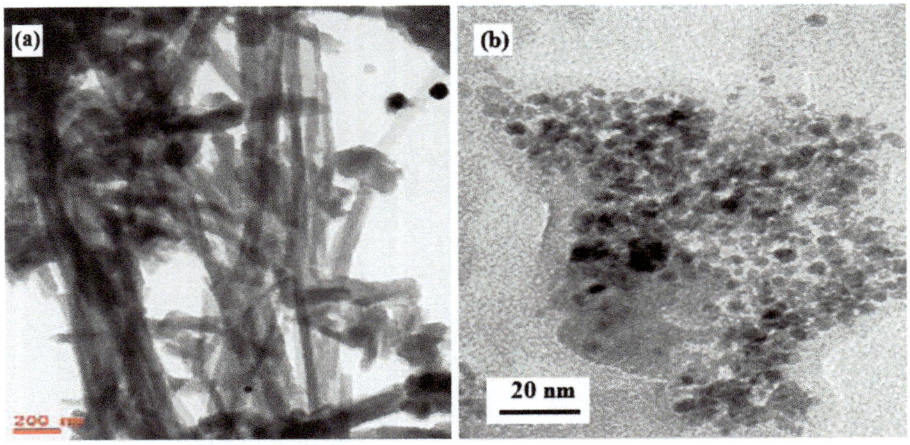

Figure 6 TEM images of the PTSA-doped TANI/Fe_3O_4 nanocomposites. **(a)** Crude and **(b)** dispersed nanocomposites.

Fe_3O_4 NPs were coated on the surface of the TANI nanostructures. The morphology, crystallinity, magnetic properties and thermal stability of the nanocomposites were found to be dependent on the concentration and molar ratio of PTSA to NPPD. Magnetic studies show superparamagnetic behaviour, with saturation magnetization values much lower than those for bulk magnetite. The TANI/Fe_3O_4 nanocomposites showed higher thermal stability as compared to pure TANI, indicating the interaction between Fe_3O_4 NPs and TANI in the nanocomposites.

Methods

Materials

N-phenyl-1,4-phenylediamine (NPPD) and PTSA were obtained from Aldrich Chemical Company (Milwaukee, USA) and were used as received without further purification. APS ($(NH_4)_2S_2O_8$), $FeCl_3{\cdot}6H_2O$, $FeSO_4{\cdot}7H_2O$, ammonia (NH_3-H_2O), sodium hydroxide and methanol were purchased from Merck & Co., Inc. (Bengaluru, India) and were used as received without any purification. All other reagents were of analytical grade and used without further purification. Milli-Q (EMD Millipore Corporation, Billerica, MA, USA) water with resistance greater than 18 MΩ was used for the preparation of all solutions.

Synthesis of TANI nanorods/ Fe_3O_4 NPs composites

PTSA-doped TANI/Fe_3O_4 nanocomposites were prepared via *in situ* oxidative chemical polymerization method using APS as an oxidant. In a typical synthesis process, 0.1840 g of NPPD was dissolved in 50 mL of Milli-Q water and then 1.9105 g of PTSA was added in under vigorous stirring. The reaction mixture was quickly cooled at 0°C to 5°C using ice-water bath with constant stirring for 2 h. Then, a pre-cooled solution of 1 mL aqueous APS (1 mol L^{-1}) was added dropwise to the above under vigorous stirring. The colour of the

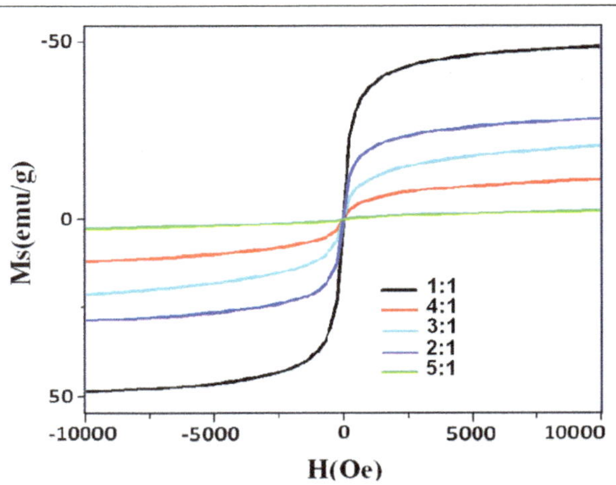

Figure 7 Variation of magnetization as a function of field for PTSA-doped TANI/Fe_3O_4 nanocomposite samples.

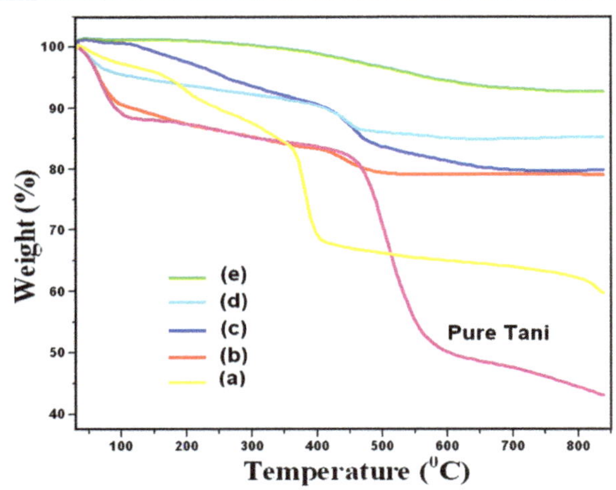

Figure 8 TGA curve of the pure TANI and PTSA-doped TANI/Fe_3O_4 nanocomposites.

reaction mixture slowly turns from colourless to light blue and finally to dark green. The reaction is allowed to proceed for 5 h at 0°C to 5°C before allowing it to attain room temperature. To the above reaction mixture, 10 mL of 1 mol L^{-1} $FeCl_3{\cdot}6H_2O$ and 10 mL of 0.5 mol L^{-1} $FeSO_4{\cdot}7H_2O$ aqueous solutions were added at 75°C under stirring and then 5 mL of 25% NH_4OH solution was added. The colour of the reaction mixture turns black, indicating the formation of magnetite. The reaction was allowed to proceed for 5 h under vigorous stirring until TANI/Fe_3O_4 dispersion was obtained. The as-synthesized nanocomposites were centrifugated and washed with a large amount of distilled water and then with methanol. Finally, the product was dried under vacuum at 50°C for 5 h. The same procedure has been adopted for the preparation of all PTSA-doped TANI/Fe_3O_4 composites while varying the molar ratio of PTSA to NPPD from 5:1 to 1:1.

Characterization

Powder XRD patterns were recorded using a Siemens AXS D5005 X-ray diffractometer (Mumbai, India) using CuKα radiation (λ = 1.54 Å) at 5°/min. Fourier transform infrared (FTIR) spectra were recorded on a PerkinElmer SPECTRUM 1000 FTIR Spectrometer (Thane, India) over the range of 400 to 4,000 cm^{-1}. The powder samples were mixed thoroughly with KBr and pressed into transparent pellets. The morphology is investigated with a JEOL JSM6610 LV scanning electron microscope (SEM, Tokyo, Japan) equipped with an electron probe microanalyzer and a FEI TECNAI G2 S-Twin transmission electron microscope (Hillsboro, OR, USA) at accelerating voltages of 120 and 200 kV. The samples for SEM were mounted on aluminium studs and gold coated prior to imaging. EDAX was also performed with the same equipment to determine elemental analysis. For the TEM measurements, the powders were ultrasonicated in DMSO, and a drop of the suspension was dried on a carbon-coated microgrid. For UV-visible absorption spectra, the samples were dissolved in DMSO, and the spectra were recorded on a PerkinElmer double beam LS-50 spectrophotometer. The room temperature magnetization in the applied magnetic field was performed using vibrating sample magnetometer. Thermogravimetric analysis (TGA) was conducted with a Cahn TG131 TGA system (Cerritos, CA, USA) at a heating rate of 20°C/min under nitrogen atmosphere from room temperature to 850°C.

Additional file

Additional file 1 Figure S1 and S2. TEM images and SAED patterns of the PTSA-doped TANI/Fe_3O_4 nanocomposites.

Competing interests

The authors declare that they have no competing interests.

Authors' contributions

KB and YP carried out the synthesis and characterization of the nanocomposites. KB and AVPR prepared the manuscript. YP carried out the magnetic and thermal studies. All authors read and approved the final manuscript.

Authors' information

AVPR is a professor at the Department of Inorganic & Analytical Chemistry, Andhra University, India. He received his PhD degree in Solid State Chemistry from Pennsylvania State University, USA. His current research has focused on solid state synthesis of various nanostructural materials. KB is an assistant professor at the Department of Inorganic & Analytical Chemistry, Andhra University, India. He obtained his MS degree from the University of Hyderabad in 2004. He received his PhD degree in organic-inorganic nanohybrid materials for catalysis and antimicrobial application. Currently, he is working on nanocomposites of one-dimensional conducting polymer and metal, metal oxide nanoparticle for various technological applications. YP received his MS degree in Analytical Chemistry from Andhra University in 2010. Currently, he is working on semiconductor nanostructure materials for environmental applications.

Acknowledgements

This work was supported by the University Grant Commission, India (UGC, India) under major research project. YPK thanks UGC, India for the award of research fellowship. The authors are grateful to Prof. TP Radhakrishnan and Muvva D. Prasad, Centre for Nanotechnology, University of Hyderabad, for providing the TEM facility.

References

1. Lee, K, Cho, S, Park, SH, Heeger, AJ, Lee, CW, Lee, SH: Metallic transport in polyaniline. Nature **441**, 65–68 (2006)
2. Sfez, R, De-Zhong, L, Turyan, I, Mandler, D, Yitzchaik, S: Polyaniline Monolayer Self-Assembled on Hydroxyl-Terminated Surfaces. Langmuir **17**, 2556 (2001)
3. Choi, J, Kim, SJ, Lee, J, Lim, JH, Lee, SC, Kim, KJ: Controlled self-assembly of Nanoporous alumina for the self-templating synthesis of polyaniline nanowires. Electrochem Commun **9**, 971–975 (2007)
4. Li, GR, Feng, ZP, Zhong, JH, Wang, ZL, Tong, YX: Electrochemical Synthesis of Polyaniline Nanobelts with Predominant Electrochemical Performances. Macromolecules **43**, 2178–2183 (2010)
5. Chen, J, Chao, D, Lu, X, Zhang, W: Novel interfacial polymerization for radially Oriented polyaniline nanofibers. Mater Lett **61**, 1419–1423 (2007)
6. Nastase, F, Stamatin, I, Nastase, C, Mihaiescu, D, Moldovan, A: Synthesis and characterization of PAni-SiO_2 and PTh-SiO_2 nanocomposites thin films by plasma polymerization. Prog Solid State Chem **34**, 191199 (2006)
7. Cruz-Silva, R, Ruiz-Flores, C, Arizmendi, L, Romero-Garci, J, Arias-Marin, E, Moggio, I, Castillon, FF, Farias, MH: Enzymatic synthesis of colloidal polyaniline particles. Polymer **47**, 1563–1568 (2006)
8. Hou, H, Jin, Z, Reuning, A, Schaper, A, Wendorff, JH, Greiner, A: Poly(p-xylylene) Nanotubes by Coating and Removal of Ultrathin Polymer Template Fibers. Macromolecules **119**, 27 (2002)
9. Huang, JX, Kaner, RB: Nanofiber Formation in the Chemical Polymerization of Aniline: A Mechanistic Study. Angew Chem Int Ed **43**, 5817–5821 (2004)
10. Zhang, X, Manohar, SK: Polyaniline nanofibers: chemical synthesis using surfactants. Chem Commun **20**, 2360–2361 (2004)
11. Chiou, NR, Lee, LJ, Epstein, AJ: Porous membrane controlled polymerization of nanofibers of polyaniline and its derivatives. J Mater Chem **18**, 2085–2089 (2008)
12. Gangopadhyay, R, De, A: Conducting Polymer Nanocomposites: A Brief Overview. Mater **12**, 608–622 (2000)
13. Meldrum, FC, Heywood, BR, Mann, S: Magnetoferritin: in vitro synthesis of a novel magnetic protein. Science **257**, 522–523 (1992)
14. Martin, CR, Mitchell, DT: Nanomaterials in Analytical Chemistry. Anal Chem **70**, 322A–327A (1998)
15. Murthy, N, Thng, YX, Schuck, S, Xu, MC, Frechet, JMJ: A Novel Strategy for Encapsulation And Release of Proteins: Hydrogels and Microgels with Acid-Labile Acetal Cross-Linkers. J Am Chem Soc **124**, 12398–12399 (2002)
16. Li, XG, Li, A, Huang, MR: Facile High-Yield Synthesis of Polyaniline Nanosticks with Intrinsic Stability and Electrical Conductivity. Chem Eur J **14**, 10309–10317 (2008)

17. Abdiryim, T, Xiao-Gang, Z, Jamal, R: Comparative studies of solid-state synthesized Polyaniline doped with inorganic acids. Mater Chem Phys **90**, 367–372 (2005)
18. Waldron, RD: Infrared spectra of ferrites. Phys Rev **99**, 1727–1735 (1955)
19. Popple Well, J, Sakhnini, L: The dependence of physical and magnetic properties of Magnetic fluids on particle size. J Magn Magn Mater **142**, 72–78 (1995)

14

Comparison of drug delivery systems: Nanotube and *p*- Sulphonatocalix[4]arene, by Density Functional Theory

Karim Zare[1,2] and Nasim Shadmani[3*]

Abstract

Nowadays, the use of calixarenes and nanotubes are widely spread in the field of drug delivery. In this work, interaction between *p*-sulphonatocalix[4]arene and nanotube(6,6)with Fluorouracil drug is investigated. Density function theory (DFT) calculations have been performed using the Gauss view and Gaussian03 by B3LYP (beck three-parameter hybrid functional combined with Lee-Yang-Parr correlation functional) method and 6-31G (*d*) standard basis set. The length bond (Å), bond angel (degree), dihedral angel (in degree), hyperconjucation energy, and total energy (KJ mol^{-1}), formation energy (KJ mol^{-1}), moment dipole (in Debye) and occupancy of between nanotube (6,6) and *p*-sulphonatocalix[4]arene with anti-cancer drug are calculated by B3LYP/6-31G (d) method. The results show that composite 2 is more stable than the other composite. The hyperconjucation energy (E^2) in the *p*-sulphonatocalix[4]arene-fluorouracil is higher than that in the other composite. The standard formation energy (ΔE_f°) in the the *p*-sulphonatocalix[4]arene-fluorouracil is more negative value (lower) to composite 1.

Keyword: Single-walled carbon nanotube; DFT; 5-Fluoropyramidin-2; 4(1H, 3H)-dion; *p*-Sulphonatocalix[4]arene; Drug design; NBO; C_v; Energy stability

Background

The extensive research on carbon nanotube done to their structural characterization and discovered by Iijima in 1991 year [1,2]. Among the numerous delivery systems currently under investigations, single-walled carbon nanotubes seem to embody a promising option [3]. Single-walled carbon nanotubes are made up of carbon atoms arranged in a series of condensed benzene rings and wrapped into a tubular form (Figure 1b). Nanotubes are two types of: single- walled carbon nanotube (SWCNT) and multi-walled carbon nanotube (MWCNT), respelsitvly. In particular, the application of single-walled carbon nanotube , calixarene, fulouren and cytosine as new carrier for drug delivery, properties became doable soon after the demonstration of cellular uptake of this new material [4-10].

In the paper, the composite of nanotube (6,6)-fluorouracil and *p*-Sulphonato-calix[4]arene)-fluorouracil are investigated as a drug delivery system. Calixarenes are a group of supramolecule that have basket-like shape which are easily available through the cyclocondensation of para-substituted phenols with formaldehyde [11-15]. The *p*-Sulphonatocalx[4]arene have received considerable attention in recent years because of their selective metal ion binding properties in aqueous solution, the formation of basket- like be layer structures in the solid state, and the observation of H-bonding with aromatic rim. Calixarenes invested in the context of electrochemical selective, sensor, stationary phases, solid phase extraction phases and drug delivery in five decade ago [16-20]. One way to increase the aqueous solubility of drugs is to use complexing agents to form hast- guest complex [21]. *P*-sulphonatocalix[4]arene are react with types of drugs , metals and other gust organic and inorganic from upper rim that inclusion SO_3H group and lower rim that inclusion OH group is not reactivity. They have shown interesting properties of activity against bacteria, fungi, cancerous cells and enveloped viruses, but also against thrombosis or fibrosis diseases [22-25]. Anti-cancer drugs act in a dominant fashion:

* Correspondence: n.shadmani@gmail.com
[3]Young Researchers and Elites Club, Rasht Branch, Islamic Azad University, Rasht, Iran
Full list of author information is available at the end of the article

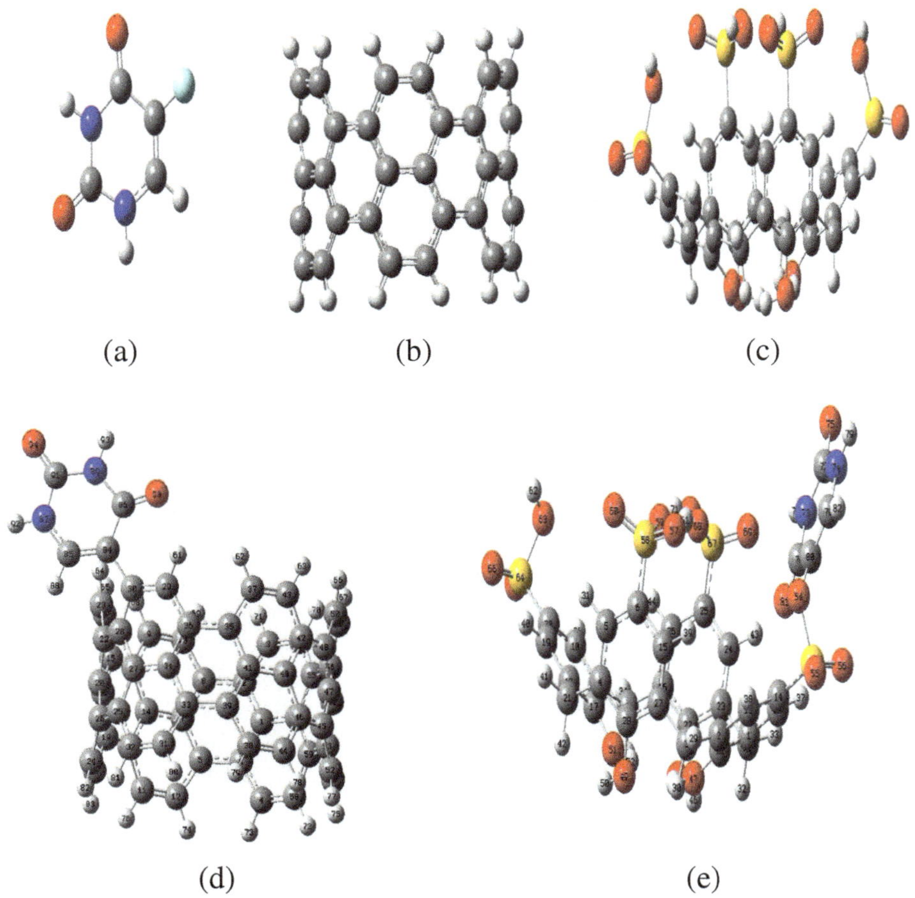

Figure 1 The structures of optimized molecular by B3LYP/6-31G (*d*) method at 298.15 K. (a) Fluorouracil, **(b)** Nanotube(6,6), **(c)** p-Sulphonatocalix[4]arene, **(d)** Composite1, **(e)** Composite 2.

when ectopically over expressed they specifically destroy tumor cells without harming normal cells. Anticancer genes have only recently emerged from studies on cancer cells [26,27]. Fluorouracil or 5-Fluoropyramidin-2, 4 (1H, 3H)-dion is used as anti- cancer drug. Fluorouracil derived of pyrimidine that used as an anti-cancer drug for years ago. The structure of Fluorouracil is observed in Figure 1a. It is anti-metabolite drug and acts in several ways, but principally as synthesis inhibitor [28]. These days there are ways to deliver drug system in the body without side effects. In this paper, we reported types of drug delivery system like single-walled nanotube, p-Sulphonatocalix[4]arene.

Results and discussion

Density Functional Theory (DFT), hybrid QM/MM and Car-Parrinello molecular dynamics simulations can be used in theory investigation. The using computer for drug design and analized interaction between drug and receptors (for example: nanotube, fullerenes and calixarene) and give optimal fit [29-32]. The more detailed information on structure, charge distribution, and energetic of the base pair are study by DFT method. At present, quantum chemical is extensive applicable to the interpretation of physical and chemical properties of drugs. The cellular receptors of many drugs used for medical treatment are react with proteins. By binding to the receptor, drugs either enhance or inhibit its activity. The biological mechanism of a disease then often suggests the types of molecules needed for nanodrugs [33-35]. Figure 1a, b, c, d, e shows the optimized composites fluorouracil, nanotube(6,6), p-Sulphonatocalix[4]arene, and nanotube(6,6)-fluorouracil (composite1), p-Sulphonatocalix[4]arenes-fluorouracil (composite2), respectively, by DFT method at B3LYP/6-31G(*d*) level.

In Table 1, it is obvious that composite 2 has formed higher hyperconjugation energy than the composite 1. Also, the results show by increasing *P* share in hybrid of atoms, the occupancy decreases. The *S* orbital share in hybrid of carbon in composite 2 is more than that in composite 1. If combined with the most stable hyperconjugation energy, the occupancy coefficient becomes smaller. Composite 2 is more stable than the composite 1. The hyperconjugation energy composite 2

Table 1 Parameters of NBO analysis of composite 1and 2 by B3LYP/6-31G (d) method at 298.15 K

Agent	Donor	Occupancy	Acceptor	Occupancy	Hybrid	E^2	ΣE^2
Composite 1	BD(1)C30- C84	1.96631	BD*(1)C27-C28	0.02586	$SP^{1.98}$	1.48	
			BD*(1)C28-C30	0.02920	$Sp^{2.06}$	1.70	
			BD*(1)C29-C30	0.02033	$Sp^{1.79}$	3.12	
			BD*(1)C29-C36	0.02164	$SP^{1.96}$	2.51	
			BD*(1)C84-C85	0.01852	$SP^{1.88}$	3.55	35.5
			BD*(1)C84-C86	0.06682	$SP^{2.28}$	1.82	
	BD*(2)C84-C85	0.23733	BD*(1)C85-N87	0.01939	$SP^{2.51}$	3.76	
			BD*(2)C29-C30	0.2195	$SP^{99.99}$	17.59	3
Composite 2	BD(1)O54- C80	1.98809	BD*(1)N73-C76	0.07349	$SP^{1.77}$	1.32	
			BD*(1)N74-C78	0.01797	$SP^{1.69}$	2.00	
			BD*(1)C78-C80	0.02461	$SP^{1.58}$	1.19	
	LP(1)O54	1.95049	BD*(1)C78- C80	0.024241	$SP^{1.58}$	2.71	
			BD*(2)C78- C80	0.024241	$SP^{99.99}$	3.67	46.3
	LP(2)O54	1.90929	BD*(1)C76-C80	0.07057	$SP^{1.72}$	8.76	
			BD*(1)C78-C80	0.02461	$SP^{1.58}$	5.14	
	BD*(2)C78-C80	0.38376	BD*(1)S53-O54	0.24241	$SP^{4.81}$	21.57	9

(46.39) is larger than that of composite1. If the S orbital hybrid of a compound is lower, the occupancy becomes larger. In composite 2, hybrid and occupancy are $SP^{99.99}$and 0.024241(acceptor) and 1.90929(donor), respectively. For example, in composite 2, hybrid and occupancy are $SP^{4.81}$ and 0.24241(acceptor) and 0.38278 (donor), respectively. In Table 1, the Mulliken charges of electronegative O_{84} donor atoms and C_{30} acceptor atom are negative and positive, respectively. The energy gap of composite 2 is larger than that of composite1; therefore, composite 2 is more stable.

We measured the parameters, such as bond length (Å), natural bond orbital (NBO) and bond angle (degree), dihedral angle (degree), distances of analysed models, of the single-walled nanotube in this paper, and the end of the nanotube that was saturated by hydrogen atoms in them were examined by DFT at the level of B3LYP and 6-31G (*d*) standard basis set and shown in Table 2. The DFT-calculated geometric parameters for composite 1 and 2 are compared in Table 2. The bond lengths C_{30}-C_{84}/O_{84} calculated for composite 1and 2 by the DFT method are 1.47 and 1.37 Å, respectively, at the B3LYP/6-31G (*d*) level. The bond distance of composite 2 decreased toward that of composite 1 because the O atom in composite 2 is more electronegative than the C atom.

The bond lengths calculated for $C_{36}\overset{\cdots}{-}C_{29}$ in composite 1 (in the nanotube ring) is 1.44 Å and that in composite 2(in the calixarene ring) is 1.39 Å. The bond lengths calculated for $C_{85}\overset{\cdots}{-}C_{87}$ in composite 1, 1.37 Å and that in composite 2 is 1.36 Å. The $C_{88}\overset{\cdots}{-}N_{92}$ bond length is lower than the $C_{36}\overset{\cdots}{-}C_{29}$ bond length in composite1and 2. That is because nitrogen is more electronegative nitrogen than carbon.

Table 2 The parameters of bond and angles propertie in B3LYP/6-31G(*d*) method for composites at 298.15 K

Agent	Composite 1	Composite 2
Bond length(Å)		
$C_{30}-C_{84}/O_{84}$	1.47	1.37
$C_{30}=C_{20}$	1.37	1.39
$C_{36}\overset{\cdots}{-}C_{29}$	1.43	1.39
$C_{28}-C_{30}$	1.47	1.51
$C_{85}\overset{\cdots}{-}C_{87}$	1.37	1.36
$N_{87}-C_{91}$	1.39	1.38
$C_{91}=O_{94}$	1.22	1.21
Bond angles (degree)		
C/S_{30}-$C_{28}\overset{\cdots}{-}C_{27}$	116.02	118.56
$C_{84}=C_{85}-H_{88}$	121.36	121.65
$C_{85}\overset{\cdots}{-}N_{87}-H_{92}$	120.82	120.99
$N_{87}-C_{91}=O_{94}$	123.42	124.32
$C_{91}-N_{89}-H_{93}$	115.60	114.99
$O_{90}=C_{86}-C_{84}$	126.90	123.98
Dihedral angles (degree)		
C_{30}-$C_{84}=C_{85}\overset{\cdots}{-}N_{87}$ [1]	178.86	−0.063
H_{87}-C_{84}-$N_{86}\overset{\cdots}{-}H_{91}$ [2]	0.79	−0.37
$C_{90}-N_{88}-C_{85}=O_{89}$ [3]	−178.64	−178.95
$H_{61}-C_{29}=C_{30}-C_{28}$ [4]	172.74	−179.95
$O_{94}=C_{91}-N_{89}-C_{86}$ [5]	179.16	−178.95
C/H_{30}-$C_{28}\overset{\cdots}{-}C_{27}\overset{\cdots}{-}C_{34}$ [6]	−26.01	−178.72

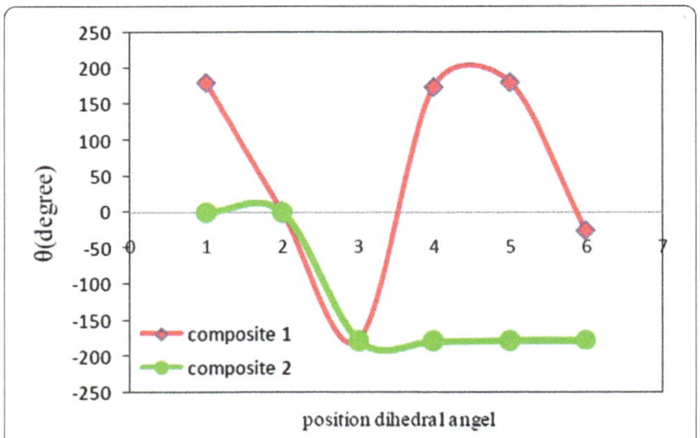

Figure 2 Pole of dihedral angel changes of composites 1 and 2 in b3lyp/6-31G (d) method at 298.15 K. Position 1: C_{30}-C_{84} = C_{85} $\overset{\cdots}{-}N_{87}$, position 2: H_{87}-C_{84}-$N_{86}\overset{\cdots}{-}H_{91}$, position 3: C_{90} - N_{88} - C_{85} = O_{89}, position 4: H_{61} - C_{29} = C_{30} - C_{28}, position 5: O_{94} = C_{91} - N_{89} - C_{86}, position 6: C/H_{30}-$C_{28}\overset{\cdots}{-}C_{27}\overset{\cdots}{-}C_{34}$.

The bond angle of N_{87}-C_{91} = O_{94} are 123.41° and 124.32° for composites 1 and 2, respectively. The bond angles of O_{90} = C_{86}-C_{84} are 126.9° and 123.9° for composite 1and 2, respectively. The bond angles C_{84}-C_{86} = O_{90} are higher than those of N_{87}-C_{91} = O_{94} in composite 1 and 2. Thus, the interaction of between nonbonding and bonding pairs on the nitrogen atom of the angle is reduced in N_{87}-C_{91} = O_{94}.

The dihedral angels of C_{30}-C_{84} = $C_{85}\overset{\cdots}{-}N_{87}$ and C/H_{30}-$C_{28}\overset{\cdots}{-}C_{27}\overset{\cdots}{-}C_{34}$ for composite 1and 2 are 178.48° and −0.06.99° (position a), -26.01° and −178.72° (position f), respectively. The different dihedral angel in position a, b, c, d, e and f, are observed in Table 2 and Figure 2.

The electron that is given by the composites in a reaction should be placed on the highest occupied molecular orbital (HOMO), with those which are captured by the composites must be placed on the lowest unoccupied molecular orbital (LUMO) [36], so the atom on which the HOMO is mainly scattered should be able to separated electrons, while the atom by holding the LUMO should achieve electrons on this basis. The energy gap is usually associated with chemical stability against electronic excitation with a larger value like greater stability [37]. Table 3 shows the HOMO and LUMO energy gaps for two composites. By evaluating HOMO/LUMO energy gaps, it is obvious that if that the gap increase, the composite will become stable; therefore, composite 2 is more stable than composite. But, in composite 2 have S and O atoms that including non bonding pair interaction with C atom fluorouracil, thus gap energies lower than composite 1.

The results of the present work were obtained by DFT optimization and formation energy (ΔE°_{f} in $KJmol^{-1}$) calculation at the B3LYP/6-31G (*d*) level. ΔE°_{f} is calculated using formula $\sum E^{\circ}_{product} - \sum E^{\circ}_{reactant}$. The value of ΔE°_{f} in composite 1and 2 are −13.275 and −38.892 $KJmol^{-1}$, respectively; therefore, composite 2 has lower formation energy than composite 1. The energy gap, total energy, $\sum E^{2}$, HOMO and LUMO of composites 1 and 2 were also calculated with B3LYP method and 6-31G (*d*) basis set. The energy (kJ mol^{-1}) and dipole moments (Debye) indicate the consistency between the two composite calculations in the DFT method. The total energy, the sum of is transitional energy, rotational energy and vibration energy, in B3LYP/6-31G (d) level for composite 1and 2 was calculated. The obtained results are shown in Table 3.

Heat capacity is a measurable physical quantity that specifies the amount of heat required to change the temperature of an object by a given amount. Translation energy, rotation energy, and a combination of the two types of energy in vibration (kinetic and potential) of atoms represent the degrees of freedom of motion which classically contribute to the heat capacity of matter, but loosely bound electrons may also participate. For quantum mechanical reasons, at any given temperature, some of these degrees of freedom may be unavailable, or only partially available, to store thermal energy. Quantum theory can be used to quantitatively predict the specific heat capacity of simple systems.

Conclusions

In this paper, the results show that the complex of *p*-Sulphonatocalix[4]arene and fluorouracil is more stable than nanotube-fluorouracil. Composite 2 is a better conditioner of drugs than composite 1. NBO analysis shows larger energy gap in composite2. Composite2 has lower formation energy, and it is more stable than composite1.

Methods

The investigation is carried out using a personal computer which has Intel ®Pentium ®dual CPU with 2-GB RAM. Nanotube-fluorouracil (composite 1) and *p*-Sulphonatocalix

Table 3 Parameters of stability composites are calculated by B3LYP/ 6-31G (d) method at 298.15 K

Agent	E_{total} ($KJmol^{-1}$)	ΔE°_{f} ($KJmol^{-1}$)	HOMO ($KJmol^{-1}$)	LUMO ($KJmol^{-1}$)	Gap ($KJmol^{-1}$)	μ (Debye)	C_v
Composite 1	465.73	−13.275	−0.15877	−0.08918	−0.0696	0.527	191.03
Composite 2	396.25	−38.897	−0.2601	−0.07261	−0.18749	−1.137	186.77

[4]arene-fluorouracil (composite 2) included a calixarene derivation (with different atom number) which reacts with the anti-cancer drug. In this paper, drug delivery properties are investigated by NBO analysis and DFT method. DFT calculations have been performed using the nanotube modeler [38] Gaussview [39] and Gaussian 03 [40] by B3LYP method and 6-31G (*d*) standard basis set. NBO analysis [41] calculations have also been performed for all composites by B3LYP method and 6-31G(*d*) standard basis set. Then complexes of the calix[n]arenes derivative and single-walled nanotube with fluorouracil are optimized. The bond length (Å), bond angel (degree), dihedral angel (degree), hyperconjucation energy(E^2), and total energy (E_t in KJ mol^{-1}), moment dipole (μ in Debye), occupancy, and HOMO/LUMO are investigated in the composites by B3LYP/6-31G (*d*) method.

The nanotube ($C_{60}H_{24}$) and *p*-sulphonatocalix[4] arenes ($C_{28}H_{24}O_{16}S_4$) and fluorouracil ($C_4H_3FN_2O_2$) are reacted by Equations 1and 2, thus is formed products nanotube- fluorouracil ($C_{64}H_{26}N_2O_2$), *p*-sulphonatocalix [4]arenes- fluorouracil ($C_{32}H_{26}N_2O_{18}S_4$) and HF in phases gas.

$$C_4H_3FN_2O_2 + C_{60}H_{24} \rightarrow C_{64}H_{26}N_2O_2 + HF\ (\text{bond C-C}) \tag{1}$$

$$C_4H_3FN_2O_2 + C_{28}H_{24}O_{16}S_4 \rightarrow C_{32}H_{26}N_2O_{18}S_4 + HF\ (\text{bond C-O}) \tag{2}$$

Competing interests

The authors declare that they have no competing interests.

Authors' contributions

KZ and NSH both contributed to this research work. Both authors read and approved the final manuscript.

Acknowledgements

Both of authors are grateful to University of Science and Research Branch, Islamic Azad University, Tehran, Iran. We are grateful to Shahid Beheshti University, Tehran, Iran. We are grateful to Young Researchers and Elites Club, Rasht Branch, Islamic Azad University, Rasht, Iran for financial support.

Author details

[1]Department of Chemistry, Science and Research Branch, Islamic Azad University, Tehran, Iran. [2]Department of Chemistry, Shahid Beheshti University, Evin, Tehran, Iran. [3]Young Researchers and Elites Club, Rasht Branch, Islamic Azad University, Rasht, Iran.

References

1. Iijima, S: Helical microtubules of graphitic carbon. Nature **354**, 56–58 (1991)
2. Colomer, JF, Henrard, L, Lambin, P, Van Tendeloo1, G: Electron diffraction study of small bundles of single-wall carbon nanotubes with unique helicity. Phys Rev B **64**, 1–7 (2001)
3. S Berber, S, Kwon, YK, Tom´anek, D: Unusually high thermal conductivity of carbon nanotubes. Phys. Rev. Lett **84**, 4613–4616 (2000)
4. Singh, R, Pantarotto, D, Lacerda, L, Pastorin, G, Klumpp, C, Prato, M, Bianco, A, Kostarelos, K: Tissue biodistribution and blood clearance rates of intravenously administered carbon nanotube radiotracers. Proc Natl Acad Sci USA **103**, 3357–3362 (2006)
5. White, CT, Tocorov, TN: Carbon nanotubes as long ballistic conductors. Nature **393**, 240–241 (1998)
6. Taft, BJ, Lazareck, AD, Withey, GD, Yin, A, Xu, JM, Kelley, SO: Site-specific assembly of DNA and appended cargo on arrayed carbon nanotubes. J. Am. Chem. Soc. **126**, 12750–12751 (2004)
7. Wong Shi Kam, N, Jessop, TC, Wender, PA, Dai, H: Nanotube molecular transporters: internalization of carbon nanotube-protein conjugates into mammalian cells. J. Am. Chem. Soc. **126**, 6850–6851 (2004)
8. Pantarotto, D, Singh, R, McCarthy, D, Erhardt, M, Briand, JP, Prato, M, Kostarelos, K, Bianco, A: Functionalized carbon nanotubes for plasmid DNA gene delivery. Angew. Chem., Int. Ed. **43**, 5236–5242 (2004)
9. Petersen, JL, Egan, DM: Small security: nanotechnology and future defense. Defense Horizons **8**, 1–6 (2002)
10. Zhdanov, VP, Kasemo, B: Simulations of the reaction kinetics on nanometer supported catalyst particles. Surf. Sci. Rep. **39**, 25–104 (2000)
11. Gutsche, CD, Bauer, LJ: The conformational properties of calix[4]arenes, calix[6]arenes, calix[8]arenes, and oxacalixarenes. J. Am. Chem. Soc. **107**, 6052–6059 (1985)
12. Gungor, O, Memon, S, Yilmaz, M: Synthesis of alkyl nitrile and alkyl benzonitrile derivatives of calix[4]arene and their polymer supported analogues: a comparative study in two-phase extraction systems. J. Funct. Polym **63**, 1–9 (2005)
13. Asfari, Z, Böhmer, V, Harrowfield, J, Vicens, J: Calixarenes. ED; Kluwer, Dordrecht (2001)
14. Faraji, M, Zare, K, Aghaei, H, Farajtabar, A, Asfari, Z, Gharib, F: Complexation of p-Sulphonato-calix[6]arene by Glycine, Glycyl-glycine, and Glycyl-glycyl -glycine in Aqueous Solution. J. Solution Chem. **41**, 2074–2081 (2012)
15. Gharib, F, Zare, K, Taghvaei, S, Monajjemi, M, Amiri, A: Complexation of tri-o -propoyl-p-t-butyl calix[4]arene with alkali metal cations in carbon tetrachloride solvent. Main Group Met. Chem. **26**, 255–258 (2003)
16. Gharib, F, Zare, K, Amiri, A: Spectrophotometric studies of alkali metal ion complexes of p-tetrabutyl- calix[4]arenes. Russ. J. Inorg. Chem. **49**, 1511–1514 (2004)
17. Gharib, F, Osouleddini, N, Zare, K, Taghvaei-Ganjali, S: Spectrophotometric studies of alkali and alkali earth metal ions complexes of mono amino derivative of calix[4]arene. Russ. J. Inorg. Chem. **55**, 434–437 (2010)
18. Brunink, JAJ, Verboom, W, Engbersen, JFJ, Reinhoudt, DN, Harkema, S: Synthesis and cation complexation selectivity of bis(syn-proximally) functionalized calix[4]arenes. Recl. Trav. Chim. Pays-Bas **111**, 511–516 (1992)
19. Shirakawa, S, Shimizu, S: Dehydrative amination of alcohols in water using a water-soluble calix[4]resorcinarene sulfonic acid. Syn. lett **10**, 1539–1542 (2008)
20. Liu, YL, Liu, L, Wang, YL, Han, YC, Wang, D, Chen, YJ: Calix[*n*]arene sulfonic acids bearing pendant aliphatic chains as recyclable surfactant-type Brønsted acid catalysts for allylic alkylation with allyl alcohols in water. Green Chem. **10**, 635–640 (2008)
21. Zhang, YJ, Cao, WX, Xu: Interaction of sulfonated calix[*n*]arenes with Rhodamine B and its application to determine acetylcholine in a real neutral aqueous medium J. Chin. Chem **20**, 322–326 (2002)
22. Guo, D, Liu, Y: Calixarene-based supramolecular polymerization in solution. Chem. Soc. Rev. **41**, 5907–5921 (2012)
23. Dalili, N, Zare, K, Gharib, F, Shadmani, N: Calixarenes or nanotubes: which are better in drug delivery? JNSC **2**, 213–217 (2011)
24. Dings, RP, Miller, MC, Nesmelova, I, Astorgues-Xerri, L, Kumar, N, Serova, M, Chen, X, Raymond, E, Hoye, TR, Mayo, KH: Antitumor agent calixarene 0118 targets human galectin-1 as an allosteric inhibitor of carbohydrate binding. J. Med. Chem. **14**, 5121–5129 (2012)
25. Fátima, AD, Fernandes, SA, Sabino, AA: Calixarenes as new platforms for drug design. Curr. Drug Discovery Technol. **6**, 1–20 (2009)
26. Moore, MJ, Erichman, C, Kaizer, L, Fine, S: A phase II study of 5-fluorouracil, leucovorin and interferon alpha in advanced pancreatic cancer. Anticancer Drugs **4**, 555–557 (1993)
27. Cornell, GN, Cahow, CE, Frey, C, McSherry, C, Beal, JM: Clinical experience with 5-fluorouracil (NSC-19 893) in the treatment of malignant disease. Cancer Chemother Rep **9**, 23–30 (1960)
28. Jemal, A, Murray, T, Samuels, A, Ghafoor, A, Ward, E, Thun, MJ: Cancer statistics. CA A Cancer Journal for Clinicians **53**, 5–26 (2003). 2003
29. Fernandez, L, Viruela-Martin, P, Latorre, J, Guillem, C, Beltrán, A, Amorós, P: Molecular precursors of mesostructured silica materials in the atrane route: a DFT/GIAO/NBO theoretical study. J. Mol. Stru: THEOCHEM **822**, 89–102 (2007)
30. Monajjemi, M, Chahkandi., B, Zare, K, Amiri, A: Study of the hydrogen bond in different orientations of adenine-thymine base pairs: an ab initio study. Biochemistry **70**, 366–376 (2005)

31. Tirado-Rives, J, Jorgensen, WL: Performance of B3LYP density functional methods for a large set of organic molecules. J. Chem. Theory Comput. **4**, 297–306 (2008)
32. Zeevaart, JG, Wang, L, Thakur, VV, Leung, CS, Tirado-Rives, J, Bailey, CM, Domaoal, RA, Anderson, KS, Jorgensen, WL: Optimization of azoles as anti-human immunodeficiency virus agents guided by free-energy calculations. J. Am. Chem. Soc. **130**, 9492–9499 (2008)
33. Dolg, M: Recent progress in understanding chemical shifts. J. Chem. Phys. **115**, 7348–7355 (2009)
34. Basilio, N, Francisco, V, Garcia-Rio, L: Aggregation of *p*-Sulfonatocalixarene-based amphiphiles and supra-amphiphiles. Int. J. Mol. Sci **14**, 3140–3157 (2013)
35. Casabianca, LB, Kallgren, JB, Natarajan, JK, Alumasa, JN, Roepe, PD, Wolf, C, de Dios, AC: Antimalarial drugs and heme in detergent micelles: an NMR study. Inorg. J. Biochem **103**, 745–748 (2009)
36. Abul Haider Shipar, M: DFT studies on fructose and Glycine Maillard reaction: formation of the Heyns rearrangement products in the initial stage. J Iran Chem Soc **8**, 433–448 (2011)
37. Vassil, B: Delchev: Computational (DFT and TD DFT) study of the electron structure of the tautomers / conformers of uridine and deoxyuridine and the processes of intramolecular proton transfers. J Mol Model **16**, 749–757 (2010)
38. Nanotube Modeler. version 1.2.4 (2004–2005).. www.jcrystal.com/products/wincnt/Nanotube
39. Frisch, A, Nielsen, AB, Holder, AJ: GaussView User Manual. Gaussian Inc., Wallingford (2000)
40. Frisch, MJ, Trucks, GW, Schlegel, HB, Scuseria, GE, Robb, MA, Cheeseman, JR, Zakrzewski, VG, Montgomery, JA, Stratmann, RE, Burant, JC, Dapprich, S, Millam, JM, Daniels, AD, Kudin, KN, Strain, MC, Farkas, O, Tomasi, J, Barone, V, Cossi, M, Cammi, R, Mennucci, B, Pomelli, C, Adamo, C, Clifford, S, Ochterski, J, Petersson, GA, Ayala, PY, Cui, Q, Morokuma, K, Malick, DK, Rabuck, AD, Raghavachari, K, Foresman, JB, Cioslowski, J, Ortiz, JV, Baboul, AG, Stefanov, BB, Liu, G, Liashenko, A, Piskorz, P, Komaromi, I, Gomperts, R, Martin, RL, Fox, DJ, Keith, T, Al-Laham, MA, Peng, CY, Nanayakkara, A, Gonzalez, C, Challacombe, M, Gill, PMW, Johnson, B, Chen, W, Wong, MW, Andres, JL, Gonzalez, C, Head Gordon, M, Replogle, ES, Pople, JA: GAUSSIAN 03, revision B.02. Gaussian Inc, Pittsburgh, PA (2003)
41. Glendening, ED, Reed, AE, Carpenter, JE, Weinhold, F: NBO version 3.1. Gaussian Inc, Pittsburg, PA (2003)

15

DFT/NBO study of Nanotube and Calixarene with anti-cancer drug

Karim Zare[1,2], Nasim Shadmani[3*] and Elham Pournamdari[4]

Abstract

Nowadays use of calixarenes and nanotubes are widely spread in the pharmaceutical industry. In this work, interaction of between calix[4]arene and nanotube (6, 6) with Fluorouracil drug are investigated. The DFT calculations have been performed using the Gauss view and Gaussian98 in B3LYP method and 6-31G (d) standard basis set at 298.15K. There are calculated length bond (Å), bond angel (deg), dihedral angel (deg), energy hyperconjucation, and total energy (KJ mol^{-1}), moment dipole (Debye), occupancy between nanotube (6, 6) and calix[4]arene with anticancer drug in B3LYP/6-31G (*d*) method. These cases and medicines show that complex1 is more stable than complex1. The parameter of E^2, gap energy and $\Delta E°$ in composite of nanotube- Fluorouracil are higher than calix[4]arene-Fluororacil; therefore, complex1is more stable.

Keyword: Drug delivery; Nanotube; DFT; Fluorouracil; Calix[n]arene; NBO; Capacity heat; Formation energy

Introduction

The application of nanotechnology in disease treatment, diagnosis, monitoring, and in the control of biological systems at the single molecule or molecular assembly level is referred to as nanomedicine. The major goal of nanomedicine is the design of material capable of delivery and targeting of pharmaceutical, therapeutic, and diagnostic agents [1-5]. The type of drug delivery system carbon nanotube discovery by Iijma in 1990_S [6]. A nanotube can be thought of as a hexagonal network of carbon atoms that has been rolled up to make aseamless cylinder [7,8]. Carbon nanotubes exhibit superior thermal [9], mechanical [10], and electrical properties [11] and are considered the most promising building block for manufacturing low-cost, high-performance nanostructured composite materials [12].

In the paper, complexation of between fluorouracil and nanotube (6,6)/calix[4]arene are investigated as drug delivery system. Calixarenes, crown ether, and cyclodextrin are group of organic macrocyclic agents that have cup like shape which are easily available through the cyclocondensation of para-substituted phenols with formaldehyde [13,14]. One way to increase the aqueous solubility of drugs is to use complexing agents to form hast- guest complexes [15-17]. Calixarenes are promising materials for nanomedicine application in drug delivery systems. For example, hydrophilic derivatives have shown interesting levels of activity against bacteria [19], fungi, cancerous cells and enveloped viruses, but also against thrombosis or fibrosis diseases [10,20-22]. Anticancer genes act in a dominant fashion: when ectopically over expressed, they specifically destroy tumor cells without harming normal cells. This cell destruction can come in various modes such as apoptosis, mitotic catastrophe followed by apoptosis or necrosis, and autophage. Anti-cancer genes have only recently emerged from studies on cancer cells [23-26]. Fluorouracil or 5-Fluoropyramidin-2,4(1H, 3H)-dion is used as anticancer drug. Fluorouracil is an analog of pyrimidine which has been used as an anti-cancer drug for 40 years. The structure of fluorouracil is observed in Figures 1a and 2. It is anti metabolite drug and acts in several ways, but principally as synthesis inhibitor. These days there are ways to deliver a drug in the body without side effects [27]. In this paper, we reported types of drug delivery system such as nanotube and calix[4]arene.

Results and discussion

The several computational tools of, such as Density Functional Theory (DFT), Car-Parrinello molecular dynamics

* Correspondence: n.shadmani@gmail.com
[3]Young Researchers and Elites Club, Rasht Branch, Islamic Azad University, Rasht, Iran
Full list of author information is available at the end of the article

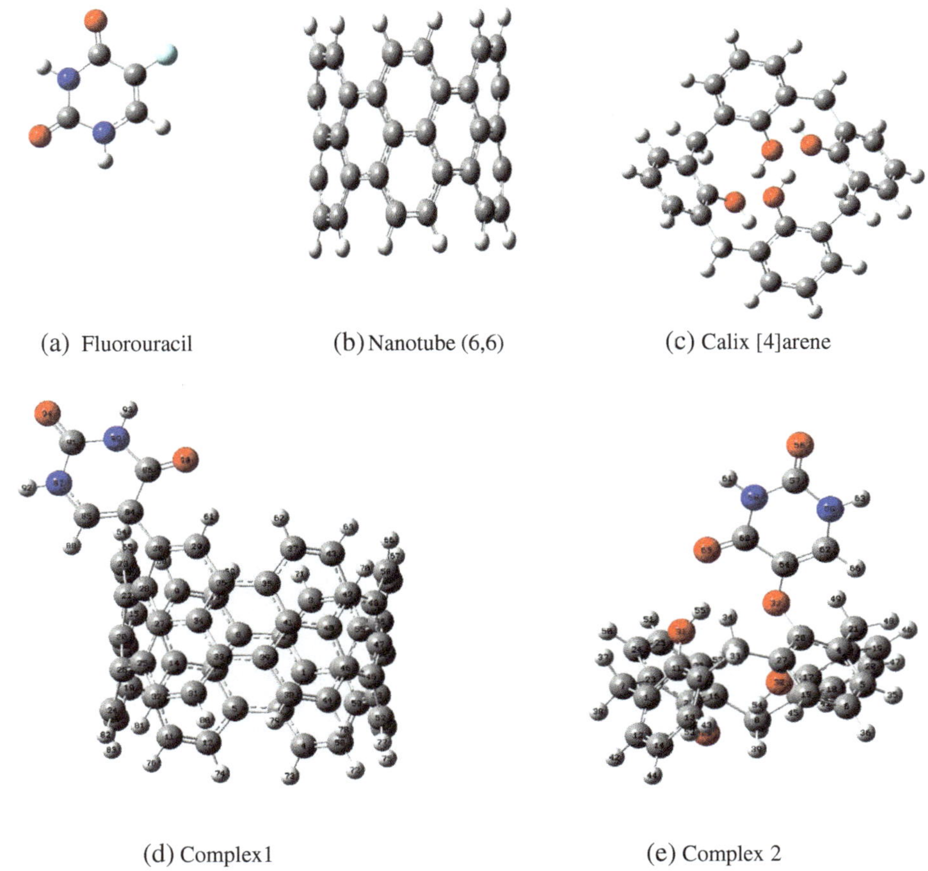

Figure 1 The structures of optimized (a, b, c, d, e) using B3LYP/6-31G (*d*) method at 298.15K.

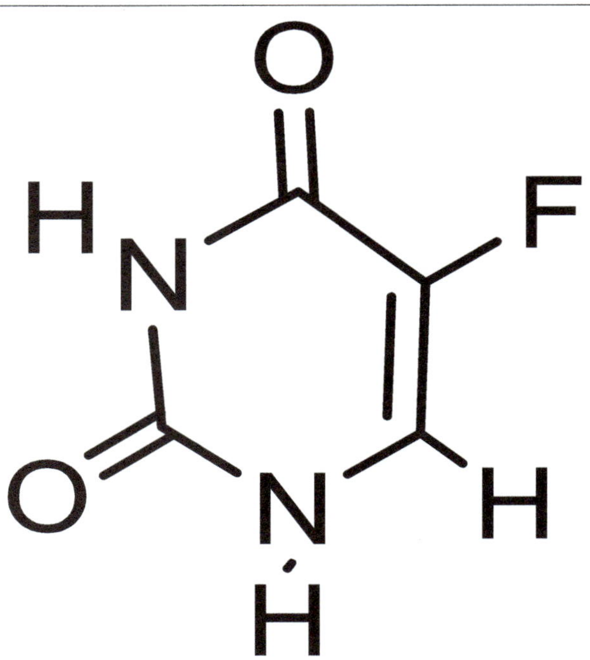

Figure 2 The structure of 5-fluoropyramidin-2,4(1H,3H)-dione.

simulations, and hybrid QM/MM approaches, can be used for calculations. Density functional quantum chemical calculations have recently provided a relatively consistent picture on base pair interaction energies and geometrics. This can lead to more detailed information on structure, charge distribution, and energetic of the base pair [28-31]. At present, quantum chemical is almost universally applicable to the interpretation of physical and chemical properties of various compounds [32]. Understanding the biochemical mechanism of a disease usually suggests the types of molecules required for new drugs. In all cases, the aim of using the computer for drug design is to analyze the interactions between the drug and receptor sites and to design drugs that give an optimal fit [33-35]. Figure 1 shows the optimized compound calix[4]arene, nanotube (6,6), fluorouracil, nanotube (6,6)-fluorouracil (complex 1), and calix[4]arene-fluorouracil (complex 2) by DFT method in level B3LYP/6-31G (*d*).

The highest occupied molecular orbital (HOMO) and the lowest unoccupied molecular orbital (LUMO), [36], the HOMO-LUMO bond gap have been found as a measure of the structural stability properties [37]. The parameters of bond length (Å), natural bond orbital (NBO) and bond angle (deg), dihedral angle (deg),

Table 1 Parameters of in B3LYP/6-31G (*d*) method optimized by complexes 1 and 2 at 298.15K

Agent	Complex 1	Complex 2
$C_{30}-C_{84}/O_{84}$	1.47	1.40
$C_{30}=C_{20}$	1.37	1.35
$C_{36}\overset{\cdots}{-}C_{29}$	1.43	1.40
$C_{28}-C_{30}$	1.47	1.46
$C_{85}\overset{\cdots}{-}N_{87}$	1.37	1.38
$N_{87}-C_{91}$	1.39	1.40
$C_{91}=O_{94}$	1.22	1.23
$C_{30}-C_{28}\overset{\cdots}{-}C_{27}$	116.02	121.62
$C_{84}=C_{85}-H_{88}$	121.36	122.70
$C_{85}\overset{\cdots}{-}N_{87}-H_{92}$	120.82	120.48
$N_{87}-C_{91}=O_{94}$	123.42	123.77
$C_{91}-N_{89}-H_{93}$	115.60	115.47
$O_{90}=C_{86}-C_{84}$	126.90	124.39
$C_{30}-C_{84}=C_{85}\overset{\cdots}{-}N_{87}$ [1]	178.86	-
$H_{87}-C_{84}-N_{86}\overset{\cdots}{-}H_{91}$ [2]	0.79	0.52
$C_{90}-N_{88}-C_{85}=C_{89}$ [3]	−178.64	−179.12
$H_{61}-C_{29}=C_{30}-C_{28}$ [4]	172.74	-
$O_{94}=C_{91}-N_{89}-C_{86}$ [5]	179.16	173.95
$C/H_{30}-C_{28}\overset{\cdots}{-}C_{27}\overset{\cdots}{-}C_{34}$ [6]	−26.01	35.84

Bond distances (Å), Bond Angles (deg) and dihedral Angles (deg).
1: $C_{30}-C_{84}=C_{85}\overset{\cdots}{-}N_{87}$ dihedral angles, 2: $H_{87}-C_{84}-N_{86}\overset{\cdots}{-}H_{91}$ dihedral Angles, 3: $C_{90}-N_{88}-C_{85}=O_{89}$ dihedral Angles, 4: $H_{61}-C_{29}=C_{30}-C_{28}$ dihedral angles, 5: $O_{94}=C_{91}-N_{89}-C_{86}$ dihedral angles, 6: $C/H_{30}-C_{28}\overset{\cdots}{-}C_{27}\overset{\cdots}{-}C_{34}$ dihedral angles.

distances of analysed models of the nanotube (6, 6) and calix[4]arene are calculated by DFT at the level of B3LYP and 6-31G (*d*) standard basis set and are shown in Table 1. The DFT calculated geometric parameters for complex 1, and 2 are compared in Table 1. The bond lengths C_{30}-C_{84}/O_{84} calculated for complex 1, 2 at the DFT level range from 1.47 to 1.40 Å, at the B3LYP/6-31G (*d*) level.

The bond lengths calculated at $C_{36}\overset{\cdots}{-}C_{29}$ for complex 1 (in ring nanotube), 1.44 Å and, for complex 2, 1.40 Å, are within the range (in ring calixarene). The bond lengths calculated $C_{85}\overset{\cdots}{-}N_{87}$ for complex 1, 1.37 Å, and complex 2 are ranged 1.38 Å. The $C_{88}\overset{\cdots}{-}N_{92}$ bond length is lower than $C_{36}\overset{\cdots}{-}C_{29}$ bond length in complex 1 and 2. With that reason, there is more electronegativity of nitrogen than carbon.

The angles for N_{87}-$C_{91}=O_{94}$ are 123.41° and 123.77° for complexes 1 and 2. The angles for C_{84}-$C_{86}=O_{90}$ are 126.9° and 124.4° for complex 1, 2. The angles for C_{84}-$C_{86}=O_{90}$ is larger than the angel for N_{87}-$C_{91}=O_{94}$ in complexes 1 and 2. The interaction of between nonbonding and bonding pairs on the nitrogen atom of the angle are reduced in N_{87}-$C_{91}=O_{94}$.

The dihedral angles for C/H_{30}–$C_{28}\overset{\cdots}{-}C_{27}\overset{\cdots}{-}C_{34}$ for complex 1 and 2 range from −26.01° to 35.84° (in dihedral 6). The dihedral angles 1,2,3,4,5 and 6, are observed in Table 1 and Figure 3.

The calculations of the total energies, hyperconjoucation energy (E^2) of the optimized structures, dipole moments (μ), occupancy and hybrid at B3LYP/6-31G (*d*) levels are presented in Tables 2 and 3. In Table 2, the Mulliken charges in donor atoms electronegative O_{84} and acceptor

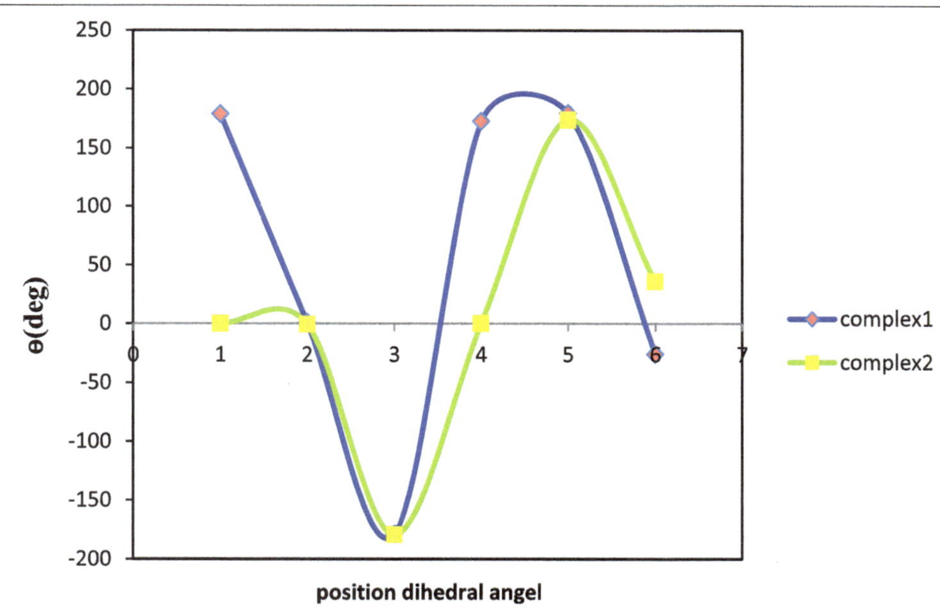

Figure 3 Pole of changed dihedral angel of complexes 1, and 2 using method B3LYP/6-31G (*d*) at 298.15 K.

Table 2 The NBO parameters of, Complex 1 and 2 are calculated in B3LYP/6-31G (*d*) method at 298.15K

Agent	Donor	Occupancy	Acceptor	Occupancy	hybrid	E^2	ΣE^2
Complex1	BD(1)C30- C84	1.96631	BD*(1)C27-C28	0.02586	$SP^{1.98}$	1.48	35.53
			BD*(1)C28-C30	0.02920	$Sp^{2.06}$	1.70	
			BD*(1)C29-C30	0.02033	$Sp^{1.79}$	3.12	
			BD*(1)C29-C36	0.02164	$SP^{1.96}$	2.51	
			BD*(1)C84-C85	0.01852	$SP^{1.88}$	3.55	
			BD*(1)C84-C86	0.06682	$SP^{2.28}$	1.82	
			BD*(1)C85-N87	0.01939	$SP^{2.51}$	3.76	
	BD*(2)C84-C85	0.23733	BD*(2)C29-C3	0.2195	$SP^{99.99}$	17.59	
Complex2	BD(1)C28-O32	1.94416	BD*(1)C4-C5	0.0219	$SP^{1.9}$	1.48	22.77
			BD*(1)C15-C27	0.02108	$SP^{1.85}$	1.6	
			BD*(1)C60-C64	0.07182	$SP^{1.71}$	1.93	
			BD*(2)C62-C64	0.24487	$SP^{1.00}$	0.66	
			BD*(1)C4-C28	0.03327	$SP^{1.98}$	1.88	
	LP(1)O32	1.93171	BD*(2)C27-C28	0.37632	$SP^{1.00}$	5.18	
			BD*(1)C4-C28	0.03327	$SP^{1.98}$	3.86	
	LP(2)O32	1.87683	BD*(1)27-C28	0.03258	$SP^{1.95}$	6.15	

C_{30} are negative and positive, respectively. Complex 1 has gap of energy that is larger than complex 2; therefore, complex 1 is stable. In Table 2, it becomes obvious that the complex 1 has formed higher hyperconjugation energy than complex 2. Also, the results show that by increasing P part in hybrid of atoms, the occupancy decreases. The S orbital part in hybrid of carbon in complex 2 is more than the S orbital part in hybrid of in complex 1. Combined with the most of hyperconjugation energy is stable. The occupancy coefficient is smaller. Complex 1 is more stable than the complex 2. The hyperconjugation energy complex 1 at 37.34 is larger than that in complex 2. The hybrid orbital S of a compound is lower. Table 2 shows the HOMO and LUMO energies for complexes. By evaluating HOMO/LUMO gap energies, it is obvious that if the gap becomes bigger, the complex will be stable; therefore, complex 1 is more stable than the complexes. The results of the present work were obtained using DFT optimization and formation energy (ΔE°_f in $KJmol^{-1}$) calculation at the B3LYP/6-31G (*d*) level. ΔE°_f is calculated using the formula $\Sigma E^{\circ}{}_{\text{product}} - \Sigma E^{\circ}{}_{\text{reactant}}$. ΔE°_f values in complexes 1 and 2 are in the range of −13.275 and +46.902 $KJmol^{-1}$; therefore, complex1 has lower formation energy than the others. The energy (kJ mol^{-1}) and dipole moments (Debye) indicate the consistency between the two complex calculations in DFT method. The gap energies and total energy, ΣE^2, HOMO and LUMO complexes 1, 2 were calculated using the B3LYP method and 6-31G (*d*) basis set. The total energy sum of energy transitional, energy rotational and energy vibration in level B3LYP/6-31G (*d*) for complexes 1, 2 was calculated. The obtained results are shown in Table 3.

Heat capacity is the measurable physical quantity that specifies the amount of heat required to change the temperature of body by a given amount. Translation, rotation, and a combination of the two types of energy in vibration (kinetic and potential) of atoms represent the degrees of freedom of motion which classically contribute to the heat capacity of matter, but loosely bound electrons may also participate. On a microscopic scale, each system particle absorbs thermal energy among the few degrees of freedom available to it, and at sufficient temperatures, this process contributes to the specific heat capacity that classically approaches a value per mole of particles that is set by the Dulong-Petit law. For quantum mechanical reasons, at any given temperature, some of these degrees of freedom may be unavailable, or only partially available, to store thermal energy. Quantum theory can be used to quantitatively predict the specific heat capacity of simple systems.

Table 3 Formation energy, total energy, HOMO, LUMO, Gap of energy, moment dipole and heat capacity are calculated in B3LYP/6-31G (*d*) method at 298.15 K

Agent	E_{total}	ΔE°_f	HOMO	LUMO	Gap	μ	C_V
Complex 1	465.73	−13.275	−0.15877	−0.08918	−0.0696	0.527	191.03
Complex 2	348.03	+46.902	−0.19989	−0.05481	−0.14508	0.591	130.52

Conclusions

In this paper, the result shows that complex 1 between Nanotube (6,6) and fluorouracil is more stable than complex 2. Thus, complex 1 is a better conditioner for drugs than complex 2. NBO analysis shows larger gap energy in complex 1. Complex 1 has lower formation energy and is more stable than complex 2.

Methods

Investigation is carried out using a personal computer (Intel (R) Pentium (R) dual CPU with 2GB RAM). Nanotube-fluorouracilarene (with different atom number) which reacts with anti-cancer drug. In this paper, the drug delivery properties are investigated by NBO analysis and DFT method. The DFT calculations have been performed using the Nanotube modeler [38], Gaussview [39] and Gaussian 03 [40] using B3LYP method and 6-31G (*d*) standard basis set. NBO analysis [41,42] calculations have been also performed for all composites using B3LYP method and the standard 6-31G (*d*) basis set. Complexes between calix[4]arenes and nanotube (6,6) with Fluorouracil drug are optimized, then bond length (Å), bond angel (deg), dihedral angel (deg), hyperconjucation energy, as well as total energy (KJmol^{-1}), moment dipole (Debye), occupancy, total energy and HOMO/LUMO are investigated between nanotube (6, 6) and calix[4]arene with anti-cancer drug using B3LYP/6-31G (*d*) method. Nanotube/calix[4]arenes and fluorouracil reaction are shown in Equations 1 and 2.

$$\text{Fluorouracil} + \text{Nanotube}\ (6,6) \rightarrow \text{Fluorouracil–Nanotube}\ (6,6) + \text{HF (bond C-C)} \tag{1}$$

$$\text{Fluorouracil} + \text{Calix[4]arene} \rightarrow \text{Fluorouracil–Calix[4]arene} + \text{HF (bond C-O)} \tag{2}$$

Competing interests

The authors declare that they have no competing interests.

Authors' contributions

KZ and NSH both contributed to this research work. Both authors read and approved the final manuscript.

Acknowledgements

Both of authors are grateful to University of Science and Research Branch, Islamic Azad University, Tehran, Iran. We are grateful to Shahid Beheshti University, Tehran, Iran. We are grateful to Young Researchers and Elites Club, Rasht Branch, Islamic Azad University, Rasht, Iran for financial support.

Author details

[1]Department of Chemistry, Science and Research Branch, Islamic Azad University, Tehran, Iran. [2]Department of Chemistry, Shahid Beheshti University, Evin, Tehran, Iran. [3]Young Researchers and Elites Club, Rasht Branch, Islamic Azad University, Rasht, Iran. [4]Department of Chemistry, Islamshahr Branch, Islamic Azad University, Islamshahr, Iran.

References

1. Wong Shi Kam, N, Jessop, TC, Wender, PA, Dai, H: Nanotube molecular transporters: Internalization of carbon nanotube-protein conjugates into mammalian cells. J. Am. Chem. Soc. **126**, 6850–6851 (2004)
2. Colomer, JF, Henrard, L, Lambin, P, Van Tendelool, G: 2001. Phys. Rev. B **64**, 1–7 (2001)
3. Moradi, O, Zare, K: Adsorption of Pb(II), Cd(II) and Cu(II) Ions in Aqueous Solution on SWCNTs and SWCNT–COOH Surfaces: Kinetics Studies. Fullerenes, Nanotubes, and Carbon Nanostructures **19**, 628–652 (2011)
4. Pouton, CW, Seymour, L: Key issues in non-viral gene delivery. Adv. Drug. Deliv. Rev. **46**, 187–203 (2001)
5. Singh, R, Pantarotto, D, Lacerda, L, Pastorin, G, Klumpp, C, Prato, M, Bianco, A, Kostarelos, K: Tissue biodistribution and blood clearance rates of intravenously administered carbon nanotube radiotracers. Proc. Natl. Acad. Sci. U.S.A. **103**, 3357–3362 (2006)
6. Iijima, S: Helical microtubules of graphitic carbon. Nature **354**, 56–58 (1991)
7. Yan, CS, Mao, HK, Li, W, Qian, J, Zhao, Y, Hemley, RJ: Ultrahard diamond single crystals from chemical vapor deposition. Physica Status Solidi (A) **201**(4), 25–27 (2004)
8. Kalele, S, Gosavi, SW, Urban, J, Kulkarni, SK: Nanoshell particles: synthesis, properties and applications. Curr. Sci. **91**, 8 (2006)
9. Wong Shi Kam, N, Jessop, TC, Wender, PA, Dai, H: Nanotube molecular transporters: Internalization of carbon nanotube-protein conjugates into mammalian cells. J. Am. Chem. Soc. **126**, 6850–6851 (2004)
10. Pastorin, G: Crucial Functionalizations of Carbon Nanotubes for Improved Drug Delivery: A Valuable Option? Pharmaceut. Res. **26**, 746–769 (2009)
11. Taft, BJ, Lazareck, AD, Withey, GD, Yin, A, Xu, JM, Kelley, SO: Site-specific 258 assembly of DNA and appended cargo on arrayed carbon nanotubes. J. Am. Chem. Soc. **126**, 12750–12751 (2004)
12. Pantarotto, D, Singh, R, McCarthy, D, Erhardt, M, Briand, JP, Prato, M, Kostarelos, K, Bianco, A: Functionalized carbon nanotubes for plasmid DNA 262 gene delivery. Angew. Chem. Int. Ed. **43**, 5242–5236 (2004)
13. Asfari, Z, Böhmer, V, Harrowfield, J, Vicens, J: Calixarenes 2001. ED; Kluwer, Dordrecht (2001)
14. Gungor, O, Memon, S, Yilmaz, M: Synthesis of Alkyl Nitrile and Alkyl Benzonitrile Derivatives of Calix[4]Arene and Their Polymer Supported Analogues: A Comparative Study in Two-Phase Extraction Systems. J. Funct. Polym. **63**, 1–9 (2005)
15. Gutsche, CD, Bauer, LJ: The conformational properties of calix[4]arenes, calix [6]arenes, calix[8]arenes, and oxacalixarenes. J. Am. Chem. Soc. **107**, 6052–6059 (1985)
16. Faraji, M, Zare, K, Aghaei, H, Farajtabar, A, Asfari, Z, Gharib, F: Complexation of p-Sulphonato-calix[6]arene by Glycine, Glycyl-glycine, and Glycyl-glycyl-glycine in Aqueous Solution. J. Solution. Chem. (2012). doi:10.1007/s10953-012-9924-2
17. Gharib, F, Zare, K, Taghvaei, S, Monajjemi, M, Amiri, A: Complexation of tri-o-propoyl-p-t-butyl calix[4]arene with alkali metal cations in carbon tetrachloride solvent. Main. Group. Met. Chem. **26**, 255–258 (2003)
18. Gharib, F, Zare, K, Amiri, A: Spectrophotometric studies of alkali metal ion complexes of p-tetrabutyl- calix[4]arenes. Russ. J. Inorg. Chem. **49**, 1511–1514 (2004)
19. Gharib, F, Osouleddini, N, Zare, K, Taghvaei-Ganjali, S: Spectrophotometric studies of alkali and alkali earth metal ions complexes of mono amino derivative of calix[4]arene. Russ. J. Inorg. Chem. **55**, 434–437 (2010)
20. Brunink, JAJ, Verboom, W, Engbersen, JFJ, Reinhoudt, DN, Harkema, S: Synthesis and cation complexation selectivity of bis (syn-proximally) functionalized calix[4]arenes. Recl. Trav. Chim. Pays-Bas **111**, 511–516 (1992)
21. Shirakawa, S, Shimizu, S: Dehydrative amination of alcohols in water using a water-soluble calix[4]resorcinarene sulfonic acid. Syn. Lett. **10**, 1539–1542 (2008)
22. Liu, YL, Liu, L, Wang, YL, Han, YC, Wang, D, Chen, YJ: Calix[*n*]arene sulfonic acids bearing pendant aliphatic chains as recyclable surfactant-type Brønsted acid catalysts for allylic alkylation with allyl alcohols in water. Green. Chem. **10**, 635–640 (2008)

23. Zhang, YJ, Cao, WX, Xu: Interaction of Sulfonated Calix[*n*]arenes with Rhodamine B and Its Application to Determine Acetylcholine in a Real Neutral Aqueous Medium J. Chin. Chem. **20**, 322–326 (2002)
24. Dolg, M: Recent progress in understanding chemical shifts. J. Chem. Phys. **115**, 7348–7355 (2009)
25. de Dios, AC, Oldfield, E: Antimalarial drugs and heme in detergent micelles: an NMR study. Solid State NMR **6**, 101–125 (1996)
26. Whitehead, RP, Benedetti, JK, Abbruzzese, JL, Ardalan, B, Goodwin, JW, Balcerzak, SP, Samlowski, WE, Lenz, H-Z, Macdonald, JS: A phase II study of high-dose 24 hour continuous infusion 5-FU and leucovorin and low-dose PALA for patients with advanced pancreatic adenocarcinoma: A Southwest Oncology Group Study. Inves. New. Drug. **22**, 335–341 (2004)
27. Cornell, GN, Cahow, CE, Frey, C, McSherry, C, Beal, JM: Clinical experience with 5-fluorouracil (NSC-19 893) in the treatment of malignant disease. Cancer Chemother. Rep. **9**, 23–30 (1960)
28. Monajjemi, M, Chahkandi, B, Zare, K, Amiri, A: Study of the hydrogen bond in different orientations of adenine-thymine base pairs: an ab initio study. Biochem **70**, 366–76 (2005)
29. Guo, D, Liu, Y: Calixarene-based supramolecular polymerization in solution. Chem. Soc. Rev. **41**, 5907–5921 (2012)
30. Tirado-Rives, J, Jorgensen, WL: Performance of B3LYP Density Functional Methods for a Large Set of Organic Molecules. J. Chem. Theory. Comput. **4**, 297–306 (2008)
31. Zeevaart, JG, Wang, L, Thakur, W, Leung, CS, Tirado-Rives, J, Bailey, CM, Domaoal, RA, Anderson, KS, Jorgensen, WL: Optimization of azoles as anti-human immunodeficiency virus agents guided by free-energy calculations. J. Am. Chem. Soc **130**, 9492–9499 (2008)
32. Casabianca, LB, Kallgren, JB, Natarajan, JK, Alumasa, JN, Roepe, PD, Wolf, C, de Dios, AC: Antimalarial drugs and heme in detergent micelles: An NMR study. Inorg, J: Biochem. **103**, 745–748 (2009)
33. Abul Haider Shipar, M: DFT Studies on Fructose and Glycine Maillard Reaction: Formation of the Heyns Rearrangement Products in the Initial Stage. J. Iran. Chem. Soc. **8**, 433–448 (2011)
34. Delchev, VB: Computational (DFT and TD DFT) study of the electron structure of the tautomers/conformers of uridine and deoxyuridine and the processes of intramolecular proton transfers. J. Mol. Model **16**, 749–757 (2010)
35. Ramachandran, KI, Deepa, G, Namboori, K: Research on computational chemistry and molecular modeling. In: Computational Chemistry and 344 Molecular Modeling: Principles and Applications, pp. 297–310. Springer, Heidelberg (2008)
36. Miehlich, B, Savin, A, Stoll, H, Preuss, H: Results obtained with the correlation energy density functionals of Becke and Lee–Yang and Parr. Chem. Phys. Lett. **157**, 200–206 (1989)
37. Dalili, N, Zare, K, Gharib, F, Shadmani, N: Calixarenes or Nanotubes: Which are better in drug delivery? JNSC **2**, 213–217 (2011)
38. www.jcrystal.com/products/wincnt/Nanotube. Nanotube Modeler. version 1.2.4 (2004–2005)
39. Frisch, A, Nielsen, AB, Holder, AJ: GaussView User Manual. Gaussian Inc., Wallingford (2000)
40. Frisch, MJ, Trucks, GW, Schlegel, HB, Scuseria, GE, Robb, MA, Cheeseman, JR, Zakrzewski, VG, Montgomery, JA, Stratmann, RE, Burant, JC, Dapprich, S, Millam, JM, Daniels, AD, Kudin, KN, Strain, MC, Farkas, O, Tomasi, J, Barone, V, Cossi, M, Cammi, R, Mennucci, B, Pomelli, C, Adamo, C, Clifford, S, Ochterski, J, Petersson, GA, Ayala, PY, Cui, Q, Morokuma, K, Malick, DK, et al: GAUSSIAN 03, revision B.02. Gaussian Inc, Pittsburgh, PA (2003)
41. Glendening, ED, Reed, AE, Carpenter, JE, Weinhold, F: NBO version 3.1. Gaussian Inc, Pittsburg, PA (2003)
42. Fernandez, L, Viruela-Martin, P, Latorre, J, Guillem, C, Beltrán, A, Amorós, P: Molecular precursors of mesostructured silica materials in the atrane route: A DFT/GIAO/NBO theoretical study. J. Mol. Struc: THEOCHEM **822**, 89–102 (2007)

Annealing temperature effect on the mechanical and tribological properties of molybdenum nitride thin films

Kaykhosrow Khojier[1*], Mohammad Reza Karami Mehr[2] and Hadi Savaloni[3]

Abstract

Mo thin films with 100-nm thickness were deposited on silicon substrates using DC magnetron sputtering method. Mo thin films were subsequently annealed at different temperatures (400°C to 900°C) with flow of nitrogen. The crystallographic structure of the samples was obtained using X-ray diffraction method. Atomic force microscopy and scanning electron microscopy were used for surface morphology investigation. Nano-indentation and scratch tests were performed to obtain the surface hardness and friction coefficient of the samples, respectively. Results show that the γ-Mo_2N(111) phase of molybdenum nitride with face-centered cubic structure and higher hardness, elastic modulus, and lower coefficient of friction and scratch volume is formed when the sample is annealed at 650°C, while the Mo_2N phase with tetragonal structure and lower hardness, elastic modulus, and higher scratch volume and friction coefficient is formed at higher temperatures of 775°C and 900°C. It is found that increasing the annealing temperature causes an increase of the grain size and film surface roughness. From the mechanical results, it may be deduced that 650°C is a critical temperature for variation of mechanical and tribological properties.

Keywords: Molybdenum nitride, Nanostructure, Hardness, Friction coefficient, Scratch volume

Background

Studies of the transition metal nitrides, especially in the form of thin films, have both scientific and technological significance. Molybdenum nitride is one of these materials that exhibit interesting properties. Molybdenum nitride thin films show high electrical conductivity and chemical stability and are promising candidates for use as Cu diffusion barrier and electrodes in microelectronics [1-4]. Furthermore, molybdenum nitrides have excellent hardness, catalytic ability, and superconductivity effects [5-9]. Hence, many researchers have produced molybdenum nitride thin films using different methods such as DC and RF magnetron sputtering, chemical vapor deposition, pulsed laser deposition, ion-assisted deposition, reactive magnetron sputtering, and heavy nitrogen ion-implantation and reported their characterization results [7-18].

In this work, we have prepared molybdenum nitride thin films using different techniques by post-annealing Mo/Si thin films with 100-nm thickness at different annealing temperatures (400°C–900°C) in the nitrogen constant flow and have investigated the influence of these variables on the nanostructure and mechanical and tribological properties of these films.

Experimental details

Film preparation

Molybdenum (99.998% purity) thin films of 100-nm thicknesses were deposited by means of a DC magnetron sputtering system using a circular sputtering target of 76-mm diameter and 1-mm thickness. The target-to-substrate distance was 10 cm. A DC power supply of 750 V and 125 mA was used as power source for sputtering. The film thickness and deposition rate were checked *in situ* by a quartz crystal monitor (6 MHz gold, Inficon Company, East Syracuse, NY, USA) positioned close to the substrate

* Correspondence: khojier@iauc.ac.ir
[1]Department of Physics, Chalous Branch, Islamic Azad University, Chalous 46615-397, Iran
Full list of author information is available at the end of the article

holder during the sputtering process. The molybdenum thin films were deposited at a deposition rate of 3 Å/s. The base pressure was 1×10^{-5} mbar, achieved with a diffusion pump coupled with a rotary pump that was changed to 3.2×10^{-2} mbar after the presence of Argon. The purity of argon gas in this work was 99.998% and controlled by a mass flow controller. The substrates for the deposition were 20×20 mm^2 of single crystal Si (400) and were ultrasonically cleaned in heated acetone then etalon. The substrates were not subjected to any heating treatment before and during the sputtering process. Post-annealing of Mo/Si films were performed at five different temperatures (400°C, 525°C, 650°C, 775°C, and 900°C) with a 200-sccm nitrogen flow. A 5°C/min temperature rise was used to reach the required annealing temperature, and the samples were kept at their annealing temperature for 120 min, then gradually cooled down to room temperature with flowing nitrogen gas.

Film characterization

The nanostructure and crystallographic orientation of the samples were obtained using a Philips XRD X'pert MPD diffractometer (Philips Analytical, Almelo, The Netherlands) (Cu Kα radiation) with a step size of 0.02° and a count time of 1 s per step. A Hysitron TriboScope® nanomechanical test instrument with 2D transducer (Hysitron Inc., Eden Prairie, MN, USA), complete software, and Berkovich diamond indenter were used for mechanical test and surface imaging. The AFM part was a NanoScope E from Digital Instruments (Santa Barbara, CA, USA). Over four indentation tests were performed on all samples, and the average of the obtained data is presented. The scan size and scan rate were 5×5 μm^2 and 1.001 Hz, respectively. Surface images, roughness parameters, nano hardness, elastic modulus, scratch volume, and friction coefficient are obtained from this analysis. A scanning electron microscope (Leo-440i, Leo Electron Microscopy Ltd., Cambridge, UK) was also used for the surface morphology study.

Results and discussion

Crystallographic structure

Figure 1 depicts the X-ray diffraction (XRD) pattern of Mo/Si as-deposited and annealed thin films with 100-nm thickness prepared in this work. In the XRD pattern of the as-deposited molybdenum thin film, two peaks (except the substrate peak) can be observed at 40.51° and 87.57° that can be related to Mo(110) and Mo(220) crystallographic orientations, respectively (with reference to JCPDS card no. 42–1120, 2θ = 40.516° and 2θ = 87.598°, system: cubic, and space group: 229). In the X-ray diffraction pattern of the annealed sample at a temperature of 400°C with flow of nitrogen, the intensity of the Mo peaks decreased, and a new peak appeared at 37.39° that can be attributed to the γ-Mo_2N(111)

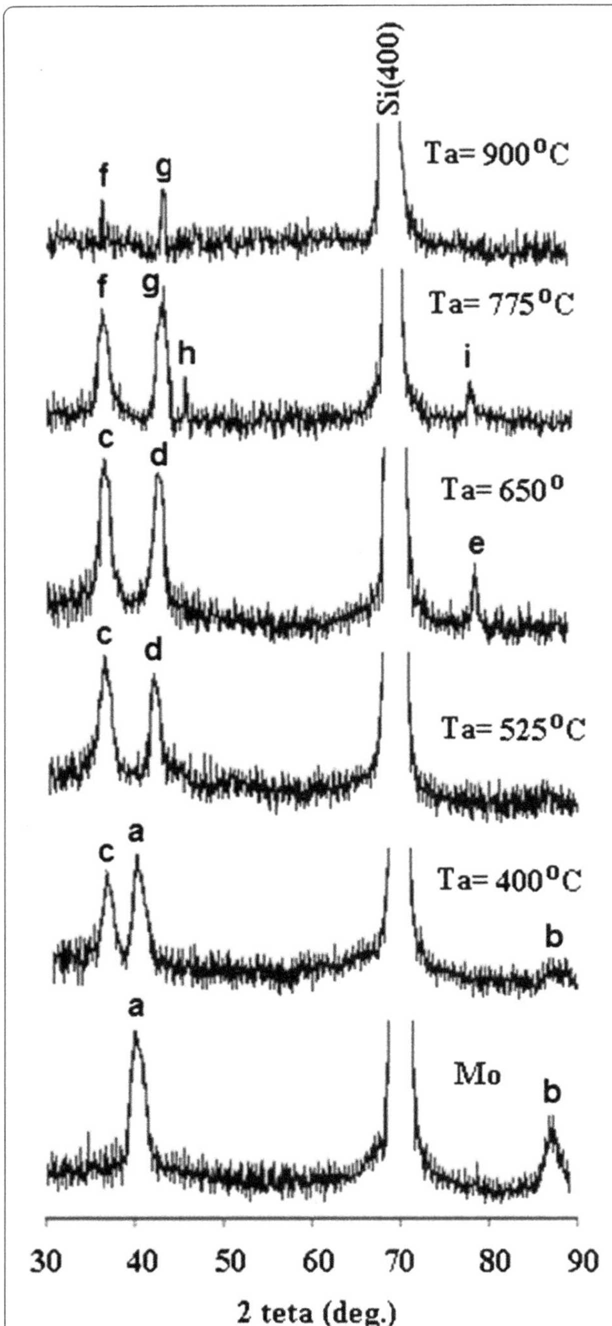

Figure 1 XRD of unheated and annealed Mo/Si thin films. (**a**) Mo(110), (**b**) Mo(220), (**c**) γ-Mo_2N(111), (**d**) γ-Mo_2N(200), (**e**) γ-Mo_2N(222), (**f**) Mo_2N(112), (**g**) Mo_2N(200), (**h**) Mo_2N(004), and (i) Mo_2N(116).

crystallographic orientation (with reference to JCPDS card no. 25–1366, 2θ =37.376°, system: cubic, and space group: 221). By increasing the annealing temperature to 525°C, the Mo peaks disappeared, and the intensity of the γ-Mo_2N(111) peak increased. In addition, a new peak of the γ-Mo_2N phase also appeared at 43.49°, which may be related to the γ-Mo_2N(200) crystallographic orientation (with reference to JCPDS card no. 25–1366,

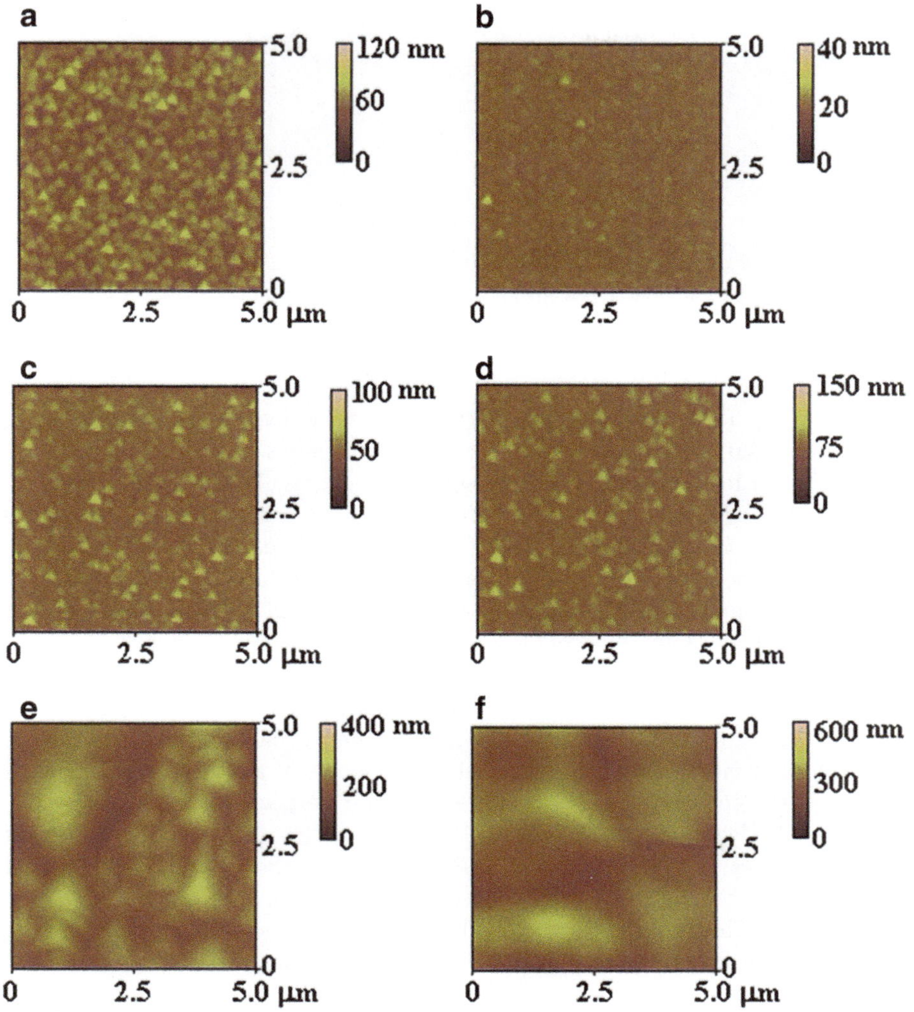

Figure 2 2D AFM images of Mo/Si thin films produced in this work. (**a**) Unheated (Mo) and annealed at (**b**) 400°C, (**c**) 525°C, (**d**) 650°C, (**e**) 775°C, and (**f**) 900°C.

2θ = 43.450°, system: cubic, and space group: 221). When the annealing temperature increased to 650°C, the intensities of the γ-Mo_2N(111) and (200) peaks increased relative to the lower annealing temperature, while a weak γ-Mo_2N(222) diffraction line at 79.73° also appeared in the XRD pattern (with reference to JCPDS card no. 25–1366, 2θ = 79.708°, system: cubic, and space group: 221).

Table 1 Results of AFM analysis

Sample no.	T_a(°C) ±5	Grain diameter (μm) ±0.02	Roughness	
			rms (nm)	R_a
K-1	as-deposited	0.19	8.02	6.48
K-2	400	0.05	1.64	1.22
K-3	525	0.07	5.14	3.78
K-4	650	0.09	6.48	4.80
K-5	775	0.70	54.7	44.1
K-6	900	1.25	78.5	64.1

At an annealing temperature of 775°C, the diffraction lines belonging to the γ-Mo_2N phase of molybdenum nitride with cubic structure disappeared, and a new phase of molybdenum nitride as a tetragonal structure is formed with the following crystallographic orientations: Mo_2N(112), Mo_2N(200), Mo_2N(004), and Mo_2N(116) at 2θ = 37.69°, 2θ = 43.19°, 2θ = 45.05°, and 2θ = 78.19°, respectively (with reference to JCPDS card no. 25–1368; 2θ = 37.685°, 43.166°, 45.020°, and 78.151° system: tetragonal; and space group: 141). At the highest annealing temperature of 900°C, the XRD pattern of the sample shows that the intensity of the molybdenum nitride peaks decreased, so the Mo_2N(004) and Mo_2N(116) peaks almost disappeared. This observation may indicate that at high temperature, reaction between nitrogen atoms/molecules and molybdenum is not favored. The tendency to grow in the γ-Mo_2N(111) crystallographic orientation at temperatures below 700°C is due to the (111) planes which are closest

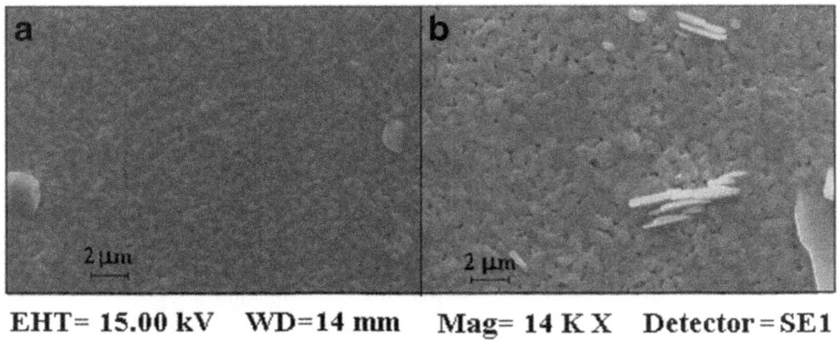

Figure 3 **SEM images of annealed Mo/Si thin films at (a) 775°C and (b) 900°C.**

packed in the face-centered cubic (FCC) molybdenum nitride structure with the lowest surface energy, and is consistent with the results of Shen for sputtered molybdenum nitride thin films [19]. In summary, the results show a mixed structure of Mo and molybdenum nitride for the annealed sample at 400°C, a polycrystal of γ-Mo_2N with FCC structure for annealed samples at 525°C and 650°C, and a Mo_2N phase of molybdenum nitride for samples which were annealed at 775°C and 900°C.

Surface physical morphology

Surface morphology of as-deposited and annealed samples was studied by atomic force microscopy (AFM) and scanning electron microscopy (SEM). Figure 2 depicts the two-dimensional (2D) AFM images of these samples, while Figure 3a,b shows SEM images of the annealed samples at 775°C and 900°C. The grain diameter (calculated from 2D AFM pictures using the Jmicrovision code) and surface roughness parameters are also given in Table 1. The results show that the grain size and surface roughness decreased by annealing at 400°C. This reduction in grain size and surface roughness can be related to the reaction of Mo surface atoms with nitrogen atoms. An increase in annealing temperature to a higher temperature (>400°C) has an inverse effect, so increasing the annealing temperature causes an increase of the grain size and film surface roughness. The increase of the grain size and surface roughness is due to the increased surface diffusion (mobility) which causes the coalescence of the grains which in turn produces larger and deeper valleys between these newly formed grains; hence, larger grains and higher surface roughness are obtained.

Mechanical properties

Figures 4 and 5 illustrate the variation of hardness and elastic modulus of all samples as a function of annealing temperature, respectively, while Figure 6a shows the 2D AFM image of the selected sample after nano-indentation test. The results show that both these quantities increase by increasing the annealing temperature up to 650°C, and then decrease. These observations can be explained as in the following paragraphs.

Film hardness and mechanical properties can be affected by different parameters such as grain size, crystallographic orientations, film density, lattice parameters, and stoichiometry [20-23]. On the other hand, the close-packed structures and small grains (with more grain boundaries) result in higher hardness. Grain boundaries

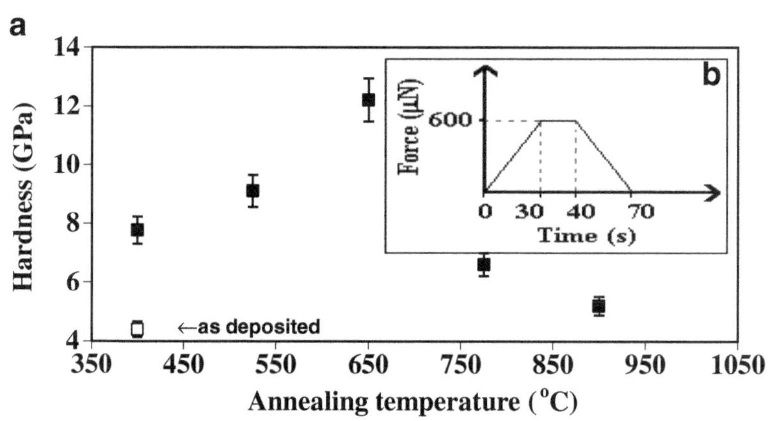

Figure 4 **Variation of hardness and loading force.** (**a**) Variation of hardness of as-deposited and annealed Mo/Si samples as a function of annealing temperature and (**b**) variation of loading force as a function of time at nano-indentation test.

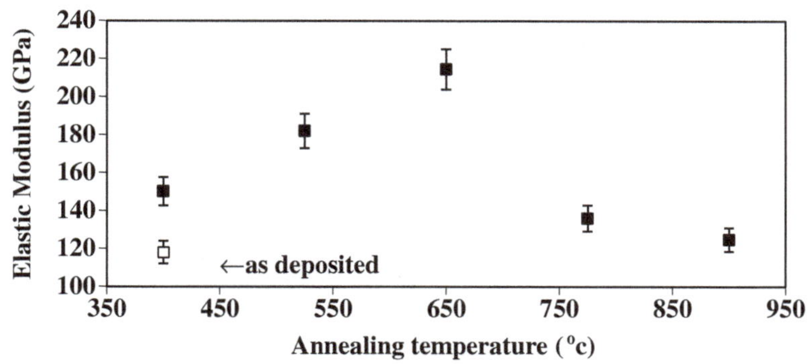

Figure 5 Variation of elastic modulus of as-deposited and annealed samples as a function of annealing temperature.

in polycrystal act as an impediment to the dislocation motion for the following two reasons: (1) dislocation must change its direction of motion due to the different orientations of grains; (2) discontinuity of slip planes from one grain to another [23].

The increasing film hardness before 650°C can be attributed to the change of the BCC structure of molybdenum to the FCC structure of γ-Mo_2N and small grains. However, an increase in annealing temperature to higher temperatures (i.e., 775°C and 900°C) has an inverse effect. This reduction is related to change of the γ-Mo_2N phase with FCC structure to the Mo_2N phase with tetragonal structure and larger size of the grains which form in higher temperature. Variation for elastic modulus in the samples is due to the ceramic nature of molybdenum nitride and larger elasticity of γ-Mo_2N relative to Mo_2N.

Tribological properties

Variations of the coefficient of friction obtained from the scratch test are plotted in Figure 7, while the 2D AFM image of the selected sample surface after the scratch test is shown in Figure 6b. The variation of (lateral force/normal force) of the selected sample as a function of time for obtaining the coefficient of friction is also shown in Figure 8. As can be seen, the friction coefficient of the molybdenum thin film decreases by annealing at the presence of nitrogen flow and increasing the annealing temperature up to 650°C. However, an increase in annealing temperature to higher temperatures (i.e., 775°C and 900°C) decreases the coefficient of friction. In the steady-state phase, the friction coefficient of the films depended mainly on their microstructure and had no direct relation to their surface topography [24]. On the other hand, the denser microstructure, namely, the smaller grain size and harder films, results in a lower friction coefficient. The variation of scratch volume of all samples as a function of annealing temperature is shown in Figure 9. As can be seen, the scratch volume decreases by increasing the annealing temperature up to 650°C, and then increases. This behavior of scratch volume is similar to the variation of friction coefficient and depends on the film microstructure. In fact, denser or harder microstructures have lower scratch volume. In summary, the results showed that an annealing temperature of 650°C is a critical temperature for mechanical and tribological properties that are due to transition of the FCC structure to the tetragonal structure and increase of the grain size which formed at a higher temperature.

Conclusions

Molybdenum nitride thin films prepared by post-annealing of MO thin films with 100-nm thickness at

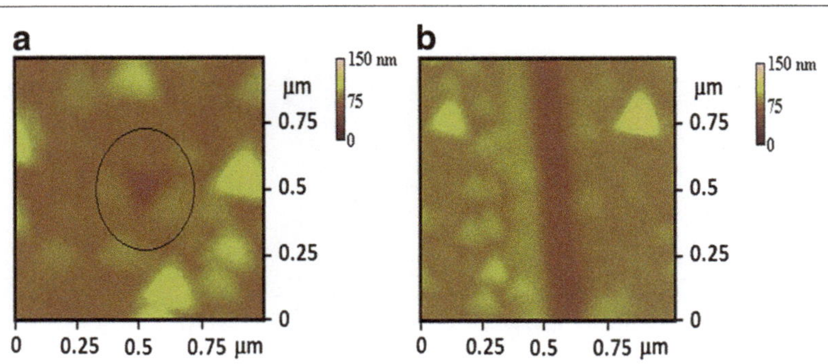

Figure 6 2D AFM pictures of a selected sample (with 100-nm thickness and annealed at 650°C). (**a**) After nano-indentation and (**b**) after scratch tests.

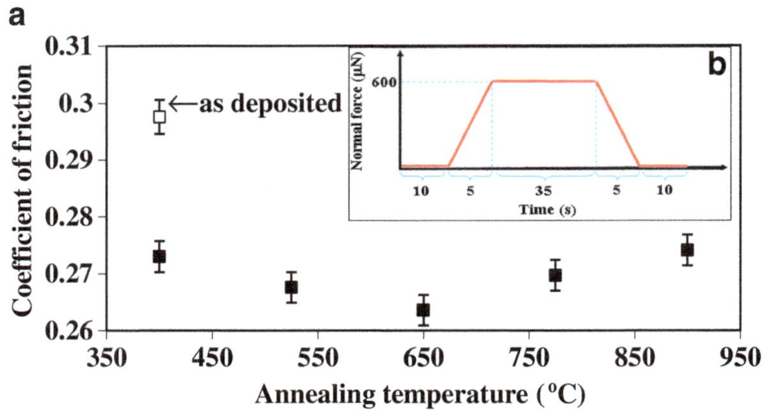

Figure 7 Variation of friction and normal force. (**a**) Variation of friction coefficient of as-deposited and annealed Mo/Si samples as a function of annealing temperature and (**b**) variation of normal force as a function of time at scratch test.

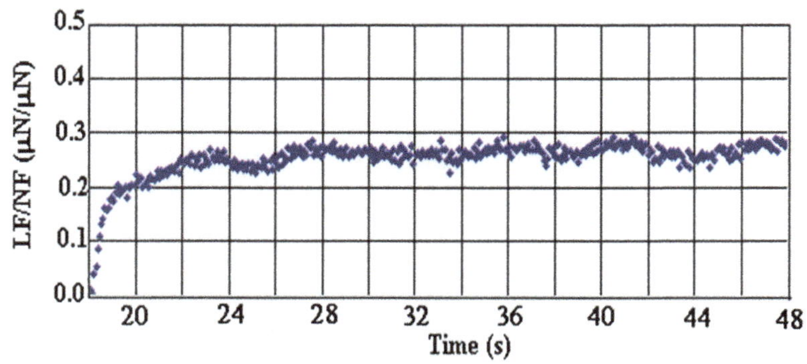

Figure 8 Variation of lateral force/normal force as a function of time of the selected sample. With 100-nm thickness and an annealing temperature of 650°C (K-4) in scratch test.

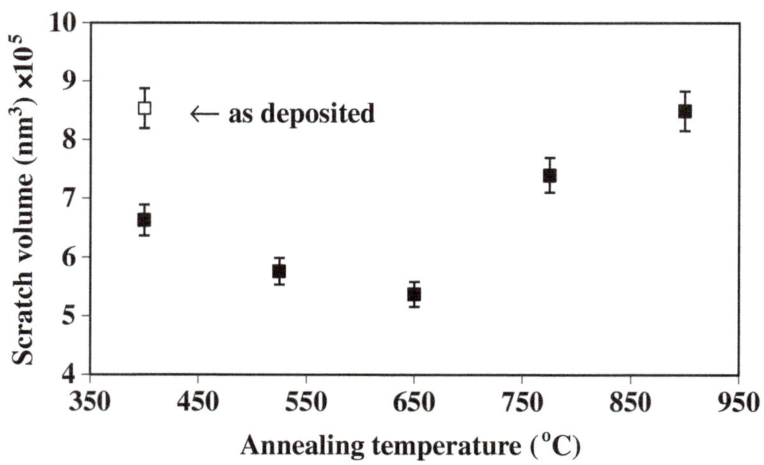

Figure 9 Variation of scratch volume of as-deposited and annealed Mo/Si samples as a function of annealing temperature.

different annealing temperatures in the range of 400°C to 900°C were studied. XRD, AFM, and SEM were employed for the study of the nanostructure and surface morphology. The mechanical and tribological properties of these layers were also considered using nano-indentation and scratch tests. The results showed that an annealing temperature of 650°C was a critical temperature for structural, mechanical, and tribological properties. The annealed sample at a lower temperature of 650°C had a γ-Mo_2N phase with FCC structure, smaller grains, and better mechanical properties, and the annealed sample at 775°C and 900°C had a Mo_2N phase with tetragonal structure and larger grains. The scratch test results also showed that scratch volume and friction of coefficient decreased by increasing the annealing temperature up to 650°C, and then increased.

Competing interests
The authors declare that they have no competing interest.

Authors' contributions
All authors provided the same contributions in this article. All authors read and approved the final manuscript.

Acknowledgments
This work was carried out with the support of the Islamic Azad University, Chalous Branch and Central Tehran Branch. The authors are also grateful to Dr. Zare.

Author details
[1]Department of Physics, Chalous Branch, Islamic Azad University, Chalous 46615-397, Iran. [2]Department of Physics, Faculty of Science, Central Tehran Branch, Islamic Azad University, Tehran 13185-786, Iran. [3]Department of Physics, University of Tehran, North Kargar Street, Tehran 14395-515, Iran.

References
1. Chuang, JC, Tu, SL, Chen, MC: Sputter-deposited Mo and reactively sputter-deposited Mo-N films as barrier layers against Cu diffusion. Thin Solid Films **346**, 299–306 (1999)
2. Alen, P, Ritala, M, Arstila, K, Keinonen, J, Leskela, M: Atomic layer deposition of molybdenum nitride thin films for Cu metallizations. J. Electrochem. Soc. **152**(5), 361–366 (2005)
3. Tewg, JY, Kuo, Y, Lu, J: Zirconium-doped tantalum oxide gate dielectric films integrated with molybdenum, molybdenum nitride, and tungsten nitride gate electrodes. J. Electrochem. Soc. **152**(8), 643–650 (2005)
4. Tsui, BY, Huang, CF, Lu, CH: Investigation of molybdenum nitride gate on SiO_2 and HfO_2 for MOSFET application. J. Electrochem. Soc. **153**(3), 197–202 (2006)
5. Hones, P, Martin, N, Regula, M, Levy, F: Structural and mechanical properties of chromium nitride, molybdenum nitride, and tungsten nitride thin films. J. Phys. D.: Appl. Phys. **36**, 1023–1029 (2003)
6. Gulbinski, W, Suszko, T: Thin films of Mo_2N/Ag nanocomposite – the structure, mechanical and tribological properties. Surf. Coat. Technol. **201**, 1469–1476 (2006)
7. Savvides, N: High Tc superconducting B1 phase MoN films prepared by low-energy ion-assisted deposition. J. Appl. Phys. **62**, 600–611 (1987)
8. Shi, YH, Zhao, BR, Zhao, YY, Li, L, Liu, JR: Superconducting and normal-state properties of MoNx thin films. Phys. Rev. B **38**, 4488–4491 (1988)
9. Inumaru, K, Baba, K, Yamanaka, S: Synthesis and characterization of superconducting b-Mo2N crystalline phase on Si substrate: an application of pulsed laser deposition to nitride chemistry. Chem. Mater. **17**, 5935–5940 (2005)
10. Inumaru, K, Baba, K, Yamanaka, S: Structural distortion and suppression of superconductivity in stoichiometric B1-MoN epitaxial thin films. Phys. Rev. B. **73**, 052504–052508 (2006)
11. Kendig, LP, Rek, ZU, Yalisove, SM, Bilello, JC: The role of impurities and microstructure on residual stress in nanoscale Mo films. Surf. Coatings Technol. **132**, 124–129 (2000)
12. Windt, D, Vac, J: Stress, microstructure and stability of Mo/Si, W/Si, and Mo/C multilayer films. Sci. Technol. A **18**, 980–991 (2000)
13. Roberson, SL, Finello, D, Davis, RF: Growth of MoxN films via chemical vapor deposition of $MoCl_5$ and NH_3. Surf. Coat. Tech. **102**, 256–259 (1998)
14. Roberson, SL, Finello, D, Davis, RF: Electrochemical evaluation of molybdenum nitride electrodes in H2SO4 electrolyte. Thin Solid Films **324**, 75–80 (1998)
15. Lee, JY, Jeon, SR, Park, JW: Effect of deposition conditions on the physical and electrical properties of reactive sputtered molybdenum nitride film. J. Mater. Sci. Lett. **15**(17), 1495–1497 (1996)
16. Kattelus, H, Koskenala, J, Nurmela, A, Niskanen, A: Stress control of sputter-deposited Mo–N films for micromechanical applications. Microelec. Eng **60**, 97–105 (2002)
17. Anitha, VP, Major, S, Chandrashekharam, D, Bhatnagar, M: Deposition of molybdenum nitride thin films. Surf. Coat. Technol. **79**, 50–54 (1996)
18. Savaloni, H, Motmaen-Dadgar, M, Ghoranneviss, M, Hantehzadeh, MR: Temperature and N+ energy dependence on nano-structural modifications and characteristics of Mo surface. Appl. Surf. Sci. **253**, 2915–2923 (2006)
19. Shen, YG: Effect of deposition conditions on mechanical stresses and microstructure of sputter-deposited molybdenum and reactively sputter-deposited molybdenum nitride films. Mater. Sci. Engin. A. **359**, 158–167 (2003)
20. Hainsworth, SV, Soh, WC: The effect of the substrate on the mechanical properties of TiN coating. Sur. Coat. Technol. **163–164**, 515–520 (2003)
21. Holleck, H: Material selection for hard coatings. J. Vac. Sci. Technol. A. **4**, 2661–2669 (1986)
22. Patsalas, P, Charitidis, C, Logothetidis, S: The effect of substrate temperature and biasing on the mechanical properties and structure of sputtered titanium nitride thin films. Sur. Coat. Tech. **125**, 335–340 (2000)
23. Wang, TG, Jeong, D, Liu, Y, Wang, Q, Iyengar, S, Melin, S, Kim, KH: Study on nanocrystalline Cr2O3 films deposited by arc ion plating: II. Mechanical and tribological properties. Surf. Coat. Tech **206**(10), 2638–2644 (2012)
24. Harlin, P, Carlsson, P, Bexell, U, Olsson, M: Influence of surface roughness of PVD coatings on tribological performance in sliding contacts. Surf. Coat. Technol. **201**, 4253–4259 (2006)

Size-dependent *in vivo/in vitro* results of homoeopathic herbal extracts

Rajesh Barve[1] and Ramesh Chaughule[2*]

Abstract

This paper presents evidence-based results of clinical trial on the use of nanosized particles in homoeopathic *Terminalia arjuna* tincture. Dynamic light scattering technique demonstrates that homoeopathic tincture is a mixture of particles varying from micron to nanometre sizes. The process of succussion, homogenization and filtration brings down the particles of source material to nanosize as a 'top-down' process. As compared to the conventionally used micron-sized *T. arjuna* powder, the results show significant clinical improvement due to the presence of nanoparticles in tincture by the above method. Similar results were reconfirmed on experiments with bacterial cultures. From clinical trials, it has been confirmed that the dose of medicine required by the patient is significantly reduced with the reduction in particle size. No toxicity was observed even after giving medicine for more than 8 weeks. Size-dependent sensitivity of the drug can help to tackle the problem of drug resistance in bacteria.

Keywords: Lipid profile, *Terminalia arjuna*, Homoeopathy, Clinical trials, DLS method, *Holarrhena antidysenterica*, *E. coli* cultures, Zone of inhibition

Background

Holarrhena antidysenterica is a tropical plant which grows up to 5 m in height. The bark extract of this plant is used in ayurvedic and homoeopathic systems of medicine for the treatment of tropical diarrhoeas and dysentery. It belongs to Apocynaceae family. Around 30 alkaloids have been isolated from the plant, mostly from the bark. These include conessine, kurchine, kurchicine, holarrhimine, conarrhimine, conaine, conessimine, isoconessimine, conimine, holacetin and conkurchin [1]. It is a well-known drug for amoebic dysentery and other gastric disorders. It is also indicated for the treatment of diarrhoea, indigestion, flatulence and colic [2]. Homoeopathic system of medicine uses ethanolic extracts, called as mother tincture, and their dilutions (potencies). As per the Homoeopathic Pharmacopoeia of India standard, the drug strength is 1/10. Since there is no standardization criterion for the particle size of the tincture, the experimentation was carried out by changing the particle size of the tincture, and size-related effects were compared *in vitro*.

Terminalia arjuna is known over 2,000 years as good cardioprotective medicine. Many useful phytoconstituents have been isolated from *T. arjuna* which includes triterpenoids for cardiovascular properties, tannins and flavonoids for its anticancer properties and so on. The bark leaves and fruits of *T. arjuna* have been used in the indigenous system of medicine for different ailments [3]. The bark contains a very high level of flavonoids compared to other commonly used plant items. Flavonoids detected from its bark are arjunolone, flavones, bicalein, quercetin, kempferol and pelorgonidin [4-6]. Traditionally in ayurvedic medicine, *T. arjuna* powder with a dose of 1 to 2 g/day and an average powder size of about 50 μm is used for the patients. *T. arjuna* is also known to reduce lipid levels. This was verified by *in vivo* studies on 32 hyperlipidemic patients by giving them this mother tincture which has much lesser particle size.

Results and discussions

Table 1 shows the results of ten patients who took *T. arjuna* dose of 0.083 g for 8 weeks or more. The patients showed significant reduction in low-density lipoprotein (LDL), very low-density lipoprotein (VLDL), triglyceride (TG) and total cholesterol as compared to the traditional dose of 1 to 2 g/day of *T. arjuna* powder

* Correspondence: rsctifr@yahoo.com
[2]Department of Chemistry, Ramnarain Ruia College, Matunga, Mumbai 400019, India
Full list of author information is available at the end of the article

Table 1 ***In vivo*** **experiment of** ***T. arjuna*** **tincture for certain hyperlipidemic patients taking drug for 8 weeks or more**

Patient number	Total cholesterol reduction (%)	HDL increase (%)	LDL reduction (%)	VLDL reduction (%)	TG reduction (%)
1	27	36	35	45	46
5	25.8	-	41.8	19	18.9
8	37	50	30	-	23.2
11	42	21	37	-	5.2
13	23	5	26.1	24.2	23.9
20	28	7.1	39	10	31.5
25	15.6	14.2	25.2	50	46.9
27	13.8	33.3	24.8	41.6	41.6
29	25.4	21	46.2	33.3	12.2
30	17.7	4.6	11.1	42.3	48.7

with a mean particle size of 50 μm available in the market. There is also an increase in high-density lipoprotein (HDL) levels by about 20% which is hardly seen in any other anti-lipid drug as HDL is cardioprotective. Figure 1 represents *Escherichia coli* M200 culture with 2.67-μm particle size of *Holarrhena* tincture on LHS plate showing no zone of inhibition. However, with 13.8-nm particle size of the same tincture on RHS plate, *E. coli* M200 gives 10 mm of zone inhibition. Figure 2 represents *E. coli* 10536 culture with *Holarrhena* tincture of particle size 2.67 μm, and Figure 3 shows *E. coli* 10536 culture with *Holarrhena* tincture having particle size of 13.8 nm.

Conclusions

This study clearly demonstrates particle size-dependent variation in the drug action. Reduction in the particle size of the drug improves the sensitivity of the drug as seen in *in vivo* experiment of *T. arjuna* drug. Reduction of particle size from 50 μm to 23 nm showed drastic reduction in dose requirement, i.e. from 1 to 2 g/day to 0.083 g/day. No toxicity was observed with this nanodose even after giving for 2 months or more.

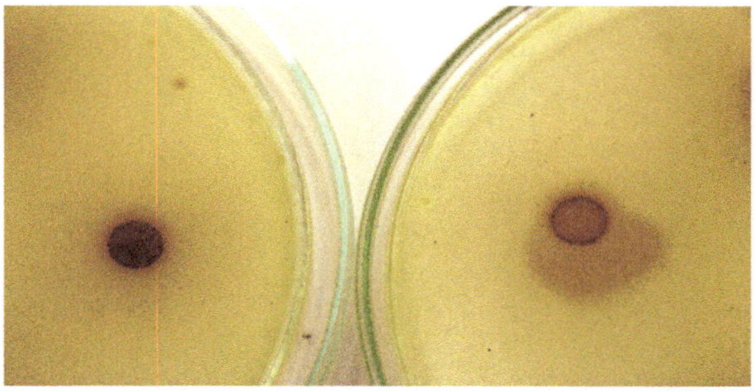

Figure 1 *E. coli* **M200 culture with** *Holarrhena* **tincture of 2.67 μm and 13.8 nm in size.**

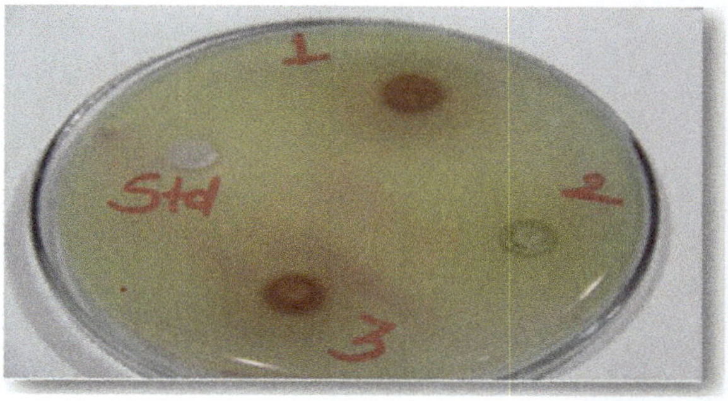

Figure 2 *E. coli* **10536 culture with** *Holarrhena* **tincture of 2.67-μm size (1 and 3), alcohol (2) and blank disc (Std).**

From *in vitro* experiments with *E. coli*, it is clear that reduction of particle size of tincture improved the culture results by 15% or more. Interestingly, *E. coli* M200 showed resistance to *Holarrhena* tincture at 2.67-μm particle size but showed sensitivity at 13.8-nm particle size tincture. This phenomenon is reported for the first time and can help to reduce drug resistance problem for the patients in the future.

Methods

T. arjuna tincture manufactured by Bakson, India was used for *in vivo* clinical trials using the dynamic light scattering (DLS) system of Aimil Malvern Instruments from Baroda. Its mean particle size was 23 nm (Figure 4). *Holarrhena* tincture also manufactured by Bakson, India was used for bacterial studies. The particle size determined by DLS was found to be 2.67 μm (Figure 5). Further reduction in particle size of this tincture to 13.8 nm (Figure 6) was achieved by standard succussion, homogenization and filtration methods.

Four pathogenic strains of *E. coli* (11100, 11105, 10536, M200) were collected and inoculated in culture plates (Table 2). Nutrient agar was used as culture media. Sterile Whatman paper discs of 5-mm-diameter size were impregnated with 0.1 ml *Holarrhena antidysenterica* tincture with particle sizes of 2.67 μm and 13.8 nm, respectively. Discs

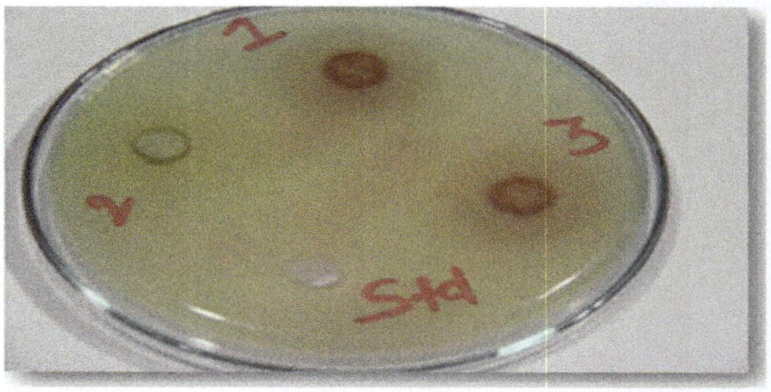

Figure 3 *E. coli* **10536 culture with** *Holarrhena* **tincture of 13.8-nm size (1 and 3), alcohol (2) and blank disc (Std).**

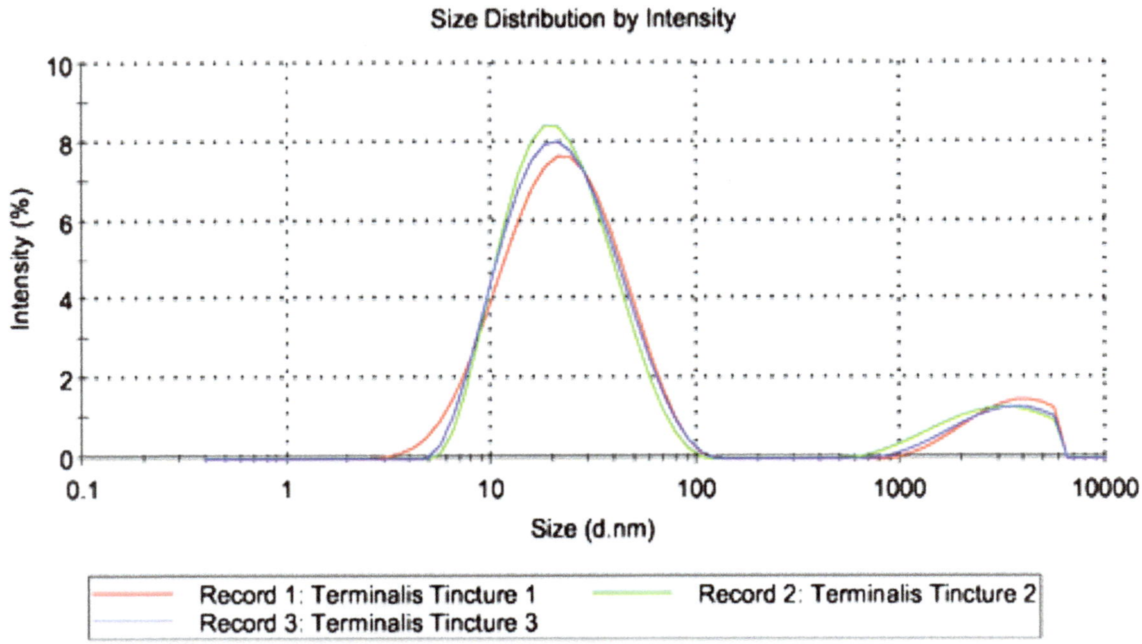

Figure 4 Intensity distribution of arjuna nanoparticles using dynamic light scattering (DLS) system.

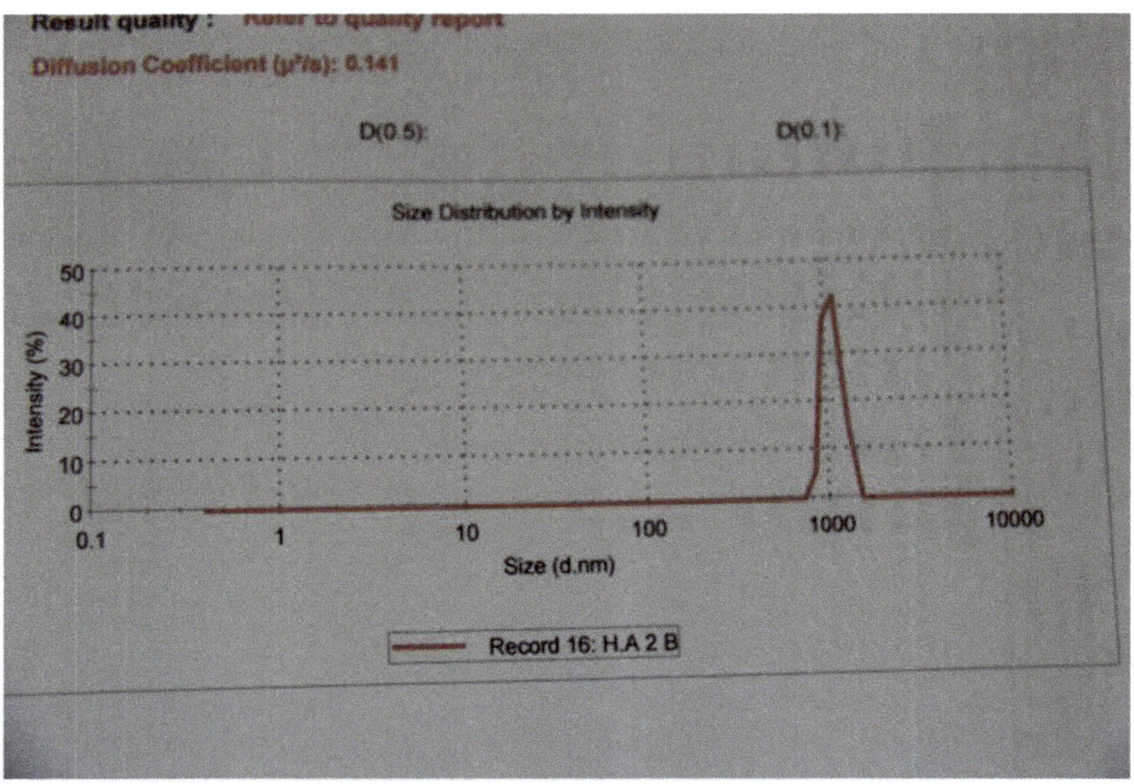

Figure 5 *Holarrhena* tincture particle size (2.67 μm) using DLS method.

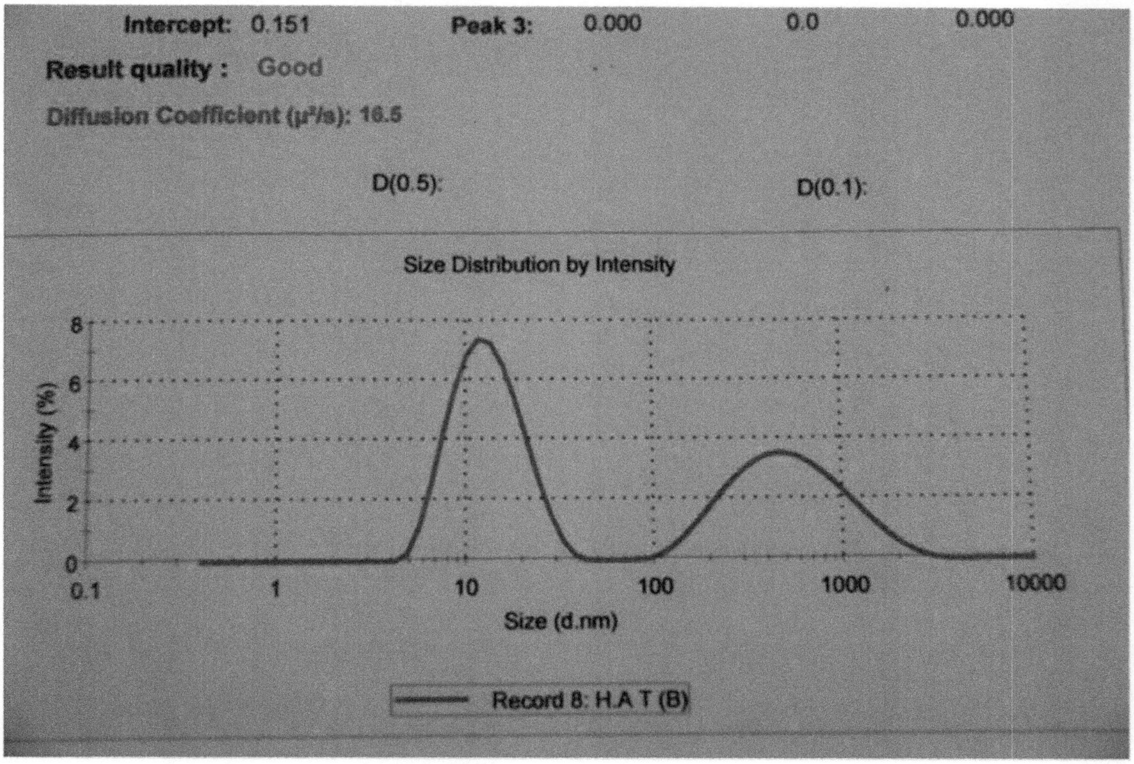

Figure 6 Reduction of *Holarrhena* tincture particle size (13.8 nm).

were placed on agar media inoculated with bacterial cultures. The disc with 0.1 ml ethanol was kept as a control since tinctures contain 60% ethanol. Culture plates were incubated at 32°C for 36 h to get good bacterial colonies. Finally, zones of inhibitions were measured and compared to know the efficacy of the drug and effect of particle sizes (Figures 1, 2, 3).

All 32 hyperlipidemic patients, who were not taking any anti-lipid medicines, were randomly selected. Their vital data regarding lipid levels, diet, habits, type of work, stress level and history of present- and past-associated illnesses were noted. Every patient was given oral dose of 10 drops of *T. arjuna* tincture with water at bedtime. The calculated weight of this 23-nm particle size tincture dose is 0.083 g. Dose was given for 8 weeks and more to compare the changes in lipid levels (Table 1).

Table 2 ***E. coli*** **and** ***H. antidysenterica*** **tincture**

	Holarrhena tincture zone of inhibition (mm)	
	2.67 μm	**13.8 nm**
E. coli M200	No zone	10
E. coli 10536	24	26
E. coli 11105	22	25
E. coli 11100	24	25

Competing interests

The authors declare that they have no competing interests.

Authors' contributions

RB and RC carried out the *T. arjuna* and *Hollarrhena antidysenterica* studies and participated in the experimental work on the *in vitro* studies of bacterial cultures. RB treated the specific patients in his clinic by giving nanodose of *T. arjuna* tincture. Both the authors provided the same contributions in this article and read and approved the final manuscript.

Acknowledgements

The authors gratefully thank Dr. Medha Shette, Dr. Vivek Athavale and Dr. Pusalkar for their help in the pathological work.

Author details

[1]Department of Repertory, Virar Homoeopathic Medical College, Virar, 401305, Maharashtra, India. [2]Department of Chemistry, Ramnarain Ruia College, Matunga, Mumbai 400019, India.

References

1. Kumar, N, Singh, B, Bhandari, P, Gupta, AP, Kaul, VK: Steroidal alkaloids from *Holarrhena antidysenterica* (L.) Wall. Chem. Pharm. Bull (Tokyo) **55**(6), 912 (2007)
2. Singh, KP: Clinical studies of amoebiasis and giardiasis evaluating the efficacy of kutaja (*Holarrhena autidysenterica*) in Eutamoeta histotylica cyst passes. Ancient Science of Life **5**, 228–231 (1986)
3. Warrier, PK, Nambiar, VPK, Ramankutty, C: Indian Medicinal Plants: A Compendium of 500 Species, p. 592. Orient Blackswan, India (1996)
4. Sharma, PN, Shoeb, PN, Kapil, RS, Popli, SP: Arjunolone – a new flavone from stem bark of *Terminalia arjuna*. Indian Journal of Chemistry **21B**, 263 (1982)

5. Nema, R, Jain, P, Khare, S, Pradhan, A, Gupta, A, Singh, D: Preliminary phytochemical evaluation and flavonoids quantification of *Terminalia arjuna* leaves extract. Int. J. Pharm. Phytopharmacol. Res. **1**(5), 283 (2012)
6. Nema, R, Jain, P, Khare, S, Pradhan, A, Gupta, A, Singh, D: Antibacterial and antifungal activity of *Terminalia arjuna* leaves extract with special reference to flavonoids. Basic Research J. Med. Clin. Sc. **1**(5), 63 (2012)

18

Investigation of a QM/MM study on interaction of a carbon nanotube with cytarabine drug in various solvents and temperatures

Fahimeh Najafi

Abstract

At the present time, usage of nanotubes in medicinal science and biology is further studied. Nanotubes can pass through cell walls, transfer, and liberate drugs in particular tissues. The goal of this article is to survey the interaction of a nanotube with an anticancer drug. In this study, transferring of an anticancer drug called cytarabine with zigzag (5, 0) carbon nanotube is studied. All computations are surveyed by quantum mechanics and molecular mechanics (MM)/Monte Carlo simulation (in various force fields such as MM+, assisted model building with energy refinement, and optimized potential for liquid simulations) in various temperatures, and their results are achieved in gas phase, water solvent, and methanol solvent, respectively.

Keywords: Solvent effect, Quantum mechanics, Cytarabine, Zigzag (5, 0) carbon nanotube, Force field

Background

Cytarabine is a chemotherapy agent utilized chiefly in the therapy of hematological malignancies like non-Hodgkin lymphoma and acute myeloid leukemia [1]. In addition, cytarabine has been known as Ara-C (arabinofuranosyl cytidine) [2]. In two randomized trials, the utilization of interferon and intermittent parenteral cytarabine has been estimated [3,4]. This combination raises the cytogenetic response rate further, and in one trial [3], it enhanced overall survival compared with interferon alone. However, subcutaneously administered cytarabine has some defects. Following parenteral administration by the ubiquitous enzyme deoxycytidine deaminase, which results in a comparatively short serum half-life, it quickly undergoes systemic deamination [5]. It needs daily subcutaneous injection to render the drug and is also associated with upper gastrointestinal symptomatology, particularly mucositis and nausea [3]. In addition, carbon nanotubes (CNTs) have attracted an uninterrupted interest in different areas since their discovery because of their unique structure and electrical/mechanical characteristics [6,7] which make them interesting for many potential usages in electronic devices, sensors, structural integrity [8], biomedical engineering [9-11], nanoinjectors [12], drug delivery [13], gene therapy [14], and in catalysis supports. We can categorize CNTs as single-wall carbon nanotubes (SWCNTs) and multiwall carbon nanotubes (MWCNTs) [15]. In fact, SWCNTs are grapheme cylinders with a wall thickness of 1 atomic layer and a diameter of around 1 nm. High aspect ratio, enormous surface, hollow geometry [6] are some their exciting characteristics. Other characteristics include high-current carrying capacity, superior thermal properties, fast response, and good reversibility [7,16]. SWCNTs have three configurations. It will be in the form of an armchair if the nanotubes $n = m$ and whenever $n = 0$ or $m = 0$, it will be in zigzag form with any other n and m, it will be in chiral variety. Each of these configurations has its own characteristics and properties [17,18]. MWCNTs are made of multiple rolled layers (concentric tubes) of graphite. Using two models, we can depict the configurations of multiwall nanotubes. In one of these models, the Russian model, sheets of graphite are set in concentric cylinders, e.g., (0, 8) SWCNTs are within a larger (0, 17) single-wall nanotube. In the second model, the parchment model, a single sheet of graphite is rolled around itself, just like a scroll of parchment or a rolled newspaper. In multiwall nanotubes, the interlayer distance is somehow equal to the distance between grapheme layers in graphite, which is nearly 3.4 Å. Two concentric cylindrical grapheme layers conform to double-wall carbon

Correspondence: fahimnajafi@Gmail.com
Department of Chemistry, Roudehen Branch, Islamic Azad University, Roudehen, Iran

nanotubes. Accordingly, they blend with the interesting electronic characteristic of SWCNTs with the superior mechanical strength of MWCNTs [19].

This paper is concentrated on the interaction of a carbon nanotube with an anticancer drug named cytarabine. Nowadays, in the universe of new medicine, carbon nanotubes have experimentally showed their capacity in passing through the cell shell. Accordingly, scientists believe that they can apply them in releasing active drug molecules in the cell, particularly in the most sensitive and essential molecules for special diseases such as cancer, AIDS. The physical and chemical nature of these nanotubes has been studied by many scientists in order to prepare this material for such a significant task. Scientists pay special attention to carbon nanotubes consisting of fullerenes [20] because of their unequaled electrical, optical, and thermal characteristics. Scientists are very hopeful to develop and utilize carbon nanotubes in releasing vaccines by conducting basic and fundamental projects. Releasing drugs in cancer cells without harming healthy cells of tissues under study is very significant. After a number of research, we can understand that nanotubes can do this task correctly [21,22]. With the use of molecular mechanics (MM)/ Monte Carlo (MC) simulation (in various force fields such as MM+, assisted model building with energy refinement (AMBER), and optimized potential for liquid simulations (OPLS)), the interaction of cytarabine with zigzag (5, 0) carbon nanotube was studied. Furthermore, we have studied the effects of solvents on their interaction by means of quantum mechanics (QM). We applied the self-consistent reaction field method (SCRF) for the simulation of solvent effects by QM method. Thus, it can be seen that we described the solvent effects on the thermodynamic values of this interaction using QM and MM methods.

Results and discussion

Here, the combination schema of cytarabine anticancer drug and zigzag (5, 0) carbon nanotube is shown in Figure 1. We have examined the thermodynamic values of the interaction of cytarabine with zigzag (5, 0) carbon nanotube in different dielectric constants.

In this work, we have done the said calculation by studying the following reaction:

$$\text{Cytarabine} + \text{SWCNT} \rightarrow \text{Cytarabine–SWCNT} \quad (1)$$

Thus, in the first stage, we studied various force fields to estimate potential energy (kcal mol^{-1}) and other kinds of geometrical parameters on the interaction of cytarabine with the zigzag (5, 0) carbon nanotube. By utilizing three various force fields, the potential energy of molecules will not be similar. Thus, to compare the calculated potential energy of one specific molecule with a special force field and with the potential energy of another molecule computed in another force field is neither reasonable nor possible. In this examination, we revealed the differences in force fields by comparing the calculated potential energy utilizing different force fields such as MM+, AMBER, and OPLS that are show in (Figures 2, 3, and 4). Thus, it can be observed which water solvent has the lowest value of potential energy alterations. Also, it can be seen why OPLS is the best force field in the comparison of potential energy alterations in the water solvent (Figure 5, Table 1). Using different force fields, we can obtain the theoretical potential energy values; these are done by the joining of or attraction of the Van Der Waals force caused by dipole-dipole interactions and the experimental repelling forces caused by Pauli repulsion. In the opposite direction of the encircling medium, a dipole moment is made by the

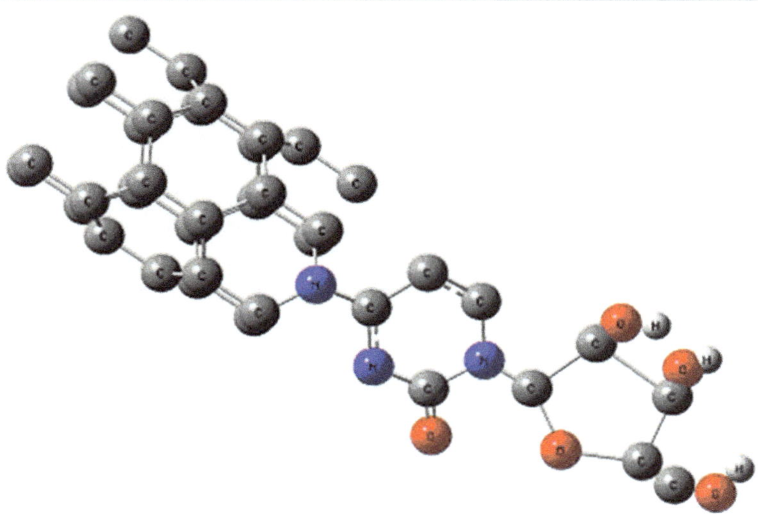

Figure 1 Shape of (cytarabine and SWCNT) system.

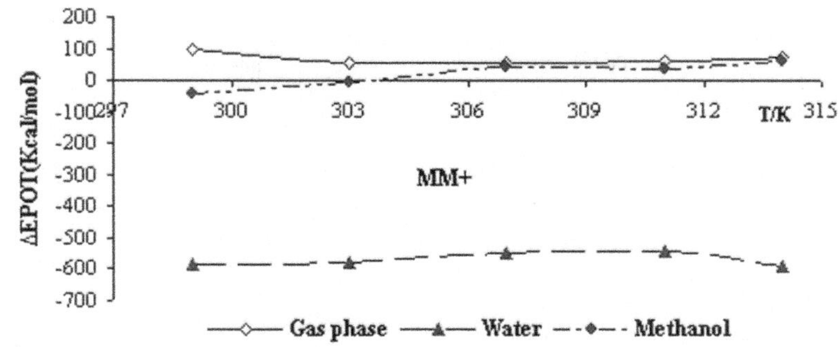

Figure 2 **Potential energy alterations versus temperature in different solvents and MM+ force field for reaction 1.**

thermodynamic values of solute dipole moment. Conversely, the medium polarizes the charge distribution in the solvent. The dipole moment values of cytarabine and zigzag (5, 0) carbon nanotube, (cytarabine/zigzag (5, 0) carbon nanotube) system in gas phase, and water and methanol at various temperatures have been related. It is revealed that dipole moment values in gas phase, water, and methanol increase with the increasing dielectric constant; we demonstrate these results in Table 2. A continuous dielectric constant signifies the presence of a solvent with a given dielectric constant in Onsager reaction field template. We must consider that the solute to be fixed into the cavity (usually with a spherical shape), with its radius in the molecule, and the electric field, which is related to the solute by the solvent dipole, will interact with the molecular dipole and will be directed to net stabilization. Now, it must be noted that salvation energy computed by SCRF method is the electrostatic distribution to the free energy of salvation. In this study, the cavity radii of cytarabine, zigzag (5, 0) carbon nanotube, and (cytarabine/zigzag (5, 0) carbon nanotube) system were calculated and were then substituted while efficiency was improving. We have also achieved the volumes of cytarabine, zigzag (5, 0) carbon nanotube, and (cytarabine/zigzag (5, 0) carbon nanotube) system. After computing the values of cytarabine, zigzag (5, 0) carbon nanotube, and (cytarabine/zigzag (5, 0) carbon nanotube) system, we came up with these outcomes: 8.70, 5.06, 5.73 Å, respectively. We have reached the point wherein, along the increasing solvent dielectric constant, the dipole moment for the anticancer drug, cytarabine, zigzag (5, 0) carbon nanotube, and (cytarabine/zigzag (5, 0) Carbon nanotube) system in different ranges of temperatures (between 299 and 314 K) also increased, which is based on the dipole moment data in Table 2. After computing the values for the Onsager template, the fact that Gibbs free energy, energy, enthalpy, and entropy values of cytarabine, zigzag (5, 0) carbon nanotube, and (cytarabine/zigzag (5, 0) carbon nanotube) system are sensitive to the polarity of the encircling solvent. Thus, the Gibbs free energy, enthalpy, entropy, and energy values are revealed according to this approach in Table 3. All of these values have relationships with the dielectric constant. According to Table 3, everybody can understand that by increasing the temperature from 299 to 314K, the thermodynamic values of Gibbs free energy of cytarabine, zigzag (5, 0) carbon nanotube, and (cytarabine/zigzag (5, 0) carbon nanotube) system in gas phase, water, and methanol will decrease. We can get the lowest amount of Gibbs free energy at 314 K. The achieved Gibbs free energy values decrease, in all inspected temperature, if the solvent dielectric constant increases. For example, Gibbs free energy obtains the most negative value whenever cytarabine, zigzag (5, 0) carbon nanotube, and (cytarabine/zigzag (5, 0) carbon nanotube) system are released in water solvent. More or less, Table 3 shows the obtained Gibbs free energy, enthalpy, energy, and entropy values in different

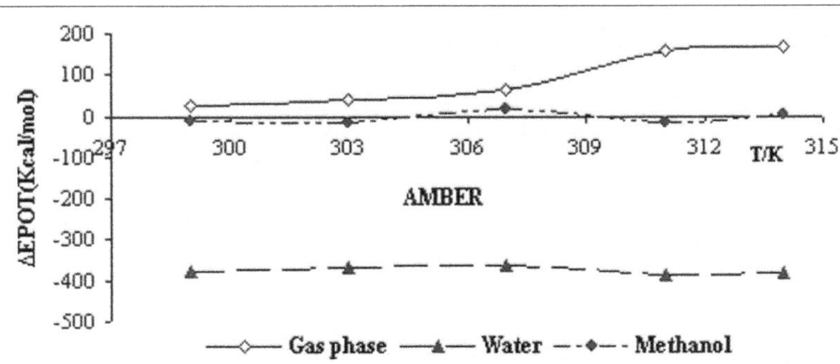

Figure 3 **Potential energy alterations versus temperature in different solvents and AMBER force field for reaction 1.**

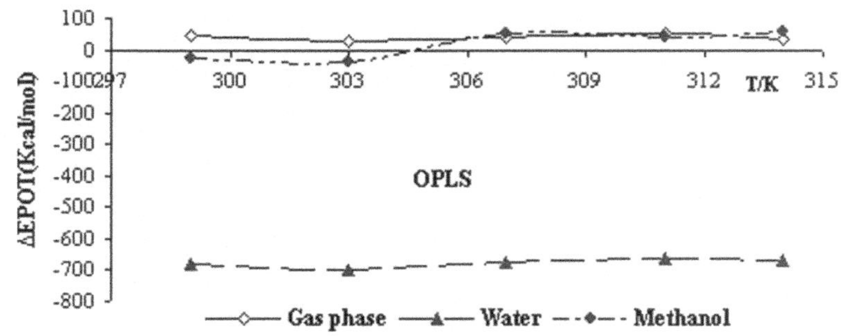

Figure 4 Potential energy alterations versus temperature in different solvents and OPLS field for reaction 1.

solvents. According to the temperature that changed between 299 and 314 K, the changes of energy, enthalpy, and entropy of cytarabine, zigzag (5, 0) carbon nanotube, and (cytarabine/zigzag (5, 0) carbon nanotube) system in gas phase, water, and methanol can be observed. Thus, we get to this point wherein the most positive value of energy, enthalpy, and entropy is 314 K if the temperature increases from 299 to 314 K. The other important point is that, after computing the values of energy, enthalpy, and entropy based on the solvent dielectric constant and increased values in temperature between 299 and 314 K, the value of energy and enthalpy becomes negative. The lowest values of energy and enthalpy and positive values of entropy can be obtained when cytarabine, zigzag (5, 0) carbon nanotube, and (cytarabine/zigzag (5, 0) carbon nanotube) system are in water solvent. In such cases, we can observe the most positive value of entropy.

Study the following reaction:

$$\text{Cytarabine} + \text{SWCNT} \rightarrow \text{Cytarabine–SWCNT}$$

In Table 4, we can observe the facts about the (cytarabine/zigzag (5, 0) carbon nanotube) system. If the solvent dielectric constant increases, the Gibbs free energy values of the interaction between solvent and solute molecules decrease. Noticeably, in Table 4, we can understand that the increase of solvent dielectric constant makes the enthalpy values of interaction between solvent and solute molecules move toward negative amounts. It reveals that the interaction between solvent and solute molecules is strong. According to Equation 1, the achieved entropy values are reasonable. In fact, the above reaction suggests that by coupling the anticancer drug of cytarabine with zigzag (5, 0) carbon nanotube, the entropy lessens. Now, we estimate and survey the data from another viewpoint on Equation 1:

We can see diagrams of $\Delta\mu_{\text{reaction}}$, $\Delta G_{\text{reaction}}$, $\Delta H_{\text{reaction}}$, and $\Delta S_{\text{reaction}}$ according to the temperature changes in Figures 6, 7, 8, and 9, respectively, in which the values of $\Delta\mu_{\text{reaction}}$, $\Delta G_{\text{reaction}}$, $\Delta H_{\text{reaction}}$, and $\Delta S_{\text{reaction}}$ are obtained on the basis of following function:

$$\Delta J = \Delta J_{\text{cytarabine-SWCNT}} - \left(\Delta J_{\text{cytarabine}} + \Delta J_{\text{SWCNT}}\right)$$

In this function, the J parameter is equal to the dipole moment values μ for Figure 6, Gibbs free energy values G for Figure 7, enthalpy values H for Figure 8, and entropy values for Figure 9. In Figure 6, we can recognize which of the values will change by the increasing solvent constant, $\Delta\mu_{\text{reaction}}$, in Figure 1 if the temperature is increased up to 314 K. Moreover, Figure 7 elaborates that the increase of temperature will also heighten the value

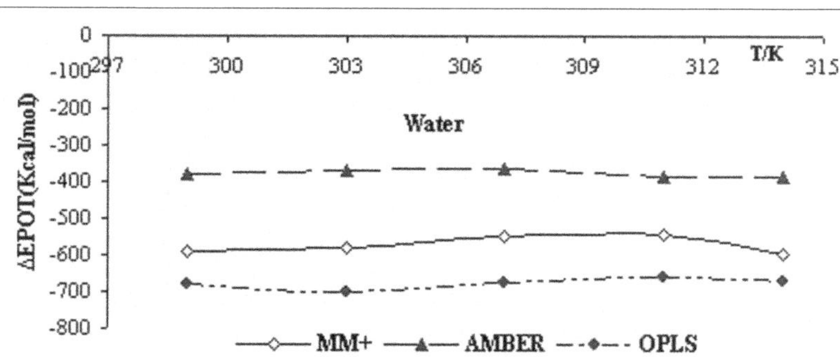

Figure 5 Potential energy alterations versus temperature for water solution in different force fields for reaction 1.

Table 1 Theoretical potential energy (kcal mol^{-1}) values in different solvents and same force fields

Media	Temperature (K)	$E_{Potential\ energy}$ (kcal mol^{-1})									$\Delta E_{Potential\ energy}$ (kcal mol^{-1})		
		Cytarabine			SWCNT			Cytarabine-SWCNT					
		MM+	AMBER	OPLS	MM+	AMBER	OPLS	MM+	AMBER	OPLS	MM+	AMBER	OPLS
Gas phase	299	44.73204	141.94	48.44767	213.1887	162.8669	213.1573	352.448	332.8886	306.8255	94.52726	28.0817	45.22053
	303	45.02634	171.42	66.91479	212.2316	136.945	224.6387	315.1821	346.3216	319.7139	57.92416	37.9566	28.16041
	307	32.37732	131.4695	54.23368	220.7707	167.3872	231.0695	309.8467	363.5419	326.0575	56.69868	64.6852	40.75432
	311	55.4327	44.88987	57.82224	218.8069	153.9778	214.3011	335.407	356.1648	323.0684	61.1674	157.29713	50.94506
	314	40.52354	43.73027	57.3092	218.3201	164.461	219.1929	332.468	374.97	308.1852	73.62436	166.77873	31.6831
Methanol	299	122.7491	96.822	94.18385	209.3418	182.3241	187.874	287.887	269.6387	255.4425	−44.2039	−9.5074	−26.6154
	303	97.2437	91.61169	133.0873	202.8216	188.161	193.7327	295.4447	262.1135	290.104	−4.6206	−17.65919	−36.716
	307	42.40936	99.70736	49.09336	206.8541	170.8445	195.9675	289.2411	288.5759	297.5085	39.97764	18.02404	52.44764
	311	43.54883	86.28603	40.9453	200.088	181.7183	189.4955	282.5071	251.2833	269.6509	38.87027	−16.72103	39.2101
	314	45.18951	93.74397	43.58532	205.6906	178.9124	180.3274	311.3402	274.331	280.4594	60.46009	1.67463	56.54668
Water	299	503.1399	274.194	526.6146	305.8046	268.0094	298.9091	220.7991	162.7351	146.6212	−588.145	−379.4683	−678.903
	303	503.894	272.4241	531.7231	304.791	244.0615	317.1772	228.5874	150.1159	150.7781	−580.098	−366.3697	−698.122
	307	500.529	275.7412	525.9601	291.5029	234.2437	295.3419	244.3345	144.6781	146.1943	−547.697	−365.3068	−675.108
	311	500.506	275.4447	526.4454	300.6973	273.7059	284.4701	257.0042	164.1786	151.0472	−544.1929	−384.972	−659.868
	314	501.2367	270.8107	531.7561	313.3305	262.9936	284.886	221.697	149.4395	147.7274	−592.870	−384.3648	−668.915

$\Delta E_{Potential\ energy}$ (kcal mol^{-1}) = Difference of energy for (cytarabine-SWCNT) system + Sum of cytarabine and SWCNT for reaction 1 in different dielectric constants.

Table 2 Theoretical dipole moment μ (Debye) and $\Delta\mu$ (Debye) values in various temperatures for three phases

Temperature (K)	μ (Debye)			$\Delta\mu$ (Debye)	μ (Debye)			$\Delta\mu$ (Debye)	μ (Debye)			$\Delta\mu$ (Debye)
	Cytarabine	SWCNT	Cytarabine-SWCNT		Cytarabine	SWCNT	Cytarabine-SWCNT		Cytarabine	SWCNT	Cytarabine-SWCNT	
	Gas phase				Water				Methanol			
299	9.0282	0.0035	1454.3548	1445.3231	9.2747	0.0049	3048.6991	3039.4195	9.2579	0.0045	2876.4738	2867.2114
303	9.0279	0.0039	1454.3541	1445.3223	9.2756	0.0049	3048.6995	3039.419	9.2585	0.0042	2876.4735	2867.2108
307	9.0293	0.0036	1454.3549	1445.322	9. 2746	0.0047	3048.6989	3039.4196	9.2578	0.0040	2876.4737	2867.2119
311	9.0284	0.0037	1454.3551	1445.323	9. 2761	0.0048	3048.6993	3039.4184	9.2687	0.0041	2876.4732	2867.2004
314	9.0297	0.0039	1454.3546	1445.321	9. 2763	0.0046	3048.6990	3039.4181	9.2683	0.0045	2876.4735	2867.2007

$\Delta\mu$ (Debye) = Difference of dipole moment for (cytarabine-SWCNT) system + Sum of cytarabine and SWCNT for reaction 1 in different temperatures and three phases.

Table 3 Theoretical thermodynamic parameters in various temperatures for three phases at HF/3-21G level of theory

System	Media	Quantities	Temperature (K)				
			299	303	307	311	314
			Theoretical thermodynamic values				
Cytarabine	Gas phase	*G*	−548150.9401	−548150.9481	−548151.1686	−548151.296	−548151.4095
		H	−548119.9476	−548119.4974	−548119.2811	−548118.9685	−548118.7193
		E	−548120.0178	−548119.8632	−548119.5651	−548119.1063	−548118.8972
		S	0.103654	0.103794	0.103868	0.103947	0.104109
	Water	*G*	−548151.7088	−548151.9786	−548152.0916	−548152.3529	−548152.5122
		H	−548120.5483	−548120.2354	−548119.8861	−548119.6926	−548119.4326
		E	−548120.8651	−548120.4968	−548120.1091	−548119.8125	−548119.6061
		S	0.104599	0.104763	0.104904	0.105017	0.105318
	Methanol	*G*	−548151.6849	−548151.7821	−548151.8596	−548152.0382	−548152.5035
		H	−548120.5451	−548120.0847	−548119.6799	−548119.386	−548119.3087
		E	−548120.7163	−548120.4109	−548120.0616	−548119.6959	−548119.501
		S	0.104496	0.104612	0.104820	0.104991	0.105079
SWCNT	Gas phase	*G*	−707511.0458	−707511.5212	−707511.887	−707512.6031	−707512.9225
		H	−707476.036	−707475.791	−707475.433	−707475.299	−707475.175
		E	−707476.565	−707476.283	−707475.927	−707475.792	−707475.669
		S	0.117088	0.117918	0.118736	0.119948	0.120213
	Water	*G*	−707511.4476	−707511.9512	−707512.3825	−707512.7784	−707513.275
		H	−707476.1402	−707475.816	−707475.625	−707475.368	−707475.268
		E	−707476.7112	−707476.391	−707476.198	−707475.94	−707475.843
		S	0.118084	0.119256	0.11973	0.120451	0.121035
	Methanol	*G*	−707511.3785	−707511.8259	−707512.246	−707512.7319	−707513.0681
		H	−707476.101	−707475.805	−707475.601	−707475.3026	−707475.197
		E	−707476.658	−707476.312	−707476.155	−707475.853	−707475.751
		S	0.117984	0.11888	0.119364	0.120350	0.120608
Cytarabine - SWCNT	Gas phase	*G*	−710372.6617	−710372.8408	−710373.0034	−710373.1847	−710373.3096
		H	−710347.9077	−710347.9296	−710347.411	−710347.2389	−710347.084
		E	−710348.1863	−710347.8016	−710347.654	−710347.437	−710347.229
		S	0.083067	0.083205	0.083363	0.083427	0.083509
	Water	*G*	−806458.3497	−806458.3631	−806458.5791	−806458.7369	−806458.9894
		H	−806431.3989	−806431.0059	−806430.8517	−806430.6162	−806430.5546
		E	−806431.5072	−806431.217	−806430.9681	−806430.7761	−806430.6375
		S	0.090135	0.090287	0.090316	0.090420	0.0 90555
	Methanol	*G*	−806300.2457	−806300.3181	−806300.5183	−806300.8488	−806301.0853
		H	−806273.5288	−806273.3328	−806272.9297	−806272.8613	−806272.7854
		E	−806273.7822	−806273.5086	−806273.0515	−806272.9147	−806272.8582
		S	0.089654	0.089721	0.089865	0..089992	0.090127

Theoretical thermodynamic parameters *G, H, E* (kcal mol^{-1}), and *S* (kcal mol^{-1} K^{-1}) for cytarabine, SWCNT, and (cytarabine-SWCNT) system.

of $\Delta G_{reaction}$. Thus, $\Delta G_{reaction}$ has the lowest value at 299 K, and we then can reach to the point where the highest resistance for Equation 1 will occur. Also, Figure 9 reveals that the value of $\Delta S_{reaction}$ declines with increasing temperature. The most positive value, $\Delta S_{reaction}$, in Equation 1 can be observed, and it is the more resistant rate. In Table 4, we must notice that the alteration of $\Delta G_{reaction}$, $\Delta H_{reaction}$, and $\Delta S_{reaction}$, in terms of solvent dielectric constant, will decline the value of the noted $\Delta G_{reaction}$ in as much as the negative value of $\Delta G_{reaction}$ is

Table 4 Theoretical thermodynamic parameter alterations in various temperatures for three phases at HF/3-21G level of theory

Media	Quantities	Temperature(K)				
		299	303	307	311	314
		Theoretical thermodynamic values				
Gas phase	ΔG	545289.3242	545289.6285	545290.0522	545290.7144	545291.0224
	ΔH	545248.0759	545247.3588	545247.3031	545247.0286	545246.8103
	ΔE	545248.3965	545248.3446	545247.8381	545247.4613	545247.3372
	ΔS	−0.137675	−0.138507	−0.139241	−0.140468	−0.140813
Water	ΔG	449204.8067	449205.5667	449205.895	449206.3944	449206.7978
	ΔH	449165.2896	449165.0455	449164.6594	449164.4444	449164.146
	ΔE	449166.0691	449165.6708	449165.339	449164.9764	449164.8116
	ΔS	−0.132548	−0.133732	−0.134318	−0.135048	−0.135798
Methanol	ΔG	449362.8177	449363.2899	449363.5873	449363.9213	449364.4863
	ΔH	449323.1173	449323.2143	449322.3512	449321.8273	449321.7203
	ΔE	449323.5921	449322.9755	449322.1651	449322.6342	449322.3938
	ΔS	−0.132826	−0.133771	−0.134319	−0.135349	−0.13556

Theoretical thermodynamic parameter alterations ΔG, ΔH, ΔE (in kcal mol^{-1}), and ΔS (kcal mol^{-1} K^{-1}) for interaction of cytarabine with and zigzag (5, 0) carbon nanotube. Parameter alterations Δ (kcal mol^{-1}) = Difference of energy for (cytarabine-SWCNT) system + Sum of cytarabine and SWCNT for reaction 1 in three phases.

for water solvent; the water solvent is the best environment for Equation 1. According to Table 4, $\Delta H_{\text{reaction}}$ and $\Delta S_{\text{reaction}}$ values are reduced by the increase of solvent dielectric constant in temperature between 299 and 314 K. If Table 4 is considered, the fact that $\Delta S_{\text{reaction}}$ at 299 K gains the most positive value is attained; thus, the temperature 299 K will be the choice for Figure 1.

Conclusions

In this article, cytarabine, zigzag (5, 0) carbon nanotube, and (cytarabine/zigzag (5, 0) carbon nanotube) system were studied by two methods, quantum mechanics and molecular mechanic/MC simulation. Diagrams were showed that ΔE_{POT} and thermodynamic values of the reaction were dependent to the solvent dielectric constant. The calculated ΔE_{POT} values for the reaction in water are more stable because the potential energy alterations become lower. The computed values of ΔE_{POT} from force fields MM+, AMBER, and OPLS in water solvent were compared, and they showed that the OPLS force field is the best force field for the study of Equation 1. In the next calculation, we investigated the obtained results by the HF/3-21G and Gaussian 98 showing that water is the best solvent when cytarabine, zigzag (5, 0) carbon nanotube, and (cytarabine/zigzag (5, 0) carbon nanotube) system are placed in it because they have the most negative Gibbs free energy and the most positive entropy in this solvent.

Methods

Computational method

We conducted all computations by quantum mechanics and molecular mechanics/Monte Carlo simulation (in

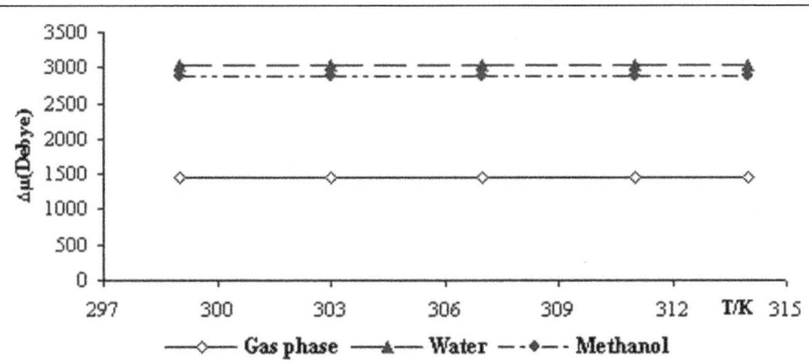

Figure 6 Dipole moment alterations calculated versus temperature for reaction 1.

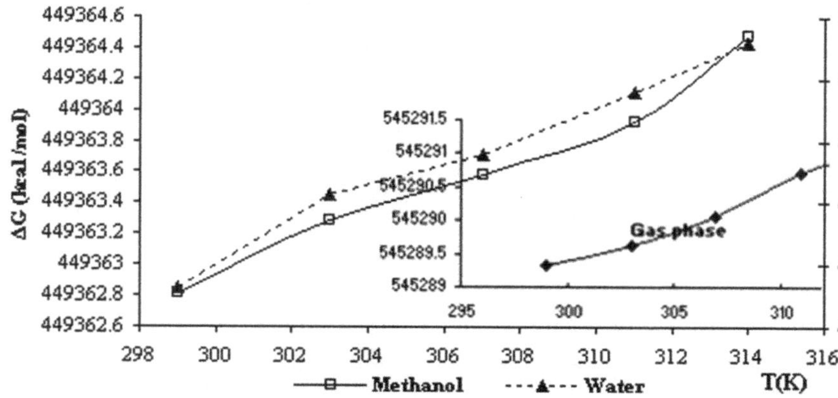

Figure 7 Gibbs free energy alterations (kcal mol^{-1}) calculated versus temperature for reaction 1 for three phases.

various force fields such as MM+, AMBER, and OPLS) in various temperatures. At first level, we carried out all calculations using the program HYPER CHEM (Hypercube, Inc., Gainesville, FL, USA). In this study, different force fields were used to determine the potential energy. Thus, we can observe the difference in force field by comparing the computed potential energy with the applied force fields namely MM+, AMBER and OPLS. At the next level, using the Hartree-Fock (HF) method, we conducted calculations of the theoretical level. It is important to choose a level which is well matched with the molecular system being studied. Accurate geometry coordinates and energy parameters control conformational interconversions; in molecular systems, they are particularly important. Low-energy structures found on each surface were closed and exposed to unlimited quantum mechanical minimization using HF/3-21G SCRF [23]. To determine thermodynamic values of cytarabine, zigzag (5, 0) carbon nanotube, and (cytarabine/zigzag (5, 0) carbon nanotube) system, we performed a theoretical analysis at the HF/3-21 level. By means of the Onsager method as implemented in the Gaussian 98 program [24], we modeled the solvent effects. The geometries of the cytarabine and single-wall carbon nanotube and the interaction of a nanotube with an anticancer drug in various solvents were optimized using the Onsager model at the Hartree-Fock level of theory. We investigated the cytarabine single-wall carbon nanotube and interaction of a nanotube with an anticancer drug in gas phase (ε = 1) and various solvent media and dielectric constants: water (ε = 78.39) and methanol (ε = 32.63), at various temperatures.

Theoretical background

SCRF method

In using the SCRF model in the quantum chemical theory, the form and volume of the solute molecule needs to be determined specifically for any set of compounds; we know some of the approaches to calculate these properties, but no nonempirical method for their study had been developed. However, we can conclude from

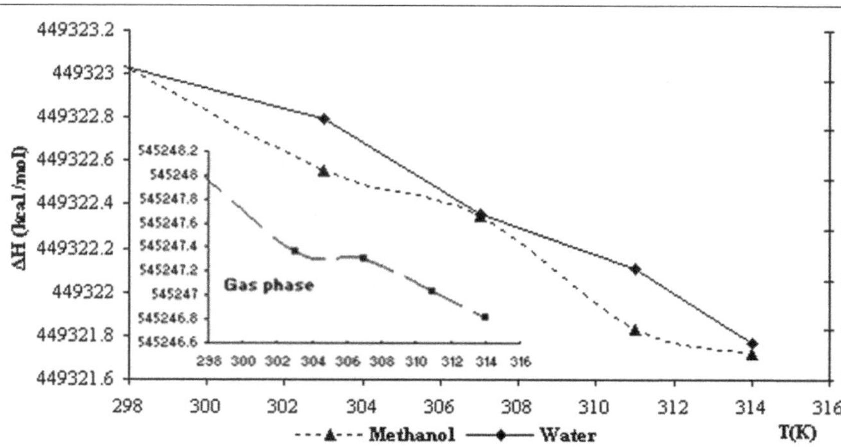

Figure 8 Enthalpy alterations (kcal mol^{-1}) calculated versus temperature for reaction 1 for three phases.

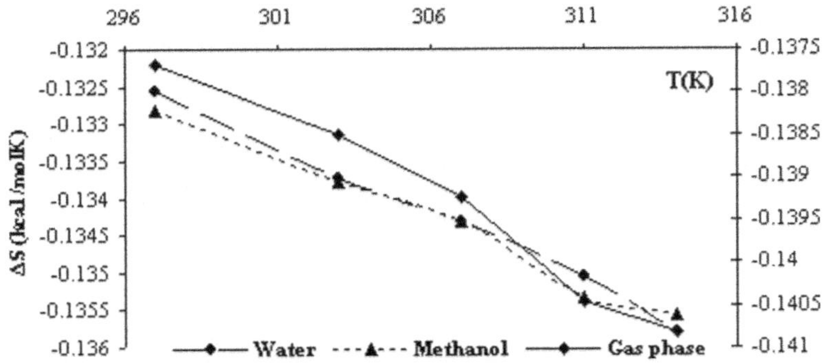

Figure 9 Entropy alterations (kcal mol^{-1} K^{-1}) calculated versus temperature for reaction 1 for three phases.

the results of the model calculations that the simple model that assumed a spherical or an ellipsoidal shape of the cavity for the solute molecule is likely satisfactory for comparatively tiny and strong molecules. We then selected this method in our computations [25]. In the past years, the Onsager-SCRF code elaborated by Wiberg and co-workers [26,27] for the Gaussian computational code has been famous. The Onsager model describes the system as a molecule with a multipole moment inside a spherical cavity surrounded by a continuous dielectric. In some programs, only the dipole moment is used, and therefore, we cannot apply them for molecules with zero dipole moment. Qualitatively, the results from using the Onsager model and HF calculations are correct rules.

Molecular mechanics (Monte Carlo simulation)

By performing statistical sampling experiments, the Monte Carlo method provides approximate solutions to a variety of mathematical problems. Where statistical simulation is defined in quite general terms as a method that utilizes sequences of random numbers to conduct the simulation, we can loosely determine them as a statistical simulation method. We then can declare that the Monte Carlo method is a collection of various methods and unique processes. This process includes performing many simulations using random numbers and probabilities to get approximations of the answer to the problem. Our notion about defining the characteristic of the Monte Carlo method is its use of random numbers in its simulation. We apply the metropolis algorithm more than any other algorithm because of its simplicity in the Monte Carlo method [28]. By random displacement, we can determine the accuracy of the algorithm in small displacement; all moves are acceptable, but in huge cases, only the small moves are accepted. In this study, by comparing the computed potential energy using the force field MM+, AMBER, and OPLS in the Monte Carlo simulation, we illustrate the chemical calculation. Scientists use the HyperChem professional 7.1 in this investigation. Using this software, geometry optimization and Monte Carlo simulation have been performed. They have performed the quantum chemical calculations on the interaction of cytarabine with open-end SWCNT by means of the Gaussian 98 program package [29] using the standard 3-21G basis set. We apply a simple approximation in the Hartree-Fock level and in the Gaussian program, and in order to calculate the radius of cavity which forms the hypothetical surface of the molecule [30,31]. We also utilize the volume of the solute. Both the solute and solvent are placed in the same cavity. How the SCRF approaches define the cavity and reaction field differ with each other. We take into account and consider the effect of the solvent in accordance with the self-consistent reaction field method. This method is based on the Onsager reaction field theory of electrostatic solvations. We consider the solvent as a uniform dielectric with a given dielectric constant, especially in this model. We then investigate solvent effects on the values of Gibbs free energy, enthalpy, entropy, energy, dipole moment of water, and methanol surrounding cytarabine open-end of the (SWCNT) system within the Onsager self-consistent reaction field model using the Hartree-Fock method as well as the various temperature effects on cytarabine open-end of SWCNT. We can compare the obtained thermodynamic values of cytarabine open-end of SWCNT in gas phase and in various solvents such as water and methanol.

Competing interests

The author declares that she has no competing interests.

Acknowledgments

The author would be like to thank Islamic Azad University for the financial and other supports.

References

1. Wang, WS, Tzeng, CH, Chiou, TJ, Liu, JH, Hsieh, RK, Yen, CC, Chen, PM: High-dose cytarabine and mitoxantrone as salvage therapy for refractory non-Hodgkin's lymphoma. Jpn. J. Clin. Oncol. **27**(3), 154–157 (1997)
2. Ogbomo, H, Michaelis, M, Klassert, D, Doerr, HW, Cinatl, J: Resistance to cytarabine induces the up-regulation of NKG2D ligands and enhances natural killer Cell lysis of leukemic cells. Neoplasia **10**(12), 1402–1410 (2008)
3. Guilhot, F, Chastang, C, Michallet, M, Guerci, A, Harousseau, JL, Maloisel, F, Bouabdallah, R, Guyotat, D, Cheron, N, Nicolini, F, Abgrall, JF, Tanzer, J: Interferon alfa-2b combined with cytarabine versus interferon alone in chronic myelogenous leukemia: French Chronic Myeloid Leukemia Study Group. N. Engl. J. Med. **337**, 223–229 (1997)
4. Baccarani, M, Rosti, G, de Vivo, A, Bonifazi, F, Russo, D, Martinelli, G, Testoni, N, Amabile, M, Fiacchini, M, Montefusco, E, Saglio, G, Tura, S: A randomized study of interferon-α versus interferon-α and low-dose arabinosyl cytosine in chronic myeloid leukemia. Blood **99**, 1527–1535 (2002)
5. Capizzi, RL, White, JC, Powell, BL, Perrino, F: Effect of dose on the pharmacokinetic and pharmacodynamic effects of cytarabine. Semin. Hematol. **28**(3 Suppl 4), 54–69 (1991)
6. Iijima, S: Helical microtubules of graphitic carbon. Nature **354**, 56–58 (1991)
7. Collins, PG, Zettle, A, Bando, H, Smalley, RE: Nanotube nanodevice. Science **278**, 100–103 (1997)
8. Liu, JZ, Zheng, QS, Wang, LF, Jiang, Q: Mechanical properties of single walled carbon nanotube bundles as bulk materials. J Mech Phys Solids. **53**, 123–142 (2005)
9. Bianco, A, Kostarelos, K, Partidos, CD, Prato, M: Biomedical applications of functionalized carbon nanotubes. Chem. Commun. **5**, 571–577 (2005)
10. Balasubramanian, K, Burghard, M: Chemically functionalized carbon nanotubes. Small **2**, 92–180 (2005)
11. Harrison, BS, Atala, A: Carbon nanotube applications for tissue engineering. Biomaterials **28**, 344–53 (2007)
12. Chen, X, Kis, A, Zettl, A, Bertozzi, CR: A cell nanoinjector based on carbon nanotubes. Proc. Natl. Acad. Sci. **104**, 8218–8222 (2007)
13. Bianco, A: Carbon nanotubes for the delivery of therapeutic molecules. Expert Opin Drug Delivery **1**, 57–65 (2004)
14. Yang, R, Yang, X, Zhang, Z, Zhang, Y, Wang, S, Cai, Z, Jia, Y, Ma, Y, Zheng, C, Lu, Y, Roden, R. Chen, Y: Single walled carbon nanotubes-mediated in vivo and *in vitro* delivery of siRNA into antigen-presenting cells. Gene Ther. **13**, 1714–2310 (2006)
15. Dresselhaus, MS, Dresselhaus, G, Avouris, P: Carbon Nanotubes: Synthesis, Structure, Properties, and Applications. Springer, Heidelberg (2001)
16. Dresselhaus, MS, Dresselhaus, G, Eklund, PC: Science of Fullerenes and Carbon Nanostructure. Academic Press, San Diego (1996)
17. Cuniberti, G, Yi, J, Porto, M: Pure-carbon ring transistor: role of topology and structure. Appl. Phys. Lett. **81**, 850–852 (2002)
18. Dekker, C: Carbon nanotubes as molecular quantum wires. Physics Today **52** (5), 22–28 (1999)
19. Yamada, T, Namail, T, Hata, K, Futaba, DN, Mizuno, K, Fan, J, Yudasaka, M, Yumura, M, Iijima, S: Size-selective growth of double-walled carbon nanotube forests from engineered iron catalysts. Nat. Nanotechnol. **1**, 131–136 (2006)
20. Lacerda, L, Raffa, S, Prato, M, Bianco, A, Kostarelos, K: Cell-penetrating CNTs for delivery of therapeutics. Nano Today **2**, 38–43 (2007)
21. Shoeb, M: Anticancer agents from medicinal plants. Bangladesh J. Pharmacol **1**, 35–41 (2006)
22. Rida, SM, Ashour, FA, El-Hawash, SAM, ElSemary, MM, Badr, MH, Shalaby, MA: Synthesis of some novel benzoxazole derivatives as anticancer,anti-HIV-1 and antimicrobial agents. Eur. J. Med. Chem. **40**(9), 949–959 (2005)
23. Parish, C, Szczepanska, A, Espartero, JL, Moreno-Vargas, AJ, Carmona, AT, Robina, I, Remmert, S: Synthesis and conformational analysis of novel trimetric maleimide cross-linking reagents. J. Org. Chem. **72**, 6776–6785 (2007)
24. Metropolis, N, Rosenbluth, AW, Rosenbluth, MN, Teller, AH, Teller, E: Equations of state calculations by fast computing machines. J. Chem. Phys. **21**, 1087–1091 (1953)
25. Karelson, M, Tamm, T, Zerner, MC: Multicavity reaction field method for the solvent effect description in flexible molecular systems. J. Phys. Chem. **97**, 11901–11907 (1993)
26. Wong, MA, Frisch, MJ, Wiberg, KB: Solvent effects: 1 mediation of electrostatic effects by solvent. J. Am. Chem. Soc. **113**, 4776–4782 (1991)
27. Wong, MA, Frisch, MJ, Wiberg, KB: Solvent effects: 2 medium effect on the structure, energy, charge density, and vibrational frequencies of sulfamic acid. J. Am. Chem. Soc. **114**, 523–529 (1992)
28. Gelman, A, Roberts, GO, Gilks, WR: Efficient metropolis jumping rules. In: Bernardo, M. (ed.) Bayesian Statistics V, pp. 599–607. Oxford University Press, Oxford (1996)
29. Frisch, MJ, Trucks, GW, Schlegel, HB, Scuseria, GE, Robb, MA, Cheeseman, JR, Zakrzewski, VG, Montgomery, JA, Stratmann, RE, Burant, JC, Dapprich, S, Millam, JM, Daniels, AD, Kudin, KN, Strain, MC, Farkas, O, Tomasi, J, Barone, V, Cossi, M, Cammi, R, Mennucci, B, Pomelli, C, Adamo, C, Clifford, S, Ochterski, J, Petersson, GA, Ayala, PY, Cui, Q, Morokuma, K, Malick, DK, et al.: Gaussian 98 Revision A.6. Gaussian, Inc, Pittsburgh PA (1998)
30. Witanowski, M, Biedrzycka, Z, Sicinska, W, Grabowski, Z: A study of solvent polarity and Hydrogen bonding effects on the nitrogen NMR shieldings of N-nitramines and *ab initio* calculations of the nitrogen shieldings of C-nitro and O-nitro systems. J. Mol. Struct. **602–603**, 199–207 (2002)
31. Mora-Diez, N, Senent, ML, Garcia, B: *Ab initio* study of solvent effects on the acetohydroxamic acid deprotonation processes. Chemical Physics **324**(2–3), 350–358 (2006)

Nanocrystalline titania-based sulfonic acid (TiO_2-Pr-SO_3H) as a new, highly efficient, and recyclable solid acid catalyst for preparation of quinoxaline derivatives

Seyyed Vahid Atghia* and Somayyeh Sarvi Beigbaghlou

Abstract

Nanocrystalline titania-based sulfonic acid (TiO_2-Pr-SO_3H) was used as an efficient and reusable catalyst for the synthesis of quinoxalines. The clean and mild acidity condition, quantitative yields of products, short reaction time, and low reaction temperature are attractive features of this method, making it suitable for heat- or acid-sensitive substrates, particularly in drug synthesis. In practice, this method affords an advantageous combination of satisfactory yields, easy product isolation, and purification.

Keywords: Nanocrystalline titania, 1,2-Dicarbonyls, Quinoxaline, Solid acid catalyst

Background

Among various classes of nitrogen-containing heterocyclic compounds, quinoxaline derivatives are important components of several pharmacologically active compounds [1,2]. These compounds are widely used as anticancer and anthelmintic agents [3], antiviral, and antibacterial [4]. Also, quinoxalines have been reported for their applications as dyes [5] and building blocks in the synthesis of organic semiconductors [6], and they also serve as useful rigid subunits in macrocyclic receptors or molecular recognition [7] and chemically controllable switches [8]. Therefore, the preparation of these compounds has gained considerable attention in recent years.

Improved methods have been reported for the synthesis of quinoxalines; some of the newly used reagents include alumina [9], montmorillonite K-10 [10], sulfated TiO_2 [11-13], clayzic [14], zirconium (IV)-modified silica gel [15], PEG-400 [16], heteropoly acid [17], ZrO_2/M_xO_y/MCM-41 [18], cellulose sulfuric acid [19], and $Ga(OTf)_3$ [20]. These procedures, although effective, have various drawbacks such as requirement of long reaction times (24 h [14]), potential hazards, difficulty in catalyst preparation (e.g., preparation of ZrO_2/M_xO_y/MCM-41 involves a complex procedure and calcination at 500°C) and low yields. Hence, introduction of new methods to circumvent these problems is still in demand.

In recent years, there has been considerable growth of interest in the catalysis of organic reactions by solid acid catalysts. Solid acid catalysts provide numerous opportunities for recovering and recycling the catalysts from reaction mixtures. These features can lead to improved processing steps, better process economics, and environmentally friendly industrial manufacturing.

In the present study, the synthesis of chemically bound adsorbed sulfonic acid on TiO_2 (TiO_2-Pr-SO_3H) by the reaction of (3-mercaptopropyl) trimethoxysilane and TiO_2 and oxidation of the thiol groups with H_2O_2 is reported. The characterization of TiO_2-Pr-SO_3H was performed by means of various techniques such as Fourier transform infrared spectroscopy (FT-IR), X-ray diffraction (XRD), scanning electron microscopy (SEM), transmission electron microscopy (TEM), and thermogravimetric analysis (TGA). The acidic property of the synthesized catalyst was evaluated by the calculation of Hammett acidity function. The assessment of the catalyst was carried out in the preparation of quinoxalines under mild conditions at room temperature (Scheme 1).

* Correspondence: atghia@yahoo.com
Young Researchers Club, Rasht Branch, Islamic Azad University, Rasht, Iran

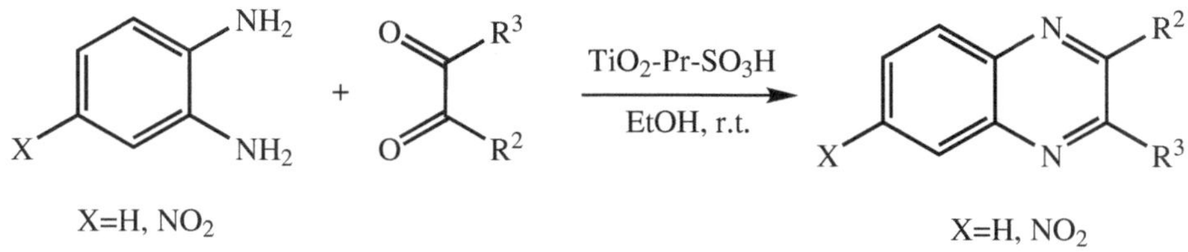

Scheme 1 Preparation of quinoxalines in the presence of TiO_2-Pr-SO_3H.

Results and discussion

In an initial study, in order to examine the catalytic activity of catalyst, we examined the reaction of 1,2-phenylenediamine (1 mmol) with benzil (1 mmol) in various solvents (EtOH, THF, MeCN, EtOAc, and toluene) and also under solvent-free classical heating conditions in the presence of different amounts of the catalyst. The best result was achieved by carrying out the reaction in the presence of 10 mg of TiO_2-Pr-SO_3H in EtOH (Table 1, entry 9). The feasibility of the reaction was successfully tested by producing good yields of the products using various 1,2-dicarbonyl compounds and O-phenylenediamines (Table 2). The present protocol also works well for condensation of aliphatic 1,2-diamines and 1,2-dicarbonyl compounds to give the corresponding quinoxaline derivatives in moderate yields.

Another advantage of the present methodology is the reusability of the catalyst. After completion of the reaction, the catalyst was removed by simple filtration, then washed with acetone, dried at 100°C for 2 h, and reused for three consecutive reactions without any appreciable change in its catalytic activity (Table 2, entry 1). The infrared (IR) spectra and TGA of the recovered catalyst after the third run showed the same spectra of the unused catalyst, which show that the catalyst remained unchanged after three usages.

Table 1 Optimization of the reaction conditions

Entry	Amount of catalyst (mg)	Solvent	Time (min)	Yield (%)
1	20	-	60	60[a]
2	20	THF	60	70
3	20	MeCN	60	81
4	20	EtOH	60	93
5	20	EtOAc	60	75
6	20	Toluene	60	85
7	18	EtOH	30	94
8	15	EtOH	10	95
9	10	EtOH	10	95
10	5	EtOH	20	90

All reactions carried out at room temperature. [a]Carried out at 120°C.

The possible mechanism for the preparation of quinoxalines in the presence of TiO_2-Pr-SO_3H as a promoter is shown in Scheme 2. On the basis of this mechanism, TiO_2-Pr-SO_3H catalyzes the reaction by the electrophilic activation of 1,2-dicarbonyls,making these compounds susceptible to nucleophilic attack by the diamine. Successive elimination of H_2O results in the formation of quinoxaline derivatives and regenerates TiO_2-Pr-SO_3H in the reaction mixture. A similar mechanism has been proposed for this reaction with the sulfated TiO_2catalyst [10].

In order to show the efficiency of this method, Table 3 compares the results from the preparation of 2,3-diphenylquinoxaline in the presence of TiO_2-Pr-SO_3H and some of the other catalysts. In comparison with previously reported methods, our catalyst promotes the reaction very effectively and gives the desired product in a very short time at high yield, and very low amount of the catalyst is needed. Also, our synthesized catalyst can be recovered simply by filtration and can be reused in the next runs without a significant yield decrease of the products. Moreover, our procedure is environmentally friendly as it does not use any toxic auxiliary or solvent.

Conclusions

In conclusion, we have described a novel and efficient method for the preparation of quinoxalines catalyzed by TiO_2-Pr-SO_3H, as a new and efficient reagent, under mild conditions. High yields of the products, short reaction times, ease of the preparation and recyclability of the reagent, solvent-free nature of the reaction, and easy work-up are among the other advantages of this method which make this procedure an alternative to the useful available methods.

Methods

General

Chemicals were purchased from Merck (Darmstadt, Germany) and Sigma-Aldrich (St. Louis, MO, USA) chemical companies. All of the yields refer to the isolated products. Identification of the products was confirmed by comparison of their physical constants with those of authentic samples. The purity determination of

Table 2 Synthesis of quinoxalines in the presence of TiO_2-Pr-SO_3H

Entry	Amine	Carbonyl compound	Time (min)	Yield[a] (%)	Product
1	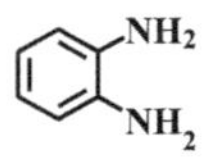	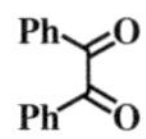	10 (10, 14)[b]	95 (94, 90)[b]	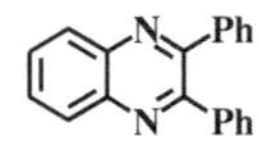
2	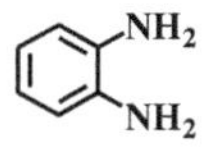	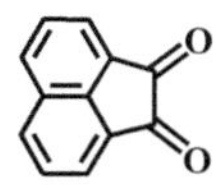	5	91	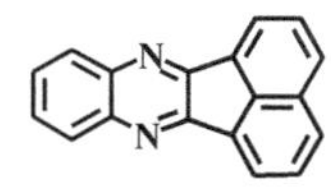
3	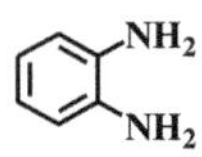	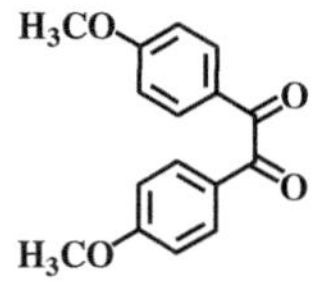	4	96	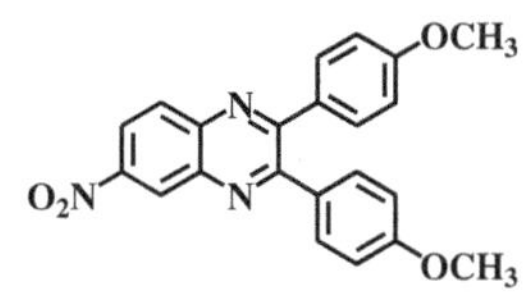
4	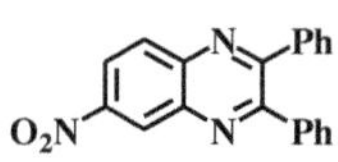	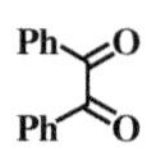	10	90	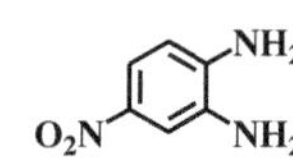
5	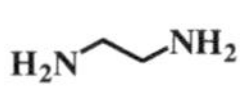	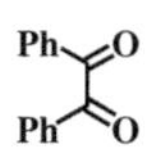	25	85	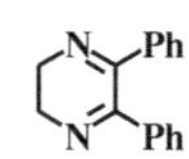
6[c]	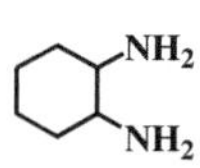	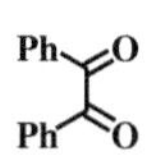	30	80	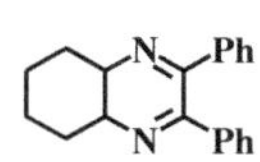

[a]Isolated yields; [b]Times and yields in successive reusing runs 1 to 3; [c]Isolated as cis-/*trans*-mixture.

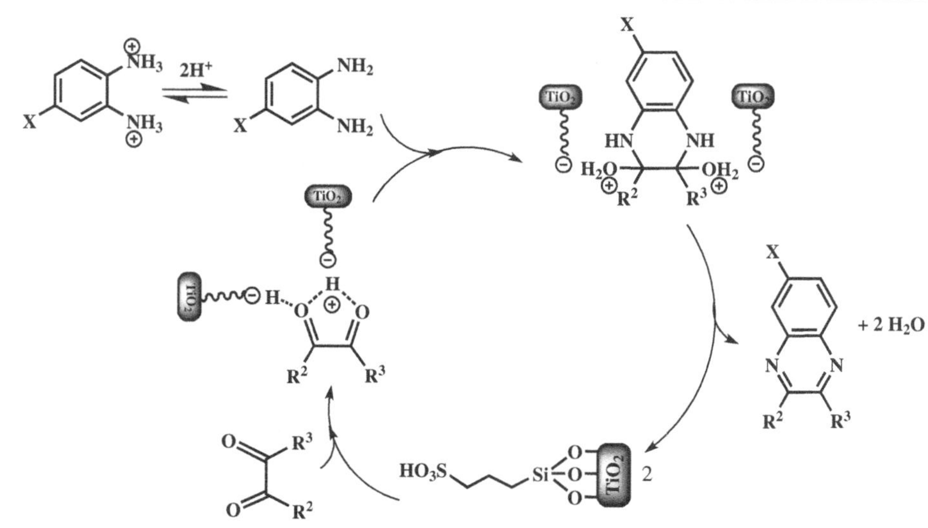

Scheme 2 Proposed mechanism.

Table 3 Comparison of the efficiency of TiO_2-Pr-SO_3H in the synthesis of 2,3-diphenylquinoxaline with other reported methods

Entry	Reagent	Solvent	Amount of catalyst (mg)	Time (min)	Yield (%)[a]
1	TiO_2-Pr-SO_3H (this work)	EtOH	10	10	95
2	Montmorillonite K-10 [10]	H_2O	21	150	100
3	Clayzic [14]	H_2O-EtOH	100	24	89
5	Heteropoly acid [17]	Acetic acid	30	5	98
6	ZrO_2/M_xO_y/MCM-41 [18]	MeCN	40	120	97
7	Cellulose sulfuric acid [19]	EtOH	10	60	93
8	Cellulose sulfuric acid [19]	H_2O	10	150	80

[a]Isolated yields.

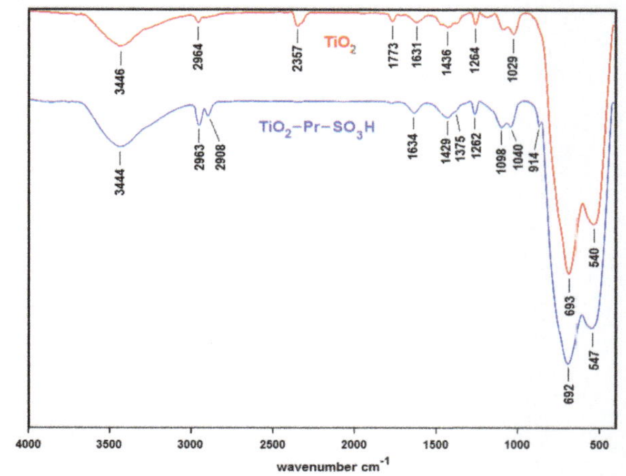

Figure 1 FT-IR spectra of TiO_2 and TiO_2-Pr-SO_3H.

the substrate and reaction monitoring were accompanied by thin-layer chromatography (TLC) on silica gel POLYGRAM SILG-UV_{254} (Carl Roth GmbH, Karlsruhe, Germany) plates.

Instrumentation

Thermogravimetric analyses were conducted using a TGA PYRIS 1 (PerkinElmer Instruments, Waltham, MA, USA) thermoanalyzer instrument. Samples were heated from 25°C to 600°C at ramp 10°C/min under N_2 atmosphere. Wide-angle XRD measurements were performed at room temperature on a Siemens D-500 X-ray diffractometer (Munich, Germany), using Ni-filtered Cu-Kα radiation ($\lambda = 0.15418$ nm). IR and FT-IR spectra were obtained in KBr wafers on Shimadzu IR-470 (Kyoto, Japan) and PerkinElmer RX1 (Waltham, MA, USA) spectrophotometers respectively. TEM analysis was performed on a Philips model CM 10 (FEI Co., Hillsboro, OR, USA) instrument. Sample was prepared by sonicating a small amount of powder in methanol and then placing a drop of the mixture on a 3-mm TEM Cu grid having a lacey carbon support film. Scanning election microphotographs were obtained on a LEO 1430VP instrument. The absorption spectra in the UV-visible regions were recorded by a PerkinElmer LAMBDA 25 recording spectrophotometer. ^1H and ^{13}C NMR spectra were measured in $CDCl_3$ with a Brucker DRX-400AVANCE (Karlsruhe, Germany) spectrometer at 400 and 100 MHz, respectively.

Catalyst preparation

Synthesis of 3-mercaptopropyltitania

To 20 g of TiO_2 (anatase, from Sigma-Aldrich) in dry toluene, 25 mL of (3-mercaptopropyl) trimethoxysilane was added, and the reaction mixture was refluxed for 24 h. After this period, the mixture was filtered to obtain 3-mercaptopropyltitania (MPT), which was washed with acetone and dried.

Oxidation of 3-mercaptopropyltitania

MPT was oxidized with 10 wt.% H_2O_2 in methanol (20 mL) for 24 h at room temperature. The prepared sample was then treated with 1 N H_2SO_4 at ambient temperature for complete protonation, and then the mixture was filtered and washed with H_2O and acetone to obtain TiO_2-Pr-SO_3H catalyst (Scheme 3).

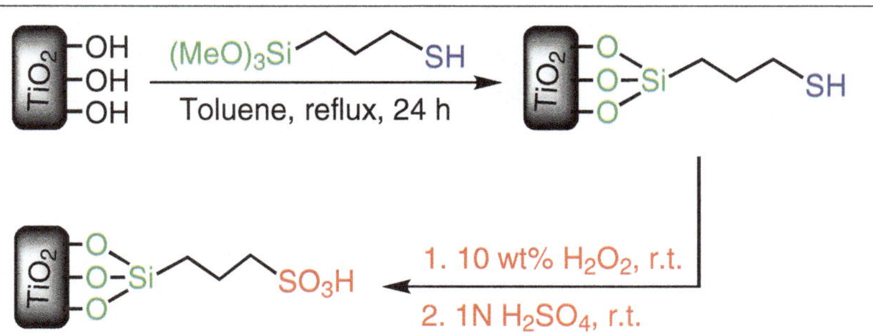

Scheme 3 Preparation of nanoporous titania-based sulfonic acid (TiO_2-Pr-SO_3H).

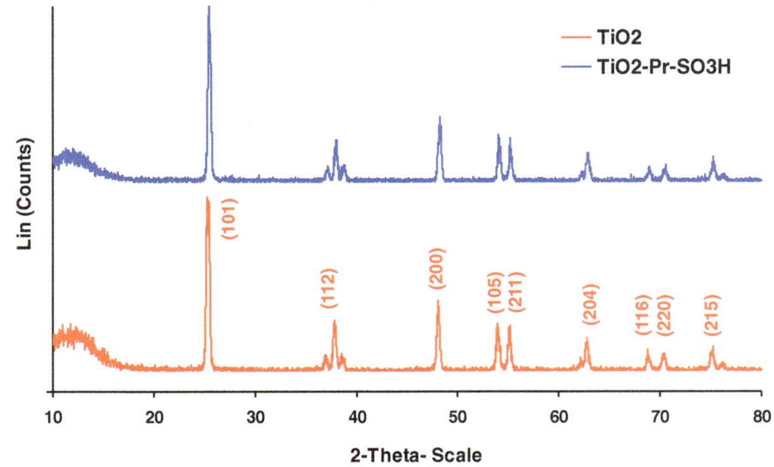

Figure 2 XRD patterns of TiO_2 and TiO_2-Pr-SO_3H.

Catalyst characterization

FT-IR analysis

FT-IR spectra for the pure TiO_2 and TiO_2-Pr-SO_3H samples are shown in Figure 1. In the case of TiO_2, the peaks at 3,446 and 1,773 cm^{-1} correspond to the -OH stretching and bending vibrations of the adsorbed water, respectively. The spectrum of functionalized TiO_2 by sulfonic acid displays almost the same pattern as that of pristine TiO_2, but the band at 3,000 to 3,600 cm^{-1} that centered at 3,444 cm^{-1} is flattened in sulfonated TiO_2 which can be attributed to the modification of TiO_2. Also, CH stretching vibrations of silylating agent was observed at 2,963 and 2,908 cm^{-1}, and the bands at 1,040 and 914 cm^{-1} can be collectively attributed to Si-O stretching vibrations. Furthermore, the band at 1,375 cm^{-1}corresponds to the stretching frequency of S = O in SO_3H.

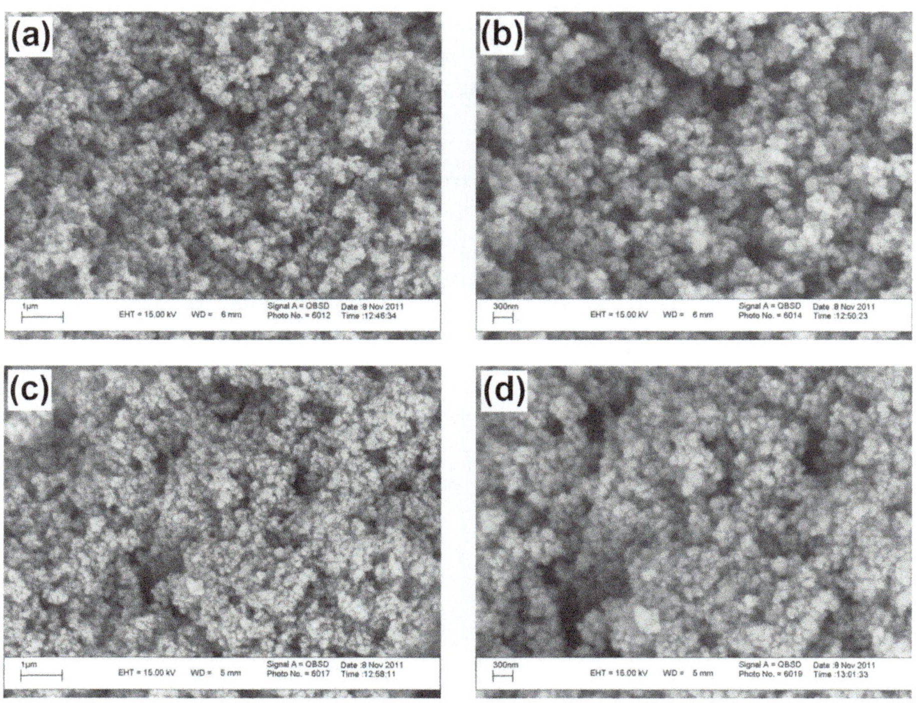

Figure 3 SEM images of TiO_2 and TiO_2-Pr-SO_3H. TiO_2 (**a**) ×30,000 and (**b**) ×50,000 magnifications and TiO_2-Pr-SO_3H (**c**) ×30,000 and (**d**) ×50,000 magnifications.

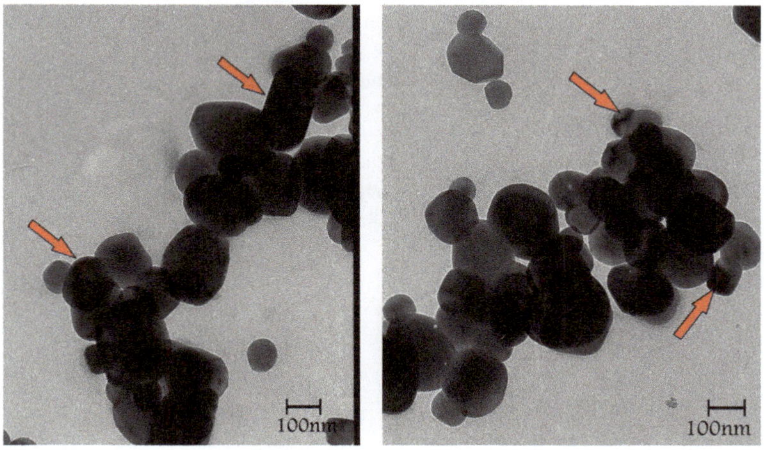

Figure 4 TEM images of TiO_2-Pr-SO_3H. Red arrow, corroded.

Wide-angle X-ray diffraction

XRD patterns of TiO_2 and TiO_2-Pr-SO_3H are given in Figure 2. XRD patterns clearly show anatase lines. It seems that the peak intensities of TiO_2-Pr-SO_3H are almost the same as those of TiO_2, and the sulfate modification does not change the phase of TiO_2. The average crystallite sizes of TiO_2 and TiO_2-Pr-SO_3H, determined using Debye-Scherrer equation, are 15.3 and 20 nm, respectively, which demonstrated that sulfate modification increases the size of TiO_2.

SEM analysis

The representative SEM images of TiO_2 and TiO_2-Pr-SO_3H are shown in Figure 3a,b,c,d with ×30,000 and ×50,000 magnifications. All of the SEM images exhibit a cloud-like structure and small spherical-shaped particles. In TiO_2-Pr-SO_3H, aggregation of TiO_2 nanoparticles was clearly seen. However, the SEM micrographs of the catalyst show some modifications with respect to TiO_2such that the primary surface structure of TiO_2 has changed.

TEM analysis

TEM images of TiO_2-Pr-SO_3H are shown in Figure 4. TiO_2-Pr-SO_3H particles are seen in different sizes ranging from 20 to 100 nm, and the average crystallite size of TiO_2-Pr-SO_3H, determined using XRD, was 20 nm. As can be seen in Figure 4, the particles are slightly corroded in the modification path. This is indicated in the figure by the arrow.

Thermal analysis

TGA was performed for the characterization of TiO_2-Pr-SO_3H in comparison with TiO_2 (Figure 5). The TGA curve of TiO_2 displays a weight loss (4 wt.%) below 100°C which corresponds to the loss of the physically adsorbed water. Also, there is a slight weight loss (3 wt.%) in the range of 100°C to 600°C, which possibly corresponds to the dehydroxylation of TiO_2.

TGA of the catalyst shows an initial weight loss (4 wt.%) due to the desorption of water below 100°C. This is followed by a second weight loss that started

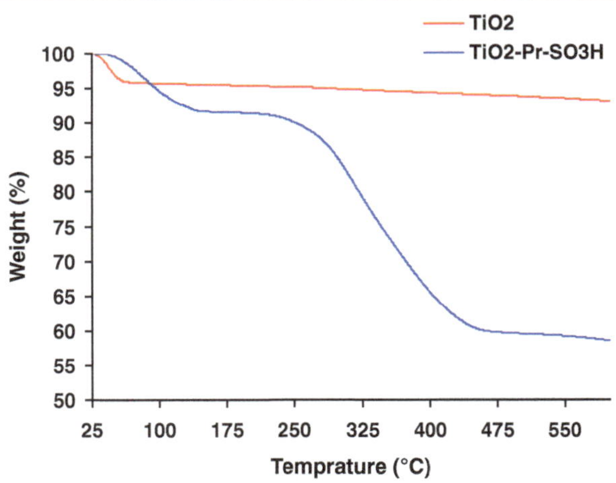

Figure 5 TGA curves of $TiO_2$and TiO_2-Pr-SO_3H.

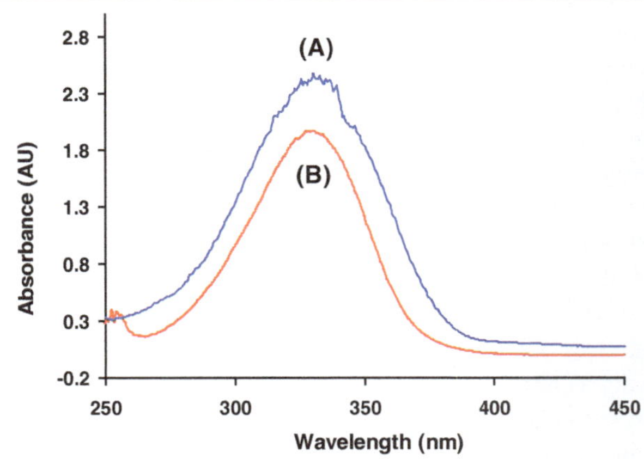

Figure 6 Absorption spectra of 4-nitroaniline (indicator; curve A) and TiO_2-Pr-SO_3H (catalyst; curve B) in CCl_4.

Table 4 Calculation of Hammett acidity function (H_0) of TiO_2-Pr-SO_3H

Entry	Catalyst	A_{max}	$[I]_s$ (%)	$[IH^+]_s$ (%)	H_0
1	-	2.4113	100	0	-
2	TiO_2-Pr-SO_3H	1.1862	49.19	50.80	0.98

Condition for UV-visible spectrum measurement: solvent, CCl_4; indicator, 4-nitroaniline ($pK(I)_{aq} = 0.99$), 1.44×10^{-4} mol/L; catalyst, TiO_2-$HClO_4$, 20 mg, 25°C.

at 245°C, corresponding to the loss of the covalently bound organic group. From this weight loss, it is calculated that the loading of the organic group bound to the titania surface was 2.52 mmol g^{-1}. Also, from the TGA, we understood that TiO_2-Pr-SO_3H has a greater thermal stability (until 245°C) relative to TiO_2.

Surface acidity studies

The Hammett acidity function (H_0) can effectively express the acidity strength of an acid in organic solvents. It can be calculated using the following equation:

$$H_0 = pK(I)_{aq} + \log([I]_s/[IH^+]_s),$$

Here, 'I' represents the indicator base (mainly substituted dinitroanilines), and $[IH^+]_s$ and $[I]_s$ are respectively the molar concentrations of the protonated and unprotonated forms of the indicator. The $pK(I)_{aq}$ values are already known (for example the $pK(I)_{aq}$ value of 4-nitroaniline is 0.99) and can be obtained from many references. According to the Lambert-Beerlaw, the value of $[I]_s/[IH^+]_s$ can be determined and calculated using the UV-visible spectrum.

In our experiment, 4-nitroaniline was chosen as the basic indicator, and $CCl_4$was chosen as the solvent because it is aprotic. The maximal absorbance of the unprotonated form of 4-nitroaniline was observed at 329 nm in CCl_4. As Figure 6 shows, the absorbance of the unprotonated form of the indicator in TiO_2-Pr-SO_3H was weak as compared to the sample of the indicator in CCl_4, which indicated that the indicator was partially in the form of $[IH^+]$. The results obtained are listed in Table 4, which shows the acidity strength of TiO_2-Pr-SO_3H.

General procedure

To a mixture of a substituted 1,2-diketone (1 mmol) and 1,2-diamine (1 mmol) in ethanol (5 mL), TiO_2-Pr-SO_3H (10 mg) was added. The mixture was stirred at room temperature and the reaction was monitored by TLC. After completion, the reaction mixture was heated for a few minutes. The catalyst was removed by filtration, the filtrate was concentrated, and the solid residue was recrystallized from ethanol to give the pure product.

Competing interests

The authors declare that they have no competing interests.

Authors' contributions

Both authors - SVA anc SSB - contributed equally in this article. Both authors read and approved the final manuscript.

Acknowledgments

The authors acknowledge the editor who made the significant revision and contribution towards improving the standard of our article.

References

1. Sarges, R, Howard, HR, Browne, RC, Label, LA, Seymour, PA: 4-Amino[1,2,4]triazolo[4,3-a]quinoxalines. A novel class of potent adenosine receptor antagonists and potential rapid-onset antidepressants. J Med Chem **33**, 2240–2254 (1990)
2. Brown, KL: Chemistry and enzymology of vitamin B_{12}. Chem Rev **105**, 2075–2150 (2005)
3. Sakata, G, Makino, K, Kuraswa, Y: Recent progress in the quinoline chemistry: synthesis and biological activity. Heterocycles **27**, 2481–2515 (1998)
4. Ali, MM, Ismail, MMF, El-Gabby, MSA, Zahran, MA, Ammar, TA: Synthesis and antimicrobial activities of some novel quinoxalinone derivatives. Molecules **5**, 864–873 (2000)
5. Sonawane, ND, Rangnekar, DW: Synthesis and application of 2-styryl-6,7-dichlorothiazolo[4,5-*b*]-quinoxaline based fluorescent dyes: part 3. J Heterocyc Chem **39**, 303–308 (2002)
6. Dailey, S, Feast, JW, Peace, RJ, Sage, IC, Till, S, Wood, EL: Synthesis and device characterisation of side-chain polymer electron transport materials for organic semiconductor applications. J Mater Chem **11**, 2238–2243 (2001)
7. Elwahy, AHM: Synthesis of new benzo-substituted macrocyclic ligands containing quinoxaline subunits. Tetrahedron **56**, 897–907 (2000)
8. Crossley, JC, Johnston, LA: Laterally-extended porphyrin systems incorporating a switchable unit. Chem Commun (2002). doi:10.1039/B111655J
9. Jafarpour, M, Rezaeifard, A, Danehchin, M: Easy access to quinoxaline derivatives using alumina as an effective and reusable catalyst under solvent-free conditions. Appl Catal A: Gen **394**, 48–51 (2011)
10. Huang, TK, Wang, R, Shi, L, Lu, X: Montmorillonite K-10: an efficient and reusable catalyst for the synthesis of quinoxaline derivatives in water. Catal Commun **9**, 1143–1147 (2008)
11. Krishnakumar, B, Swaminathan, M: Solvent free synthesis of quinoxalines, dipyridophenazines and chalcones under microwave irradiation with sulfated Degussa titania as a novel solid acid catalyst. J Mol Catal A: Chem **350**, 16–25 (2011)
12. Krishnakumar, B, Swaminathan, M: A recyclable and highly effective sulfated TiO_2-P25 for the synthesis of quinoxaline and dipyridophenazine derivatives at room temperature. J Organomet Chem **695**, 2572–2577 (2010)
13. Krishnakumar, B, Velmurugan, R, Jothivel, S, Swaminathan, M: An efficient protocol for the green synthesis of quinoxaline and dipyridophenazine derivatives at room temperature using sulfated titania. Catal Commun **11**, 997–1002 (2010)
14. Dhakshinamoorthy, A, Kanagaraj, K, Pitchumani, K: Zn^{2+}-K10-clay (clayzic) as an efficient water-tolerant, solid acid catalyst for the synthesis of benzimidazoles and quinoxalines at room temperature. Tetrahedron Lett **52**, 69–73 (2011)
15. Sharma, RK, Sharma, C: Zirconium(IV)-modified silica gel: preparation, characterization and catalytic activity in the synthesis of some biologically important molecules. Catal Commun **12**, 327–331 (2011)
16. Zhang, XZ, Wang, JX, Sun, YJ, Zhan, HW: Synthesis of quinoxaline derivatives catalyzed by PEG-400. Chin Chem Lett **21**, 395–398 (2010)
17. Heravi, MM, Bakhtiari, K, Bamoharram, FF, Tehrani, MH: *Wells-Dawson* type heteropolyacid catalyzed synthesis of quinoxaline derivatives at room temperature. Monatsh Chem **138**, 465–467 (2007)
18. Ajaikumar, S, Pandurangan, A: Efficient synthesis of quinoxaline derivatives over ZrO_2/M_xO_y (M = Al, Ga, In and La) mixed metal oxides supported on MCM-41 mesoporous molecular sieves. Appl Catal A Gen **357**, 184–192 (2009)

19. Shaabani, A, Rezayan, AH, Behnam, M, Heidary, M: Green chemistry approaches for the synthesis of quinoxaline derivatives: comparison of ethanol and water in the presence of the reusable catalyst cellulose sulfuric acid. C R Chimie **12**, 1249–1252 (2009)
20. Cai, JJ, Zou, JP, Pan, XQ, Zhang, W: Gallium(III) triflate-catalyzed synthesis of quinoxaline derivatives. Tetrahedron Lett **49**, 7386–7390 (2008)

Ab initio simulations of p-type porous silicon nanostructures

Emilye Rosas Landa Loustau[1,2*], Jesús A del Río[1,2], Julia Tagüeña-Martínez[1], Luis E Sansores[3] and Rocío Nava[1]

Abstract

The morphology of porous silicon (p-Si) depends on several parameters such as the doping type and the carriers' concentration of the crystalline silicon substrate. The electrolytes used in the p-Si fabrication also have an important role. The final structure determines if p-Si is luminescent or suitable for photonic applications. Experimental results on p-Si produced by electrochemical etching show that although the carriers are greatly reduced by the etching process, boron atoms remain in the bulk. The study of p-type porous silicon nanostructures by means of an *ab initio* computational simulation might help to understand how boron atoms influence the p-Si final structure. Here, we report electronic and topological properties of ten p-type porous silicon structures as an extension of our previous paper on p-type crystalline silicon. Our results suggest that the boron atoms can not remain bonded on the porous surface but do so in the bulk. The presence of impurities changes the bond distance of their neighbors within a radius of 5 Å. The energy of the models is essentially the same for all the boron positions in the silicon backbone. The high electronic density around the boron impurity could influence the trajectory of an HF ion entering a p-Si pore during the fabrication process.

Keywords: Porous silicon, Nanomaterials, Computational simulations

Background

We have extensively studied porous silicon, both theoretically [1-3] and experimentally [4,5], and successfully implemented different applications such as biosensors [6], luminescent structures [7], one-dimensional photonic crystals [5], mirrors for solar concentration devices [8], and filters [5]. These diverse applications are possible in varying the structure of the porous silicon samples by changing the crystalline silicon substrate and the electrochemical etching conditions.

In order to understand the role of boron in the structure of porous silicon (p-Si) fabricated from p-type silicon wafers, we have recently studied some properties of crystalline p-type silicon models by means of *ab initio* computational simulations [9]. This article is another step in this direction. Here, we simulate ten p-type porous silicon models (PSiHB), with the highest number of silicon atoms used until now on a periodic solid configuration by an *ab initio* code. Our models incorporate experimental conditions like the concentration of boron atoms (corresponding to the resistivity of the p-type silicon wafers used in the fabrication of p-Si for photonic applications (NR, JdR, and PPM, unpublished work)) and dangling bonds passivated with hydrogen. The technical features of the *ab initio* software employed in our work and the procedure used to generate the PSiHB models are described in the 'Methods' section. In the 'Results and discussion' section, we report the total radial distribution functions (RDFs), topological radius influence of boron atoms, plane angle distributions, and relative electronic densities (RelD(E)s) along planes which contain the first Si neighbors of the B impurities.

Our results suggest that boron atoms cannot remain bonded on the surface of the pore but are bound in the bulk. The boron impurities favor the crystallinity of the PSiHB models because they reduce the interatomic distances among second silicon neighbors. Also, we find a large electronic density around the boron atom with a short-range influence of about two atomic layers. We

*Correspondence: emiloustau@gmail.com
[1]Centro de Investigación en Energía, Universidad Nacional Autónoma de México (UNAM), Temixco, CP. 62580, México
[2]Centro de Ciencias de la Complejidad (UNAM), Mexico, CP 04510, Mexico
Full list of author information is available at the end of the article

speculate that due to this electronic density, during the electrochemical fabrication of p-Si, the HF ion avoids the boron atoms. This ion movement could determine the p-Si final topology.

Experimental facts

There are some experimental facts that support our PSiHB theoretical models. It is known that the HF electrochemical attack of the silicon substrate begins at surface irregularities due to a tip-type effect [10]. Then, the porous formation continues vertically due to the presence of holes on the pore tips as an electrical current is applied. In our model, the porous silicon structure remains crystalline, as it happens in real samples (see Figure 1a, b). We use the fact that few boron atoms remain in the p-Si structure after the electrochemical etching of the silicon substrate [11]. Another important fact is that the final topology of the samples strongly depends on the silicon substrate electrical resistivity. In low-resistivity (10^{-3} Ω cm^{-1}) substrates, porous silicon shows a column-like structure, see Figure 2a, and photonic features (NR, JdR, and PPM, unpublished work). On the other hand, high-resistivity (10 Ω cm^{-1}) substrates produce a sponge-like structure with luminescent properties, as can be observed in Figure 2b. Although, in our simulations, we do not include the applied electric field used to produce p-Si by electrochemical etching, it is possible to fabricate p-Si by chemical etching with HF without any electrical current. Figure 3 shows a porous sample obtained by electrochemical etching with 40 mA cm^{-2} current in only 60 s and then chemically etched by the sample in the HF solution during 68 h without any current. As it can be observed, chemical etching is a very slow process but capable of growing porous in a silicon substrate.

Results and discussion

To analyze the PSiHB models from a topological point of view, we begin by studying their relaxed structures and the behavior of a B atom localized in the bulk (Figure 4b) and on the pore surface of the PSiHB models (Figure 4c). Considering that the length of the Si-B bond is 2.00 [12] or 2.05 Å [13] and that bond lengths Si-H and B-H are 1.66 and 1.20 Å [12], respectively, we can see from the figures that when the B impurity is in the bulk, it is bonded to three Si atoms at less than 2.05 Å, which is in agreement with [12] and [13]. However, when the boron atom is on the pore surface, it has only one bond with a Si atom at 2.05 Å. This suggests that under room temperature conditions, porous silicon will not have boron atoms on the pore surface.

We also report the superposition of the total RDFs of each PSiHB model and the RDF of the PSiH model (Figure 5a). The RDFs of the PSiHB models present sharper peaks for the first and second silicon neighbors than the PSiH RDF, which indicates that the PSiHB structures are more ordered than the PSiH structure [14]. Between the first and second silicon neighbors of the PSiHB RDFs, a peak appears, unlike the classical three broad peaks of the RDF of the amorphous silicon (a-Si) network [15]. For the PSiHB models, the mean interatomic distance between first silicon neighbors is 2.35 Å, the second silicon neighbors are distributed in two peaks at 3.25 and 3.85 Å, and the third neighbors are at a distance of 4.85 Å. For the PSiH model, its first, second, and third silicon neighbors are at 2.45, 3.85, and 5.05 Å respectively. Now, remembering that the crystalline silicon interatomic distance is 2.35 Å and that its second and third silicon neighbors are at 3.85 and 4.50 Å, respectively, we conclude that the boron impurities of the PSiHB models attract their silicon neighbors within an approximate radius of 5.00 Å, causing a crystalline ordering between silicon first neighbors. This brings some of the second neighbors from 3.85 to 3.25 Å and moves the third neighbors distance from 5.05 to 4.85 Å.

In Figure 5b, we present the superposition of the PSiHB boron radii of influence (r_B) defined in the previous section, and we compare it with the effect of one B atom in a p-type silicon crystal. The value of the boron radii of influence is bigger than the 10% for the atoms that are at less than 5 Å from the B impurities. The boron radii of influence coincide with the approximate boron effect observed in the PSiHB RDFs and with the contractions of the first and second Si nearest neighbors toward the substitutional impurity calculated by [13,16] and observed by [17].

Besides the RDFs, we have obtained the plane angle distributions of the PSiH and PSiHB models with the purpose of understanding the topological B effect in the p-type p-Si structure. The plain angle distribution of the PSiH model is showed in Figure 6a, and the distribution of one of the PSiHB models is presented in Figure 6b. Both distributions are broad and exhibit two peaks around 40° and 110°. The median of the PSiHB and PSiH bond angle

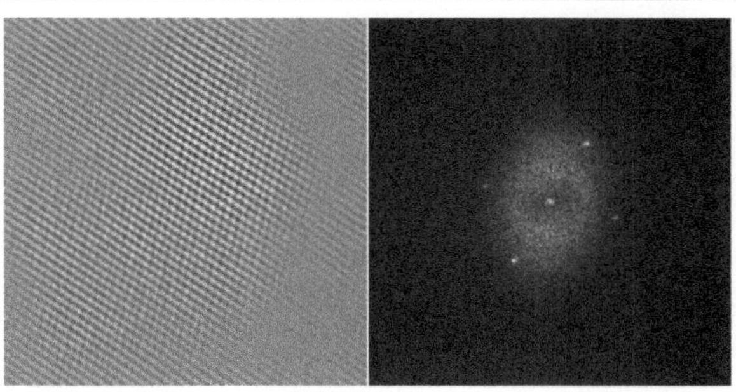

Figure 1 Crystallinity of the p-Si structure. TEM images of a p-Si sample prepared from a boron p-type silicon wafer with a resistivity of 0.001 to 0.005 Ω cm^{-1}. The sample was placed on an amorphous carbon grid for measurement. From (**a**) the Fourier-filtered diffraction lattice image and (**b**) the transmission electron diffraction pattern, it can be observed that the p-Si final structure remains crystalline.

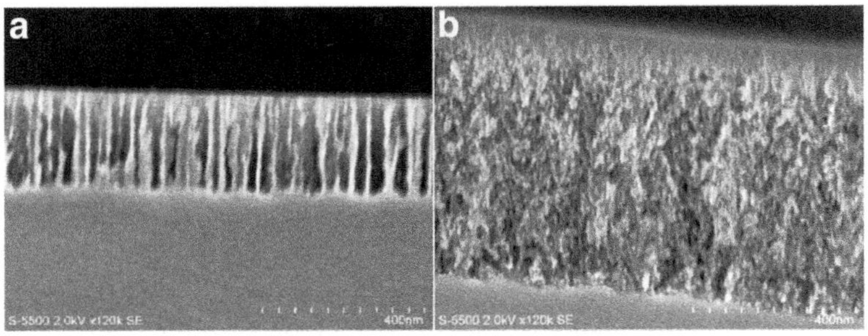

Figure 2 Porous silicon topology. Porous silicon produced by electrochemical etching on a silicon substrate in an electrolyte composed of ethanol, HF and glycerol. (**a**) Low electrical-resistivity substrate (10^{-3} Ω cm^{-1}), (**b**) high-resistivity substrate (10 Ω cm^{-1}).

distributions is 107.62° and 106.93° respectively. Recalling that the crystalline bond angle is 109.47°, we conclude that for both structures, most of the tetrahedral bonds are a bit distorted from the crystalline value, but their distributions are different from the a-Si ones [18], thus remain crystalline. Because the PSiH and PSiHB are similar, we are not able to distinguish the effect of the B atom in the bond angle distortion.

We have calculated the PSiHB RelD(E)s and PSiH D(E) on the planes where the B atom and its Si neighbors lie. The PSiH D(E) is presented in Figure 7a. The electronic densities of the silicon atoms are represented as little hills and, as we can notice, there are open spaces because of the pore. The RelD(E) of one of the PSiHB models is presented in Figure 7b; a huge electronic charge appears around the B impurity, and the hills surrounding the B atom represent the redistribution of the electronic density, caused by the impurity present in the plane, beyond their silicon neighbors. The ten PSiHB RelD(E)s are similar; their maximum, minimum, and mean heights are 1.90, 1.61 and 1.75 electrons/Bohr^2, respectively. The height of the PSiHB RelD(E) model with the B atom on the pore surface is 1.77 electrons/Bohr^2 near the mean value of the RelD(E) height distribution; we have not observed any peculiarity in this RelD(E)s with respect to the RelD(E)s of the rest of the PSiHB models with impurity in the bulk. Considering the plane of the impurity of first neighbors, the presence of the B atom modifies the silicon electronic charge distribution in a range less than two atomic layers; this result agrees with [16].

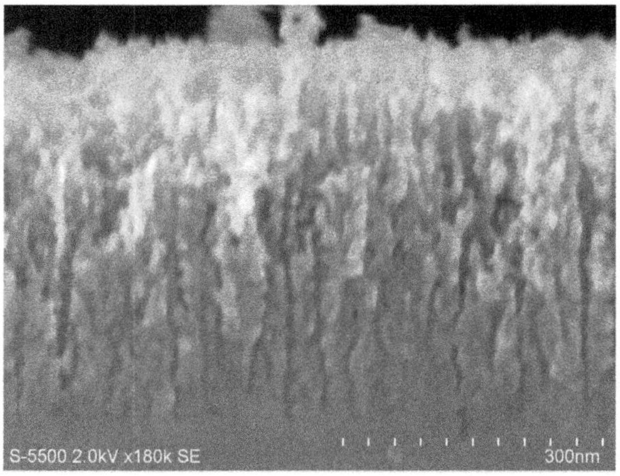

Figure 3 Electrochemical etching without an electrical current. Porous silicon produced first by electrochemical etching of a low-resistivity substrate (applying a 40-mA cm^{-2} current for only 60 s). The sample was chemically etched in HF solution for 68 h without an electrical current. The electrolyte is composed of ethanol, HF and glycerol (volume ratio 7:3:1).

Conclusions

We conclude that the presence of boron atoms in the porous silicon samples favors crystallinity since it reduces the interatomic distance among its first silicon neighbors around 10% with respect to pure silicon increasing the short range order, which is observed through the sharp peaks in the RDFs. Also, the plane angle distribution around the impurity is close to the crystalline value. Another interesting result of our simulations is that boron atoms do not remain on the pore surface because of weak bonding to the silicon network. We find that boron atoms present a bigger electronic density than silicon atoms which could indicate that during chemical etching, the HF ions will tend to avoid the impurities at a distance of at least 5 Å (two atomic layers). This might be the reason why p and $p+$ porous silicon samples present different topologies. When more boron atoms are present, there are not many possible trajectories for the HF and a more regular structure is obtained, while when a small number of impurities are present, the etching could be in many directions, generating a sponge-like structure.

Methods

We generated ten PSiHB models with 255 Si atoms, 154 hydrogen atoms, and one B atom. For nine of the ten

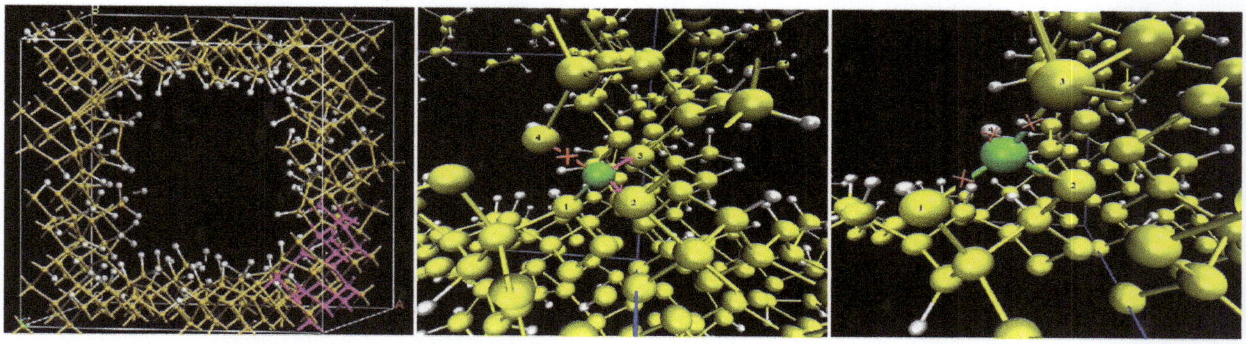

Figure 4 Porous silicon models. (**a**) Boron positions (magenta) in the PSiHB non-relaxed models. (**b**) PSiHB-relaxed structure with the boron atom (green) in the bulk and three bonds with the Si atoms 1, 2 and 3. The impurity is no longer bonded with Si atom number 4. (**c**) PSiHB-relaxed structure with the boron atom (green) on the pore surface. The impurity has just one bond with Si atom number 2.

models, the B atom is placed on different bulk locations, and for one of the models, B is on the pore surface. The PSiHB models have one central and regular pore along the *z*-axis of the supercell. The optimized structures were calculated using ABINIT that is an *ab initio* code based on total energy pseudopotential methods [19]. This code is based on the density functional theory [20,21] and uses plane waves to expand the electronic wave functions.

To construct the PSiHB models, we followed the next steps:

1. The diamond structure Si cell was replicated four times on each axis to obtain a crystalline supercell with 512 Si atoms, whose edge length is 21.72 Å, with a density of 2.33 g/cm^3.
2. The central pore was constructed eliminating 50% of the silicon atoms and with a regular transverse section.
3. The silicon dangling bonds were passivated with hydrogen atoms.
4. For each PSiHB model, one Si atom was substituted by a B one. The position of the impurity is different for each model. Just one of the PSiHB structures has a B atom bonded on the inner pore surface, while the other structures have their impurities in the bulk (Figure 4a). In this step, the B atom is bonded to four atoms, but this is an artificial state because B has just three valence electrons.
5. Once the PSiHB models have been constructed, an optimization of their geometries is implemented in the ABINIT code to obtain the lowest energy configuration. The VMD code was used to visualize the resulting optimized structures.
6. The Si and H atoms were simulated with a local type Troullier-Martins pseudopotential; the B atoms, with Troullier-Martins-Fermi. The exchange-correlation energy functional used for the local density

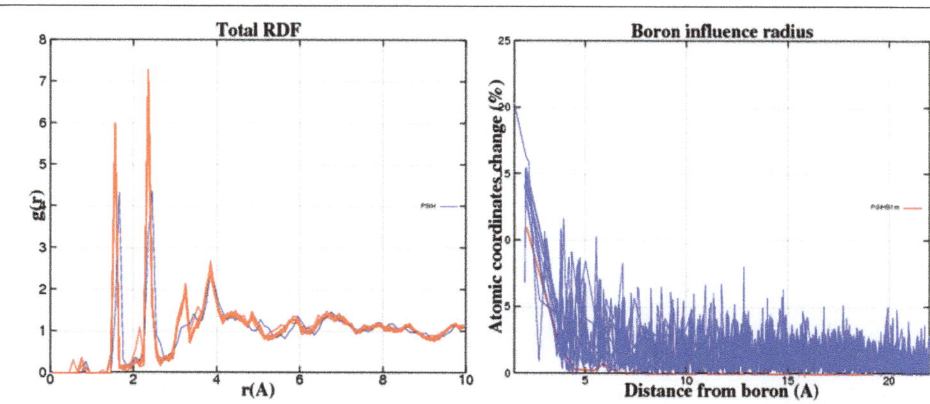

Figure 5 Radial distribution functions. (**a**) Radial distribution functions of the PSiHB models (red line) compared with the RDF of the hydrogenated porous silicon model (blue line). (**b**) Boron radii of influence of the PSiHB models (in blue) compared with the atomic coordinate change produced in a p-type silicon crystal structure due to the presence of a boron impurity (red line).

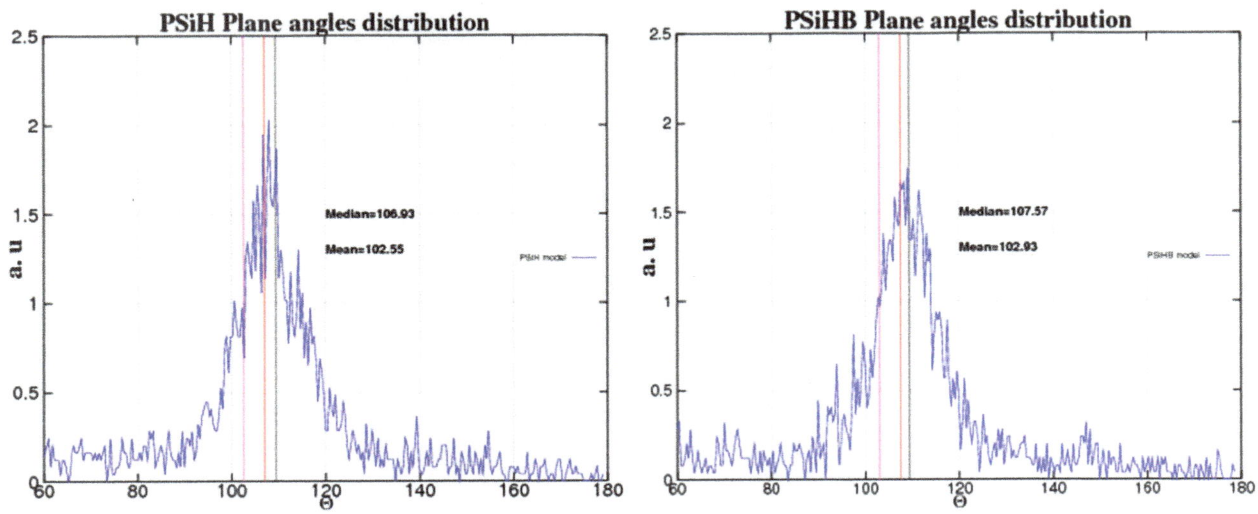

Figure 6 Plane angle distributions. (**a**) Plane angle distribution of the PSiH and (**b**) plane angle distribution of one of the PSiHB models. The plane angle value for the silicon crystal structure is in black. The median and mean of the distributions are in red and magenta, respectively.

approximation was the one proposed by Goedecker et al. [22].

7. For geometry optimization, a cutoff energy radius of 10 Ha and a tolerance force of 5×10^{-5} Ha/Bohr were used.
8. To reproduce the doping level of the experimental p-type silicon wafers, the PSiHB resistivity (ρ) was set as $\rho = 1 \times 10^{-3}\ \Omega\ \text{cm}^{-1}$ corresponding to a 1×10^{20} carriers/cm^3 concentration [23].

We are interested in the changes on the topology and electronic densities due to the presence of a B atom impurity in the PSiHB models. We have obtained their RDFs with the VMD code and plane angles analyzing the ABINIT files. We define a boron radius of influence (r_B) as follows:

$$r_B = |r_{PSiH} - r_{PSiHB}|, \qquad (1)$$

where r_{PSiH} and r_{PSiHB} are the distance vectors of each Si atom from the B impurity, at the origin of the coordinate system, on a pure porous silicon structure passivated with hydrogen (PSiH) and on a PSiHB structure, respectively. To study the electronic behavior around the B impurity, we calculate the relative electronic densities (RelD(E)) of the PSiHB structures as follows:

$$\text{RelD(E)} = \text{D(E)PSiHB} - \text{D(E)PSiH}, \qquad (2)$$

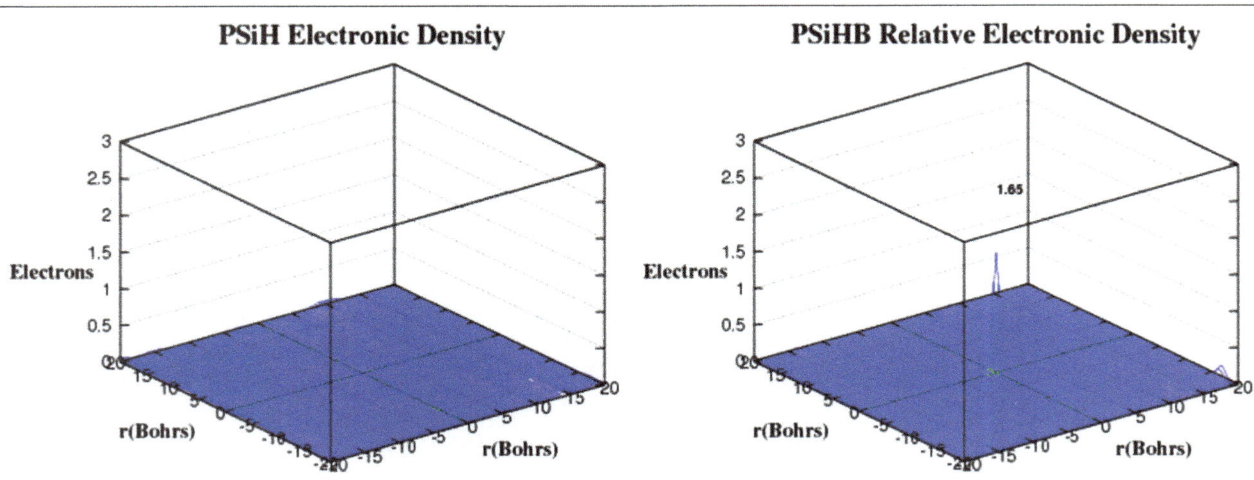

Figure 7 Electronic densities. (**a**) PSiH D(E). The hills are the Si atoms electronic densities. (**b**) RelD(E) of a PSiHB model. The boron impurity is in the bulk and at the center of the plane showed.

where D(E)PSiHB is the electronic density of a PSiHB model, and D(E)PSiH is the electronic density of a pure porous silicon model passivated with hydrogen. We have obtained the RelD(E) of each PSiHB model on a plane determined by a B atom and two of its first Si atom neighbors. In order to visualize the local changes around the boron, we consider $r_B > 1\%$ and RelD(E) > 0.01 electrons/Bohr^2.

Competing interests
The authors declare that they have no competing interests.

Authors' contributions
ERLL carried out the computational simulations and result analysis and participated in the manuscript elaboration. JAdR and JT coordinated the project, discussed the results, and helped to draft the manuscript. LES examined the results and computational simulation methodology and helped to draft the manuscript. RN carried out the experiments, obtained most of the experimental images, and participated to draft the manuscript. All authors read and approved the final manuscript.

Acknowledgements
The authors thank Dr. Maria Beatriz de la Mora and Orlando Hernández Cristobal for the SEM and TEM images. Support from the National University of México (UNAM) through a Posdoctoral Grant to E. R. L. Loustau is acknowledged. We thank the Computing and Information Technology Division of UNAM and the National Supercomputing Center (CNS) for the computer resources. This work was supported in part by DGAPA-UNAM under grant PAPIIT IN109812.

Author details
[1]Centro de Investigación en Energía, Universidad Nacional Autónoma de México (UNAM), Temixco, CP. 62580, México. [2]Centro de Ciencias de la Complejidad (UNAM), Mexico, CP 04510, Mexico. [3]Instituto de Investigaciones en Materiales (UNAM), Mexico, CP 04510, Mexico.

References
1. JE Lugo, JA del Río, JT Martínez, Influence of surface coverage on the effective optical properties of porous silicon modeled as a Si-wire array. J. Appl. Phys. **81**, 1923–1928 (1997)
2. J Tagüeña-Martínez, YG Rubo, M Cruz, MR Beltran Wang, Tight-binding description of disordered nanostructures: an application to porous silicon. Appl. Surf. Sci. **142**, 564–568 (1998)
3. M Cruz, CWJTnM Beltrán, R Yuri, G Rubo, PRB: Supercell approach to the optical properties of porous silicon. Phys. Rev. B. **59**, 1–15387 (1999)
4. V Agarwal, G Malpuech, M Zamfirescu, A Kavokin, D Coquillat, D Scalbert, M Vladimirova, Gil, B, Photon bloch oscillations in porous silicon optical superlattices. Phys Rev. Lett. **92**, 097401–097406 (2004)
5. V Agarwal, J del Río, Filters, mirrors and microcavities from porous silicon. Int. J. Mod. Phys. B. **20**, 99–110 (2006)
6. AG Palestino, JA del Río, C Gergely, MB de la Mora, E Pérez, Three-dimensional spatial resolution of the nonlinear photoemission from biofunctionalized porous silicon microcavity. Appl. Phys. Lett. **91**, 091–1219093 (2007)
7. R Nava, Assisted electrochemical etching of photonic luminescent multilayers of porous silicon. J. Phys. D: Appl. Phys. **43**, 455102–455107 (2010)
8. MB de la Mora, R Nava, J Tagüeña-Martínez, JA del Río, OA Jaramillo, Viability study of porous silicon photonic mirrors as secondary reflectors for solar concentration systems. Sol. Energy Mater. Sol. Cells. **93**, 1218–1224 (2009)
9. ERL Loustau, JA del Río, J Tagüeña-Martínez, LE Sansores, R Nava, Mora de la M B, Ab initio simulation of p-type silicon crystals. Solid State Commun. **152**, 1619–1624 (2012)
10. V Lehmann, H Föll, Formation Mechanism and Properties of Electrochemically Etched Trenches in n-Type Silicon. J. Electrochem. Soc. **137**, 653-659 (1990)
11. G Polisski, GDTSFK, D Kovalev, Boron in mesoporous Si — Where have all the carriers gone?. Physica B. **273-274**, 951–954 (1999)
12. J Gilbert, Bock Mains C W, M Trachtman, Theoretical structures for boron-silicon fluorohydrides. J. Phys. Chem. **93**, 1745–1752 (1989)
13. J Fritsch, KE Scmidt, JB Page, GB Adams, First-principles local-orbital study of the boron-induced reconstruction of Si(001). Phys. Rev. B. **57**, 9745–9756 (1998)
14. ERL Loustau, *Imulación ab initio del silicio amorfo poroso Saarbrücken, Germany.* (Editoral Académica, Española, 2012)
15. F Alvarez, CC Diaz, AA Valladares, Radial distribution functions of ab initio generated amorphous covalent networks. Phys. Rev. B. **65**, 08–113112 (2002)
16. J Chang, MJ Stott, Si($\mathbf{001}/B$) surface reconstruction. Phys Rev. B. **53**, 0–13704 (1996)
17. Y Wang, RJ Hammers, E Kaxiras, Atomic structure and bonding of boron-induced reconstructions on Si(001). Phys. Rev. Lett. **74**(3), 403–406 (1995)
18. F Alvarez, AA Valladares, Ab initio generation of amorphous semiconducting structures. The case of a-Si. Boron in mesoporous. **299**, 259–264 (2002)
19. MC Payne, DCATAA, MP Teter, JD Joannopoulos, Iterative minimization techniques for ab initio total-energy calculations: molecular dynamics and conjugate gradients. Rev. Mod. Phys. **64**, 1045–1096 (1992)
20. P Hohenberg, W Kohn, Inhomogeneous electron gas. Phys. Rev. B. **136**, 864–871 (1964)
21. W Kohn, LJ Sham, Self-consistent equations including exchange and correlation effects. Phys. Rev. A. **140**, 1133–1138 (1965)
22. S Goedecker, JHM Teter, Separable dual-space Gaussian pseudopotentials. Phys. Rev. B. **54**, 1703–1710 (1996)
23. WR Thurber, YML, RL Mattis, J Filliben, Resistivity-dopant density relationship for phosphorus-doped silicon Solid State Science and Technology. J. Electrochem. Soc. **127**, 2291–2294 (1980)

Phytosynthesis of silver nanoparticles using *Pterocarpus santalinus* leaf extract and their antibacterial properties

Kasi Gopinath, Shanmugam Gowri and Ayyakannu Arumugam*

Abstract

Green synthesis is one of the rapid, reliable, and best routes for the synthesis of silver nanoparticles (Ag NPs). The current study revealed that the aqueous leaf extract of *Pterocarpus santalinus*, which contains steroids, saponins, tannins, phenols, triterpenoids, flavonoids, glycosides, and glycerides, is found to be responsible for bioreduction during the synthesis of spherical Ag nanoparticles. The formed Ag NPs were characterized by ultraviolet-visible (UV-vis), Fourier transform-infrared (FT-IR), X-ray diffraction (XRD), atomic force microscopy (AFM), energy-dispersive X-ray spectroscopy (EDX), and scanning electron microscopy (SEM) analysis. UV-vis spectra of the aqueous medium containing silver nanoparticles showed a surface plasmon resonance peak at 418 nm. FT-IR analysis was performed to analyze the biomolecules responsible for the reduction of Ag NPs. XRD results confirmed the presence of silver nanoparticles with face-centered cubic structure. The EDX analysis showed the completed inorganic composition of the synthesized Ag NPs. AFM analysis exemplified the results of particle sizes (41 nm). The calculated crystallite sizes are in the range of 20 to 50 nm, and the spherical nature of the Ag NPs was ascertained by SEM. The synthesized Ag NPs exhibited good antibacterial potential against gram-positive and gram-negative bacterial strains. The zone of inhibition effect of antibacterial activity depends upon the concentration of Ag NPs.

Keywords: *Pterocarpus santalinus*; Leaf extract; Nanoparticles; Green synthesis; Antimicrobial activity

Background

The application of nanoscale materials and structures, usually ranging from 1 to 100 nm, is an emerging area of nanoscience and nanotechnology. Silver nanoparticles (Ag NPs) have rapidly increased due to their unusual optical, chemical, electronic, photo-electrochemical, catalytic, magnetic, antibacterial, and biological labeling properties [1]. Silver nanoparticles were used in broad range of applications like biomedical [2], drug delivery [3], food industries [4], agriculture [5], textile industries [6], water treatment [7] as an antioxidant [8], antimicrobial [9], anti-cancer [10], cosmetics [11], ointments [12], and larvicides [13]. Nanoparticle synthesis is usually carried out by various physical and chemical methods using various hazardous and toxic chemicals. However, green synthesis approaches of producing Ag NPs are an alternative source of conventional method and possess excellent antimicrobial activity [1]. Recently, the green synthesis of Ag NPs has been reported using the extract of plants such as *Artocarpus heterophyllus* [14], *Sesbania grandiflora* [15], *Punica granatum* [16], *Pithecellobium dulce* [17], *Malva parviflora* [18], *Iresine herbstii* [19], *Hibiscus cannabinus* [20], *Hevea brasiliensis* [21], *Euphorbia prostrata* [22] *Cissus quadrangularis* [23-25], *Catharanthus roseus* [26], *Coccinia grandis* [27], *Ixora coccinea* [28], *Lippia citriodora* [29], *Manilkara zapota* [30], *Piper pedicellatum* [31], *Prosopis juliflora* [32], and *Semecarpus anacardium* [33].

Pterocarpus santalinus belongs to Fabaceae family, and it has been used for treatment of vomiting, treating eye diseases, mental aberrations, and ulcers. In addition, it has shown to have antipyretic, anti-inflammatory, anthelmintic, tonic, hemorrhage, dysentery, aphrodisiac, and diaphoretic activities [34]. In this present study, we have reported the green synthesis and characterization of silver nanoparticles using *P. santalinus* leaf extract. To the best

* Correspondence: sixmuga@yahoo.com
Department of Nanoscience and Technology, Alagappa University, Karaikudi, Tamil Nadu 630 004, India

of our knowledge, this is the first report for the synthesis of Ag NPs using *P. santalinus* leaf extract.

Results and discussion

A reduction of Ag NPs was clearly observed when *P. santalinus* leaf extract was added with $AgNO_3$ solution within 20 min. The colorless solution was changed to brown color which indicates the formation of silver nanoparticles.

UV-vis spectroscopy and Fourier transform-infrared spectroscopy analysis

The mixture of leaf extract and $AgNO_3$ solution was subjected to ultraviolet-visible (UV-vis) spectroscopy analysis in the recorded spectra, and it showed a observable peak at 418 nm which corresponds to the wavelength of the surface plasmon resonance of Ag NPs (Figure 1). Various reports have established that the resonance peak of silver nanoparticles appears around this region [22]. Fourier transform-infrared (FT-IR) analysis was performed to identify the possible biomolecules responsible for the reduction of the Ag^+ ions and capping of the reduced Ag NPs synthesized using *P. santalinus* leaf extract. The strong IR bands were observed at 3,382, 2,922, 2,337, 1,614, 1,384, 1,070, and 590 cm^{-1}. The bands which appeared at 3,382 and 2,922 cm^{-1} correspond to -OH stretching and aliphatic -C-H stretching, respectively [35]. The bands at 2,337 and 1,613 cm^{-1} are due to the CO_2 and C=C stretching, respectively. The IR bands observed at 1,384 and 1,070 cm^{-1} may be ascribed to -C-O and -C-O-C stretching modes, respectively. The low band at 590 cm^{-1} corresponds to C-Cl stretching. Hence, the main components such as steroids, saponins, tannins, phenols, triterpenoids, flavonoids, glycosides, and glycerides present in the leaf extract of *P. santalinus* are prime responsible for the observed reduction and capping during the synthesis of Ag NPs. The two new strong bands recorded at 825 and 463 cm^{-1} in the spectra of the synthesized material were assigned to C-H bending and metal (Ag), respectively. The C-H bending peak may be raised due to the reduction of $AgNO_3$ to Ag nanoparticles (Figure 2).

X-ray diffraction and energy dispersive X-ray spectroscopy analysis

X-ray diffraction pattern (XRD) was recorded for the synthesized Ag NPs (Figure 3). Three distinct diffraction peaks at 38.04°, 44.23°, and 64.37° were indexed with the planes (111), (200), and (220) for the face-centered cubic silver as per the JCPDS card no. 89–3722 [36]. The well-resolved and intense XRD pattern clearly showed that the Ag NPs formed by the reduction of Ag^+ ions using *P. santalinus* leaf extract are crystalline in nature. Similar results were reported for Ag NPs in the literature [14-22]. The low intense peak at 77.34° belongs to (311) plane. Energy-dispersive X-ray spectroscopy (EDX) (Figure 4) illustrated the chemical nature of synthesized silver nanoparticles using *P. santalinus* leaf extract. The peak was obtained at the energy of 3 keV for silver, and also some of the weak peaks for C, O, Cl, Al, and Na were found. The emission energy at 3 keV indicates the reduction of silver ions to element of silver. Similarly, *C. quadrangularis* stem extract derived silver nanoparticles showed an EDX spectrum, emission energy at 3 keV for silver, and some of the weak signals from Cl, K, O, Ca, Mg, and S were observed [25].

Atomic force microscopy and scanning electron microscopy

Surface topology of the formulated silver nanoparticles was studied by atomic force microscopy (AFM) analysis (Figure 5a, b). The micrographs clearly indicate that the

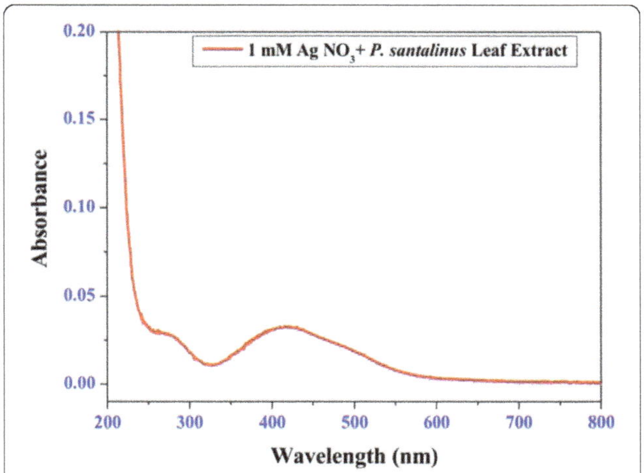

Figure 1 UV-vis spectrum of Ag NPs synthesized by *P. santalinus* leaf extract.

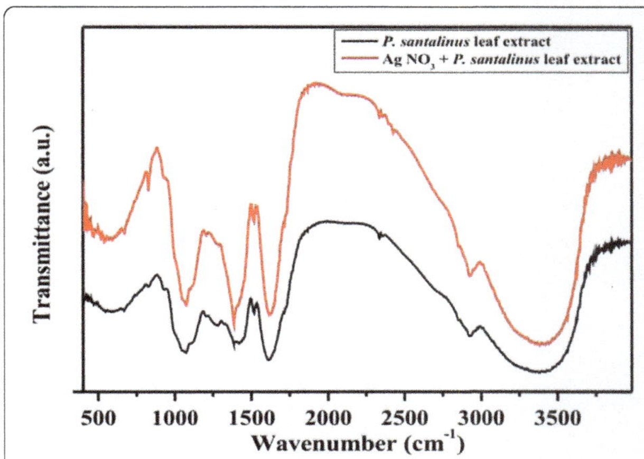

Figure 2 FT-IR spectra of silver nanoparticles synthesized using *P. santalinus* leaf extract.

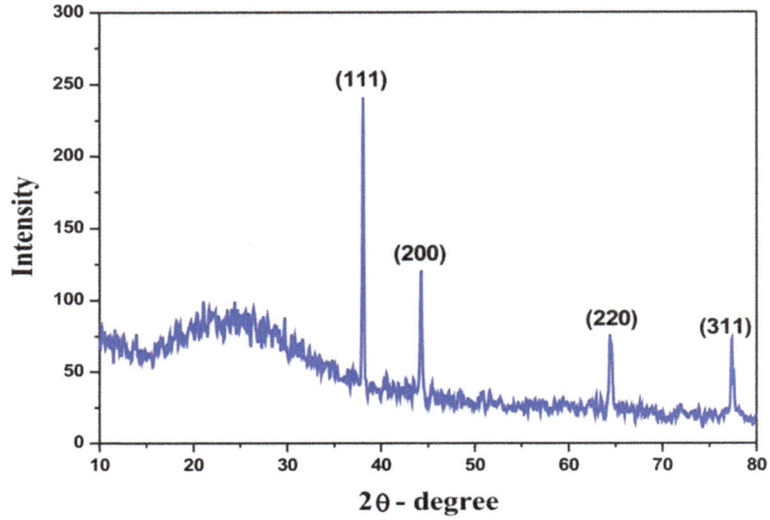

Figure 3 XRD pattern of synthesized silver nanoparticles.

formulated Ag NPs possess spherical shape and have the calculated sizes in the range of 20 to 50 nm. The scanning electron microscopy (SEM) image (Figure 6) further ascertains that the silver nanoparticles are predominantly spherical in morphology with their sizes ranging from 20 to 50 nm and have an average size of about 20 nm.

Antibacterial assay

The antibacterial assay was performed against three gram-positive and gram-negative bacterial pathogens using green-synthesized Ag NPs. The concentration of silver nanoparticles varied between 10, 50, and 100 μL (Table 1). *Staphylococcus aureus* and *Streptococcus pneumoniae* exhibit similar zone of inhibition for all three concentrations. The most significant effect of silver nanoparticles at low concentration of 10 μL per disc against *S. dysenteriae* is to produce a 2-mm inhibition zone of gram-negative bacteria, and at the same concentration, the nanoparticles did not show any significant effect on *Bacillus subtilis*, *Pseudomonas aeruginosa*, and *Proteus vulgaris* (Figure 7). The zone of inhibition increases with

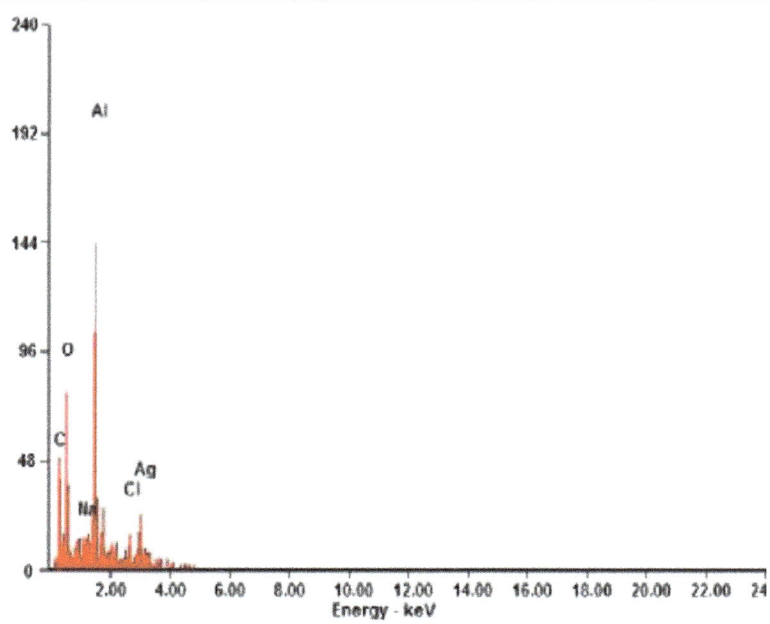

Figure 4 EDX analysis of silver nanoparticles.

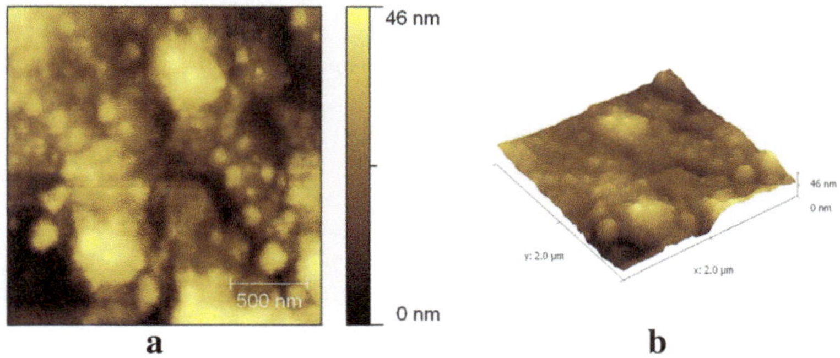

Figure 5 AFM image of synthesized silver nanoparticles (a) 2D image, (b) 3D image.

increasing concentration of silver nanoparticles as the Ag NPs bind with cytoplasmic membrane and killed the bacterial cell. Electrostatic attraction of silver nanoparticles causes damage of bacterial cell membrane to the formation of pits on the surface, and these structural changes take place due to cell expiration [37]. The prokaryotic bacteria have a mesosome cell organelle, and they are present in the inside of plasma membrane. It produced more enzymes as well as major function of cellular respiration, DNA replication, cell division, and increased the surface area of the bacterial cell membrane. Ag NPs interfere with the bacterial cell membrane and bind with mesosome cell organelle and after that reduce the mesosomal function and increase the ROS generation. Ag NPs interact with thiol groups in protein which induced the inactivation of the bacterial protein synthesis as well as DNA replication [38]. Similarly, oxygen associates with silver and reacts with the sulfhydryl (–S-H) groups on cell wall to remove the hydrogen atoms (as water), causing the sulfur atoms to form an R-S-S-R bond, blocked the respiration, and causing the lethal effect of bacterial cells [39]. Ag NPs naturally interact with the membrane of bacteria and disrupt the membrane integrity; silver ions bind to sulfur, oxygen, and nitrogen of essential biological molecules and inhibit bacterial growth [40].

Conclusions

The present work indicates the green-synthesized Ag NPs using *P. santalinus* leaf extract. The AFM and SEM images suggested that the particles are spherical shaped

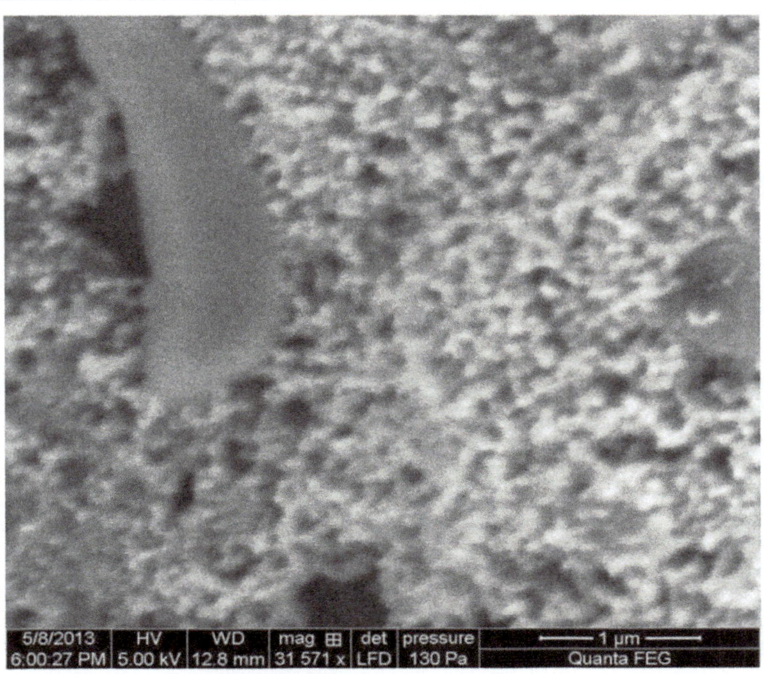

Figure 6 SEM image of Ag NPs using the leaf extract of *P. santalinus*.

Table 1 Antibacterial activity of silver nanoparticles at various concentrations against gram-positive and gram-negative bacteria

Concentration of silver nanoparticles	Zone of inhibition (in mm)					
	Bacillus subtilis	*Staphylococcus aureus*	*Streptococcus pneumoniae*	*Pseudomonas aeruginosa*	*Proteus vulgaris*	*Shigella dysenteriae*
10 μL	0	1	1	0	0	2
50 μL	1	1	1	1	2	2
100 μL	2	3	3	3	2	3

with average size of 20 nm. The antimicrobial activity depends upon the concentration of Ag NPs to produce the most significant effects against the gram-positive and gram-negative bacteria. This green-synthesized method is rapid, facile, convenient, less time consuming, environmentally safe, and can be applied in a variety of existing applications. This plant leaf extract compounds can be extended to the synthesis of other metal and metal oxide nanoparticles.

Methods

Synthesis of silver nanoparticles using *P. santalinus* leaf extract

Fresh *P. santalinus* plant leaves were collected from Karaikudi (Tamil Nadu, India) and used to retrieve their extraction. First, the leaves were cleaned with tap water, followed by distilled water and then finely cut into small pieces. Ten grams of finely cut leaves was added with 100 mL of double distilled water and boiled at 50°C to 60°C for 5 min. The obtained extraction was filtered using Whatman No. 1 filter paper, and the filtrate was collected in 250-mL Erlenmeyer flask and stored at room temperature for further usage. Then, 1 mL of *P. santalinus* leaf extract was added to 100 mL of 1 mM $AgNO_3$ solution at room temperature, and the reduction of Ag NPs was clearly observed within the next 20 min.

Characterization

The synthesized silver nanoparticles were subjected to UV-vis analysis in the wavelength range of 350 to 800 nm using Shimadzu spectrophotometer (model UV-1800, Shimadzu, Kyoto, Japan) operating at a resolution of 1 nm. Also, FT-IR analysis was carried out in the range of 400 to 4,000 cm^{-1}. XRD pattern was recorded using Cu Kα radiation (λ = 1.54060 Å) with nickel monochromator in the range of 2θ from 10° to 80°. The average crystallite size of the synthesized Ag NPs was calculated using Scherrer's formula $D = 0.9\lambda/\beta\cos\theta$. EDX analysis for a thin film of the sample prepared on a aluminum foil (1 cm × 1 cm) by dropping 100 μL of the sample on the foil and allowed to dry for 30 min for further use. AFM analysis for a thin-film sample was prepared on a glass slide (1 cm × 1 cm) by dropping 100 μL of the sample on the slide and allowed to dry for 30 min. The slides were then scanned with AFM (APE Research-model no: A100SGS). The AFM characterization was carried out in ambient temperature in non-contact mode using silicon nitrite tips with varying

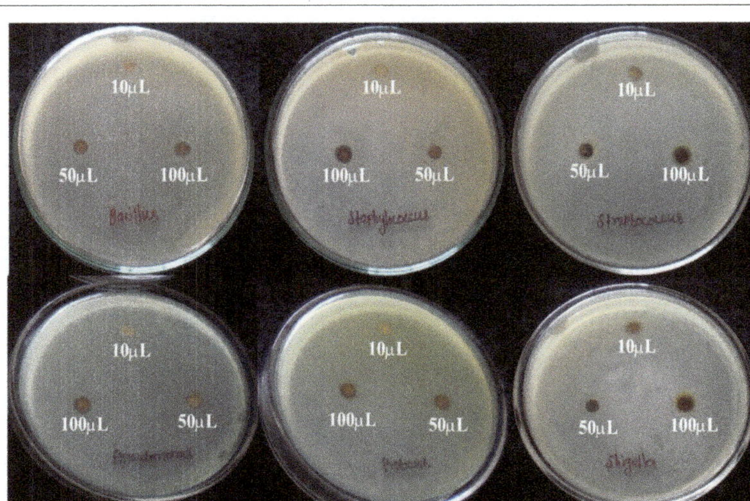

Figure 7 Antibacterial activity of Ag NPs at different concentrations against gram-positive and gram-negative bacteria.

resonance frequencies. The morphology of the synthesized Ag NPs was examined by SEM.

Antibacterial activity of Ag NPs

The minimum inhibitory concentration of the green-synthesized Ag NPs was examined using three gram-positive (*B. subtilis, S. aureus, S. pneumoniae*) and three gram-negative bacteria (*P. aeruginosa, P. vulgaris, S. dysenteriae*) were assayed by disk diffusion method. These six bacteria were grown in nutrient broth medium for 24 h. Approximately 20 mL of molten and cooled nutrient agar was poured into the petri dishes. The six tested organisms were swapped over the nutrient agar medium, and the disks containing silver nanoparticles were kept over the medium using sterile forceps. The silver nanoparticle-loaded disks were prepared at different concentrations of 10, 50, and 100 μL/6-mm disc. The plates were incubated for 24 h at 37°C, and the inhibition zone diameters were measured.

Competing interests

The authors declare that they have no competing interests.

Authors' contributions

KG and SG carried out the nanoparticles synthesis, characterization, and antimicrobial activity. AA carried out the manuscript preparation. All authors read and approved the final manuscript.

Acknowledgments

The authors gratefully thank School of Physics, Alagappa University for extending the XRD facility and also the Department of Industrial Chemistry, Alagappa University for providing the SEM with EDX facilities.

References

1. Sharma, VK, Yngard, RA, Lin, Y: Silver nanoparticles: green synthesis and their antimicrobial activities. Adv. Colloid Interface Sci. **145**, 83–96 (2009)
2. Chaloupka, K, Malam, Y, Seifalian, AM: Nanosilver as a new generation of nanoproduct in biomedical applications. Trends Biotechnol. **28**, 580–588 (2010)
3. Prow, TW, Grice, JE, Lin, LL, Faye, R, Butler, M, Becker, W, Wurm, EMT, Yoong, C, Robertson, TA, Soyer, HP, Roberts, MS: Nanoparticles and microparticles for skin drug delivery. Adv. Drug Deliv. Rev. **63**, 470–491 (2011)
4. Chaudhry, Q, Castle, L: Food applications of nanotechnologies: an overview of opportunities and challenges for developing countries. Trends Food Sci. Tech. **2011**(22), 595–603 (2011)
5. Nair, R, Varghese, SH, Nair, BG, Maekawa, T, Yoshida, Y, Sakthi Kumar, D: Nanoparticulate material delivery to plants. Plant Sci. **179**, 154–163 (2010)
6. Kelly, FM, Johnston, JH: Colored and functional silver nanoparticle wool fiber composites. ACS Appl. Mater. Interfaces **3**, 1083–1092 (2011)
7. Dankovich, TA, Gray, DG: Bactericidal paper impregnated with silver nanoparticles for point-of-use water treatment. Environ. Sci. Technol. **45**, 1992–1998 (2011)
8. Niraimathi, KL, Sudha, V, Lavanya, R, Brindha, P: Biosynthesis of silver nanoparticles using *Alternanthera sessilis* (Linn.) extract and their antimicrobial, antioxidant activities. Colloid. Surface B **102**, 288–291 (2013)
9. Sankar, R, Karthik, A, Prabu, A, Karthik, S, Shivashangari, KS, Ravikumar, V: *Origanum vulgare* mediated biosynthesis of silver nanoparticles for its antibacterial and anticancer activity. Colloid. Surface B. **108**, 80–84 (2013)
10. Boca, SC, Potara, M, Gabudean, AM, Juhem, A, Baldeck, PL, Astilean, S: Chitosan-coated triangular silver nanoparticles as a novel class of biocompatible, highly effective photothermal transducers for *in vitro* cancer cell therapy. Cancer Lett. **31**, 131–140 (2011)
11. Jain, J, Arora, S, Rajwade, JM, Omray, P, Khandelwal, S, Paknikar, KM: Silver nanoparticles in therapeutics: development of an antimicrobial gel formulation for topical use. Mol. Pharm. **6**, 1388–1401 (2009)
12. Murphy, CJ: Sustainability as a design criterion in nanoparticle synthesis and applications. J. Mater. Chem. **18**, 2173–2176 (2008)
13. Roopan, SM, Rohit, Madhumitha, G, Abdul Rahuman, A, Kamaraj, C, Bharathi, A, Surendra, TV: Low-cost and eco-friendly phyto-synthesis of silver nanoparticles using Cocos nucifera coir extract and its larvicidal activity. Ind. Crop Prod. **43**, 631–635 (2013)
14. Jagtap, U, Bapa, VA: Green synthesis of silver nanoparticles using *Artocarpus heterophyllus* Lam. Seed extract and its antibacterial activity. Ind. Crop Prod. **46**, 132–137 (2013)
15. Das, J, Paul Das, M, Velusamy, P: *Sesbania grandiflora* leaf extract mediated green synthesis of antibacterial silver nanoparticles against selected human pathogens. Spectrochim. Acta. A. **104**, 265–270 (2013)
16. Edison, TJ, Sethuraman, MG: Biogenic robust synthesis of silver nanoparticles using *Punica granatum* peel and its application as a green catalyst for the reduction of an anthropogenic pollutant 4-nitrophenol. Spectrochim. Acta. A. **104**, 262–264 (2013)
17. Raman, N, Sudharsan, S, Veerakumar, V, Pravin, N, Vithiya, K: *Pithecellobium dulce* mediated extra cellular green synthesis of larvicidal silver nanoparticles. Spectrochim. Acta. A. **96**, 1031–1037 (2012)
18. Zayed, MF, Eisa, WH, Shabaka, AA: *Malva parviflora* extract assisted green synthesis of silver nanoparticles. Spectrochim. Acta. A. **98**, 423–428 (2012)
19. Dipankar, C, Murugan, S: The green synthesis, characterization and evaluation of the biological activities of silver nanoparticles synthesized from *Iresine herbstii* leaf aqueous extracts. Colloid. Surface B. **98**, 112–119 (2012)
20. Bindhu, MR, Umadevi, M: Synthesis of monodispersed silver nanoparticles using *Hibiscus cannabinus* leaf extract and its antimicrobial activity. Spectrochim. Acta. A. **101**, 184–190 (2013)
21. Guidelli, EJ, Ramos, AP, Zaniquelli, ME, Baffa, O: Green synthesis of colloidal silver nanoparticles using natural rubber latex extracted from *Hevea brasiliensis*. Spectrochim. Acta. A. **82**, 140–145 (2011)
22. Zahir, AA, Rahuman, AA: Evaluation of different extracts and synthesised silver nanoparticles from leaves of *Euphorbia prostrata* against *Haemaphysalis bispinosa* and *Hippobosca maculata*. Vet. Parasitol. **187**, 511–520 (2012)
23. Valli, JS, Vaseeharan, B: Biosynthesis of silver nanoparticles by *Cissus quadrangularis* extracts. Mater. Lett. **82**, 171–173 (2012)
24. Santhoshkumar, T, Rahuman, AA, Bagavan, A, Marimuthu, S, Jayaseelan, C, Kirthi, AV, Kamaraj, C, Rajakumar, G, Zahir, AA, Elango, G, Velayutham, K, Iyappan, M, Siva, C, Karthik, L, Rao, KV: Evaluation of stem aqueous extract and synthesized silver nanoparticles using *Cissus quadrangularis* against *Hippobosca maculata* and *Rhipicephalus* (Boophilus) *microplus*. Exp. Parasitol. **132**, 156–165 (2012)
25. Vanaja, M, Gnanajobitha, G, Paulkumar, K, Rajeshkumar, S, Malarkodi, C, Annadura, G: Phytosynthesis of silver nanoparticles by *Cissus quadrangularis*: influence of physicochemical factors. J. Nanostructure Chem. **3**, 17 (2013)
26. Kotakadi, VS, Rao, YS, Gaddam, SA, Prasad, TNVKV, Reddy, AV, Gopal, DVRS: Simple and rapid biosynthesis of stable silver nanoparticles using dried leaves of *Catharanthus roseus*. Linn. G. Donn and its anti microbial activity. Colloid. Surface B. **105**, 194–198 (2013)
27. Arunachalam, R, Dhanasingh, S, Kalimuthu, B, Uthirappan, M, Rose, C, Mandal, AB: Phytosynthesis of silver nanoparticles using Coccinia grandis leaf extract and its application in the photocatalytic degradation. Colloid. Surface B. **94**, 226–230 (2012)
28. Karuppiah, M, Rajmohan, R: Green synthesis of silver nanoparticles using *Ixora coccinea* leaves extract. Mater. Lett. **97**, 141–143 (2013)
29. Cruz, D, Falé, PL, Mourato, A, Vaz, PD, Serralheiro, ML, Lino, ARL: Preparation and physicochemical characterization of Ag nanoparticles biosynthesized by *Lippia citriodora* (Lemon Verbena). Colloid. Surface B. **81**, 67–73 (2010)
30. Rajakumar, G, Rahuman, AA: Acaricidal activity of aqueous extract and synthesized silver nanoparticles from *Manilkara zapota* against *Rhipicephalus (Boophilus) microplus*. Res. Vet. Sci. **93**, 303–309 (2012)
31. Tamuly, C, Hazarika, M, Borah, SC, Das, MR, Boruah, MP: In situ biosynthesis of Ag, Au and bimetallic nanoparticles using *Piper pedicellatum* C.DC: Green chemistry approach. Colloid. Surface B **102**, 627–634 (2013)
32. Raja, K, Saravanakumar, A, Vijayakumar, R: Efficient synthesis of silver nanoparticles from *Prosopis juliflora* leaf extract and its antimicrobial activity using sewage. Spectrochim. Acta. A. **97**, 490–494 (2012)

33. Raju, D, Hazra, S, Hazra, S, Mehta, UJ: Phytosynthesis of silver nanoparticles by Semecarpus anacardium L. leaf extract. Mater. Lett. **102–103**, 5–7 (2013)
34. Arunakumara, KKIU, Walpola, BC, Sibasinghe, S, Yoon, MH: Pterocarpus santalinus Linn. f. (Rath handun): a review of its botany, uses, phytochemistry and pharmacology. J. Korean Soc. Appl. Biol. Chem. **54**, 495–500 (2011)
35. Kumar, KM, Mandal, BK, Sinha, M, Krishnakumar, V: *Terminalia chebula* mediated green and rapid synthesis of gold nanoparticles. Spectrochim. Acta. A. **86**, 490–494 (2012)
36. Vivek, R, Thangam, R, Muthuchelian, K, Gunasekaran, P, Kaveri, K, Kannan, S: Green biosynthesis of silver nanoparticles from *Annona squamosa* leaf extract and its *in vitro* cytotoxic effect on MCF-7 cells. Process Biochem. **47**, 2405–2410 (2012)
37. Sondi, I, Salopek-Sondi, B: Silver nanoparticles as antimicrobial agent: a case study on E.coli as a model for Gram-negative bacteria. J. Colloid Interface Sci. **275**, 177–182 (2004)
38. Feng, QL, Wu, J, Chen, GQ, Cui, FZ, Kim, TN, Kim, JO: A mechanistic study of the antibacterial effect of silver ions on *Escherichia coli* and *Staphylococcus aureus*. J. Biomed. Mater. Res. **52**, 662–668 (2000)
39. Siva Kumar, V, Nagaraja, BM, Shashikala, V, Padmasri, AH, Madhavendra, SS, Raju, BD, Rama Rao, KS: Highly efficient Ag/C catalyst prepared by electro-chemical deposition method in controlling microorganisms in water. J. Mol. Catal. A. Chem. **223**, 313–319 (2004)
40. Juan, L, Zhimin, Z, Anchun, M, Lei, L, Jingchao, Z: Deposition of silver nanoparticles on titanium surface for antibacterial effect. Int. J. Nanomed. **5**, 261–267 (2010)

22

Preparation and study of vitamin A palmitate microemulsion drug delivery system and investigation of co-surfactant effect

Solmaz Maleki Dizaj[1,2]

Abstract

The oral route is the main route of drug delivery in many diseases. A major problem in oral oil-soluble drug administration is poor aqueous solubility. One way to deliver oil-soluble drugs is to incorporate the drug into an inert vehicle, such as microemulsions. The purpose of this research was to develop the oil-in-water (o/w) microemulsion areas of pseudo-ternary phase diagrams for a drug delivery system. Our systems consist of sunflower oil + surfactant (Tween 80) + vitamin A palmitate + different co-surfactants + water. Vitamin A palmitate is an oil-soluble drug with poor aqueous solubility. Our goal is to increase its aqueous solubility. The effects of different co-surfactants on the o/w microemulsion region were investigated. The co-surfactants studied were anhydrous glycerol, sucrose, ethanol, and 1-propranol. A titration technique was employed for the preparation of the samples. The phases were identified by visual inspection and polarized microscopy. Samples of the microemulsion area were separated, and pseudo-ternary phase diagram was plotted for them. The system sunflower oil + Tween 80 + sucrose + water+ model drug showed the largest o/w microemulsion region. Consequently, sucrose was selected as the best co-surfactant. Finally, some samples of this system were selected for particle size measurement and stability testing. Results showed that more than 90% of the samples in different temperatures (4°C, 45°C, and room temperature) are stable.

Keywords: Drug delivery system; Pseudo-ternary phase diagram; Co-surfactants; Microemulsion

Background

Water-soluble drugs can diffuse easily in the body, but oil-based drugs have poor aqueous solubility and a low bioavailability [1]. Microemulsions are suitable systems for the delivery of oil-soluble drugs. Microemulsions are thermodynamically stable systems, having unique properties. They have a small particle size (nano size), high stability, large interfacial area, and low interfacial tension and form spontaneously. This spontaneous formation in the gastrointestinal tract presents the drug in a solubilized form, and the small size of the formed droplet provides a large interfacial surface area for drug absorption [2]. When the droplet size of a microemulsion is decreased, the emulsion stability against sedimentation is being increased [3]. In microemulsion systems, solubilization may be defined as the phenomenon whereby substances that are otherwise insoluble in a given medium are brought into solution by incorporating them into, or upon, aggregates of colloidal dimensions, termed micelles [4].

The dispersed phase of microemulsion can act as a potential drug reservoir, and the drug is partitioned between the dispersed phase and dispersion medium and can get transported through the membrane, e.g., skin, mucous membrane, etc. Another important point about using microemulsion is that the same medium could be used both for hydrophilic and lipophilic drugs [5].

There are three types of microemulsion: bicontinuous, oil-in-water, and water-in-oil. Oil-in-water (o/w) microemulsions are more important in delivering oil-soluble drugs. Generally, a microemulsion contains oil, surfactant, cosolvent/co-surfactant, and other components. For a drug delivery system, formulations are isotropic mixtures of an oil, a surfactant, a co-surfactant (or solubilizer), and a drug. The basic principle of this system is its ability to form fine

Correspondence: solmazmaleki64@yahoo.com
[1]Faculty of Pharmacy, Tabriz University of Medical Science, Tabriz, Iran
[2]Student Research Committee, Tabriz University of Medical Science, Tabriz, Iran

o/w microemulsions under gentle agitation following dilution by aqueous phases [2].

The hydrophilic-lipophilic balance of a surfactant system is a good guide for mixing water and oil into a microemulsion [6]. Nonionic surfactants with high hydrophilic-lipophilic balance (HLB) values are used in the formulation of drug delivery systems [7]. On the other hand, from the pharmacological point of view, surfactants with low critical micelle concentration (cmc) value have more stable micelles. Examples of nonionic surfactants with high HLB and low cmc values are Span 80, Tween 80, and Tween 20. Surfactants with a high cmc value may dissociate into monomers, their content may precipitate in the blood [8], and they are not suitable for drug delivery.

Often the use of co-surfactants is required for the optimal formation of a microemulsion. The co-surfactant is often the second surfactant but may also refer to a low molecular weight amphiphile, such as an alcohol. Co-surfactants increase the flexibility of the surfactant film around the microemulsion droplet [9]. The role of the co-surfactant is to overcome the repulsive forces of similar phases and fluidity of the oil and water to increase the permeability of two phases to form a microemulsion [10]. Short- and medium-chain alcohols, such as butanol, pentanol, ethanol, isopropanol, or propylene glycol, are commonly added as co-surfactants [9].

The use of pseudo-ternary phase diagram is required to map the optimal composition range for three excipients. This technique is mainly used to map the microemulsion areas [7]. Pseudo-ternary phase diagrams can be used to show the influence of changes in the volume fractions of the different phases on the phase behavior of the system [11].

Microemulsions are characterized using dynamic light scattering, polarized light microscopy, electrical conductivity, and rheology [9]. Dynamic light scattering is used to measure nanoscale particles of liquid mediums such as microemulsions. Polarized light microscopy can distinguish between isotropic and anisotropic materials. Microemulsion is an isotropic system, and birefringence is not found in it. A black background under polarized light is a key distinctive property of microemulsion. The conductometery method is used to detect boundary of different areas in the ternary phase diagram.

Results and discussion

Identification of microemulsions

The phases were identified by visual inspection and polarized microscopy. A microemulsion is optically clear and transparent. The samples with transparent appearance were separated for further investigations. Polarized light microscopy can distinguish between isotropic and anisotropic materials [9]. Microemulsion is an isotropic system, and birefringence is not found in it. The samples exhibited no birefringence, and a black background under polarized light is a microemulsion. Figure 1 shows our microemulsion picture under polarized light microscopy.

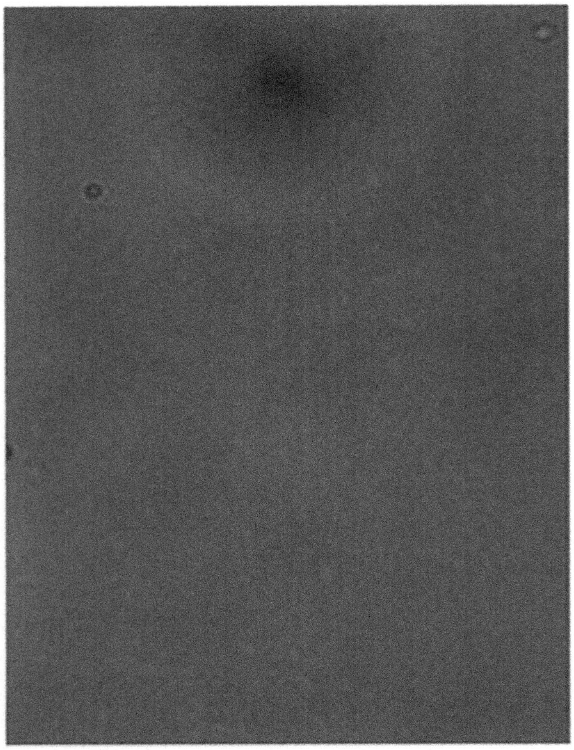

Figure 1 Black background of vitamin A palmitate microemulsion under polarized light microscopy.

Choosing the best system

The o/w area of the pseudo-ternary phase diagrams for our different systems is shown in Figures 2, 3, 4, 5, 6. Results show that with short-chain alcohols, such as ethanol and 1-propanol, microemulsions are very difficult to form even when the surfactant is in high concentration. Systems with anhydrous glycerol and sucrose form relatively good o/w microemulsion with Tween 80, but this region for sucrose is larger than for anhydrous glycerol. The system sunflower oil + Tween 80 + sucrose + water + vitamin A showed the largest microemulsion region. This system has higher solubility compared with the other systems, and our choice for drug formulation is increased in this system. Consequently, sucrose was selected as the best co-surfactant for this system, and the system was investigated for particle size measurement and stability studies.

Particle size measurement

A key distinctive property of microemulsion is its nanoscale particle size [2]. For the selected system, particle size analysis was performed by dynamic light scattering.

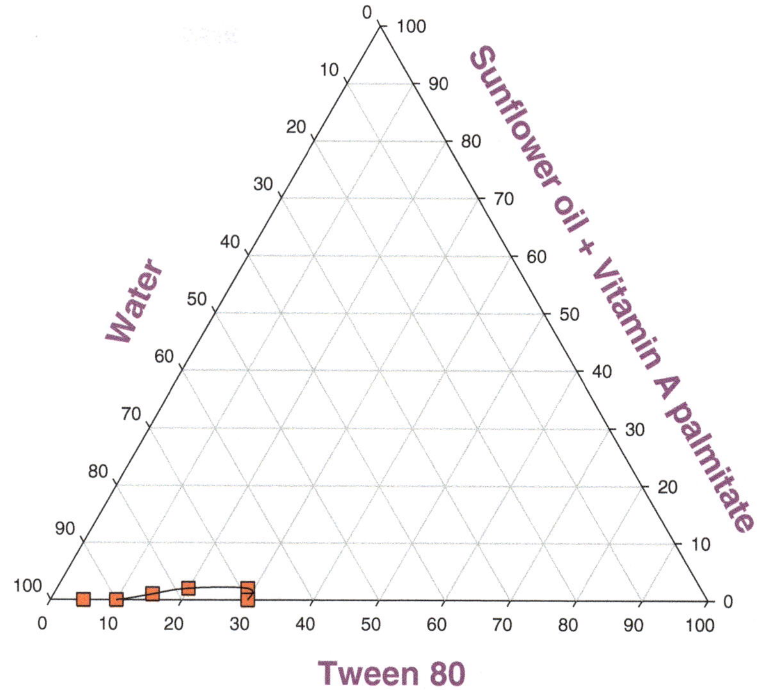

Figure 2 Pseudo-ternary phase diagram of the vitamin A palmitate microemulsion region without co-surfactant.

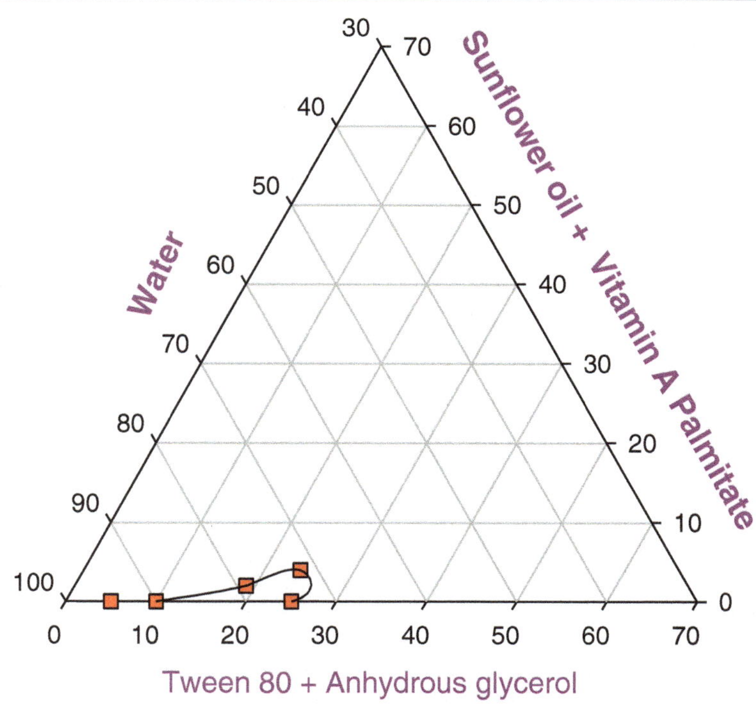

Figure 3 Pseudo-ternary phase diagram of the vitamin A palmitate microemulsion region with anhydrous glycerol co-surfactant.

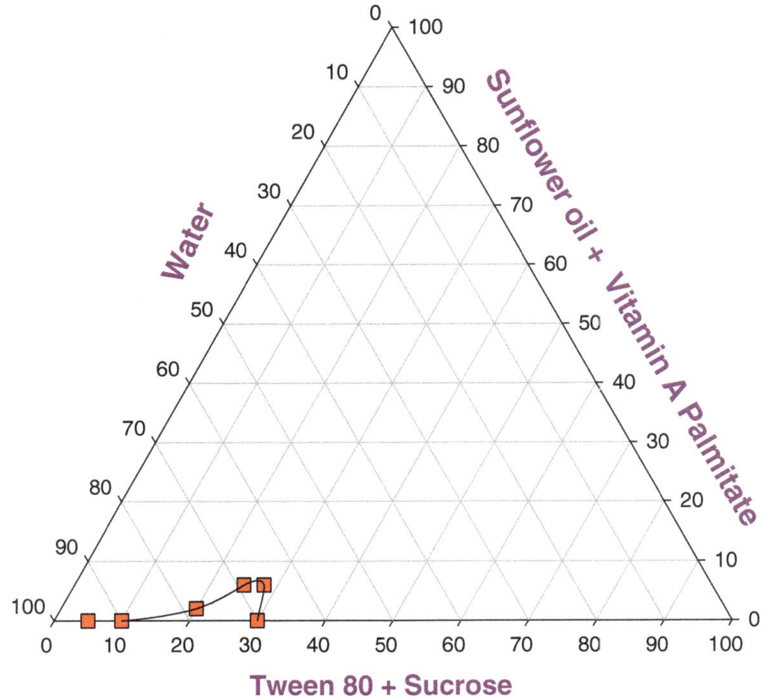

Figure 4 Pseudo-ternary phase diagram of the vitamin A palmitate microemulsion region with sucrose co-surfactant.

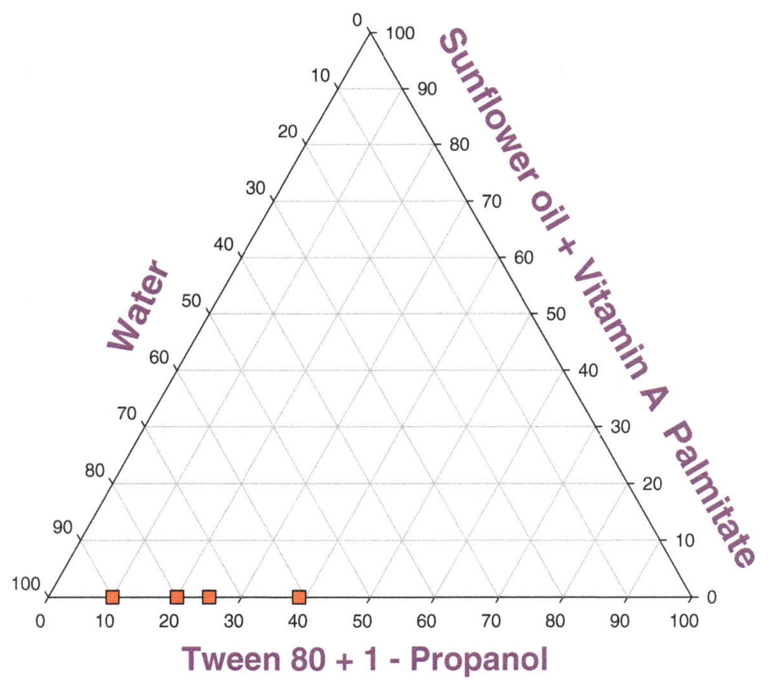

Figure 5 Pseudo-ternary phase diagram of the vitamin A palmitate microemulsion region with 1-propanol co-surfactant.

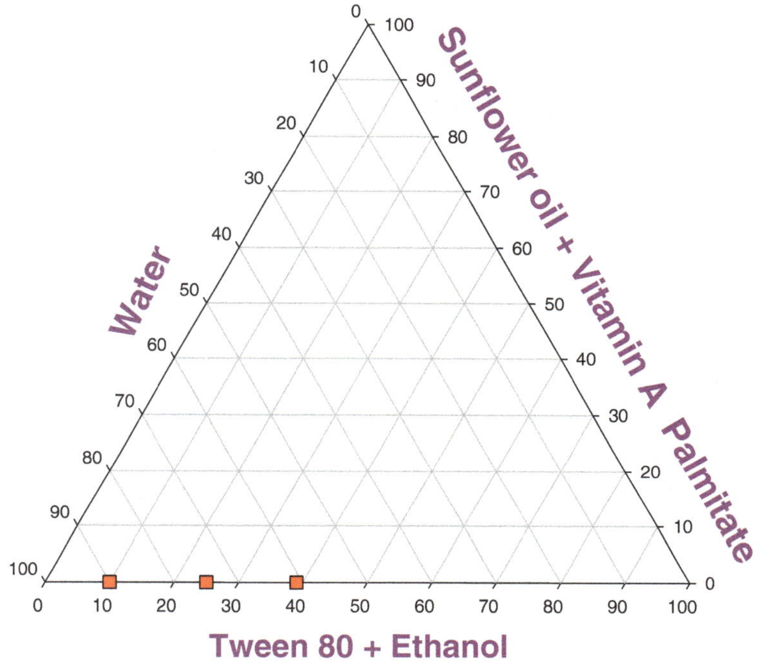

Figure 6 **Pseudo-ternary phase diagram of the vitamin A palmitate microemulsion region with ethanol co-surfactant.**

Dynamic light scattering method, sometimes called photon correlation spectroscopy, is used to measure particles in a liquid medium. Recently, particles with a diameter of less than 1 nm can also be measured with this method. Figure 7 shows the particle size distributions of our microemulsions.

Stability studies

Some samples were selected for stability testing. The samples were kept at different temperatures (4°C, 45°C, and room temperature) for 48 h and were evaluated periodically. In some cases, sugar crystallization or phase separation was observed. Of the samples which were kept at 4°C, 93% remained stable and 6% of them showed phase separation or sugar crystallization. The samples at room temperature showed the highest stability (96%). The stability percentage obtained for samples which were kept at 45°C was 94%.

Conclusions

We develop the o/w microemulsion areas of pseudo-ternary phase diagrams for a drug delivery system consisting of sunflower oil, surfactant (Tween 80), vitamin A palmitate, different co-surfactants, and water. The effects of different co-surfactants on the vitamin A palmitate microemulsion drug delivery system region were investigated. The co-surfactants studied were anhydrous glycerol, sucrose, ethanol, and 1-propranol. In conclusion,

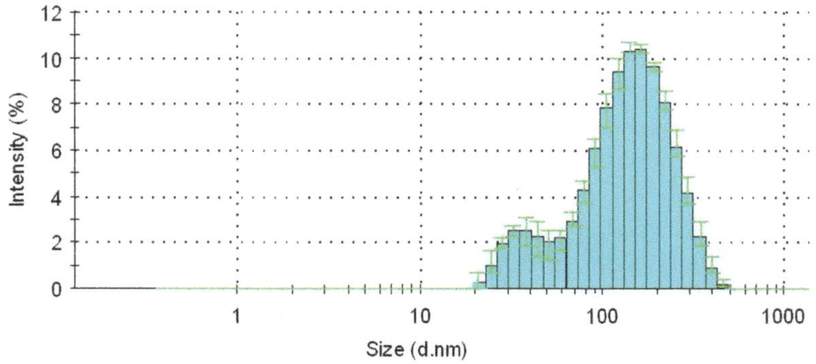

Figure 7 **Particle size distribution of vitamin A palmitate microemulsion with DLS.**

the system sunflower oil + Tween 80 + sucrose + water + vitamin A showed the largest microemulsion region. This system has higher solubility compared with the other systems, and our choice for drug formulation is increased in this system. Consequently, sucrose was selected as the best co-surfactant for this system.

Methods

Materials

The materials used were as follows: Tween 80 (polysorbate 80) surfactant (AppliChem, Darmstadt, Germany), sucrose (AppliChem), anhydrous glycerol, 1-propanol and ethanol (Merck, Darmstadt, Germany), sunflower oil (Faravardehaye Roghani Iran Co., Tehran, Iran), and vitamin A palmitate (USP30, Germany).

Methods

From the stability point of view, the ratio of surfactant to co-surfactant was fixed at 1:1 on the weight basis. In the oil phase, the ratio of sunflower oil to vitamin A was fixed at 9:1 in all of the samples. A titration technique was employed for the preparation of the samples of the five systems below:

Tween 80 + sunflower oil + water + vitamin A palmitate (without co-surfactant)
Tween 80 + sunflower oil + water + vitamin A palmitate + anhydrous glycerol as a co-surfactant
Tween 80 + sunflower oil + water + vitamin A palmitate + sucrose as a co-surfactant
Tween 80 + sunflower oil + water + vitamin A palmitate + 1-propanol as a co-surfactant
Tween 80 + sunflower oil + water + vitamin A palmitate + ethanol as a co-surfactant

Deionized water was added in different volumes to the mixture of sunflower oil, surfactant, co-surfactant, and model drug at room temperature. All of the samples were stirred for 24 h and 1,200 rpm using a magnetic stirring plate. After phase preparation, we identify the microemulsion area by visual inspection (a microemulsion is optically clear and transparent) and polarized microscopy (Olympus, BX51TRF Model, Tokyo, Japan). In each system, samples of the microemulsion area were separated, and the o/w area of the pseudo-ternary phase diagram using Sigma plot (12) software was plotted for them. The system sunflower oil + Tween 80 + sucrose + water + vitamin A showed the largest microemulsion region. Then sucrose was selected as the best co-surfactant, and further studies such as particle size analysis and stability studies were performed for it. Particle size analysis was performed by dynamic light scattering (NANOPHOX PCCS with WINDOX 5 software, Sympatec, Clausthal-Zellerfeld, Germany).

Competing interests
The author declares that has no competing interests.

Acknowledgements
We would like to thank Urmia Payam Noor University and Dr Alireza Salehi Sadaghiani for his valuable guidance.

References

1. Wang, Z: Pseudo ternary phase diagram of a drug delivery system. University of Waterloo, Dissertation (2009)
2. Taha, E, Al-Saidan, S, Samy, AM, Khan, MA: Preparation and in vitro characterization of self-nanoemulsified drug delivery system of all-trans-retionol acetate. Int. J. Pharm. **285**, 109–119 (2004)
3. Najafi, M, Kadkhodaie, R, Mortazavi, SA: Evaluation factors affecting the formation and properties of nanoemulsion stabilized with Tween 80. Iran Food SciTechnol Res. J. **6**(1), 9–16 (2010)
4. Swarbrick, J: Phase equilibrium diagrams. J. Soc. Cosmetic **19**, 187–209 (1968)
5. Moulik, SP, Raksjit, AK: Physicochemisty and applications of microemulsion. J. Surface Sci. Technol **22**(3–4), 159–186 (2006)
6. Salehi, A: Zwitterionic and catanionic surfactants. University of Lund, Division of Physical Chemistry, Dissertation (1990)
7. Kamble, V: Self micro emulsifying drug delivery system. Int. J. Pharma. and Bio. Sci. **1**, 2 (2010)
8. Rangel, C: Micellar solubilization of drugs. J. Pharmaceut. Sci. **8**(2), 147–163 (2005)
9. Zachar, C: Pharmaceutically relevant microemulsions with potential topical, ophthalmic and parenteral applications. The University of Toledo, Dissertation (2010)
10. Ghahraee, M, Khadiv-Parsi, P, Kaffashi, B, Momajjed, A, Babaee, J: Microemulsion preparation form pumpkin seed oil (*Cucurbita pepo*) as an oral drug carrier. J. Med. Plants **30**(2), 44–53 (2010)
11. Sadeghhasani, S, Samiee, L: Study of nanoparticle size measurement with dynamic light scattering. J. Nanotechnol. **7**, 130 (2009)

23

Room temperature aqueous phase synthesis and characterization of novel nano-sized coordination polymers composed of copper(II), nickel(II), and zinc(II) metal ions with *p*-phenylenediamine (PPD) as the bridging ligand

Rimpy Gupta, Sumit Sanotra, Haq Nawaz Sheikh* and Bansi Lal Kalsotra

Abstract

Nanostructured metal-organic hybrid materials composed of nickel(II), copper(II), and zinc(II) metal ions and *p*-phenylenediamine (PPD) as the organic ligand were synthesized in aqueous medium at room temperature. The synthesized compounds were characterized by elemental analyses, powder X-ray diffraction (PXRD) spectra, Fourier transform infrared spectra, nuclear magnetic resonance (^1H NMR) spectra, electronic spectra, scanning electron microscopy, N_2 adsorption-desorption isotherm, and dynamic light scattering studies. N_2 adsorption-desorption isotherm of copper(II)-PPD compound confirmed that it has mesoporous structure as it exhibits type-IV reversible isotherm with *H1* hysterisis. Steep adsorption indicated that the mesopores possessing it are of uniform order. Barrett-Joyner-Halenda model showed an average pore diameter of 5.2 nm. The PXRD patterns of all the three compounds are identical and showed well-defined and highly intense diffraction peaks, thereby suggesting their nature as crystalline. The broadness of the diffraction peaks indicated that the particles are of nanometer dimensions.

Keywords: Nanostructured, *p*-Phenylenediamine, Aqueous medium, Mesoporous, Crystalline

Background

Metal-organic frameworks (MOFs) are a new development in the interface between molecular coordination chemistry and material science. They are a class of porous polymeric material consisting of metal ions linked together by organic bridging ligands. Metal-organic hybrid materials have been used for ion exchange, gas adsorption, catalysis and enantio-selective separations [1,2]. Using different metal ions and selecting special organic ligands, a variety of metal-organic hybrid materials with significant new structures and properties have been synthesized [3,4].

So far, most studies on the synthesis of the metal-organic hybrid materials have focused on obtaining large crystal structures. In principle, if metal-organic materials can be obtained at a nanoscale, the surface area of the materials will be larger, and the quantum mechanical effects such as the 'quantum size effect' begin to play a significant role. These effects play minor role when going from macro to micro dimensions but become increasingly important when nanometer size range is reached [5-9]. Scaling down these materials to the nanoregime has enabled their use in a broad range of applications including catalysis, spin-crossover, templating, biosensing, biomedical imaging, and anticancer drug delivery [10-12].

However, researchers have not paid enough attention to the significance of syntheses of the metal-organic hybrid nanomaterials. Some hybrid nanomaterials have been reported [10,13,14]. Phenylenediamine (PPD) has been proven as a good building block for the synthesis of the inorganic–organic structures [15,16]. The $[CuCl_4-(H_2L)]$ (L=*p*-phenylenediamine) has been

* Correspondence: hnsheikh@rediffmail.com
Post-Graduate Department of Chemistry, University of Jammu, Jammu Tawi 180006, India

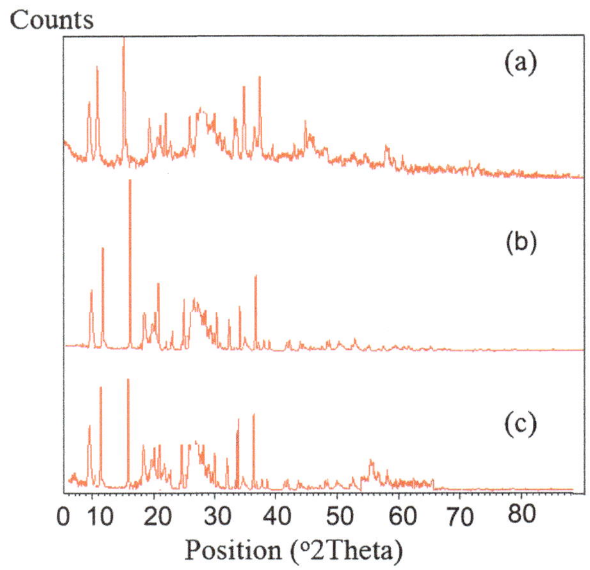

Figure 1 PXRD patterns for (a) 1, (b) 2, and (c) 3.

synthesized in strong acidic condition and identified as having a layered structure in which the hydrogen bonding plays an important role [15]. Sun and others have successfully synthesized submicrometer-scale, monodisperse spherical colloids of coordination polymers from the mixture of H_2PtCl_6 and *p*-phenylenediamine (PPD) aqueous solutions at room temperature [17]. The influence of molar ratio and concentration of reactants on such colloids was investigated, and the optimum experimental parameters for the formation of monodisperse products were proposed [17]. Based on these studies, we have tried to synthesize some functional MOF nanomaterials using PPD as the organic ligand. In this study, we report on three new hybrid nanostructures composed of nickel(II), copper(II), and zinc(II) with *p*-phenylenediamine as the organic ligand. Such hybrid nanostructures with versatile properties provoked by combining the merits of both organic as well as inorganic part may find applications in many fields [18-20].

Various attempts to obtain single crystals of synthesized polymers have so far been unsuccessful. The polymers have been characterized by elemental analyses, powder X-ray diffraction (PXRD), Fourier transform infrared (FTIR), ^{1}H nuclear magnetic resonance (NMR), electronic spectra, scanning electron micrography (SEM), N_2 adsorption-desorption isotherm, and dynamic light scattering (DLS) studies. In this paper, we present our results and propose possible structures for synthesized compounds based on several spectral data. The unique nature of the work presented is the porosity of compounds which has been explored here.

Results and discussion

Elemental analysis and solubility

The compounds are fully soluble in dimethyl sulfoxide (DMSO) but partially soluble in ethanol. The compounds are insoluble in other solvents.

PXRD analysis

The compounds were examined for crystalline/amorphous nature by PXRD in as obtained state. All the compounds are assumed to be isostructural on the basis of their powder X-ray diffraction diagrams (Figure 1). Further, all the compounds produce highly intense X-ray reflections in their corresponding PXRD pattern, indicating that all the compounds are crystalline in nature. The

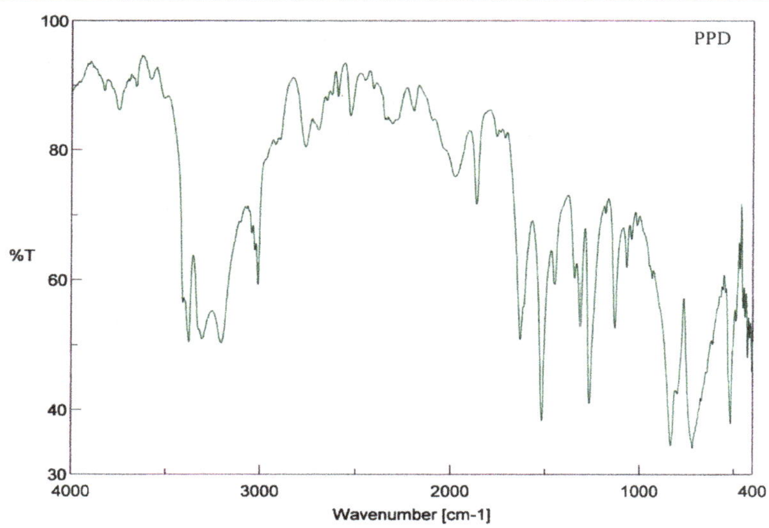

Figure 2 FTIR spectrum of PPD.

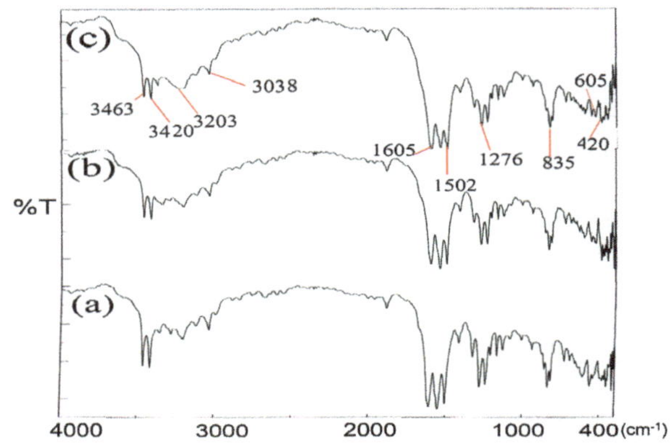

Figure 3 FTIR spectra of (a) 1, (b) 2, and (c) 3.

broadness of the diffraction peaks indicates that the particles are of nanometer dimensions. The average crystallite size of the nanoparticles was calculated using Scherrer's formula:

$$L = 0.89\lambda\ /\beta\ cos\theta,$$

where L is the average crystallite size, λ = 1.5418Å for CuKα, β is the half maximum peak width, and θ is the diffraction angle in degrees. The average crystallite size has been found to be about 40 nm.

FTIR spectral data

In PPD, two sharp bands observed at 3,498 and 3,450 cm^{-1} are due to ν_{as}(N-H) and ν_s(N-H) stretching modes of NH_2 group (Figure 2). In compounds **1**, **2**, and **3**, these bands appear at 3,463 and 3,420 cm^{-1}, respectively (Figure 3), confirming the presence of primary aromatic amine (NH_2) group [21]. A strong band near 1,605 cm^{-1} is due to N-H bending mode of NH_2 group [22]. A weaker absorption band near 3,203 cm^{-1} is the Fermi resonance band which appeared as a result of the

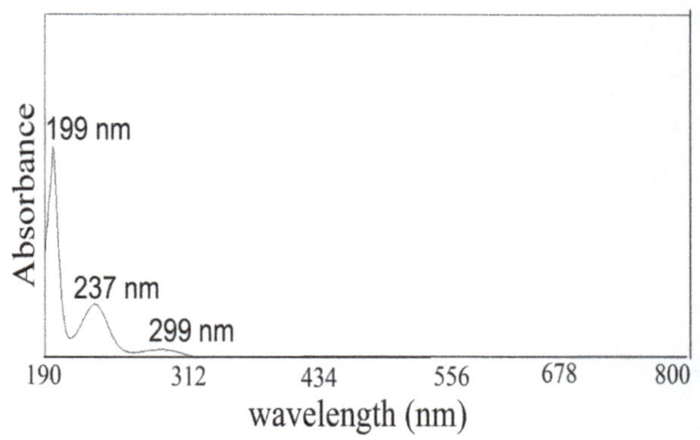

Figure 5 UV–vis spectrum of PPD.

interaction between the overtone of the 1,605 cm^{-1} band with the symmetric N-H stretching band. Further, bands at 1,276, 3,038, 835, and 1,502 cm^{-1} are assigned as C-N aromatic stretching [23], C-H aromatic stretching, C-H aromatic deformation [22,23], and C=C aromatic stretching bands [22], respectively. The new non-ligand bands observed in the region around 605 and 420 cm^{-1} are assigned to the υ(M–N) bands and υ(M–O) bands, respectively [24]. The broad band around 3,200 cm^{-1} is due to υ(OH) of coordinated water molecules [24-27].

^{1}H NMR study

The ^{1}H NMR spectra of PPD and compound **1** were recorded in deuterated DMSO. The ^{1}H NMR spectrum of PPD showed two signals (Figure 4a). The sharp signal at δ 6.31 is due to aromatic protons, whereas the broad signal at δ 3.37 is due to protons of -NH_2 group. The peak at δ 2.50 is the residual solvent peak.

The ^{1}H NMR spectrum of **1** (Figure 4b) is also characterized by the two signals which exactly correspond to two types of protons in **1**. Compared to the spectrum of PPD, broad signal at δ 4.00 in the spectrum of **1** is attributed to the -NH_2 protons of the benzenoid ring,

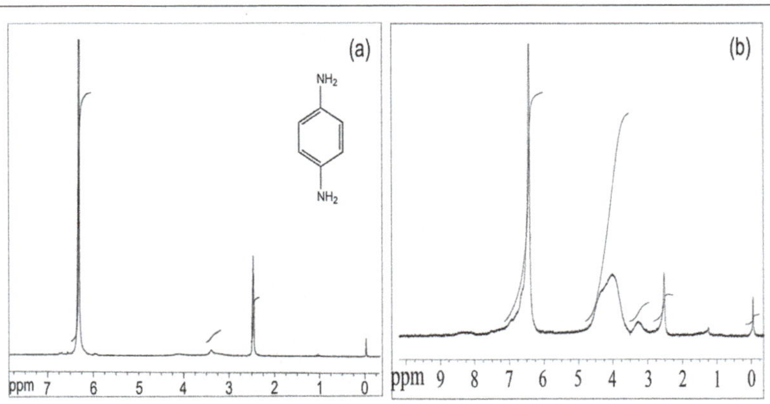

Figure 4 ^{1}H NMR spectra of (a) PPD and (b) 1.

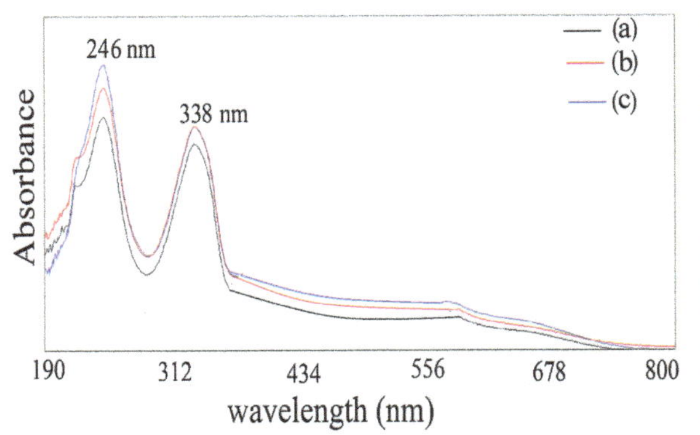

Figure 6 UV-vis spectrum of (a) **1**, (b) **2**, and (c) **3**.

whereas the aromatic protons of the benzenoid ring appear as a sharp signal at δ 6.50. Again, the peak at δ 2.50 is the residual solvent peak.

UV-vis spectral studies

The UV–vis spectra of PPD and compounds **1**, **2**, and **3** were recorded in dimethyl sulfoxide to investigate their electronic properties. The UV–vis spectrum of PPD (Figure 5) shows three narrow and intense characteristic absorption bands centered at 199, 237, and 299 nm. The bands located at 199 and 237 nm can be assigned to π-π^* transitions of the benzenoid rings, whereas the weak band centered at 299 nm is ascribed to the transition derived from a so-called π-p conjugation by the unshared electrons of NH_2 and benzenoid ring [28,29]. However, in **1**, **2**, and **3** (Figure 6), the bands due to the π-π^* transitions undergo bathochromic shift to 246 nm, and the band due to the π-p transition undergoes bathochromic shift to 338 nm when compared with the spectrum of PPD. It is probable that bathochromic shift occurs as a result of the coordination of the lone pair electrons on the N donor atoms of PPD with the metal ion site, thus stabilizing the excited state relative to the ground state, leading to longer wavelength absorption maxima [30].

Magnetic studies

The observed magnetic moment values of **1** and **2** at room temperature are 1.71 and 2.80 BM, respectively, suggesting their paramagnetic nature, whereas compound **3** was found diamagnetic in nature. The values correspond to +2 oxidation state for Cu, Ni, and Zn in the respective compounds.

Morphology and dynamic light scattering DLS studies

The morphologies of **1**, **2**, and **3** were determined by SEM. Typical SEM images of the particles of **1**, **2**, and **3** are shown in Figures 7, 8, and 9, respectively. As depicted from low magnification SEM images, only a few particles are in single state, while most of the particles are prone to condense with each other and form nanoclusters. Higher magnification SEM images reveal that these nanoclusters form a porous network structure.

We further examined the particle size of each compound by DLS technique. All the compounds were dispersed in water by sonication for 5 min. As observed from Figure 10, the DLS measurements show mean particle sizes of 400, 340, and 360 nm for **1**, **2**, and **3** respectively, with narrow size distribution.

Figure 7 Typical SEM images of the particles of 1. (**a**) Low magnification SEM image, (**b**) and (**c**) high magnification SEM images.

Figure 8 Typical SEM images of the particles of 2. (**a**) Low magnification SEM image, (**b**) and (**c**) high magnification SEM images.

N_2 adsorption-desorption isotherm: textural/structural properties

Figure 11a shows the nitrogen sorption isotherm of **1**. The figure suggests that according to IUPAC classification, compound **1** exhibits type IV reversible isotherm with *H1* hysterisis, characteristic of mesoporous materials. The inflection occurs at relative pressure 0.09. Generally, the inflection position depends on the diameter of mesopores, and sharpness usually indicates the uniformity of the mesopores. The higher the value of inflection position, the bigger the diameter of mesopores. The greater the sharpness, the more uniformity the mesopores are. The inflection position corresponds to the pronounced uptake of nitrogen and indicates the stage at which the monolayer coverage is complete and the multilayer adsorption is about to begin.

As observed from the isotherm, compound **1** shows steep adsorption. This indicates that compound **1** possesses mesopores of uniform order and has ordered structural as well as textural features [31]. In addition, the hysteresis loop observed at higher relative pressure indicates that it possesses large pores [32] which are filled at relatively higher pressures by capillary condensation of N_2. The type *H1* further suggests that the polymer consists of *agglomerates* [33].

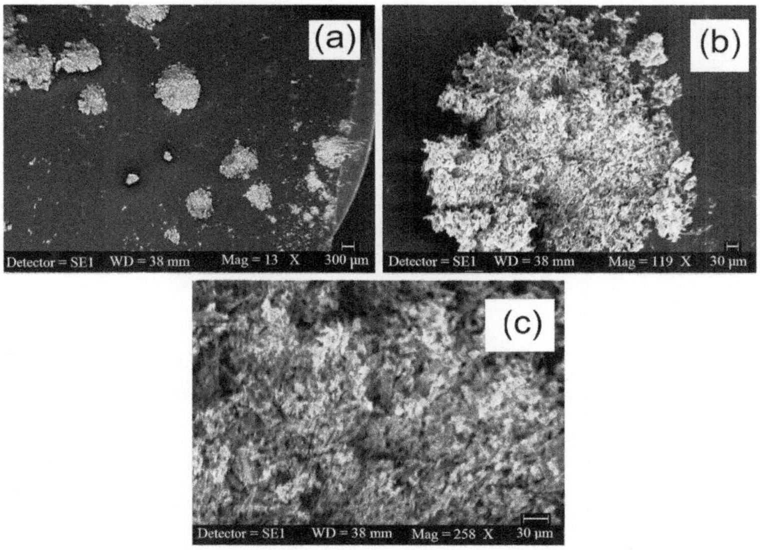

Figure 9 Typical SEM images of the particles of 3. (**a**) Low magnification SEM image, (**b**) and (**c**) high magnification SEM images.

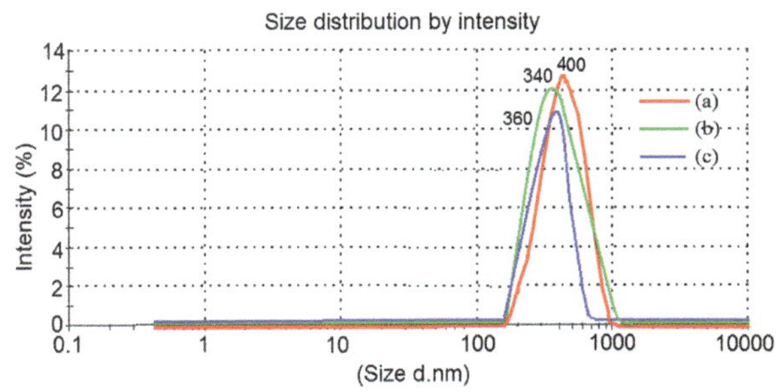

Figure 10 Particle size measurement of (a) 1, (b) 2 and (c) 3 in water by DLS technique.

In view of further verification, the pore size distribution pattern was obtained from the desorption branch by employing Barrett-Joyner-Halenda (BJH) model and is illustrated in Figure 11d. The distribution curve shows that the pores present in the polymer are of variable sizes with their diameters ranging from 1.4 to 12.4 nm. The maxima of the curve occurs at 2.3 nm, suggesting that majority of the pores have diameters of 2.3 nm. The average pore diameter was found to be 5.2 nm, whereas the total pore volume has been found to be 0.085 cc/g for pores smaller than 19 nm diameter at relative pressure P/Po = 0.89.

The specific surface area determined by Brunauer-Emmett-Teller (BET) method is found to be 65.25 m^2/g (Figure 11b), whereas the Langmuir surface area is found to be 87.3 m^2/g. The values of the micropore volume, micropore area, and external surface area obtained by the *t*-method micropore analysis (de Boer) are 0.00012 cc/g, 0.35 m^2/g, and 64.9 m^2/g, respectively (Figure 11c). Thus, total BET surface area (65.25 m^2/g) is exactly equal to the sum of the micropore area (0.35 m^2/g) and external the surface area (64.9 m^2/g). This behavior further confirms that compound **1** satisfies the BET analysis with type IV isotherm, indicating the formation of multiple layers during adsorption [33]. The surface area of the micropores is relatively low compared with the other porous crystalline structures, which might be caused by the relatively compact structure of the polymer [34].

Proposed mechanism of polymerization

A plausible formation process is briefly presented as following:

While preparing aqueous solutions of the respective metal salts, M^{2+} ions get hydrated and form hexa-aqua complexes. When these aqua complexes are mixed with aqueous solution of PPD, two water molecules on the trans positions of aqua complex are replaced by two PPD molecules, resulting in tetra-aqua PPD substituted complex. The free nitrogen atoms of the coordinated PPD can further capture hexa-aqua complex molecules by coordination interactions. This coordination-induced assembly process can proceed repeatedly until the depletion of reactants in the solution, resulting in the formation of large coordination polymers having a porous structure. The Cl^- anions are accommodated in this

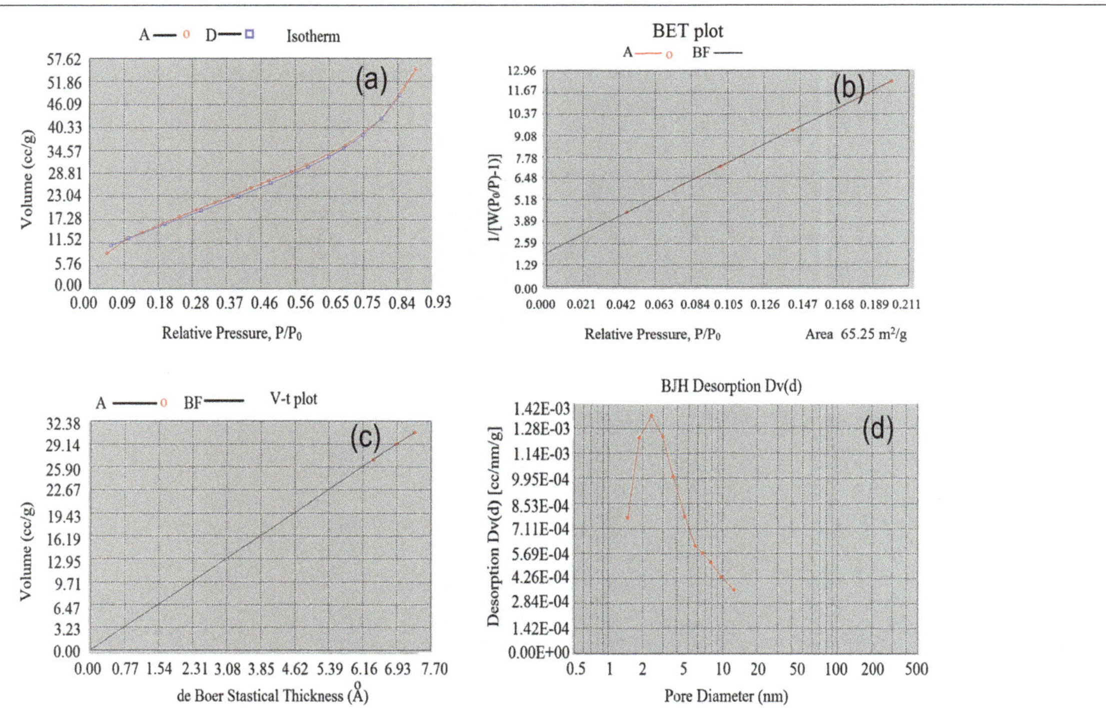

Figure 11 Surface area and porosity studies of 1. (**a**) N_2 sorption isotherm, (**b**) BET plot, (**c**) *t*-plot analysis, and (**d**) BJH pore size distribution.

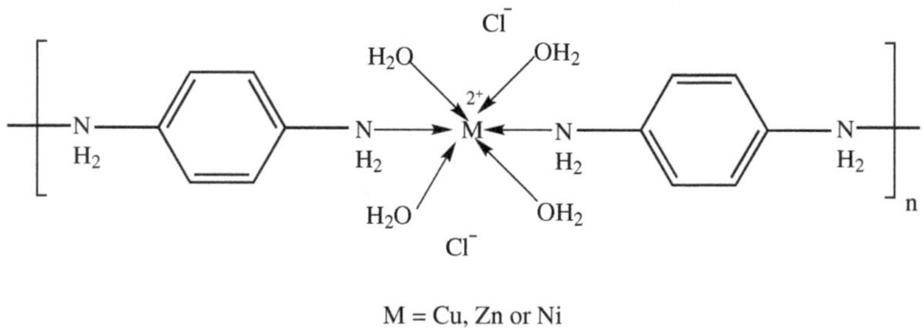

Scheme 1 One-dimensional linear coordination polymers comprising a repeat unit.

porous structure as the counter ions. Thus, we get one-dimensional linear coordination polymers comprising a repeat unit shown in Scheme 1 in which the *M*(II) ion is six- coordinated. The *M*(II) ion binds to four oxygen atoms from the four coordinated water molecules (which come from aqueous medium) and two nitrogen atoms from two different PPD ligands in the distorted octahedral coordination geometry. However, the detailed formation mechanism of such coordination polymers is not clear at present time and needs further

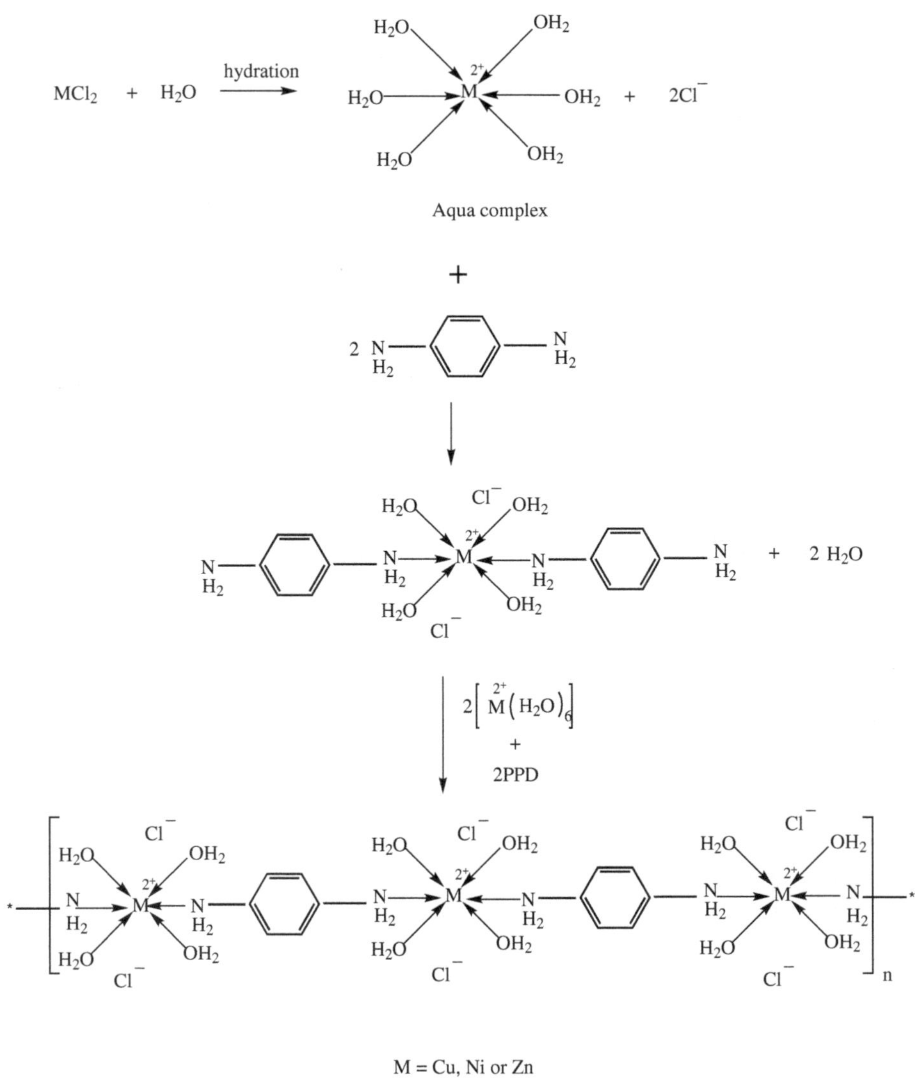

Scheme 2 Proposed mechanism.

investigation. However, the proposed mechanism is shown as Scheme 2.

Conclusions

Mesoporous nanostructured metal-organic hybrid materials composed of copper(II), nickel(II), and zinc(II) metal ions with PPD as the organic ligand were synthesized at room temperature. The proposed formula of the compounds is $\{[M(H_2O)_4(PPD)_2]^{2+}.2Cl^-\}_n$, where M = Cu, Ni or Zn. The PXRD profile shows the crystalline nature of the polymers. N_2 adsorption-desorption isotherm confirms the mesoporous nature of **1** as it exhibits type IV isotherm with *H1* hysterisis. The steepness of the inflection point shows that the mesopores formed are of uniform order. The BJH curve shows an average pore diameter of 5.2 nm. The specific surface area determined by the BET method is 65.25 m^2/g.

Methods

Materials

Doubly-distilled water was used to prepare aqueous solutions. *p*-Phenylenediamine was received from Sigma-Aldrich (MO, USA). $CuCl_2$, $NiCl_2.6H_2O$, and $ZnCl_2$ were of analytical grade and used without further purification.

Characterization techniques

Microanalyses were carried out using CHNS analyzer Leco Model-932 (LECO Corporation, St. Joseph, MI, USA). The concentrations of the metal ions were estimated by atomic absorption spectroscopy (Perkin-Elmer 2380, Waltham, MA, USA). PXRD patterns were recorded from 0° to 70° on a Rigaku Miniflex diffractometer using monochromatic CuKα radiations (The Woodlands, TX, USA). FTIR spectra from 4,000 to 400 cm^{-1} were recorded on Perkin-Elmer Spectrum RX 1 FTIR spectrophotometer using KBr disks. ^1H NMR spectra were recorded on Bruker Avance DPX 200 NMR spectrometer (200 MHz; Dubai, UAE) using tetramethyl silane as internal standard. Electronic spectra (190 to 800 nm) of 10^{-3} M ethanolic solutions of PPD and its compounds with Ni(II), Cu(II), and Zn(II) were recorded on UV-visible double beam spectrophotometer, PG Instruments Ltd. Model T90+ (Leicestershire, England). Scanning electron micrographs (SEM) were collected on Jeol T-300 scanning electron microscope with gold coating (Tokyo, Japan). The particle size of each compound was determined by DLS technique using Zetasizer Nano ZS-90 (Malvern Instruments Ltd., Worcestershire, UK).

N_2 adsorption-desorption and Brunauer-Emmett-Teller surface area studies

The specific surface area, total pore volume, and average pore diameter were measured by N_2 adsorption-desorption method using NOVA 1200 (Quanta Chrome, Boynton Beach, FL, USA) instrument. The bath temperature was maintained at 77 K. The sample was degassed at 423 K for 12 h before measurement. Pore size distribution and pore volume were obtained by applying BJH analysis to the adsorption branch of the nitrogen adsorption-desorption isotherm. The specific surface area of the sample was calculated using Brunauer-Emmett-Teller method. Micropore surface area and micropore volume were determined from *t*-plot analysis. All the flow rates were maintained at normal temperature and pressure.

Synthesis of *M*(II)-PPD compounds (*M*(II) = Cu(II), Ni(II), Zn(II))

In a typical experiment, 1.4 ml of 15 M PPD aqueous solution was introduced into 40 ml of water first, and then 5.5 ml of 3.8 M aqueous solution of the metal salt; $CuCl_2/NiCl_2.6H_2O/ZnCl_2$ with 1:1 molar ratio of PPD to metal was added into the resulting solution under vigorous stirring at room temperature. A gradual color change was observed, and a large amount of blackish-brown-colored precipitates occurred within several hours. The precipitates thus formed were collected by centrifugation, washed several times with absolute ethanol and water, and then dried in air. The yield percentage obtained in each case was around 68%.

- Elemental analysis of the compound composed of Cu(II) with PPD, 1: ($C_{12}H_{24}CuN_4O_4$, 351.89): Calc: C 40.96, H 6.87, N 15.92, Cu 18.06%; Found: C 40.85, H 6.75, N 15.78, Cu 18.25%.
- Elemental analysis of the compound composed of Ni (II) with PPD, 2: ($C_{12}H_{24}NiN_4O_4$, 347.04): Calc: C 41.53, H 6.97, N 16.14, Ni 16.91%; Found: C 41.69, H 6.66, N 16.78, Ni 16.55%.
- Elemental analysis of the compound composed of Zn(II) with PPD, 3: ($C_{12}H_{24}ZnN_4O_4$, 353.73): Calc: C 40.74, H 6.84, N 15.84, Zn 18.49%; Found: C 40.85, H 6.65, N 15.68, Zn 18.25%.

Competing interests

The authors declare that they have no competing interest.

Authors' contributions

RG, SS, HNS, and BLK provided the same contributions in this article. All authors read and approved the final manuscript.

Authors' information

RG was born in Jammu, India. She received her M.Sc. degree in Chemistry from the University of Jammu in 2008 and M. Phil. degree in Chemistry from the same university in 2009 under the supervision of Dr. Sheikh. Presently, she is doing her Ph.D. study in Chemistry from the same university, under the supervision of Dr. Sheikh. SS was born in Jammu, India. He received his M.Sc. degree in Chemistry from the University of Jammu in 2009. Presently, he is doing his Ph.D. study in Chemistry from the same university under the supervision of Dr. Sheikh. HNS was born in Jammu, India. He received his M. Sc. degree in Chemistry from the University of Jammu in 1992, and Ph.D. degree in Chemistry from the same university in 1998. He joined the Department of Chemistry, University of Jammu as Lecturer in 1999, became

Associate Professor in 2006, and Professor in 2013. He is presently working as Professor in the Department of Chemistry, University of Jammu, Jammu. At present, he is actively involved in research in coordination polymers and nanomaterials, and has M. Phil. and Ph.D. students in these fields. BLK was born in Jammu, India. He received his M.Sc. degree in Chemistry from J & K University 1969, Ph. D. degree in Chemistry from Delhi University in 1973, and Post Doctoral degree from UTEP, El Paso, Texas, USA in 1978. He retired as Dean of the Faculty of Science, University of Jammu in 2007. He has worked on much scientific research areas such as Organometallic Chemistry, Coordination Chemistry, Environmental Chemistry, and Nanomaterials.

Acknowledgments
The authors gratefully acknowledge the financial support from the Defence Research Development Organization, New Delhi. They thank the Department of Anatomy, All India Institute of Medical Sciences, New Delhi for the Electron Microscope Facility. They are also thankful to the National Chemical Laboratory, Pune for the thermal, powder X-ray diffraction, and adsorption-desorption studies.

References

1. James, SL: Metal-organic frameworks. Chem. Soc. Rev. **32**, 276–288 (2003)
2. Eddaoudi, M, Moler, DB, Li, H, Chen, B, Reineke, MR, O'Keeffe, M, Yaghi, O: Modular chemistry: secondary building units as a basis for the design of highly porous and robust metal-organic carboxylate frameworks. Accounts Chem. Res. **34**, 319–330 (2001)
3. Seo, JS, Whang, D, Lee, H, Jun, SI, Oh, J, Jeon, YJ, Kim, K: A homochiral metal-organic porous material for enantioselective separation and catalysis. Nature **404**, 982–986 (2000)
4. Fukuoka, A, Sakamoto, Y, Guan, S, Inagaki, S, Sugimoto, N, Yukushima, F, Hirahara, K, Iijima, S, Ichikawa, MJ: Novel templating synthesis of necklace-shaped mono-and bimetallic nanowires in hybrid organic–inorganic mesoporous material. J. Am. Chem. Soc. **123**, 3373–3374 (2001)
5. Shi, H, Qi, L, Ma, L, Cheng, H: Polymer-directed synthesis of penniform $BaWO_4$ nanostructures in reverse micelles. J. Am. Chem. Soc. **125**, 3450–3451 (2003)
6. Zhang, H, Yang, D, Li, D, Ma, X, Li, S, Que, D: Controllable growth of ZnO microcrystals by a capping-molecule-assisted hydrothermal process. Cryst. Growth Des. **5**, 547–550 (2005)
7. Kuang, D, Xu, A, Fang, Y, Liu, H, Frommen, C, Fenske, D: Surfactant-assisted growth of novel PbS dendritic nanostructures via facile hydrothermal process. Adv. Mater. **15**, 1747–1750 (2003)
8. Kim, F, Connor, S, Song, H, Kuykendall, T, Yang, P: Platonic gold nanocrystals. Angew. Chem. Int. Ed. **43**, 3673–3677 (2004)
9. Markovich, G, Collier, CP, Henrichs, SE, Remacle, F, Levine, RD, Heath, JR: Architectonic quantum dot solids. Acc. Chem. Res. **32**, 415–423 (1999)
10. Lin, W, Rieter, J, Taylor, KML: Modular synthesis of functional nanoscale coordination polymers. Angew. Chem. Int. Ed Engl. **48**(4), 650–658 (2009)
11. Taylor, KML, Rieter, WJ, Lin, W: Manganese-based nanoscale metal-organic frameworks for magnetic resonance imaging. J. Am. Chem. Soc. **130**, 14358–14359 (2008)
12. Rieter, WJ, Pott, KM, Taylor, KML, Lin, W: Nanoscale coordination polymers for platinum-based anticancer drug delivery. J. Am. Chem. Soc. **130**, 11584–11585 (2008)
13. Sahraneshin, A, Asahina, S, Togashi, T, Singh, V, Takami, S, Hojo, D, Arita, T, Minami, K, Adschiri, T: Surfactant-assisted hydrothermal synthesis of water-dispersible hafnium oxide nanoparticles in highly alkaline media. Cryst. Growth Des. **12**(11), 5219–5226 (2012)
14. Tsuroka, T, Kawasaki, H, Nawafune, H, Akamatsu, K: Controlled self-assembly of metal–organic frameworks on metal nanoparticles for efficient synthesis of hybrid nanostructures. Appl. Mater. Interfaces. **3**(10), 3788–3791 (2011)
15. Bourne, SA, Mangombo, Z: Phenylamines as building blocks to layered inorganic–organic structures. Cryst. Eng. Comm. **6**, 437–442 (2004)
16. Pansanel, J, Jouaiti, A, Ferlay, S, Hosseini, M: Molecular tectonics: generation of 2-D molecular networks by combination of coordination and hydrogen bonds. New J. Chem. **30**, 71–76 (2006)
17. Sun, X, Dong, S, Wang, E: Coordination-induced formation of submicrometer-scale, monodisperse, spherical colloids of organic–inorganic hybrid materials at room temperature. J. Am. Chem. Soc. **127**, 13102–13103 (2005)
18. Mederos, A, Dominguez, S, Hernandez-Molina, R, Sanchiz, J, Brito, F: Coordinating ability of phenylenediamines. Coord. Chem. Rev. **193–195**, 913–939 (1999)
19. Mederos, A, Dominguez, S, Hernandez-Molina, R, Sanchiz, J, Brito, F: Coordinating ability of ligands derived from phenylenediamines. Coord. Chem. Rev. **193–195**, 857–911 (1999)
20. Han, MG, Foulger, SH: Crystalline colloidal arrays composed of poly(3,4-ethylenedioxythiophene)-coated polystyrene particles with a stop band in the visible regime. Adv. Mater. **16**(3), 231–234 (2004)
21. Wang, JJ, Jiang, J, Hu, B, Yu, SH: Uniformly shaped poly(*p*-phenylenediamine) microparticles: shape-controlled synthesis and their potential application for the removal of lead ions from water. Adv. Funct. Mater. **18**(7), 1105–1111 (2008)
22. Cataldo, F: On the polymerization of *p*-phenylenediamine. Eur. Polym. J. **32**(1), 43–50 (1996)
23. Mallick, K, Witcomb, MJ, Dinsmore, A, Scurrell, MS: Fabrication of metal nanoparticles and polymer nanofibers composite material by an in situ chemical synthetic route. Langmuir **21**(17), 7964–7967 (2005)
24. Nakamoto, K: Infrared and Raman Spectra of Inorganic and Coordination Compounds, 5th edn. Wiley, New York (1997)
25. Ding, B, Liu, YY, Zhao, XJ, Yang, EC, Wang, XG: Hydrothermal synthesis and characterization of a novel luminescent lead(II) framework extended by novel Pb-μ1,1-(N)CS-Pb bridges. J. Mol. Struct. **920**, 248–251 (2009)
26. Luo, Y-M, Li, J, Xiao, L-X, Tang, R-R, Tang, X-C: Synthesis, characterization and fluorescence properties of Eu(III) and Tb(III) complexes with novel mono-substituted-diketone ligands and 1,10-phenanthroline. Spectrochim. Acta Part A. **72**, 703–708 (2009)
27. Xie, C, Zhang, Z, Wang, X, Liu, X, Shen, G, Wang, R, Shen, D: The synthesis and structure of a novel 1D copper(II) weak coordination polymer with 2,6-pyridinedicarboxylic acid. J. Coord. Chem. **57**, 1173–1178 (2004)
28. Min, Y-L, Wang, T, Zhang, Y-G, Chen, Y-C: The synthesis of poly(*p*-phenylenediamine) microstructures without oxidant and their effective adsorption of lead ions. J. Mater. Chem. **21**, 6683–6689 (2011)
29. Liu, Z, Zu, Y, Fu, Y, Guo, S, Zhang, Y, Liang, H: Synthesis of hybrid nanostructures composed of copper ions and poly(*p*-phenylenediamine) in aqueous solutions. J. Nanopart. Res **10**(8), 1271–1278 (2008)
30. Christie, RM: Colour Chemistry, pp. 12–44. Royal Society of Chemistry, United Kingdom (2001)
31. Jha, RK, Shylesh, S, Bhoware, SS, Singh, AP: Oxidation of ethyl benzene and diphenyl methane over ordered mesoporous M-MCM-41 (M = Ti, V, Cr): synthesis, characterization and structure–activity correlations. Micropor. Mesopor. Mater **95**(1–3), 154–163 (2006)
32. Shylesh, S, Jha, RK, Singh, AP: Assembly of hydrothermally stable ethane-bridged periodic mesoporous organosilicas with spherical and wormlike structures. Micropor. Mesopor. Mater. **94**(1–3), 364–370 (2006)
33. Sing, KSW, Everett, DH, Haul, RAW, Moscou, L, Pierotti, RA, Rouquerol, J, Siemieniewska, T: Reporting physisorption data for gas/solid systems with special reference to the determination of surface area and porosity. Pure Appl. Chem. **57**, 603–619 (1985)
34. Huo, J, Wang, L, Irran, E, Yu, H, Gao, J, Fan, D, Li, B, Wang, J, Ding, W, Amin, AM, Li, C, Ma, L: Hollow ferrocenyl coordination polymer microspheres with micropores in shells prepared by Ostwald ripening. Angew. Chem. **49**(48), 9237–9241 (2010)

24

Synthesis, characterization, and experimental investigation of surface activity of SERS substrates using neodymium oxide (Nd_2O_3)

Fatemeh Sadeghi[1], Farzad Khani[2], Abbas Hemati Azandaryani[3,4], Yadollah Mansouri[5,6*], Zahra Shaykhi Mehrabadi[3*] and Amir Nikjou[1]

Abstract

The activity of two surface-enhanced Raman scattering (SERS) substrates was investigated using neodymium oxide (Nd_2O_3) molecules. Colloidal silver solution, containing nano-sized silver particles and silver-coated filter papers were used as the SERS substrates. In order to characterize the substrates, ultraviolet–visible spectroscopy, dynamic light scattering, and scanning electron microscopy were used. Subsequently, gain enhancement factors were calculated from the measured spectra to monitor the change in effectiveness of the SERS substrates under prolonged storage time. Experimental results show that for enhancing the Raman signal, the activity of two fabricated substrates is very effective.

Keywords: Surface-enhanced Raman scattering (SERS), Colloidal silver, Silver-coated filter paper, Neodymium oxide (Nd_2O_3)

Background

Surface-enhanced Raman scattering (SERS) is a sensitive and selective technique in which Raman scattering enhancing is achieved for those adsorbed molecules on metal nano-sized particles. Using this method, not only for the molecule structure analysis but also the information about the direction of molecule absorption and interaction between the molecule and the substrates can be attained [1-3].

SERS was first observed on absorbed pyridine molecules on the surface of roughed silver electrode with chemical method by Fleischmann et al. [4]. At the beginning, an unusual increase in the intensity of Raman signals was attributed to the growth of roughed substrate area. Then in 1977, Jean Marie and his colleagues found that the growth of the Raman scattering's cross section has another reason than the increase of molecules due to being roughed [5]. Nowadays, there are different ideas to describe this effect, which two major categories are considering the intensity proportion of Raman scattering with the polarization of molecule and electric field shown as follows.

1. Electromagnetic effect: The mechanism is based on the optical properties of the noble metals and their ability to support plasmon resonances at visible wavelengths.
2. Chemical effect or charge transition: In this condition, the molecule is absorbed chemically by the metal nano-sized particles, and Raman enhancing is due to the exchange of electrons from metal to molecule and their return to the metal.

In other words, the molecule's polarization is affected by the interaction between the molecule and metal's surface. This effect is possible only for the first electronic state of the adsorbed molecule. Another important factor is the selection rules. When a molecule is adsorbed on the gold, silver, or other noble metals, its symmetry is decreased, and this leads to weaken the selection rules. As a result, more frequencies than those in the ordinary Raman spectroscopy are found [6-8].

* Correspondence: yadollamansouri@yahoo.com; sheikhizahra@yahoo.com
[5]Environmental Epidemiological Research Center, Kermanshah University of Medical Sciences, Kermanshah 6714676573, Iran
[3]Department of Applied Chemistry, Faculty of Chemistry, Razi University, Kermanshah 6715685438, Iran
Full list of author information is available at the end of the article

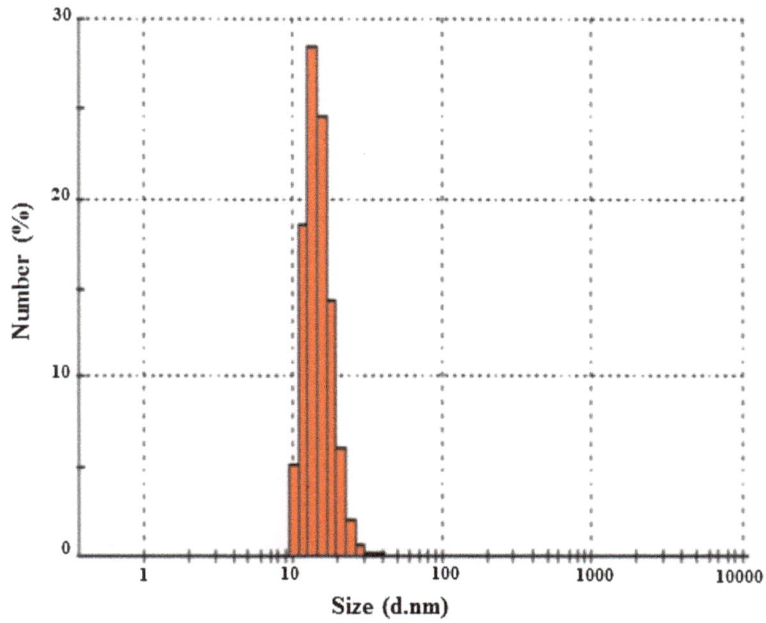

Figure 1 **The particle size distribution of a typical prepared sample.**

There are different kinds of SERS-active substrates used in the wide range of SERS applications such as metal colloids, coated filter papers with metal nanoparticles, electrochemically roughened metal electrodes, etc. [9-11].

In this study, we report a fabrication of the SERS-active substrate using colloidal silver nanoparticles and then characterized surface activity of SERS substrates by several methods such as scanning electron microscopy (SEM); dynamic light scattering (DLS) was investigated. Also, the plasmon absorption of the formed nanocrystals was monitored by ultraviolet–visible (UV–vis) spectrometry.

Results and discussions

Zetasizer Nano-Zs by Malvern Instruments (Worcestershire, UK) was used to define the size of the silver nano-sized particles. In this experiment, we dilute the collide solution with proportion of 5:1 using deionized water. Since, by increasing the average of scattering numbers of N, the number of oscillations is declined, and their effects are vanished in the large N [12]. The distribution of particles size of Ag metal that was shown in Figure 1 is relatively narrow, and the average particle size is about 16.9 nm.

In order to obtain a scanning electron microscopic image, we used freshly prepared nano-particles and micrographs of the samples (Figure 2). The size of the

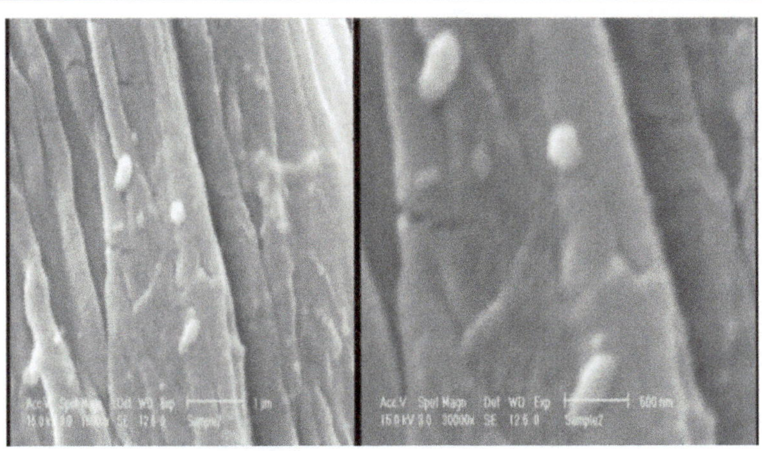

Figure 2 **SEM of filter paper coated using nano-particles of silver.**

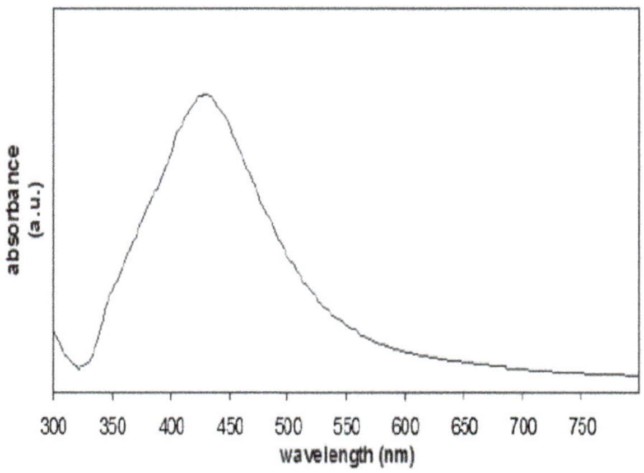

Figure 3 UV/vis spectrum of the prepared Ag colloidal sample.

nano-sized particles is estimated around 40 to 240 nm by microstructure measurement. In fact, it is supposed that the larger nano-sized particles are not, silver, perhaps they are salts existing in colloidal silver solution that are recrystallized after drying. Moreover, the micrographs demonstrated a spherical shape and a smooth surface, with a particle size in the nano-metric range [13,14].

Excitation of the localized surface plasmon resonance (LSPR) is characterized by strong, wavelength-selective absorption and enhanced electromagnetic fields at the nano-particle surface. This provides the opportunity to simply follow the optical properties of the nanostructures utilizing UV–vis spectrometry and the tremendous enhancement of the weak Raman signal, respectively. The maximum SERS enhancement is expected to take place when the max of LSPR is slightly longer than the laser (excitation) wavelength such that both the Raman scattered photon and the incident photon are strongly enhanced [11,15].

Hence, it is possible to adjust the wavelength maximum of the longitudinal resonance plasmon according to the excitation laser wavelength to obtain the largest Raman enhancement [16]. Colloidal silver solution was characterized using UV/vis/near-infrared (NIR) spectrophotometer. Figure 3 showed that the surface plasmon absorption band of colloidal silver was around 430 nm.

In Figure 4, we showed the Raman spectra and SERS of Nd_2O_3 molecules using colloidal silver: Figure 4a shows Raman spectrum in 1 M Nd_2O_3 solution without using SERS substrates. In Figure 4b, Raman spectrum in 0.01 M Nd_2O_3 solutions is shown which has lower density. According to Figure 4b, the peaks of Nd_2O_3 do not appear in this concentration. But in Figure 4c, the SERS spectrum for 0.01 M of Nd_2O_3that was calculated using the colloidal silver shows the peaks of Nd_2O_3 even better than in 1 M solution.

Figure 5 depicted the SERS spectrum of Nd_2O_3 molecules using filter paper coated with nano-silvers. Figure 5b shows that filter paper possesses high enhancing power than colloidal silver although the filter paper does not show some weak peaks. Comparison between the location of the main flow peak and their intensities is shown for both substrates in Table 1. To quantify the degradation in the SERS effectiveness of the silver nanorods affected by the storage time, we calculated the Raman enhancement factor from the SERS spectra as previously

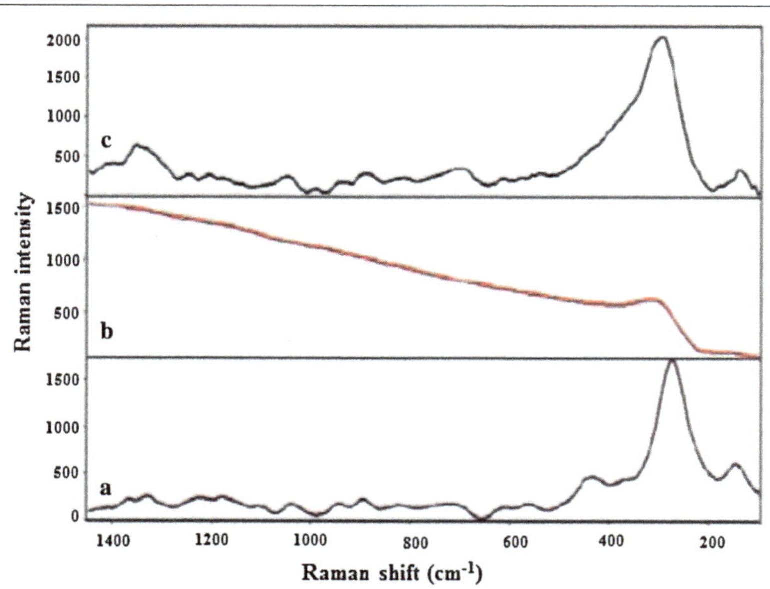

Figure 4 Raman spectra and SERS of Nd_2O_3 molecules using colloidal silver. The Raman spectrum of 1M Nd_2O_3 solution (**a**). The Raman spectrum of 0.01 M Nd_2O_3 solution (**b**). The SERS spectrum of 0.01 M Nd_2O_3 by silver colloid (**c**).

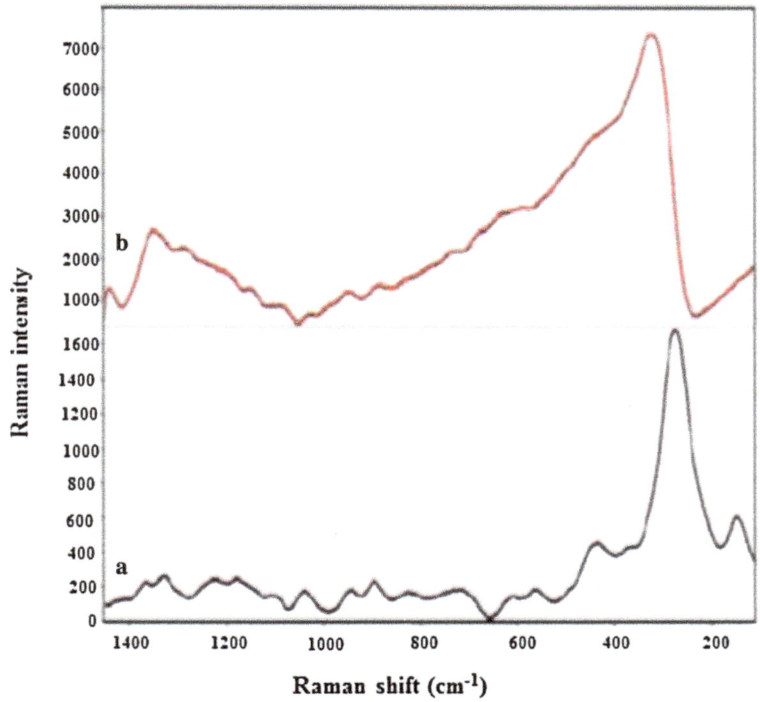

Figure 5 The Raman spectrum and SERS of Nd_2O_3 molecules using filter paper. (**a**) The Raman spectrum of 1 M Nd_2O_3 solution and (**b**) the SERS spectrum of 0.01 M of it using filter paper.

shown. The enhancement factor was defined as the ratio of inelastic scattering intensity per molecule between the presence and absence of the SERS structure. The Raman enhancement factor was calculated by following formula [17]:

$$EF = \frac{I_{SERS}}{I_{Raman}} \times \frac{C_{Raman}}{C_{SERS}} \tag{1}$$

In which, I is the intensity of the peak and C is the concentration or molarity of the solution. We use the selected solution molarity of C_{SERS}. The EF values were shown in Table 1. According to the relatively high enhancement factor (428 for filter paper and 127 for sliver colloidal, the method is practical to detect the molecules with a low concentration and even single molecules.

In addition to Raman enhancing and detecting the molecules with low concentration, we can obtain information about molecule structure of Nd_2O_3 and its absorption trait using SERS effect. The location of Raman and SERS peaks of Nd_2O_3 molecules is investigated in Table 2.

Table 1 Comparison of the intensities of the main peaks of Nd_2O_3 for two substrates

Substrate	Location of the peak (cm^{-1})	Intensity of the peak	Enhancement factor
Filter paper	322	7,327	458
Silver colloid	294	2,026	127
Nd_2O_3 molecule, 1 M solution	279	1,599	

Conclusions

In this study, SERS was used to characterize colloidal silver solution, containing nano-sized silver particles. The solution was used to make filter paper covered using nano-sized silver. Absorption spectrum of colloidal silver showed the plasmon resonance in 430-nm wavelength

Table 2 The location of Raman and SERS peaks

1 M solution of Nd_2O_3 (cm^{-1})	0.01 M solution of Nd_2O_3 in silver colloid (cm^{-1})	0.01 M solution of Nd_2O_3 on filter paper (cm^{-1})	Band
147	140		
274	294	322	Stretching vibration Nd-O
434			Stretching vibration Nd_2O_2
592		588	
721	701	722	
896	815	948	
1,180	1,209	1,154	Stretching vibration O-O
1,330	1,355	1,352	

using UV/vis/NIR. The DLS method also showed that the average size for nano-sized particles in colloid is 16.9 nm. To be certain about the existence of nano-sized silver particles on the filter paper, we used SEM imaging. SEM imaging proves the existence of the nano-sized silver particles on the filter papers, and their size is estimated about 40 to 240 nm. Moreover, the micrographs demonstrated a spherical shape and a smooth surface for particles. SERS spectrum for Nd_2O_3 was investigated for both substrates, and a satisfactory enhancement was observed.

Methods

Preparation of colloidal silver

Colloidal silver nano-particles were prepared according to Lee and Meisel's method [18]. Briefly, 18 mg of $AgNO_3$ was dissolved in 100-ml distilled water, and the solution was heated to boiling. Then, 7 ml of 1% trisodium citrate aqueous solution was added into the boiling silver nitrate solution at once, under vigorous stirring. The mixed solution was kept boiling for a further 8 min. Finally, the green-gray solution was obtained which was stable for several days or weeks.

In order to investigate rang of colloidal silver activity, we used Nd_2O_3. Nd_2O_3 is used to cover the glass for example in sunglasses, solid-state laser, and colored glasses [19]. To prepare Nd_2O_3 using colloidal silver, its molar solution with the proportion of 1:2 was mixed by colloidal silver, and the solution's pH was altered 8 up to 13 [20,21]. The peaks existing in 1 M solution were observed more clearly in pH = 12.

Preparation of the silver-coated filter papers

The filter paper used in this method has a very low filtration speed (MN 61q de 125 mm). We dropped the colloidal silver solution one by one on the filter paper, then let it to dry in the room temperature for about 10 min, and did this action several times. In this way, a group of filter papers covered using different nano-sized silver layers were achieved [22]. Finally, we drop the 0.01 M Nd_2O_3 solution drops one by one on the papers and let it to dry in the ordinary temperature.

Structural characterization and SERS detection

For characterization, the morphology of Ag-NPA-coated substrate was examined by a SEM (JSM-6700F, JEOL Ltd., Akishima-shi, Japan). The surface-enhanced Raman spectra were measured using a microscopic confocal Raman spectrometer (RM 2000, Renishaw,Wotton-under-Edge, UK), employing a Nd-YAG laser as the light source (_ = 532 nm). The beam diameter was 1 _m, and the integration time for each spectrum was set as 10 s.

Measurement of the hydrodynamic diameter of the nano-particles was accomplished using DLS (Nano ZS4700, Malvern Instruments, Worcestershire, UK). For size measurement, all paper formulations were diluted using deionized water to eliminate the effect of viscosity caused by the ingredients.

Competing interests

The authors declare that they have no competing interests.

Authors' contributions

AHA and YM conceived of the study, participated in its design and coordination, and worked on the preparation of nanoparticles, the silver-coated filter papers, and characterization with Zetasizer. FS and FK participated in the design of the study and performed the statistical analysis. ZSM drafted the manuscript, and AN rechecked the whole manuscript. All the authors read and approved the final manuscript.

Acknowledgments

We acknowledge Kenneth B Pael for his valuable editing of our article.

Author details

[1]Department of Physics, Payam Noor University, Tehran 19395-4697, Iran. [2]Department of Mathematics, Razi University, Kermanshah 6715685438, Iran. [3]Department of Applied Chemistry, Faculty of Chemistry, Razi University, Kermanshah 6715685438, Iran. [4]Nano Drug Delivery Research Center, Kermanshah University of Medical Sciences, Kermanshah 6714676573, Iran. [5]Environmental Epidemiological Research Center, Kermanshah University of Medical Sciences, Kermanshah 6714676573, Iran. [6]Department of Analytical Chemistry, Faculty of Chemistry, Razi University, Kermanshah 6715685438, Iran.

References

1. Zhang, X, Shah, NC, Van Duyne, RP: Sensitive and selective chem/bio sensing based on surfaceenhanced Raman spectroscopy (SERS). Vib. Spectrosc. **42**, 2–8 (2006)
2. Wang, LR, Fang, Y: IR-SERS study and theoretical analogue on the adsorption behavior of pyridine carboxylic acid on silver nanoparticles. Spectrochim. Acta Part A **63**, 614–618 (2006)
3. Serra, A, Manno, D, Filippo, E, Buccolieri, A, Urso, E, Rizzello, A, Maffia M: SERS based optical sensor to detect prion protein in neurodegenerate living cells. Sens. Actuators A **156**, 479–485 (2011)
4. Fleischmann, M, Hendra, PJ, McQuillan, AJ: Raman spectra of pyridine adsorbed at a silver electrode. Chem. Phys. Lett. **163**, 26–33 (1974)
5. Kneipp, K, Moskovits, M, Kneipp, H: Surface-Enhanced Raman Scattering: Physics and Applications. Springer, New York (2006)
6. Aroca, R: Surface-Enhanced Vibrational Spectroscopy. University of Windsor, Ontario, Canada (2006). Chapter 3
7. Smith, E, Dent, G: Modern Raman Spectroscopy-A Practical Approach. John Wiley and Sons, Hoboken (2005)
8. Kosuda, KM, Bingham, JM, Wustholz, KL, Van Duyne, RP: Nanostructures and surface-enhanced Raman spectroscopy. In: Andrews, DL, Scholes, GD, Wiederrecht, GP (eds.) Comprehensive Nanoscience and Technology, vol. 3, pp. 263–301. Academic, Oxford (2011)
9. Vo-Dinh, T: Surface-enhanced Raman spectroscopy using metallic nanostructures trends. Anal. Chem. **17**, 8–14 (1998)
10. Le Ru, EC, Etchegoin, PG: Principles of Surface-Enhanced Raman Spectroscopy. Elsevier, Amsterdam (2009)
11. El-Sherbiny, IM, Salih, E, Reicha, FM: Green synthesis of densely dispersed and stable silver nanoparticles using myrrh extract and evaluation of their antibacterial activity. J. Nanostruct. Chem **3**(8), 1–7 (2013)
12. Ru, ECL, Blackie, E, Meyer, M, Etchegoin, PG: Surface enhanced Raman Scattering enhancement factors: A comprehensive study. J. Phys. Chem. **111**, 13794–13803 (2007)
13. Rukiye, S, Murvet, V: Surface-enhanced Raman scattering (SERS) studies on silver nanorod substrates. Sens. Actuators B **139**, 150–155 (2009)

14. Farhadi, S, Pourzare, K, Sadeghinejad, S: Simple preparation of ferromagnetic Co_3O_4 nanoparticles by thermal dissociation of the CoII$(NH_3)_6$]$(NO_3)_2$ complex at low temperature. J. Nanostruct. Chem **3**(16), 1–7 (2013)
15. McFarland, AD, Young, MA, Dieringer, JA, Van Duyne, RP: Wavelength scanned surface-enhanced Raman excitation spectroscopy. J. Phys. Chem. B **109**, 11279–11285 (2005)
16. Banerji, B, Pramanik, SK, Pal, U, Maiti, NC: Dipeptide derived from benzylcystine forms unbranched nanotubes in aqueous solution. J. Nanostruct. Chem **3**(12), 1–6 (2013)
17. Nuntawong, N, Eiamchai, P, Wong-ek, B, Horprathum, M, Limwichean, K, Patthanasettakul, V, Chindaudom, P: Shelf time effect on SERS effectiveness of silver nanorod prepared by OAD technique. Vacuum **88**, 23–27 (2013)
18. Lee, PC, Meisel, D: Adsorption and surface-enhanced Raman of dyes on silver and gold. Sols. J. Phys. Chem. **86**, 3391–3395 (1982)
19. http://en.wikipedia.org/wiki/Nd_2O_3
20. Jing, C, Fang, Y: Simple method for electrochemical preparation of silver dendrites used as active and stable SERS substrate. J. Colloid Interface Sci. **314**, 46–52 (2007)
21. Vidhu, S, Tiwari, B, Tovmachenko, O, Krishna Darbha, G, Hardy, W, Singh, JP: Nonresonance SERS effects of colloidal silvers with different shapes. Chem. Phys. Lett. **446**, 77–83 (2007)
22. Niu, Z, Fang, Y: Surface-enhanced Raman scattering of single-walled carbon nanotubes on silver-coated and goldcoated filter paper. J. Colloid Interface Sci. **303**, 224–301 (2006)

Studies of *ab initio* and Monte Carlo simulation on interaction of fluorouracil anticancer drug with carbon nanotube

Karim Zare[1], Fahimeh Najafi[2*], Hamidreza Sadegh[3] and Ramin Shahryari ghoshekandi[3]

Abstract

In this study, we have carried out the calculation for the interaction of fluorouracil anticancer drug with carbon nanotube using Hartree-Fock theory with 3-21G basis set and molecular mechanic/Monte Carlo simulation (with MM+, AMBER, and OPLS force field) in the gas phase, methanol, and water solvents at various temperatures (298 up to 312 K). The obtained results of potential energy, Gibbs free energy, enthalpy, entropy, and dipole moment value changes are compared in gas phase, methanol, and water, and the temperature effects on the stability of the interaction between fluorouracil anticancer drug and carbon nanotube in gas phase, methanol, and water solvents.

Keywords: Hartree-Fock (HF) theory; Fluorouracil; Carbon nanotube; Force field

Background

Fluorouracil anticancer drug belongs to a bunch of medicines distinguished as antineoplastics, and it is sorted as an antimetabolite. It acts by preventing the growth of cancer cells and finally demolishing them. Fluorouracil is utilized for the treatment of the cancers such as the breast, rectum, colon, stomach, or the pancreas [1] in 1957 by Charles Heidelberger; 5-FU was designed, synthesized, and patented [2]. 5-Fluoro-2,4(1*H*, 3*H*)-pyrimidinedione is chemically named fluorouracil [3]. Three possible mechanisms of action of fluorouracil existed [4]. The first mechanism, the fluorouracil metabolite fluorodeoxyuridine monophosphate contests with uracil to bind with thymidylate synthetase and the folate cofactor [5], these consequences in declined thymidine production and hence declined DNA synthesis and repair, and eventually declined cell proliferation. Leucovorin (formyltetrahydrofolate, formyl-FH4) increases fluorouracil by stabilizing the binding of fluorodeoxyuridine monophosphate to thymidylate synthetase. The second mechanism, the fluorouracil metabolite fluorodeoxyuridine triphosphate is incorporated into DNA, thus interfering with DNA replication [4]. The final mechanism, the fluorouracil metabolite fluorouridine-5-triphosphate is being incorporated into RNA in place of uridine triphosphate, causing a fraudulent RNA and interfering with RNA processing and protein synthesis [6]. It is cell-cycle special [5]. Besides, the carbon nanotube (CNT) is a promising nanomaterial with rare physical characteristics [7]. Its geometry and chemical characteristics are ideal for an electron emitter with a low threshold voltage and long lifetime [8], and a wide range of usages have been proposed, such as gas sensors [9], transistors [10,11], thermal sensors [12,13], and pressure sensors [14]. In addition, CNTs are nanostructures which are members of the fullerene group. They are one of the most promising materials for the fabrication of nanotechnological devices in the future. Substantial study efforts are focused on the study of their structure, as well as their mechanical and electric characteristics [15,16]. Different methods have developed for the production of CNTs, for example, laser vaporization [17] or catalytic decomposition of hydrocarbons [18]. In recent times, a new production way based on the thermal decomposition of C60 in the presence of transition metals was presented [19,20]. Different conformations of carbon nanotube are three structures that the nanotube is either of the armchairs ($n = m$), zigzag ($n = 0$ or $m = 0$), or chiral (any other n and m) variety, and each of these forms has its own characteristics

* Correspondence: fahimnajafi@Gmail.com
[2]Department of Chemistry, Roudehen Branch, Islamic Azad University, Roudehen, Iran
Full list of author information is available at the end of the article

[21,22]. Single-wall nanotubes (SWNTs) are molecular scale wires with fascinating and potentially useful characteristics [15]. It is well established of which the diameter and chirality of a SWNT determine the electronic structures of the nanotube [23]. In order to achieve homogeneous nanotube materials with special physical characteristics, it is necessary to control the structures of nanotubes by chemical synthesis techniques. The prerequisite for such control is a detailed understanding of the growth mechanism of nanotubes. A double-wall carbon nanotube, which consists of two concentric cylindrical graphene layers, has been successfully synthesized by catalytic chemical vapor deposition [23-25]. Double-wall carbon nanotubes are the topic of extreme investigations versus their promising usages. In particular, the double-wall carbon nanotube with a metallic inner tube and an insulating outer tube is a model system for a molecular conductive wire covered by an insulator. In our work, the interaction of fluorouracil with CNT was studied, applying molecular mechanic/Monte Carlo (MC) simulation in several force fields. Furthermore, the influences of different environments such as gas phase, methanol, and water on this interaction by quantum mechanic (QM) have been studied too. For the simulation of the solvent influences by QM method, self-consistent reaction field (SCRF) method is mainly used. In this place, solvent effects on the potential energy changes and thermodynamic value changes of this interaction with employ QM/MM methods are depicted.

Results and discussion

All the theoretical results present in this paper were obtained by two methods, firstly, molecular mechanic (Monte Carlo simulation) and secondly, quantum mechanic.

Fluorouracil inhibits the development of cancer cells and finally destroying them, and carbon nanotube is a carbon with cylindrical outline material and a nanometric-level diameter which has a spacious range of usages such as drug delivery; thus, our purpose of this article is to survey the interaction of a nanotube with a fluorouracil anticancer drug according to reaction 1. In this work, we see the image of the combination of fluorouracil and CNT in Figure 1.

If the following reaction is considered:

$$\text{Fluorouracil} + \text{CNT} \rightarrow \text{Fluorouracil-CNT} \quad \text{reaction}$$

we can calculate the potential energy and thermodynamic value changes on the interaction of fluorouracil with CNT for reaction 1 by the following equation:

$$\Delta M = M_{\text{Fluorouracil-CNT}} - (M_{\text{Fluorouracil}} - M_{\text{CNT}}) \quad (1)$$

where M is equal to μ, G, H, E, S, and E_{POT} parameters. Therefore, first, we investigated the different force fields to determine the potential energy and other kinds of geometrical parameters on the interaction of fluorouracil and CNT. Since the three different force fields have been utilized, the calculated potential energy of the molecules will not be the same. Thus, comparing the computed potential energy of one molecule using a particular force field with the potential energy of another molecule, which is calculated by another force field, is neither rational nor possible. In this article, the difference in force fields is illustrated by comparing the computed potential energy changes using various force fields like: MM+, AMBER, and OPLS (Table 1 and Figures 2, 3, and 4). The theoretical potential energy value changes are attained using different force fields. Then, we see that water solvent has the lowest value of potential energy changes. And also, by comparing ΔE_{POT} in different force fields such as MM+, AMBER, and OPLS in water solvent, we observe that AMBER force field is best when reaction 1 occurs in the water solvent (Figure 5).

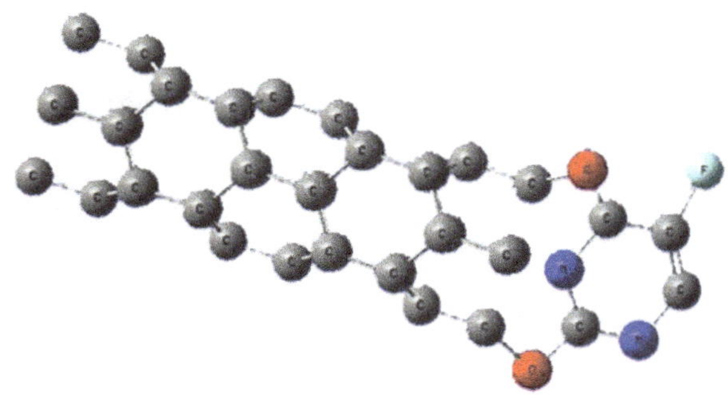

Figure 1 Studied system of fluorouracil with CNT.

Table 1 Theoretical $\Delta E_{\text{potential energy}}$ (kcal/mol) in different environments and same force fields

Environment	Temperature	$\Delta E_{\text{potential energy}}$ (kcal/mol)		
		MM+	AMBER	OPLS
Gas phase	298	8.2253	20.4087	9.673
	301	−7.9819	−231.1291	−116.5912
	304	9.1083	−230.7377	−79.0468
	308	−6.2843	−230.9912	−87.642
	312	1.8067	−236.5307	−89.6245
Methanol	298	−22.122	−22.681	−62.718
	301	−10.1062	−53.425	−149.045
	304	−29.071	−58.157	−103.659
	308	−20.622	−103.905	−123.955
	312	−24.862	−124.313	−121.214
Water	298	−29.9874	−66.1572	−156.4089
	301	−15.0026	−283.5154	−154.9751
	304	−53.4379	−313.8459	−145.4375
	308	−87.081	−327.245	−135.445
	312	−77.2617	−309.5457	−138.8755

$\Delta E_{\text{potential energy}}$ (kcal/mol), difference of energy for fluorouracil and CNT and sum of fluorouracil and CNT for reaction 1 in different dielectric constants.

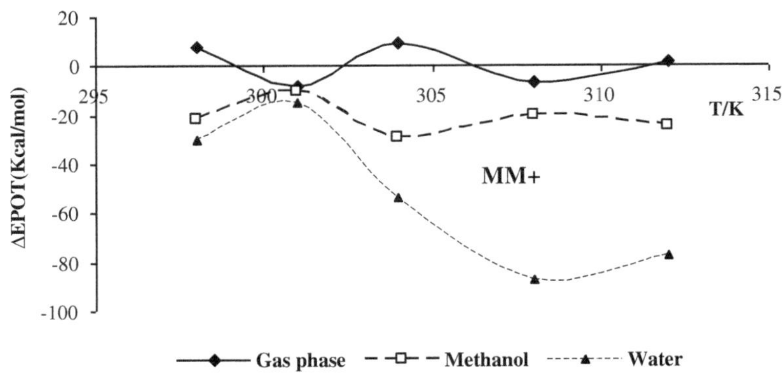

Figure 2 The potential energy changes versus temperature in different solvents and MM+ force field for reaction 1.

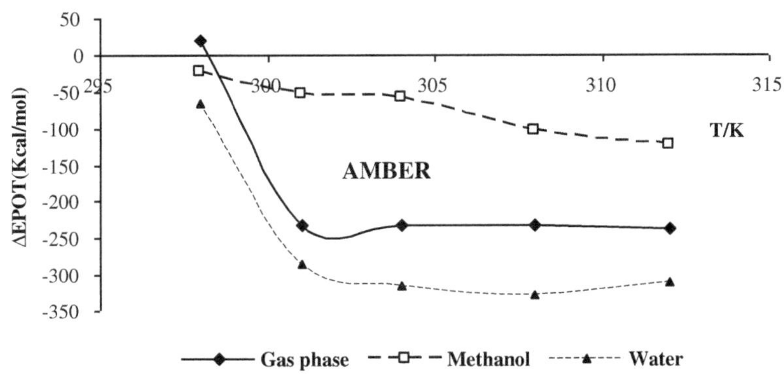

Figure 3 The potential energy changes versus temperature in different solvents and AMBER force field for reaction 1.

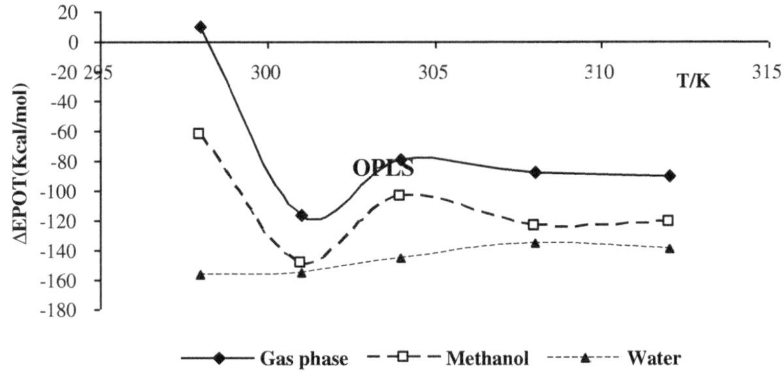

Figure 4 The potential energy changes versus temperature in different solvents and OPLS field for reaction 1.

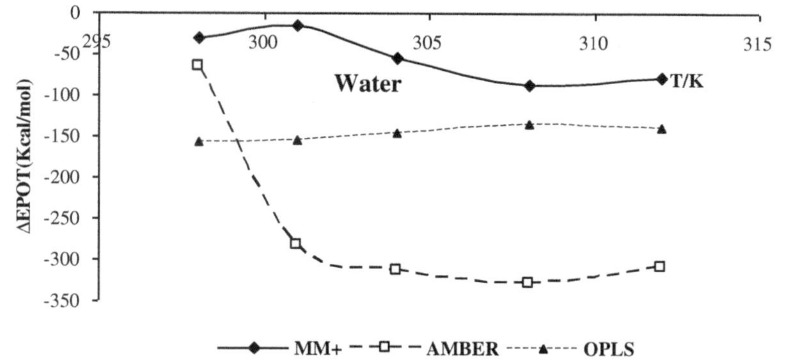

Figure 5 The potential energy changes versus temperature in water solvent and same force field for reaction 1.

In Table 2, the dipole moment values of the interaction of fluorouracil with CNT according to reaction 1 based on Equation 1 in gas phase, water, and methanol at different temperatures have been observed. Dipole moment values in gas phase, water, and methanol exhibit increasing dielectric constant; dipole moment will be increased that is demonstrated in Table 2. In the SCRF model, the solvent is signified by a continuous dielectric, recognized by a given dielectric constant. The solute supposed to be fixed into a cavity, with radius a_0 in the molecule, and the electric field which is related to the solute by the solvent dipole will interact with the molecular dipole and direct to net stabilization. It must be noted which salvation energy computed using the SCRF method is the electrostatic distribution to the free energy of salvation. In this investigation, the cavity radius of fluorouracil, CNT, and fluorouracil/CNT system (a_0) was calculated and substituted efficiency was improving. The volume of fluorouracil, CNT, and fluorouracil/open-end of CNT system was achieved that values of a_0 for fluorouracil, CNT, and fluorouracil/CNT system were 3.76, 5.64, and 10.26 Å, respectively; thus, according to Equation 1, we calculate the potential energy and thermodynamic value changes. For in the survey, you can see that the dipole moment changes Δμ (Debye) in (Table 2 and Figure 6) that by the increase of the solvent dielectric constant, the dipole moment for the interaction of fluorouracil with CNT in different temperatures such as 298, 301, 304, 308, and 312 K will be increased.

Table 2 Difference of dipole moment and sum of fluorouracil and CNT in various environments and temperatures

Temperature (K)	Δμ (Debye)		
	Gas phase	Water	Methanol
298	622.61578	935.40235	927.07144
301	622.61818	935.43612	927.06763
304	622.26749	935.862794	927.05227
308	622.44462	935.905183	927.05544
312	622.64556	935.28985	927.07670

After computing using the SCRF model, it is demonstrated which ΔG, ΔE, ΔH and ΔS values of the interaction of fluorouracil with CNT for reaction 1 are sensitive to polarity of the surrounding solvent. Thus, in Table 3, you observe that ΔG, ΔE, ΔH and ΔS values depend on the dielectric constant (ε), and Table 3 indicates which increasing temperature from 298 to 312 K and values of ΔG of interaction of fluorouracil with CNT for reaction 1 in gas phase, water, and methanol will be increased. As a result, the lowest amount of ΔG is in 298 K, and we have the highest resistance for reaction 1 in 298 K (Figure 7). Also, as the solvent dielectric constant increases, the attained ΔG values of interaction of fluorouracil with CNT decrease. So, the most negative ΔG value of this interaction is in water solvent. The water solvent is the best environment for reaction 1 as shown in Figure 7. Also, in Table 3, we can observe the attained ΔG, ΔE, ΔH, and ΔS values in various environments such as in gas phase, methanol, and water versus temperature alteration from 298 to 312 K. Thereupon, by increasing the temperature from 298 to 312 K, the most positive value of ΔE, ΔH, and ΔS is for 298 K. The other actuality that is confirmed by the calculated values of ΔE, ΔH, and ΔS versus solvent dielectric constant increase in temperature of 298 to 312 K is which of the values of ΔE and ΔH will shift toward negative values in which you can observe the most negative of these parameters in 312 K. Also, in Table 3, with the increase of solvent dielectric constant in temperature from 298 to 312 K, ΔS values increase. So, Table 3 proves which of the value of ΔS in water solvent has the most positive value (Figure 8). Then, water solvent is the best for reaction 1. So, when the studied system of interaction of fluorouracil with CNT is placed in water solvent, the lowest values of enthalpy take place (Figure 9).

Conclusion

In this work, two calculations were performed, one by HyperChem program and second by the *ab initio*

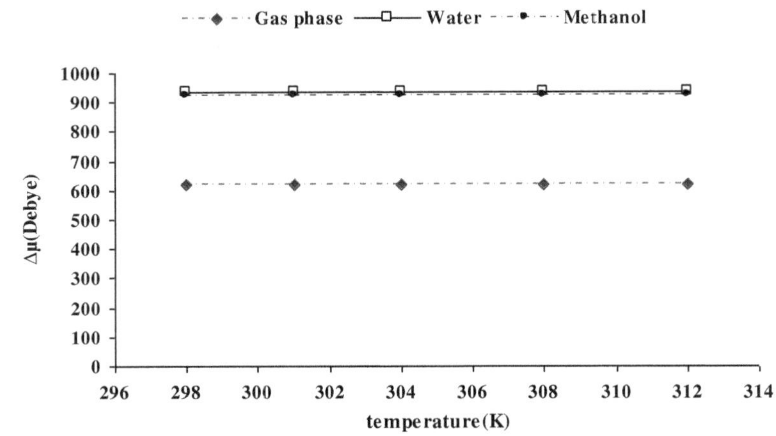

Figure 6 The dipole moment changes calculated versus temperature for reaction 1.

computations which were carried out with Hartree-Fock (HF) functional, utilizing 3-21G basis set and the Gaussian 98. According to achieved consequences, water solvent is the best environment because the studied system of the interaction of fluorouracil with CNT according to reaction 1 has the most negative values of ΔG. Also, using the SCRF model, the obtained consequences have seemed quite sensitive to the polarity of the enclosing solvent, and we can see that with the increase of solvent dielectric constant, ΔS of studied system of the interaction of fluorouracil with CNT has the most positive value. Thus, we can conclude that water solvent is best. Besides, the achieved consequences of potential energy changes from force fields MM+, AMBER, and OPLS were compared, and we observe which AMBER force field is the best for the survey of reaction 1.

Methods

The theoretical results demonstrated in this article were achieved by means of the *ab initio* and molecular mechanic/Monte Carlo simulation (MM+, AMBER, and OPLS). Geometry optimization and structural characteristics computations were done in Gaussian 98 program, Wallingford, CT, USA. Also, Monte Carlo computations were carried out by HyperChem program, Gainesville, FL, USA.

Ab initio

The most ordinary form of *ab initio* computation is named Hartree-Fock computation (abbreviated HF), as the initial approximation. This means that the coulombic electron–electron repulsion is not clearly taken into account, though its average influence is contained in the computation [26]. Quantum Monte Carlo is a technique which prevents the

Table 3 Theoretical values in different environments and temperatures at HF/3-21-3G level of theory for reaction 1

Media	Quantities	Temperature (K)				
		298	301	304	308	312
Gas phase	ΔG	−48,958.5357	−48,958.4658	−48,958.3146	−48,958.288	−48,957.9842
	ΔH	−48,980.36156	−48,980.4919	−48,980.6311	−48,980.9187	−48,980.9501
	ΔE	−48,980.40396	−48,980.52361	−48,980.969	−48,981.125	−48,981.331
	ΔS	−0.073223	−0.073273	−0.073376	−0.073496	−0.073592
Water	ΔG	−94,790.7414	−94,790.503	−94,790.4441	−94,790.1122	−94,789.8589
	ΔH	−94,812.01083	−94,812.1098	−94,812.326	−94,812.4097	−94,812.5200
	ΔE	−94,812.2387	−94,812.4763	−94,812.6035	−94,812.826	−94,812.9781
	ΔS	−0.071359	−0.071904	−0.071937	−0.072403	−0.072619
Methanol	ΔG	−93,996.591	−93,996.4897	−93,996.4098	−93,996.2923	−93,996.091
	ΔH	−94,018.06215	−94,018.3508	−94,013.5013	−94,018.6502	−94,018.8009
	ΔE	−94,018.22153	−94,018.4806	−94,013.6256	−94,018.7963	−94,018.9574
	ΔS	−0.072035	−0.072601	−0.072591	−0.072579	−0.072819

Values of ΔG, ΔH, and ΔE in kcal/mol and ΔS in kcal/mol K. Δ difference of energy for fluorouracil and CNT and sum of fluorouracil and CNT for reaction 1 in different dielectric constants.

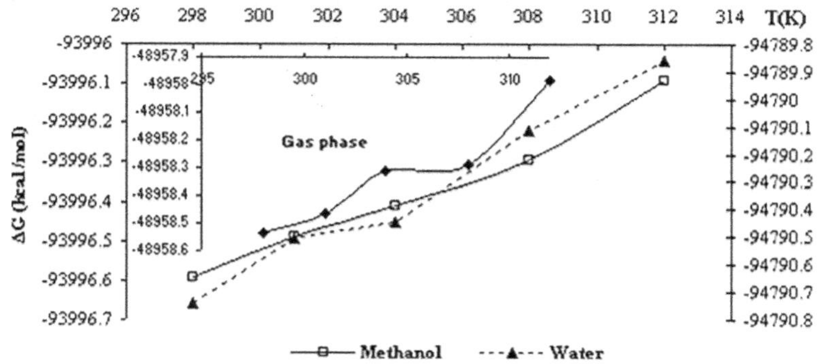

Figure 7 Calculated Δ*G* versus temperature for reaction 1 at different dielectric constants.

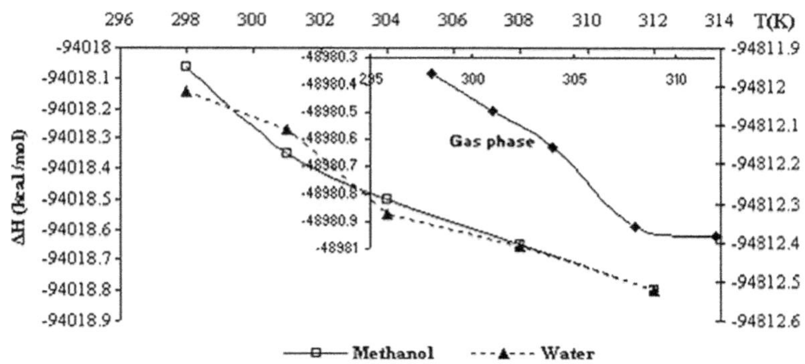

Figure 8 Calculated Δ*S* versus temperature for reaction 1 at different dielectric constants.

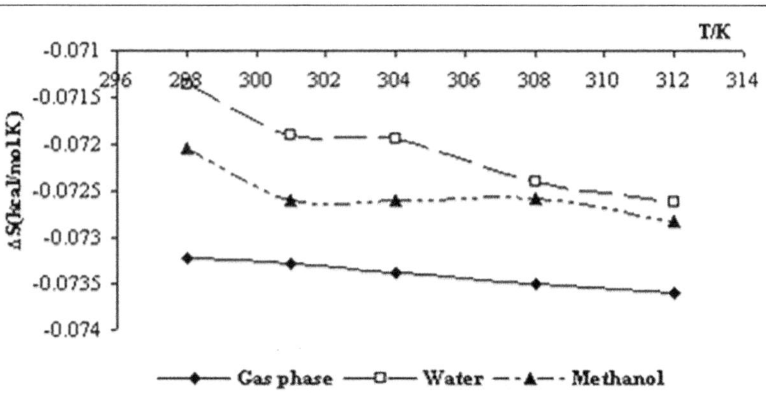

Figure 9 Calculated Δ*H* versus temperature for reaction 1 at different dielectric constants.

HF from making mistakes. There are some sorts of QMC: variation, diffusion, and Green's functions. These techniques work with an openly correlated wave function and compute integrals numerically using a Monte Carlo integration. These computations are very time-consuming, but they are the most correct techniques that are known up to now. On the whole, *ab initio* computations prepare perfect quantitative consequences and can give increasingly accurate quantitative consequences as the molecules under consideration become smaller [27].

Monte Carlo simulation

In the fields of chemistry, biology, physics, and engineering, Monte Carlo simulations are widely utilized to determine the structural and thermodynamic characteristics of complex systems at the atomic level. By Monte Carlo techniques, we can determine not only the thermodynamic averages of molecular characteristics but also minimum energy. Structures [28] in this technique [29] and a sequence of points in phase space are created from an initial geometry by adding a random 'kick' to the coordinates of a fortuitously chosen particle (atom or molecule). If the energy declines, the new computation is accepted, with a probability of $e^{-E/KT}$ if the energy heightens. The metropolis [30] procedure ensures which of the configurations in the ensemble obey a Boltzmann distribution. The probability of accepting higher energy computations permits MS technique to heighten and go over from a local minimum. In MC simulation, you need only the capability to evaluate the energy of the system if computing the first derivative is difficult or time-consuming; this capability maybe advantageous. Additionally, since only a single particle is moved in each step, the only thing that must be configurated are the energy alterations accompanied with this move, not the total energy for the complete system. The lack of the time dimension and atomic velocities are some disadvantages of MC techniques. And so, they are not appropriate for spending time and investigating characteristics depending on time-dependent phenomena or momentum [31].

In the present article, differences in force fields are demonstrated by comparing the computed potential energy changes by applying force fields MM+, AMBER, and OPLS. In this article, HyperChem program is utilized for the quantum chemical computations. We have done geometry optimization and Monte Carlo simulation by this software [32]. The QM computations on the interaction of fluorouracil with carbon nanotube have been done by Gaussian 98 program [33] utilizing the standard 3-21G basis set. Thus, solvent effects have been studied on the values of thermodynamic parameters and dipole moment of water and methanol encircling fluorouracil/CNT system within the SCRF model utilizing a Hartree-Fock method and the temperature effects on fluorouracil/CNT system. Therefore, achieved thermodynamic data of fluorouracil/CNT system are compared in gas phase and various solvent environments such as methanol and water.

Competing interests

The authors declare that they have no competing interests.

Authors' contributions

FN carried out the computational simulations and the results analysis and participated in the manuscript elaboration. KZ coordinated in the project, discussed the results, and helped to draft the manuscript. HS and RSG examined the results and helped to draft the manuscript. All authors read and approved the final manuscript.

Acknowledgements

The authors would like to thank the Islamic Azad University for financial and other supports.

Author details

[1]Department of Chemistry, Shahid Beheshti University, Evin, Tehran, Iran [2]Department of Chemistry, Roudehen Branch, Islamic Azad University, Roudehen, Iran. [3]Department of Chemistry, Shahr-e- Qods Branch, Islamic Azad University, Tehran, Iran.

References

1. Dorr, RT, Von-Hoff, DD: Cancer Chemotherapy Handbook, 2nd edn, pp. 27–28. Appleton and Lange, Norwalk (1994)
2. Chu, E: Clinical colorectal cancer: ode to 5-fluorouracil. Clin. Colorectal Cancer **6**(9), 609 (2007)
3. Longley, DB, Harkin DP, Johnston, PG: 5-fluorouracil: mechanisms of action and clinical strategies. Nat. Rev. Cancer **3**(5), 330–338 (May 2003)
4. van der Wilt, CL, Marinelli, A, Pinedo, HM, Cloos, J, Smid, K, van de Velde, CJ, Peters, GJ: The effects of different routes of administration of 5-fluorouracil on thymidylase synthase inhibition in the rat. Eur. J. Cancer **31A**(5), 754–760 (1995)
5. McEvoy, GK (ed.): AHFS 2006 Drug Information, pp. 1052–1056. American Society of Health-System Pharmacists, Bethesda (2006)
6. McEvoy, GK (ed.): AHFS 2005 Drug Information, pp. 1020–1025. American Society of Health-System Pharmacists, Bethesda (2005)
7. Dekker, C: Carbon nanotubes as molecular quantum wires. Physics Today. **52**(5), 22–28 (1999)
8. Bonard, JM, Kind, H, Stöckli, T, Nilsson, LO: Field emission from carbon nanotubes: the first five years. Solid State Electron. **45**, 893 (2001)
9. Kong, J, Franklin, NR, Zhou, C, Chapline, M, Peng, G, Cho, SK, Dai, H: Nanotube molecular wires as chemical sensors. Science **287**, 622–625 (2000)
10. Artukovic, E, Kaempgen, M, Hecht, DS, Roth, S, Gruner, G: Transparent and flexible carbon nanotube transistors. Nano Lett. **5**, 757–760 (2005)
11. Hines, DR, Mezhenny, S, Breban, M, Williams, ED, Ballarotto, VW, Esen, G, Southard, A, Fuhrer, MS: Nanotransfer printing of organic and carbon nanotube thin-film transistors on plastic substrates. Appl. Phys. Lett. **86**, 163101 (2005)
12. Selvarasah, S, Chen, C-L, Chao, S-H, Makaram, P, Busnaina, A, Dokmeci, MR: A three dimensional thermal sensor based on single-walled carbon nanotubes. In: Proceedings of the 14th International Conference on Solid-State Sensors, Actuators and Microsystems (Transducers '07), pp. 1023–1026. , Lyon (2007). 10–14 Jun
13. Fung, CKM, Wong, VTS, Chan, RHM, Li, WJ: Dielectrophoretic batch fabrication of bundled carbon nanotube thermal sensors. IEEE Trans. Nanotechnology **3**, 395–403 (2004)
14. Fung, CKM: Fabrication of CNT-based MEMS piezoresistive pressure sensors using DEP nanoassembly. In: Proceedings of the 5th International Conference on Nanotechnology (IEEENANO), pp. 199–202. Nagoya, Nagoya (2005). 11–15 Jul
15. Dresselhaus, MS, Dresselhaus, G, Eklund, PC: Science of Fullerenes and Carbon Nanotubes. Academic Press, San Diego (1996)
16. Ebbesen, TW (ed.): Carbon Nanotubes: Preparation and Properties. CRC Press, Boca Raton (1997)

17. Thess, A, Lee, R, Nikolaev, P, Dai, H, Petit, P, Robert, J, Xu, C, Lee, YH, Kim, SG, Rinzler, AG, Colbert, DT, Scuseria, GE, Tomanek, D, Fischer, JE, Smalley, RE: Crystalline ropes of metallic carbon nanotubes. Science **273**, 483 (1996)
18. Ivanov, V, Nagy, JB, Lambin, P, Lucas, A, Zhang, XB, Zhang, XF, Bernaerts, D, Van Tendeloo, G, Amellinckx, S, Van Landuyt, J: The study of carbon nanotubules produced by catalytic method. Chem. Phys. Lett. **223**, 329 (1994)
19. Ehlich, R, Biro, LP, Hertel, IV: Growth of nanotubes by decomposition of C60 on transitional metal surfaces. Synth. Met. **103**, 2486 (1999)
20. Bir´o, LP, Ehlich, R, Tellgmann, R, Gromov, A, Krawetz, N, Tschaplyguine, M, Pohl, M-M, Zsoldos, E, V´ertesy, Z, Horv´ath, ZE, Campbell, EEB: Growth of carbon nanotubes by fullerene decomposition in the presence of transition metals. Chem. Phys. Lett. **306**, 155 (1999)
21. Meng, S, Maragakis, P, Papaloukas, C, Kaxiras, E: DNA nucleoside interaction and identification with carbon nanotubes. Nano Lett. **l7**, 45 (2007). No1
22. Cuniberti, G, Yi, J, Porto, M: arXiv: Pure-carbon ring transistor: role of topology and structure. Appl. Phys. Lett. **81**, 850 (2002)
23. Flahaut, E, Peigney, A, Laurent, CH, Rousset, A: Synthesis of single-walled carbon nanotube-Co-MgO composite powders and extraction of the nanotubes. J. Mater. Chem. **10**, 249 (2000)
24. Sugai, T, Omote, H, Bandow, S, Tanaka, N, Shinohara, H: Carbon nanotubes by high-temperature pulsed arc discharge. J. Chem. Phys. **112**, 6000 (2000)
25. Zhu, H, Xu, C, Wei, B, Wu, D: A new method for synthesizing double-walled carbon nanotubes. Carbon **40**, 2021 (2002)
26. Nsangou, M, Jaidane, NE, Ben, Z: Lakhdar: Internet electronic. J. Molecular Design **5**, 89–101 (2006)
27. Young, D: Cytoclonal Pharmaceutics Inc.: Introduction to computational chemistry. Chem. Aust. **11**, 5 (1998)
28. Monajjemi, M, Chahkandi, B: Study of the hydrogen bond indifferent orientations of adenine-thymine base pairs: an ab initio study. J. Mol. Struct. (THEOCHEM) **714**, 43 (2004)
29. Jorgensen, WL: Biological reactions are slow in absence of a catalyst. Adv. Chem. Phys. **70**, 469 (1988)
30. Metropolis, NA, Rosenbluth, AW, Teller, AH, Teller, E: Equation of state calculations by fast computing machines. J. Chem. Phys. **21**, 10–87 (1953)
31. Haeri, HH, Hashemianzadeh, SM, Monajjemi, M: Temperature effects on the stochastic gating of the IP3R calcium release channel: a numerical simulation study. J. Biol. Syst. **17**, 817–852 (2009)
32. HyperChem 7.0. Hypecube Inc, Gainesville (2001)
33. Frisch, MJ, Trucks, GW, Schlegel, HB, Scuseria, GE, Robb, MA, Cheeseman, JR, Zakrzewski, VG, Montgomery Jr., JA, Stratmann, RE, Burant, JC, Dapprich, S, Millam, JM, Daniels, AD, Kudin, KN, Strain, MC, Farkas, O, Tomasi, J, Barone, V, Cossi, M, Cammi, R, Mennucci, B, Pomelli, C, Adamo, C, Clifford, S, Ochterski, J, Petersson, GA, Ayala, PY, Cui, Q, Morokuma, K, Malick, DK, et al.: Gaussian 98 Revision A.6. Gaussian, Inc, Pittsburgh (1998)

X-ray imaging technique using colloid solution of Au/silica core-shell nanoparticles

Yoshio Kobayashi[1*], Hiromitsu Inose[1], Tomohiko Nakagawa[2], Yohsuke Kubota[2], Kohsuke Gonda[2] and Noriaki Ohuchi[2]

Abstract

This work describes X-ray imaging of mouse using a colloid solution of silica-coated Au (Au/SiO_2) nanoparticles. A colloid solution of Au nanoparticles with a size of 16.9 nm was prepared using hydrogen tetrachloroaurate (III) trihydrate as Au source and sodium citrate as reducing reagent. Silica coating of the Au nanoparticles was performed by modifying the Au nanoparticle surface with (3-aminopropyl)trimethoxysilane and then by depositing silica nuclei generated through a sol–gel reaction of tetraethyl orthosilicate in water/ethanol initiated with sodium hydroxide on the surface-modified surface, which produced Au/SiO_2 particles with a size of 136.4 nm. A computed tomography value of the Au/SiO_2 colloid solution with an Au concentration of 0.036 M was as high as 1,184.8 Hounsfield units, which was quite higher than that of a commercial X-ray contrast agent with the same iodine concentration as the Au concentration. Tissues of mouse could be imaged by injecting the Au/SiO_2 particle colloid solution into them.

Keywords: Au; Silica; Core-shell; Nanoparticle; X-ray contrast agent

Background

An X-ray imaging technique is one of the quite useful medical diagnoses [1-3] because internal organs can be imaged without a surgical operation. Images taken with the technique are made clearer using contrast agents that strongly absorb X-ray [4-6]. The contrast agents, which are commercially available, are solutions that dissolve iodine compounds homogeneously in solvents at molecular level. The iodine compound molecules are not strongly dragged in fluid because of their small sizes. Consequently, they cannot stay in internal organs for a long period, which provides difficulty taking steady images. In addition, the iodine-related contrast agents cannot be used for patients in whom iodine compounds provoke adverse events like allergic reactions [7-9].

Gold is a material that has been extensively characterized [10,11]. Gold also absorbs X-ray strongly, which means that it is available for obtaining X-ray images with high contrast. From this viewpoint, studies on the X-ray imaging ability of metallic Au nanoparticles have been performed for development of new contrast agents that aid the technique to image tissues in internal organs at nanometer level [12-20]. The particles have another advantage over gold dissolved homogeneously in solvents at molecular level. The particles are strongly dragged in fluid because of their size which is larger than that of molecules. Thus, their residence time in internal organs will increase, which may allow steady imaging.

Metallic nanoparticles also have troubled us with their toxicity in internal organs [21-28]. Formation of core-shell particles containing Au nanoparticles as the core is a candidate for solving the problem since the shell can keep the particles from contacting with internal organs, which decreases the toxicity of Au nanoparticles. Materials inert for internal organs are desirable as materials for the shell. Silica is one of the promising materials for the shell because it is inert for internal organs compared to iodine compounds and metallic nanoparticles. Several research groups have proposed methods for producing core-shell particles composed of Au nanoparticle core such as spheres, rods, cages, etc. and silica shell, most of which are related to a sol–gel process [29-38]. Our research group has also studied on development of a

* Correspondence: ykoba@mx.ibaraki.ac.jp
[1]Department of Biomolecular Functional Engineering, College of Engineering, Ibaraki University, 4-12-1 Naka-narusawa-cho, Hitachi, Ibaraki 316-8511, Japan
Full list of author information is available at the end of the article

method for silica coating of Au nanoparticles [39-42]. In the developed method, the silica-coated particles were fabricated in three steps. The first, second, and third steps are (1) preparation of the Au nanoparticle colloid solution by reducing hydrogen tetrachloroaurate(III) with sodium citrate in water, (2) modification of the Au nanoparticle surface with hydrophilic groups using a silane coupling agent with an amino group as surface primer, and (3) formation of silica shells on the hydrophilic surface through the sol–gel reaction of silicon alkoxide using sodium hydroxide (NaOH) aqueous solution as a basic catalyst in ethanol, respectively.

The colloid solution of silica-coated Au (Au/SiO_2) nanoparticles with the size of 47.3 nm prepared by the developed method was examined for X-ray imaging of mouse in our previous work [42]. Properties of particles such as colloidal stability and residence time in internal organs should be dependent on the particle size. Au/SiO_2 nanoparticles with not only the size of 47.3 nm but also various sizes were successfully fabricated in another work of ours [40]. The present work examines the colloid solution of Au/SiO_2 nanoparticles with a different size for X-ray imaging of tissues in a mouse to accumulate data useful for optimizing the size of contrast agent particles.

Results and discussion

Morphology of Au/SiO_2 particles

Figure 1a shows a photograph of the concentrated Au/SiO_2 particle colloid solution. A dark red solution was obtained. No flocculation, aggregation, and precipitation were found by passing the solution through a light and then observing it with the naked eyes. This observation showed the high colloidal stability of the concentrated Au/SiO_2 particle colloid solution. A transmission electron microscopy (TEM) image of the Au/SiO_2 particles in the concentrated colloid solution is shown in Figure 1b. The average size of the particles was 136.4 ± 17.8 nm, which was close to that of the as-prepared Au/SiO_2 particles in the prior work. The Au/SiO_2 particles still had a spherical core-shell structure even after the concentrating process, which indicated that the Au/SiO_2 particles produced in the present work were mechanically stable. The inductively coupled plasma (ICP) measurement revealed that an actual Au concentration in the concentrated Au/SiO_2 particle colloid solution was 0.036 M, which indicated that 41.9% of Au contained in the Au nanoparticle colloid solution was consumed for the preparation of the concentrated Au/SiO_2 particle colloid solution.

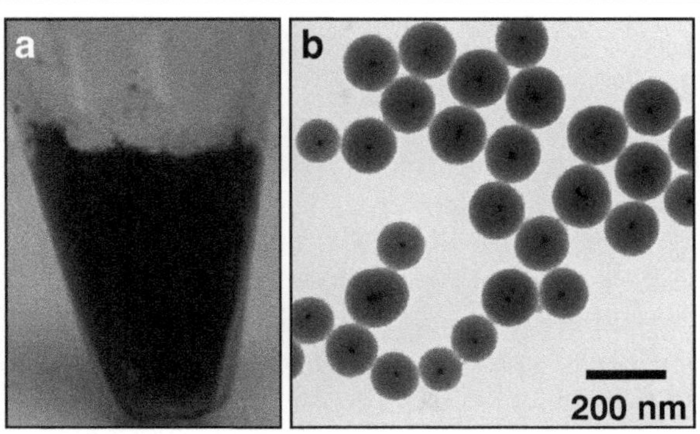

Figure 1 Concentrated Au/SiO_2 particle colloid solution (a) and TEM image of Au/SiO_2 particles in the colloid solution (b).

CT values of the Au/SiO_2 particle colloid solution

Figure 2b,c,d shows X-ray images of Au/SiO_2 particle colloid solutions with different actual Au concentrations. For comparison, an X-ray image of water is also shown. The white contrast of the image increased as the Au concentration increased.

Figure 3 shows the computed tomography (CT) value of the Au/SiO_2 particle colloid solution versus the actual Au concentration. The CT value increased quasi-linearly with an increase in Au concentration and was as high as 1,184.8 Hounsfield units (HU) at an actual Au concentration of 0.036 M. The straight line, which was drawn mathematically by the least squares method of linear regression, had a slope of 3.32×10^4 HU/M. According to our previous work [42], the slope for Iopamiron®300, which is a commercial iodine-related X-ray contrast agent purchased from Bracco Eisai (Bunkyo-ku Tokyo, Japan), was 4.76×10^3 HU/M. The slope for the Au/SiO_2 particle colloid solution was around seven times larger than that for Iopamiron®300. This tendency shown in the present work was similar to studies reported by several researchers that resulted in CT values of Au nanoparticles or clusters higher than commercial iodine compound contrast agents with the same iodine concentration as that of the Au [12,19,43]. The size of gold atom is larger than that of iodine due to its larger atom number, and the absorption coefficient for the X-ray of gold is comparable to that of iodine. This indicates that the projected area of Au nanoparticles in the colloid solution should be larger than that of the iodine compound. Consequently, such a tendency was shown for the Au/SiO_2 particle colloid solution. Accordingly, the present result on its X-ray absorption ability expected the Au/SiO_2 particle colloid solution produced in the present work to function as an X-ray contrast agent.

X-ray imaging of the mouse

Figure 4 shows X-ray images of the mouse prior to and after the injection of the concentrated Au/SiO_2 particle colloid solution. Figure 5 shows CT values of various tissues as a function of time after the injection. CT values at 0 min corresponded to those prior to the injection.

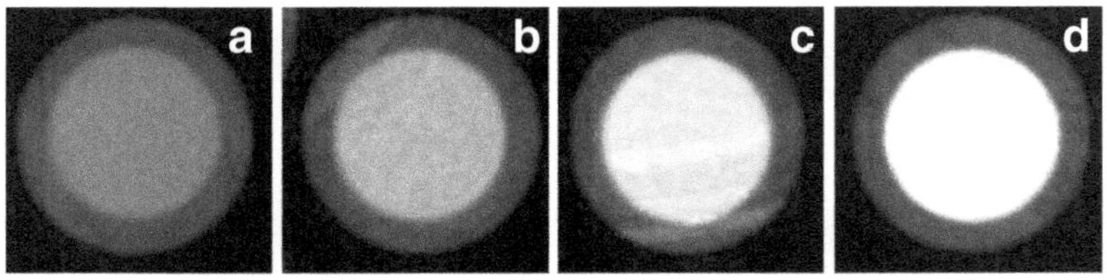

Figure 2 X-ray images of (a) water and (b, c, d) Au/SiO_2 particle colloid solutions. The Au concentrations in the colloid solutions were (b) 0.011, (c) 0.020, and (d) 0.036 M.

Prior to the injection, it was hard to recognize tissues such as the liver and the spleen in the images. CT values of the liver and the spleen were 76.7 and 96.5 HU, respectively, at 0 min. The contrasts of these tissues decreased after the injection, compared to those prior to the injection. At 5 min after the injection, their CT values increased markedly up to 115.0 and 120.2 HU, respectively. Then, the values almost leveled out in 240 min (6 h), and they did not remarkably decrease even in 2,880 min (2 days). These tendencies were similar to the results for the Au/SiO_2 particles with the size of 47.3 nm that was reported in our previous work [40]; there was no large difference in the dependence of the CT value on time between the Au/SiO_2 particle sizes examined. The present results indicated that the Au/SiO_2 particles efficiently reached the tissues through a flow in the blood vessels and were trapped in the tissues, and then the tissues could be imaged. For the kidneys, their contrasts were almost constant for all the times examined, in which their CT value was around 65 HU. This indicated that the Au/SiO_2 particles were probably recognized as alien substances and were intensively trapped in the liver and the spleen. The internal organs recognize hydrophobic materials as foreign through interaction between proteins and the hydrophilic surface of the materials [44-48]. The surface of the Au/SiO_2 particles seemed to be hydrophilic because of silanol groups derived from the silica surface of Au/SiO_2 particles. The trapping took place for the liver and the spleen with the hydrophilic surface. Accordingly, these results proved that the Au/SiO_2 particle surface was required to be more hydrophilic for improving their imaging function.

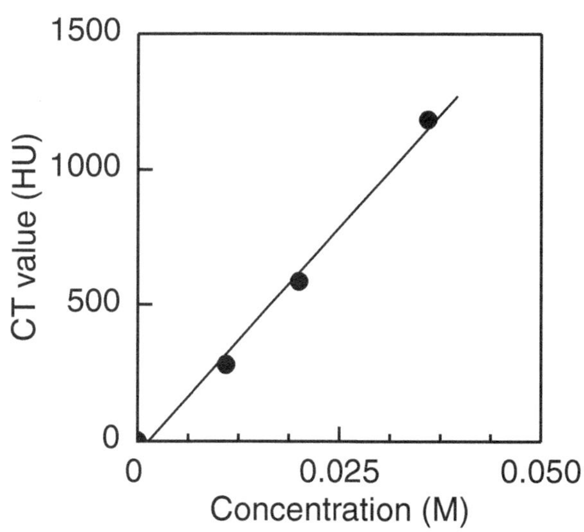

Figure 3 CT value versus Au concentration of the AuSiO$_2$ particle colloid solution.

Conclusions

The Au/SiO_2 particles with the average size of 138.9 ± 12.4 nm and the Au core size of 16.9 ± 1.2 nm that were fabricated in our previous work were applied to X-ray imaging in the present work. The actual Au concentration of the Au/SiO_2 particle colloid solution was 0.036 M after concentrated to 1/2,000 of its original volume. The concentrated Au/SiO_2 particle colloid solution had the CT value of 1,184.8 HU. This value was high compared to that for the commercial X-ray contrast agent with the same iodine concentration as the Au concentration, which indicated that the Au/SiO_2 particle colloid solutions were sensitive for absorption of X-ray. The injection of the colloid solution into a mouse made the X-ray images of its internal organs such as the liver and spleen taken clearly. Accordingly, the Au/SiO_2 particle colloid solution was proved to be able to work as an X-ray contrast agent. Further studies, in particular on the toxicity of particle colloid solutions, are in progress toward practical use.

Methods

Chemicals

Au nanoparticles were prepared from hydrogen tetrachloroaurate(III) trihydrate ($HAuCl_4{\cdot}3H_2O$, >98%, Kanto Chemical, Chuo-ku, Tokyo, Japan) as an Au source and trisodium citrate dihydrate (Na-cit, 99%, Kanto Chemical) as a reducing reagent. (3-Aminopropyl) trimethoxysilane (APMS, 97%, Sigma-Aldrich, St. Louis, MO, USA) were used as a silane coupling agent for increasing the affinity between the Au particle surface and silica shell. Tetraethyl orthosilicate (TEOS, 95%, Kanto Chemical), sodium hydroxide (NaOH) aqueous solution (1 mol/L, Kanto Chemical), and ethanol (99.5%, Kanto

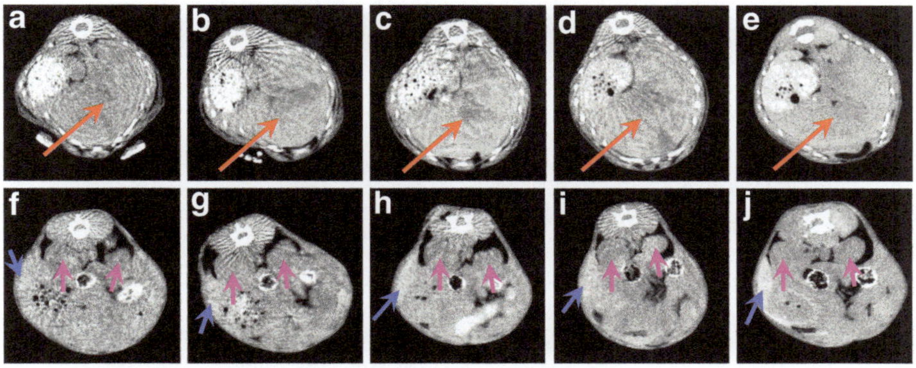

Figure 4 CT images of the internal organs of the mouse after injection. CT images of the liver (red arrows), spleen (blue arrows), and kidney (pink arrows) of the mouse after injection of Au/SiO_2 particle colloid solution. The Au concentration in the colloic solution and the volume of the injected colloid solution were 0.036 M and 80 μL, respectively. The images were taken **(a, f)** prior to injection and at **(b, g)** 5, **(c, h)** 360, **(d, i)** 1,440, and **(e, j)** 2,880 min after injection.

Chemical) were used as a silica source, a catalyst for the sol–gel reaction of TEOS, and a solvent for silica coating, respectively. Hydrofluoric acid (HF, 46.0% to 48.0%, Kanto Chemical), nitric acid (HNO_3, 69% to 70%, Kanto Chemical), and hydrochloric acid (HCl, 35.0% to 37.0%, Kanto Chemical) were used for preparing samples for ICP spectrometry. All chemicals were used as received. Water that was ion-exchanged and distilled with Yamato WG-250 (Koto-ku, Tokyo, Japan) was used in all the preparation.

Preparation of materials

A colloid solution of Au/SiO_2 particles was prepared in two steps, as reported in our previous work [40]. The first step was preparation of the Au nanoparticle colloid solution. The preparation was performed by adding a Na-cit aqueous solution to an $HAuCl_4$ aqueous solution at a constant temperature of 80°C under vigorous stirring. Initial concentrations of Au and Na-cit were 2.4×10^{-4} and 1.6×10^{-3} M, respectively. The total volume of the final reactant solution was 400 mL. Production of Au nanoparticles was implied with visual observation for color change of the solution to wine red after the addition. According to TEM, the Au nanoparticles had a volume-average size of 16.9 ± 1.2 nm [40]. The second step was silica coating of the obtained Au nanoparticles. The silica coating was performed by a sol–gel method in the presence of Au nanoparticles. At 15 min after addition of an APMS aqueous solution to the obtained Au nanoparticle colloid solution, ethanol and TEOS were successively added to the colloid solution. The following rapid injection of an NaOH aqueous solution into the Au/TEOS colloid solution initiated a sol–gel reaction of TEOS or the silica coating. The reaction temperature and time were 35°C and 24 h, respectively. The total volume of the final reactant solution was 2,000 mL. Initial concentrations of Au, APMS, NaOH, H_2O, and TEOS in the final reactant solution were adjusted to 4.3×10^{-5}, 2.0×10^{-5}, 1.0×10^{-3}, 10.7, and 1.0×10^{-2} M, respectively. TEM observation indicated that these initial concentrations resulted in production of Au/SiO_2 particles with a volume-average particle size of 138.9 ± 12.4 nm [40].

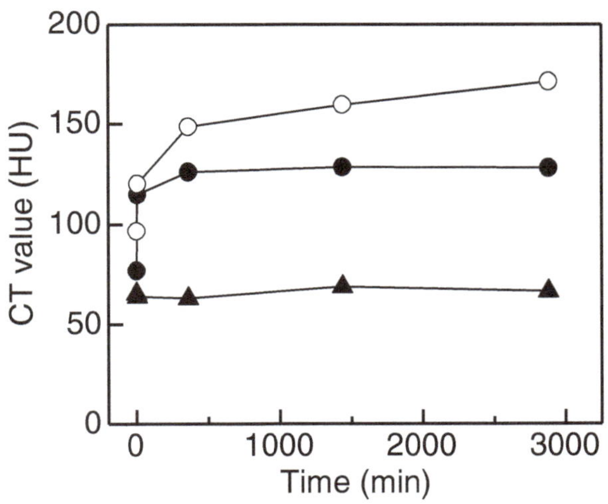

Figure 5 CT values of the internal organs of the mouse after injection. CT values of the liver (filled circle), spleen (empty circle), and kidney (triangle) of the mouse as a function of time after injection of Au/SiO_2 particle colloid solution. The Au concentration in the colloid solution and the volume of the injected colloid solution were 0.036 M and 80 μL, respectively.

Characterization

TEM was used for investigation of the morphology of the particles. TEM observation was performed with a JEOL JEM-2000FX II microscope (Akishima-shi, Japan) operating at 200 kV. To prepare samples for TEM, the particle colloid solution was dropped on a collodion-coated copper grid, and then its dispersion medium was evaporated in air followed by *in vacuo*. Volume-average particle sizes were determined by measuring dozens of particle diameters in TEM images. An actual Au concentration of the concentrated Au/SiO_2 particle colloid solution was determined by ICP. A Shimadzu ICPS-7510 emission spectrometer (Kyoto, Japan) was used for the ICP. To prepare samples for the ICP, HF and aqua regia composed of 1:3 (v/v) HNO_3/HCl were successively added to the concentrated Au/SiO_2 particle colloid solution to dissolve the silica shell and Au nanoparticles, respectively. Then, the obtained solution was diluted to adjust the Au concentration to a concentration appropriate to the ICP measurement. X-ray images and CT values of samples such as Iopamiron®300, the Au/SiO_2 particle colloid solution, and a mouse injected with the particle colloid solutions were obtained with the Aloka LaTheta LCT-200 CT system (Mitaka-shi, Japan), according to our previous work [42]. Prior to the measurement of the CT value for the Au/SiO_2 particle colloid solution, the Au/SiO_2 particles were washed by salting out the Au/SiO_2 particles with addition of a saturated NaCl aqueous solution, removing the supernatant by decantation, and then repeating a washing process composed of centrifugation, removal of the supernatant, addition of water, and shaking with a vortex mixer for several times. The Au/SiO_2 particle colloid solution was concentrated to 1/2,000 of its original volume by decreasing the amount of added water in the washing process (concentrated Au/SiO_2 particle colloid solution). Assuming complete reduction of $HAuCl_4$ to metallic Au and no loss of particles during the concentrating process, the concentrated Au/SiO_2 particle colloid solution was considered to have an initial Au concentration of 0.086 M. The samples were put into a tube with a diameter of 3.7 cm and a length of 29.5 cm. The images were taken as if the samples were cut into round slices. CT values of samples were estimated with the basis of CT values of −1,000 HU for air and 0 HU for water. The mouse used was the ICR mouse with the age of 5 to 6 weeks old. The mouse was put under anesthesia, and 80 μL of the colloid solution was injected into the mouse from its tail veins. All sugical processes were performed for the mouse under anesthesia in accordance with guidelines approved by the committee on animal experiments of Tohoku University.

Competing interests
The authors declare that they have no competing interests.

Authors' contributions
YK (Kobayashi) drafted the manuscript. KG and NO modified it. HI, TN, and YK (Kubota) participated in some practical work. All authors read and approved the final manuscript.

Acknowledgements
We express our thanks to Prof. T. Noguchi from the College of Science of Ibaraki University, Japan, for his help in the TEM observation.

Author details
[1]Department of Biomolecular Functional Engineering, College of Engineering, Ibaraki University, 4-12-1 Naka-narusawa-cho, Hitachi, Ibaraki 316-8511, Japan. [2]Division of Surgical Oncology, Graduate School of Medicine, Tohoku University, Seiryo-machi, Aoba-ku, Sendai, Miyagi 980-8574, Japan.

References
1. Bellani, G, Caironi, P: Lung imaging during acute respiratory distress syndrome: CT- and PET-scanning. Trends Anaesth Crit Care **1**, 203–209 (2011)
2. Duwek, HC, Spitzer, H, Weitzen, R, Apter, S: A biologically-based algorithm for companding computerized tomography (CT) images. Comput Biol Med **41**, 367–379 (2011)
3. Ruschin, M, Komljenovic, PT, Ansell, S, Ménard, C, Bootsma, G, Cho, YB, Chung, C., Jaffray, D: Cone beam computed tomography image guidance system for a dedicated intracranial radiosurgery treatment unit. Int J Radiat Oncol Biol Phys **85**, 243–250 (2013)
4. Lee, SJ, Jung, SY, Ahn, S: Flow tracing microparticle sensors designed for enhanced X-ray contrast. Biosens Bioelectron **25**, 1571–1578 (2010)
5. Hwang, UJ, Shin, DH, Kim, TH, Moon, SH, Lim, YK, Jeong, H, Rah, JE, Kim, SS, Kim, JY, Kim, DY, Park, SY, Cho, KH: The effect of a contrast agent on proton beam range in radiotherapy planning using computed tomography for patients with locoregionally advanced lung cancer. Int J Radiat Oncol Biol Phys **81**, e317–e324 (2011)
6. Rustighi, I, Donati, I, Ferluga, M, Campa, C, Pasqua, AE, Rossib, M, Paoletti, S: Borate complexes of X-ray iodinated contrast agents: characterization and sorption studies for their removal from aqueous media. J Hazard Mater **205–206**, 10–16 (2012)
7. Thomsen, HS: Contrast media safety - an update. Eur J Radiol **80**, 77–82 (2011)
8. Ichikawa, T, Motosugi, U, Morisaka, H, Sou, H, Onohara, K, Sano, K, Araki, T: Optimal iodine dose for 3-dimensional multidetector-row CT angiography of the liver. Eur J Radiol **81**, 2450–2455 (2012)
9. Li, X, Anton, N, Zuber, G, Zhao, M, Messaddeq, N, Hallouard, F, Fessi, H, Vandamme, TF: Iodinated a-tocopherol nano-emulsions as non-toxic contrast agents for preclinical X-ray imaging. Biomater **34**, 481–491 (2013)
10. Malarkodi, C, Rajeshkumar, S, Vanaja, M, Paulkumar, K, Gnanajobitha, G, Annadurai, G: Eco-friendly synthesis and characterization of gold nanoparticles using *Klebsiella pneumoniae*. J Nanostruct Chem **3**, 30 (2013)
11. Rajeshkumar, S, Malarkodi, C, Gnanajobitha, G, Paulkumar, K, Vanaja, M, Kannan, C, Annadurai, G: Seaweed-mediated synthesis of gold nanoparticles using *Turbinaria conoides* and its characterization. J Nanostruc Chem **3**, 44 (2013)
12. Park, JA, Kim, HK, Kim, JH, Jeong, SW, Jung, JC, Lee, GH, Lee, J, Chang, Y, Kim, TJ: Gold nanoparticles functionalized by gadolinium-DTPA conjugate of cysteine as a multimodal bioimaging agent. Bioorg Med Chem Lett **20**, 2287–2291 (2010)
13. Cho, EC, Glaus, C, Chen, J, Welch, MJ, Xia, Y: Inorganic nanoparticle-based contrast agents for molecular imaging. Trends Mol Med **16**, 561–573 (2010)
14. Wang, H, Zheng, L, Peng, C, Guo, R, Shen, M, Shi, X, Zhang, G: Computed tomography imaging of cancer cells using acetylated dendrimer-entrapped gold nanoparticles. Biomater **32**, 2979–2988 (2011)
15. Ahn, S, Jung, SY, Seo, E, Lee, SJ: Gold nanoparticle-incorporated human red blood cells (RBCs) for X-ray dynamic imaging. Biomater **32**, 7191–7199 (2011)
16. Menk, RH, Schültke, E, Hall, C, Arfelli, F, Astolfo, A, Rigon, L, Round, A, Ataelmannan, K, MacDonald, SR, Juurlink, BHJ: Gold nanoparticle labeling of cells is a sensitive method to investigate cell distribution and migration in animal models of human disease. Nanomed Nanotechnol Biol Med **7**, 647–654 (2011)
17. Roessl, E, Cormode, D, Brendel, B, Engel, KJ, Martens, G, Thran, A, Fayad, Z, Proks, R: Preclinical spectral computed tomography of gold nano-particles. Nucl Instrum Met Phys Res A **648**, S259–S264 (2011)
18. Zhang, XD, Wu, D, Shen, X, Chen, J, Sun, YM, Liu, PX, Liang, XJ: Size-dependent radiosensitization of PEG-coated gold nanoparticles for cancer radiation therapy. Biomater **33**, 6408–6419 (2012)
19. Peng, C, Zheng, L, Chen, Q, Shen, M, Guo, R, Wang, H, Cao, X, Zhang, G, Shi, X: PEGylated dendrimer-entrapped gold nanoparticles for in vivo blood pool and tumor imaging by computed tomography. Biomater **33**, 1107–1119 (2012)
20. Kim, D, Jon, S: Gold nanoparticles in image-guided cancer therapy. Inorg Chim Acta **393**, 154–164 (2012)

21. Tedesco, S, Doyle, H, Blasco, J, Redmond, G, Sheehan, D: Oxidative stress and toxicity of gold nanoparticles in *Mytilus edulis*. Aquat Toxicol **100**, 178–186 (2010)
22. Ahamed, M, AlSalhi, MS, Siddiqui, MKJ: Silver nanoparticle applications and human health. Clin Chim Acta **411**, 1841–1848 (2010)
23. Farkas, J, Christian, P, Urrea, JAG, Roos, N, Hassellöv, M, Tollefsen, KE, Thomas, KV: Effects of silver and gold nanoparticles on rainbow trout (*Oncorhynchus mykiss*) hepatocytes. Aquat Toxicol **96**, 44–52 (2010)
24. Lasagna-Reeves, C, Gonzalez-Romero, D, Barria, MA, Olmedo, I, Clos, A, Ramanujam, VMS, Urayama, A, Vergara, L, Kogan, MJ, Soto, C: Bioaccumulation and toxicity of gold nanoparticles after repeated administration in mice. Biochem Biophys Res Commun **393**, 649–655 (2010)
25. Han, X, Gelein, R, Corson, N, Mercer, PW, Jiang, J, Biswas, P, Finkelstein, JN, Elder, A, Oberdörster, G: Validation of an LDH assay for assessing nanoparticle toxicity. Toxicol **287**, 99–104 (2011)
26. Park, MVDZ, Neigh, AM, Vermeulen, JP, de la Fonteyne, LJJ, Verharen, HW, Briedé, JJ, Van Loveren, H, De Jong, WH: The effect of particle size on the cytotoxicity, inflammation, developmental toxicity and genotoxicity of silver nanoparticles. Biomater **32**, 9810–9817 (2011)
27. Schulz, M, Ma-Hock, L, Brill, S, Strauss, V, Treumann Gröters, SS, van Ravenzwaay, B, Landsiedel, R: Investigation on the genotoxicity of different sizes of gold nanoparticles administered to the lungs of rats. Mutat Res **745**, 51–57 (2012)
28. Nymark, P, Catalán, J, Suhonen, S, Järventaus, H, Birkedal, R, Clausen, PA, Jensen, KA, Vippola, M, Savolainena, K, Norppa, H: Genotoxicity of polyvinylpyrrolidone-coated silver nanoparticles in BEAS 2B cells. Toxicol in press
29. Lu, Y, Yin, Y, Li, ZY, Xia, Y: Synthesis and self-assembly of Au@SiO_2 core-shell colloids. Nano Lett **2**, 785–788 (2002)
30. Xu, J, Perry, CC: A novel approach to Au@SiO_2 core-shell spheres. J Non-Cryst Solid **353**, 1212–1215 (2007)
31. Ye, J, Broek, BV, Palma, RD, Libaers, W, Clays, K, Roy, WV, Borghs, G., Maes, G: Surface morphology changes on silica-coated gold colloids. Colloids Surf A **322**, 225–233 (2008)
32. Wu, Z, Liang, J, Ji, X, Yang, W: Preparation of uniform Au@SiO_2 particles by direct silica coating on citrate-capped Au nanoparticles. Colloids Surf A **392**, 220–224 (2011)
33. Wang, TT, Chai, F, Wang, CG, Li, L, Liu, HY, Zhang, LY: Fluorescent hollow/rattle-type mesoporous Au@SiO_2 nanocapsules for drug delivery and fluorescence imaging of cancer cells. J Colloid Inter Sci **358**, 109–115 (2011)
34. Huang, CC, Huang, CH, Kuo, IT, Chau, LK, Yang, TS: Synthesis of silica-coated gold nanorod as Raman tags by modulating cetyltrimethylammonium bromide concentration. Colloids Surf A **409**, 61–68 (2012)
35. Yu, D, Zeng, Y, Qi, Y, Zhou, T, Shi, G: A novel electrochemical sensor for determination of dopamine based on AuNPs@SiO_2 core-shell imprinted composite. Biosens Bioelectron **38**, 270–277 (2012)
36. Fülöp, E, Nagy, N, Deák, A, Bársony, I: Langmuir–Blodgett films of gold/silica core/shell nanorods. Thin Solid Films **520**, 7002–7005 (2012)
37. Shen, S, Tang, H, Zhang, X, Ren, J, Pang, Z, Wang, D, Gao, H, Qian, Y, Jiang, X, Yang, W: Targeting mesoporous silica-encapsulated gold nanorods for chemo-photothermal therapy with near-infrared radiation. Biomater **34**, 3150–3158 (2013)
38. Khlebtsov, BN, Khanadeev, VA, Panfilova, EV, Inozemtseva, OA, Burov, AM, Khlebtsov, NG: A simple Mie-type model for silica-coated gold nanocages. J Quant Spectrosc Ra **121**, 23–29 (2013)
39. Mine, E, Yamada, A, Kobayashi, Y, Konno, M, Liz-Marzán, LM: Direct coating of gold nanoparticles with silica by a seeded polymerization technique. J Colloid Interface Sci **264**, 385–390 (2003)
40. Kobayashi, Y, Inose, H, Nakagawa, T, Gonda, K, Takeda, M, Ohuchi, N, Kasuya, A: Control of shell thickness in silica-coating of Au nanoparticles and their X-ray imaging properties. J Colloid Interface Sci **358**, 329–333 (2011)
41. Kobayashi, Y, Inose, H, Nakagawa, T, Gonda, K, Takeda, M, Ohuchi, N, Kasuya, A: Synthesis of Au-silica core-shell particles by sol–gel process. Surf Eng **28**, 129–133 (2012)
42. Kobayashi, Y, Inose, H, Nagasu, R, Nakagawa, T, Kubota, Y, Gonda, K, Ohuchi, N: X-ray imaging technique using colloid solution of Au/silica/poly(ethylene glycol) nanoparticles. Mater Res Innov in press
43. Zhang, A, Tu, Y, Qin, S, Li, Y, Zhou, J, Chen, N, Lu, Q, Zhang, B: Gold nanoclusters as contrast agents for fluorescent and X-ray dual-modality imaging. J Colloid Interface Sci **372**, 239–244 (2012)
44. Hallouard, F, Anton, N, Choquet, P, Constantinesco, A, Vandamme, T: Iodinated blood pool contrast media for preclinical X-ray imaging applications - a review. Biomater **31**, 6249–6268 (2010)
45. Parveen, S, Sahoo, SK: Long circulating chitosan/PEG blended PLGA nanoparticle for tumor drug delivery. Eur J Pharmacol **670**, 372–383 (2011)
46. Dufort, S, Sancey, L, Coll, JL: Physico-chemical parameters that govern nanoparticles fate also dictate rules for their molecular evolution. Adv Drug Deliv Rev **64**, 179–189 (2012)
47. Stolnik, S, Illum, L, Davis, SS: Long circulating microparticulate drug carriers. Adv Drug Deliv Rev **64**, 290–301 (2012)
48. Bauer, S, Schmuki, P, von der Mark, K, Park, J: Engineering biocompatible implant surfaces: part I: materials and surfaces. Prog Mater Sci **58**, 261–326 (2013)

Fabrication of ZnO nanoparticles by solution-combustion method for the photocatalytic degradation of organic dye

Mohammad Amdad Ali[1*], Mahmudur Rahman Idris[2] and Md Emran Quayum[2]

Abstract

ZnO nanoparticles (NPs) were successfully fabricated by simple combustion process, in which $Zn(CH_3COO)_2$ precursors migrated with the aid of alcoholic fuel to the top of a burning lampwick and the chemical reactions occurred at the solvent-air interface of the ignited lampwick, yielding ZnO NPs. Ultraviolet-visible spectroscopic analysis and energy-dispersive spectroscopic elemental analysis revealed that the fabricated ZnO NPs are comprised mainly of Zn and O atoms. It was observed that most of the ZnO NPs exhibited an ellipsoidal shape having a particle size of 100 to 200 nm for short-length axes and 100 to 400 nm for long-length axes. To see the effectiveness of the fabricated ZnO NPs, it was employed as photocatalytic agent to degrade the organic dye, *viz.* Rhodamine B, and catalytic activities under both sunlight and UV light were analyzed and compared with respect to the commercially available ZnO (Fluka). The initial degradation rate in fabricated ZnO NPs was almost double than that of ZnO (Fluka). Furthermore, by exposing to sunlight for 1 h, the fabricated ZnO NPs degraded Rhodamine B dye more than 80%, whereas it was only approximately 35% for ZnO (Fluka). Photocatalytic activities might be attributed to porous structures with larger active surface area of the nanosized ZnO particles. With this advantage, 96% of organic contaminants were degraded by the fabricated ZnO NPs upon exposure to sunlight for 2 h.

Keywords: Nanoparticles, Photocatalyst, Organic dye, Wastewater treatment

Background

Water pollution has been a serious threat to the environment. The major sources of pollutants are from manufacturing processing industries, particularly chemical and textile industries in which organic dyes are widely used. Presently, the perilous effects of organic dyes to the environment are a particular concern. Conventional treatment methods, adsorptions for instance, are not always adequate and/or appropriate to remove organic dyes from wastewater streams in a large scale. Adsorption is a nondestructive method, which simply transfers dyes from one medium to another. In recent years, many researchers are interested in applying the 'photocatalytic decomposition process' to degrade the hard-to-biodegrade organic pollutants in industrial wastewater before discharge into mainstream water bodies [1].

Nanostructured semiconductors are a potential candidate for the mineralization of toxic organic compounds, hazardous inorganic constituent [1], and bacteria disinfection [2] owing to its strong oxidizing ability, i.e., hydroxyl radical ($\dot{O}H$). Numerous metal oxides such as zinc oxide (ZnO), titanium dioxide (TiO_2) [3], strontium titanate ($SrTiO_3$) [4], tungsten oxide (WO_3) [5], silver nanoparticle [6], and hematite (α-Fe_2O_3) [5] are turned to be potential photocatalysts. ZnO has appeared to be a prominent catalyst as far as water detoxification is concerned; on accounting that, it produces H_2O_2 more efficiently and shows high reaction and mineralization rates [7]. In addition, it was reported that ZnO showed high surface reactivity due to its higher number of active sites [7]. ZnO has been proven as a superior photocatalyst compared to commercial TiO_2 due to its higher initial rates of activities and absorption efficacy of

* Correspondence: amdad53@yahoo.com
[1]Department of Agricultural Chemistry, Bangladesh Agricultural University, Mymensingh 2202, Bangladesh
Full list of author information is available at the end of the article

solar radiations [7]. As a result, ZnO has been widely used in the treatment of wastewater owing to its excellent chemical and physical properties. A number of methods have been developed to synthesize ZnO nanoparticles (NPs), such as, the sol-gel method [8], evaporative decomposition of solution [9], hydrothermal [10], wet chemical synthesis [11], and template-assisted growth [12]. These methods are very expensive and usually employed to produce highly pure ZnO with desired shape and crystallinity.

Of these, however, it is still a challenge to find a simple fabrication process to prepare ZnO NPs with a large active surface area. In addition, the use in agricultural and environmental aspect requires a simple, cost-effective fabrication process to synthesize ZnO NPs, which can provide a high yield with reasonable purities. Solution-combustion process is simple and cost-effective, expected to provide extremely porous structure of ZnO NPs [13]. With this advantage, the solution-combustion method has been employed to fabricate ZnO NPs, with a mixture of ethanol and ethyleneglycol ($V/V = 60/40$) as the solvent and zinc acetate as the zinc source, and explore its photocatalytic degradation effects on organic dye Rhodamine B.

Results and discussion

SEM images of the microstructure ZnO NPs prepared by the combustion method is depicted in Figure 1, showing that most particles exhibited an ellipsoidal shape. Some small particles with a rounded shape and few large particles were also seen. The bigger-sized particles ought to be attributed to the aggregation or overlapping of small particles. The particle size measured from the SEM micrographs was in the range of 100 to 200 nm for short-length axes and 100 to 400 nm for long-length axes. Thus, the topological images suggested that particle sizes were not well defined; it might be resultant of uncontrolled ambient conditions.

Figure 2 shows the EDS scan profile of the fabricated ZnO NPs. The spectrum shows the presence of Zn, O, and K atom in the sample. Highly intense peaks of O and Zn were observed, indicating that the NPs were composed mainly of ZnO along with some impurities. In the scan profile, the K peak was likely to come from the cotton lampwick as cotton lampwick contains K atom. An elemental analysis on the cotton lampwick was performed to further confirm the presence of K atom.

For the characterization of the fabricated ZnO NPs, in the present study, an UV-vis spectrophotometer was employed to record the absorbance spectra in the wavelength range of 200 to 1,100 nm for dispersed ZnO NPs in distilled water. The absorbance spectrum of ZnO NPs is shown in Figure 3. A strong absorption band ranging from 200 to 400 nm was observed for ZnO NPs due to the metal ion [13]. The $n \rightarrow \sigma^*$ absorption bands appeared at approximately 295 nm. The $n \rightarrow \pi^*$ transition bands in the ZnO NPs were observed at approximately 312 nm. This spectrum was consistent with previously reported data [13]. An absorption peak centered at 374 nm (3.32 eV) was seen, thus in good agreement with previous work.

ZnO is proven to be a potential photocatalyst and has widely been used in wastewater treatment. The origin of photocatalytic process is related with the generation of electron-hole pairs with the help of bandgap radiation, thus increasing redox reactions with adsorbed species on the catalyst surface [1,13]. The hydroxyl free radical ($\dot{O}H$) is considered as a driving matter, originated from the oxidation process of OH^- or H_2O by the photogenerated electron-hole pairs in the presence of oxygen. The hydroxyl free radical ($\dot{O}H$) is a strong oxidant and can decolorize organic dyes by degrading the conjugated bonds. The possible route of hydroxyl free radical formation is shown here [2,4]:

$$ZnO + hv \rightarrow e^- + h^+$$
$$h^+ + OH^- \rightarrow \dot{O}H + H^+$$
$$h^+ + H_2O \rightarrow \dot{O}H$$

To investigate the photocatalytic activities of the fabricated ZnO NPs, an organic dye, Rhodamine B, was

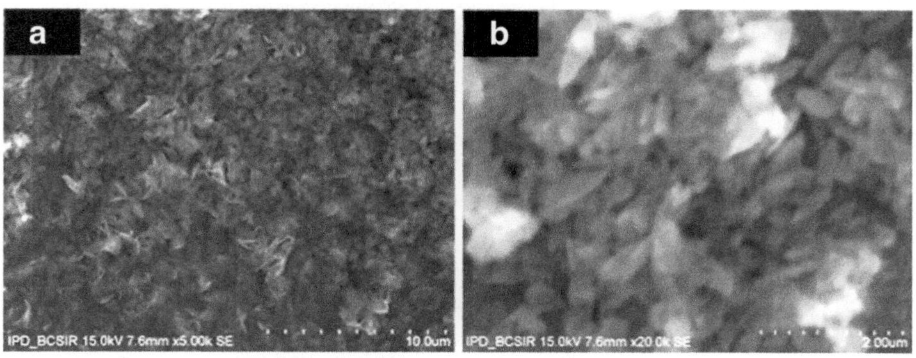

Figure 1 Typical SEM micrographs of the fabricated ZnO NPs topology. The scale bars in the micrographs represent 10 (**a**) and 2 μm (**b**).

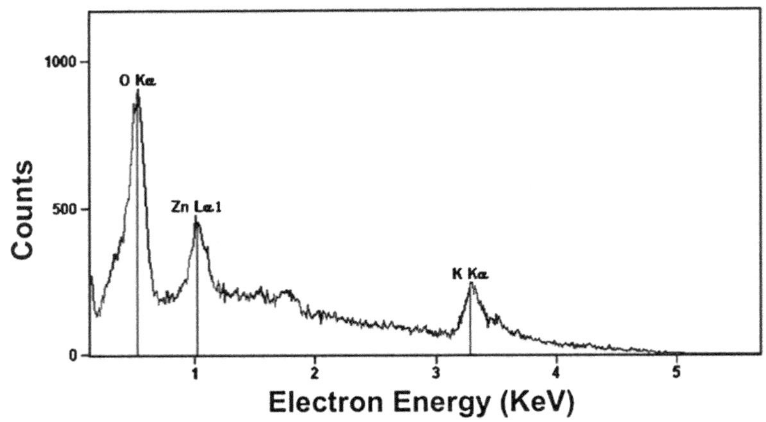

Figure 2 EDX elemental analysis of synthesized ZnO NPs.

employed. Figure 4a shows the UV-vis absorption spectra of Rhodamine B by exposure to sunlight for 30, 60, 90, 120, and 150 min, respectively, in the presence of ZnO NPs. It was seen that after irradiating for 60 min, the absorption peak intensity sharply decreased. After that, it decreased gradually with increasing exposure time. It demonstrated that the fabricated ZnO NPs has good photocatalytic degradation efficiency in sunlight. Figure 4b shows the UV-vis absorption spectra of the Rhodamine dye irradiated by UV light of 365 nm wavelength for several time intervals ranging from 20 to 130 min in the presence of ZnO NPs. A similar trend was found; the intensity of absorbance peak decreased more rapidly until 30 min compared to that with exposure to sunlight, indicating that UV light enhances the catalytic activities of ZnO NPs, which might be due to the bandgap in the ultraviolet region (approximately 3.32 eV) promoting photocatalysis upon illumination with UV radiation. After 30 min, it decreased slowly with increasing exposure time.

The maximum absorbance was observed at approximately 560 nm both in sunlight and UV light, and the percentage of degradation was calculated considering the absorbance maxima, equal to $[A_i - A_t] \times 100$, where A_i and A_t denotes the absorbance of Rhodamine B at time $t = 0$ and $t = t$, respectively. In Figure 5, the percentage of degradation with time for degradation reactions, catalyzed by commercial and nanocrystalline ZnO (Fluka) respectively, are shown and compared for both sunlight and UV light exposure. Figure 5a shows the faster degradation rate in the case of the fabricated ZnO NPs compared to that of the commercial ZnO in the early stage of reaction. Approximately 90% of the dye was degraded by the fabricated ZnO NPs, whereas it was only approximately 35% for the commercial ZnO (Fluka) after 80 min. The results of the photodegradation experiments on commercial ZnO and fabricated ZnO NPs catalyst clearly indicated that the fabricated ZnO NPs are more active than the commercial ZnO in degrading Rhodamine B. The photocatalytic activity differences can be attributed to the ZnO NPs' larger surface area with extremely porous structures, providing an elevated active surface for the fabricated ZnO NPs. Moreover, as dimension of NPs are smaller, the photogenerated electron-hole pairs can rapidly reach the catalyst surface, reducing recombination process probability. The reduced activity of ZnO NPs in the whole time course of the catalyzed reaction can be explicated by photocorrosion phenomena. This hypothesis explained that, with decreasing intensity of irradiation, a higher photocatalytic rate of the nanosized ZnO compared to the commercial ZnO was observed, thus omitting an extensive degradation of the catalyst.

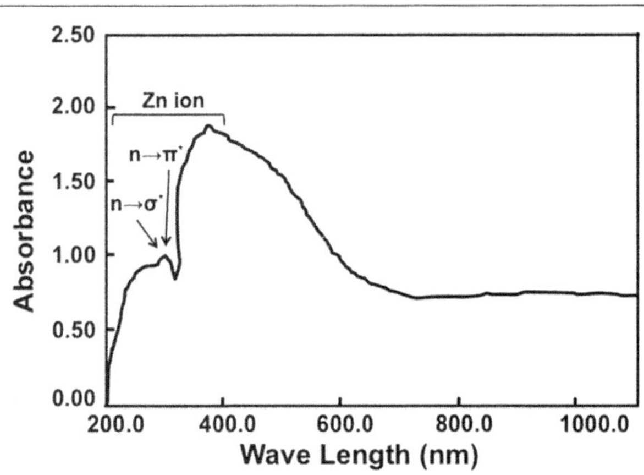

Figure 3 UV-visible spectrum of the fabricated ZnO NPs.

Conclusions

A novel, feasible, and cost-effective process of fabricating ZnO NPs was successfully demonstrated, which would be a very promising method for synthesizing other semiconductor NPs. The chemical reactions occurred at the solvent-air interface, yielding ZnO NPs via esterification

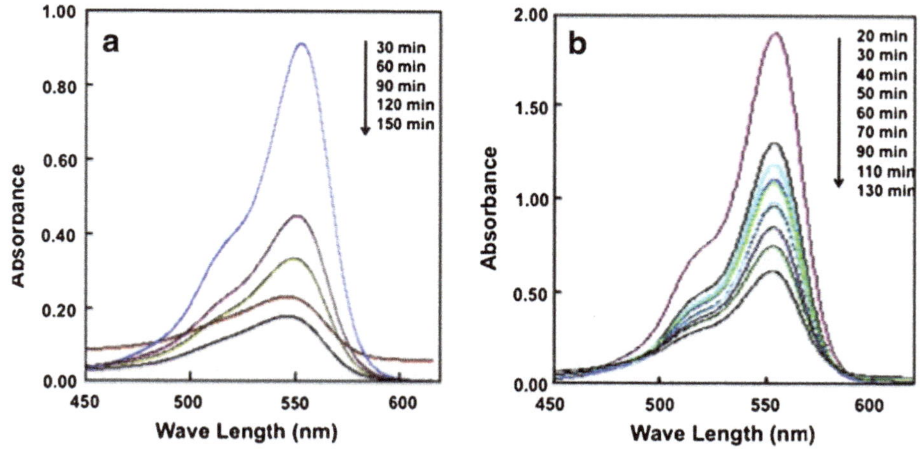

Figure 4 Absorption spectrum of Rhodamine B in the presence of the fabricated ZnO NPs. At different time intervals by exposing to (**a**) sunlight and (**b**) UV light.

of alcohol. The NPs exhibited a nonuniform size and shape which might be due to the uncontrolled growth; however, controlled processing parameters might produce a uniform size and shape of the particles. Importantly, the fabricated ZnO NPs exhibited excellent catalytic properties for the degradation of organic dyes in the presence of sunlight and UV-vis light. Porous structures provided a larger surface area with an elevated active surface, thus attributing better catalytic degradation properties of nanosized ZnO particles on the organic dye Rhodamine B over commercial ZnO (Fluka). More than 95% of Rhodamine B was degraded by exposing to sunlight for 2 h in the presence of the fabricated ZnO NPs. Therefore, it is very promising and implementable for the degradation of organic matter for wastewater treatment in small or large scale.

Methods

All chemicals were purchased from the market and used as received. In this study, $Zn(CH_3COO)_2.2H_2O$ (0.03 mol) was dissolved in 100 mL of a mixed-solvent system of ethanol-ethyleneglycol in a volume ratio of 60/40. A portion of this mixture was transferred to a spirit lamp including an absorbent cotton lampwick. The spirit lamp was ignited, and after a certain time, a yellowish-colored substance formed around the surface of the ignited lampwick. When the lampwick was quenched, the yellow-colored substance turned white. A schematic diagram of the fabrication process is shown in Figure 6. It has been reported that the interaction between zinc acetate and alcohol under solvothermal conditions follows the esterification reaction, yielding ZnO, ester, and water [13-15], according to the following reaction:

$$Zn\,(CH_3\,COO)_2 + 2R{-}OH \rightarrow ZnO + 2CH_3{-}CO{-}R + H_2O$$

The resultant white product was repeatedly peeled off and dispersed in distilled water to remove the impurity. For further purification, the synthesized ZnO NPs were dried at 500°C in a muffle furnace for 6 h (5 h) and finally preserved in an airtight desiccator. Scanning electron microscope (SEM) images were taken by a Sirion

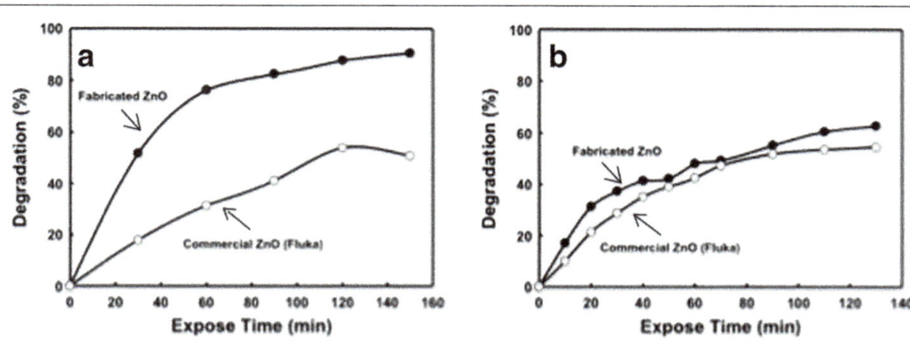

Figure 5 Comparison of dye degradation percentage with the fabricated and commercial ZnO NPs. As a function of exposure time in (**a**) sunlight and (**b**) UV-vis light.

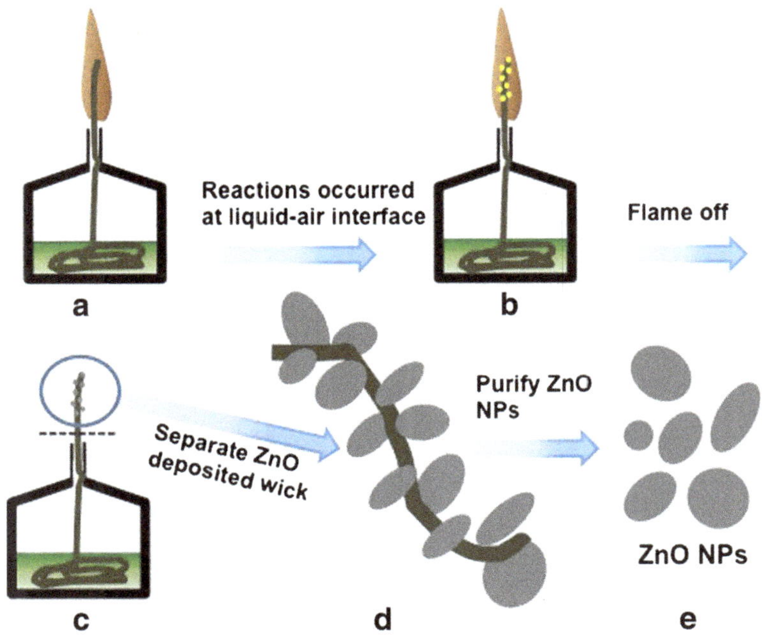

Figure 6 The fabrication process of ZnO NPs by a simple combustion method. (**a**) A spirit lamp containing $Zn(OAc)_2$ and an ethanol-ethyleneglycol mixture. (**b**) Yellow-colored ZnO deposited around the burning lampwick. (**c**) The yellow-colored ZnO turned white, indicating the formation of ZnO NPs. (**d**) A separated portion of the lampwick with ZnO NPs. (**e**) ZnO NPs peeled off, dispersed in water, and dried at 500°C.

200 field emission scanning electron microanalyzer (FEI Co., Hillsboro, OR, USA employing an operating voltage of 25 kV to investigate the particle size and shape. Energy dispersion spectroscopy (EDS) and ultraviolet-visible (UV-vis) spectroscopy using a Hitachi U-4100 spectrophotometer (Tokyo, Japan) were employed to characterize the ZnO NPs. To observe the degradation effects, 1.5 g of ZnO was dispersed in 100-mL aqueous solutions containing Rhodamine B, an organic dye, and exposed to sunlight (at a temperature around 30°C) and UV light for the desired time. Then the decolorization of the dye was measured by UV-vis spectroscopy.

Competing interests
The authors declare that they have no competing interests.

Authors' contributions
MAA analyzed, interpreted, and organized the data and prepared the manuscript. MRI carried out the experiments. MEQ designed and coordinated the overall project. All authors read and approved the final manuscript.

Authors' information
MAA is an assistant professor at the Department of Agricultural Chemistry at Bangladesh Agricultural University, Bangladesh. Presently, he is on leave and taking a Ph.D. degree at the Centre for Organic Photonics and Electronics (COPE), School of Chemistry and Molecular Biosciences, The University of Queensland, Australia. He has been awarded with an International Postgraduate Research Scholarship and the University of Queensland Centennial Scholarship for Ph.D. study. He achieved his master degree in advanced materials engineering from Kookmin University, South Korea and his bachelor degree in chemistry from the Department of Chemistry, University of Dhaka, Bangladesh. He had obtained a fellowship provided by the Ministry of Science and Technology/Korea Science and Engineering Foundation through the Center for Materials and Processes of Self-Assemble (CMPS) for a 2-year period from September 1, 2008 through August 31, 2010 to support his master degree. His research interest involves the fundamental understanding of the physical properties of optoelectronic materials to implementing the devices. He is credited with more than ten international journal publications and presented his research works in several international conferences.
MRI achieved his B.Sc. in chemistry and M.Sc. degree in physical chemistry from the University of Dhaka, Bangladesh. At present, he involves himself as a lecturer at the Department of Textile Engineering, City University, Bangladesh. He is also taking an M.Phil. degree in physical chemistry at the Department of Chemistry, University of Dhaka. His specialization is on the synthesis, characterization, and applications of nanoparticles.
MEQ is a professor of physical chemistry at the Department of Chemistry, University of Dhaka, Bangladesh. He holds a B.Sc. degree in chemistry, and M.Sc. and Ph.D. in physical chemistry. He has more than 18 years of research experience and 14 years of teaching experience at the graduate and post-graduate level. He is specialized in electrochemistry and materials chemistry. He published more than 20 international journal publications and guided many M.Phil. and Ph.D. students.

Acknowledgments
The authors thank the Bangladesh Council of Scientific and Industrial Research (BCSIR), Dhaka, Bangladesh for the SEM images.

Author details
[1]Department of Agricultural Chemistry, Bangladesh Agricultural University, Mymensingh 2202, Bangladesh. [2]Department of Chemistry, University of Dhaka, Dhaka 1000, Bangladesh.

References
1. Curri, ML, Comparelli, R, Cozzoli, PD, Mascolo, G, Agostiano, A: Colloidal oxide nanoparticles for the photocatalytic degradation of organic dye. Mater Sci Eng C **23**, 285–289 (2003)

2. Karunakaran, C, Abiramasundari, G, Gomathisankar, P, Manikandan, G, Anandi, V: Cu-doped TiO_2 nanoparticles for photocatalytic disinfection of bacteria under visible light. J Colloid Interface Sci **352**, 68–74 (2010)
3. Wu, MC, Sápi, A, Avila, A, Szabó, M, Hiltunen, J, Huuhtanen, M, Tóth, G, Kukovecz, Á, Kónya, Z, Keiski, R, Su, WF, Jantunen, H, Kordás, K: Enhanced photocatalytic activity of TiO_2 nanofibers and their flexible composite films: decomposition of organic dyes and efficient H_2 generation from ethanol–water mixtures. Nano Res **4**(4), 360–369 (2011)
4. Karimi, L, Zohoori, S: Superior photocatalytic degradation of azo dyes in aqueous solutions using TiO_2/$SrTiO_3$ nanocomposite. J Nanostructure Chem **3**, 32 (2013)
5. Bhatkhande, DS, Pangarkar, DS, Beenackers, AACM: Photo catalytic degradation for environmental applications – a review. J Chem Technol **77**, 102–116 (2001)
6. Vanaja, M, Gnanajobitha, G, Paulkumar, K, Rajeshkumar, S, Malarkodi, C, Annadurai, G: Phytosynthesis of silver nanoparticles by *Cissus quadrangularis*: influence of physicochemical factors. J Nanostructure Chem **3**, 17 (2013)
7. Ullah, R, Dutta, J: Photocatalytic degradation of organic dyes with manganese-doped ZnO nanoparticles. J Haz Mater **156**, 194–200 (2008)
8. Joo, J, Kwon, SG, Yu, JH, Hyeon, T: Synthesis of ZnO nanocrystals with cone, hexagonal cone, and rod shapes via non-hydrolytic ester elimination sol–gel reactions. Adv Mater **17**, 1873–1877 (2005)
9. Marci, G, Augugliaro, V, Lopez-Munoz, MJ, Martin, C, Palmisano, L, Rives, V, Schiavello, M, Tilley, RJD, Venezia, AM: Preparation characterization and photocatalytic activity of polycrystalline ZnO/TiO_2 systems. 1. Surface and bulk characterization. J Phys Chem B **105**, 1026–1032 (2001)
10. Ni, YH, Wei, XW, Hong, JM, Ye, Y: Hydrothermal preparation and optical properties of ZnO nanorods. Mater Sci Eng B **121**, 42–47 (2005)
11. Pal, U, Santiago, P: Controlling the morphology of ZnO nanostructures in a low-temperature hydrothermal process. J Phys Chem B **109**, 15317–15321 (2005)
12. Jie, J, Wang, G, Wang, Q, Chen, Y, Han, X, Wang, X, Hou, JG: Synthesis and characterization of aligned ZnO nanorods on porous aluminum oxide template. J Phys Chem B **108**, 11976–11980 (2004)
13. Ni, Y, Cao, X, Wu, G, Hu, G, Yang, Z, Wei, X: Preparation, characterization and property study of zinc oxide nanoparticles via a simple solution-combusting method. Nanotechnology **18**, 155603–155608 (2007)
14. Yiamsawas, D, Boonpavanitchakul, K, Kangwansupamonkon, W: Preparation of ZnO nanostructures by solvothermal method. J Micros Soc Thailand **23** (1), 75–78 (2009)
15. Bell, NS: Solvothermal routes for synthesis of zinc oxide nanorods. Symp Pros Mater Res Soc **878E**, Y2.6.1–Y2.6.11 (2005)

Optimization of ball milling parameters to produce *Centella asiatica* herbal nanopowders

Muhammad Zuhairi Borhan[1,2,3*], Rohaya Ahmad[3], Mohamad Rusop[1,2] and Saifollah Abdullah[2*]

Abstract

Nanopowders of *Centella asiatica* was produced using planetary ball mill by varying milling parameters such as milling time, mass concentration, and bead amount. Particle size analysis employing photon correlation spectroscopy was carried out to record the effect of milling parameters on the particle size produced. The morphology of milled powders was also analyzed using a field emission scanning electron microscope. The bioactive component, asiatic acid, was extracted from various sizes of *C. asiatica* powders, and its extraction yield at different powder size was calculated by high-performance liquid chromatography. Optimization of milling parameters was found to be a crucial step in determining the content of asiatic acid extracted. In this study, the highest amount of asiatic acid extracted was 25.4 mg/g, obtained at moderate conditions of the following milling parameters: 4 h of milling time, 1.2% (*w/v*) concentration of powder, and 25 g of bead load.

Keywords: Nanopowders; *Centella asiatica*; Asiaticoside; Asiatic acid; Planetary ball mill

Background

Centella asiatica is a perennial plant that mostly grows in the tropical and subtropical climate. It is known as *mandukarpani* (Indian), *gotu kala* (Chinese), or pennywort (English). It is commonly used as a wound-healing agent and for treatment of mental disorders and skin cancer. Studies also show that the herb extract could increase collagen production and strengthen tensile strength [1]. Thus, this herb is widely used in formulations of cosmetic and pharmaceutical products and possesses high commercial value. All these therapeutic effects are mainly contributed by the triterpene glycosides, mainly asiatic acid and asiaticoside (Figure 1). Asiatic acid, a pentacyclic triterpene, is the aglycone form of asiaticoside which consists of a sugar moiety: a trisaccharide containing α-L-rhamnopyranoside linked to the disaccharide gentiobiose [2]. Although the pharmacological effect of asiaticoside has been well known, some data suggest that asiatic acid is the actual component responsible for the plant's therapeutic effects since asiaticoside is being hydrolyzed to asiatic acid *in vivo* [3-5]. Thus, asiatic acid can be considered as the bioactive compound of *C. asiatica*. [4]. Recent studies showed that asiatic acid could treat liver fibrosis [6], protect primary neurons against C2-ceramide-induced apoptosis [7], inhibit acetylcholinesterase [8], and induce neurotoxicity. Furthermore, modification of asiatic acid functional groups have been carried out with the aim of improving the compound's potential as a hepatoprotective agent [9,10].

Natural products have been important sources of drugs and will continue to play an important role as a major source of new drugs in the years to come [11-13]. Normally, bioactive components exist in a low concentration; thus, an effective method is urgently needed to retrieve the analyte from their cell matrix. Many methods have been introduced such as ultrasonic extraction (UE), microwave-assisted extraction, accelerated solvent extraction, pressurized hot water extraction [14-16], and supercritical/subcritical extraction [17,18]. However, the extraction efficiency of each method is greatly dependent on the particle size of plant material itself [19,20]. Among the methods, UE has been known for their extraction efficiency with reduced processing time and solvent consumption [14].

Ball milling is a common method in the size reduction of material. The milling process produces nanopowders

* Correspondence: mzborhan@yahoo.com; saifolabdullah@yahoo.co.uk
[1]NANO-SciTech Centre (NST), Institute of Science, Shah Alam 40450, Selangor, Malaysia
[2]NANO-Innovation Centre (Nano-IC), Faculty of Applied Sciences, Shah Alam 40450, Selangor, Malaysia
Full list of author information is available at the end of the article

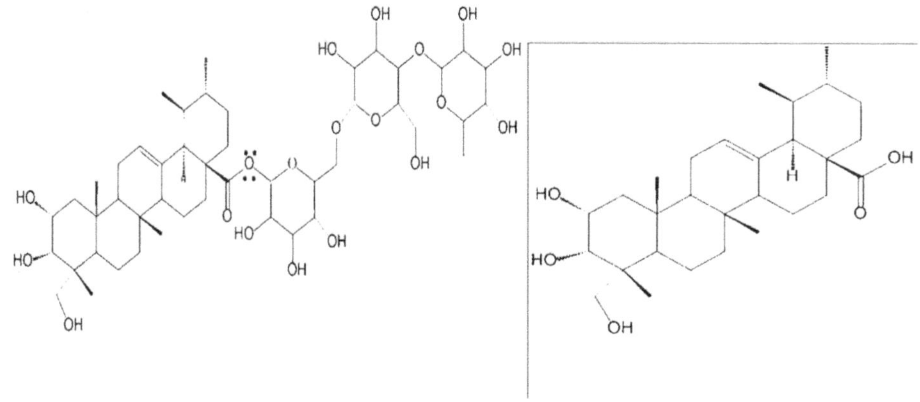

Figure 1 **Structure of asiaticoside (a) and asiatic acid (b).**

through the impact forces generated by action of centrifugal forces. Milling parameters such as milling time, mass concentration, and bead amount are important parameters that need to be considered in producing superfine powders. In this research, our aim is to investigate the effect of particle size on the amount of asiatic acid extracted employing high-performance liquid chromatography (HPLC).

Results and discussion

Optimization of milling parameters

To obtain a uniform distribution of particle size in the nanometer range, milling parameters such as milling time, mass concentration, and bead amount were varied. The results as shown in Figure 2 indicate that each set of milling condition produced different particle sizes of powder. Varying milling time, mass concentration, and bead amount from 2 to 8 h, 0.4% to 2.8% (*w*/*v*), and 5 to 100 g varied the particle size from 408.8 to 191.1, 237.1 to 581.1, and 442.6 to 212.8 nm, respectively. Statistically, increasing milling time from 2 to 4 h significantly reduced the particle size since sufficient energy was provided for size reduction. However, no significant reduction in particle size was observed after an extended milling time of 4 to 8 h. This is due to the fibrous structure of plant material itself which limited further size reduction [21]. An increase of particle size from 237.1 to 581.13 nm was observed as mass concentration is increased from 0.4% to 2.8% (*w*/*v*). Increase of mass concentration led to inefficient energy distribution in the milling process caused by the weak interaction between bead media and particles [22]. At a higher mass concentration, the viscosities of slurries increase and the distance between individual particles is reduced leading to ineffective particle capture by the grinding beads. The higher viscosities will then weaken the motion of grinding media; thus, velocity of grinding media and kinetic energy of bead decreased tremendously leading

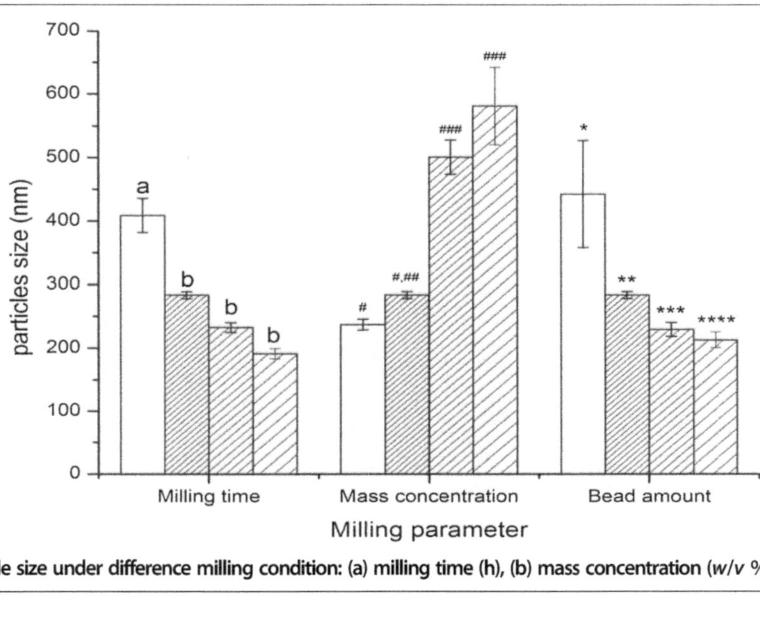

Figure 2 **Variation of particle size under difference milling condition: (a) milling time (h), (b) mass concentration (*w*/*v* %) and (c) bead amount (g).**

to ineffective milling operation [23]. The study on the effect of bead amount on particle size showed an inverse correlation. The smallest particle size (212.8 nm) was achieved with the highest amount (100 g) of bead load, and the largest particles size (442.6 nm) was found with the lowest bead amount (5 g). This can be understood by the limited amount of beads present in the reaction. With a low amount of bead load, the limiting reactant is the bead itself since the low numbers of bead present in the physical reaction will cause a reduction in the collision frequency due to limited amounts of bead per mass powder and reduced energy efficiency. Similarly, with higher bead amount, the energy efficiency would be higher, as reported by He et al. [22].

Microstructures of ground powder

As shown in Figure 3, the microstructures of coarse and milled powders showed a huge reduction of the size of nanopowders from the micropowder. It proves that the impact of collision in the milling process has tremendously broken down the micropowder to nanosize powder. The nanopowders exist in random structure in an agglomerated form. This is common since a higher surface energy possessed by the nanoparticles caused them to aggregate once it goes into a solid-state condition [24].

Effect of milling parameters on extraction yield of asiatic acid

To evaluate the amount of asiatic acid extracted, each powder from each set of milling condition was extracted and analyzed using HPLC. From the HPLC chromatogram, asiaticoside was eluted for 8 min (t_R = 8 min), while asiatic acid had a t_R of 23 min. This method shows a good separation of the two compounds for both micro- and nanosamples (Figure 4). As shown in Figure 4, all milled samples showed no presence of asiaticoside, and we believe that asiaticoside have been hydrolyzed to asiatic acid during the milling process. As shown in Figure 5, increasing the milling time, mass concentration, and bead amount from 2 to 8 h, 0.4% to 2.8% (*w*/*v*), and 5 to 100 g during grinding significantly varied the amount of asiatic acid extracted from 4.5 to 25.4 mg/g, from 2.3 to 25.4 mg/g, and from 4.1 to 25.4 mg/g, respectively. The difference in the amount of asiatic acid extracted could be caused by the degree of

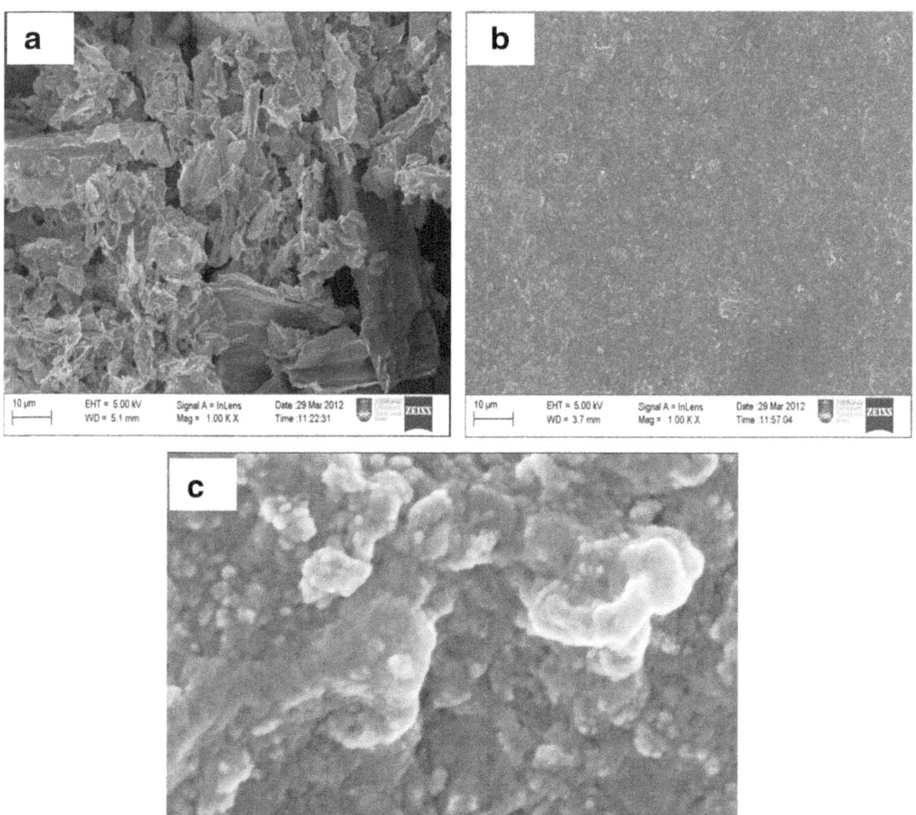

Figure 3 Microstructure of *C. asiatica* (a) micropowder (b) nanopowders, and (c) nanopowders at higher magnification.

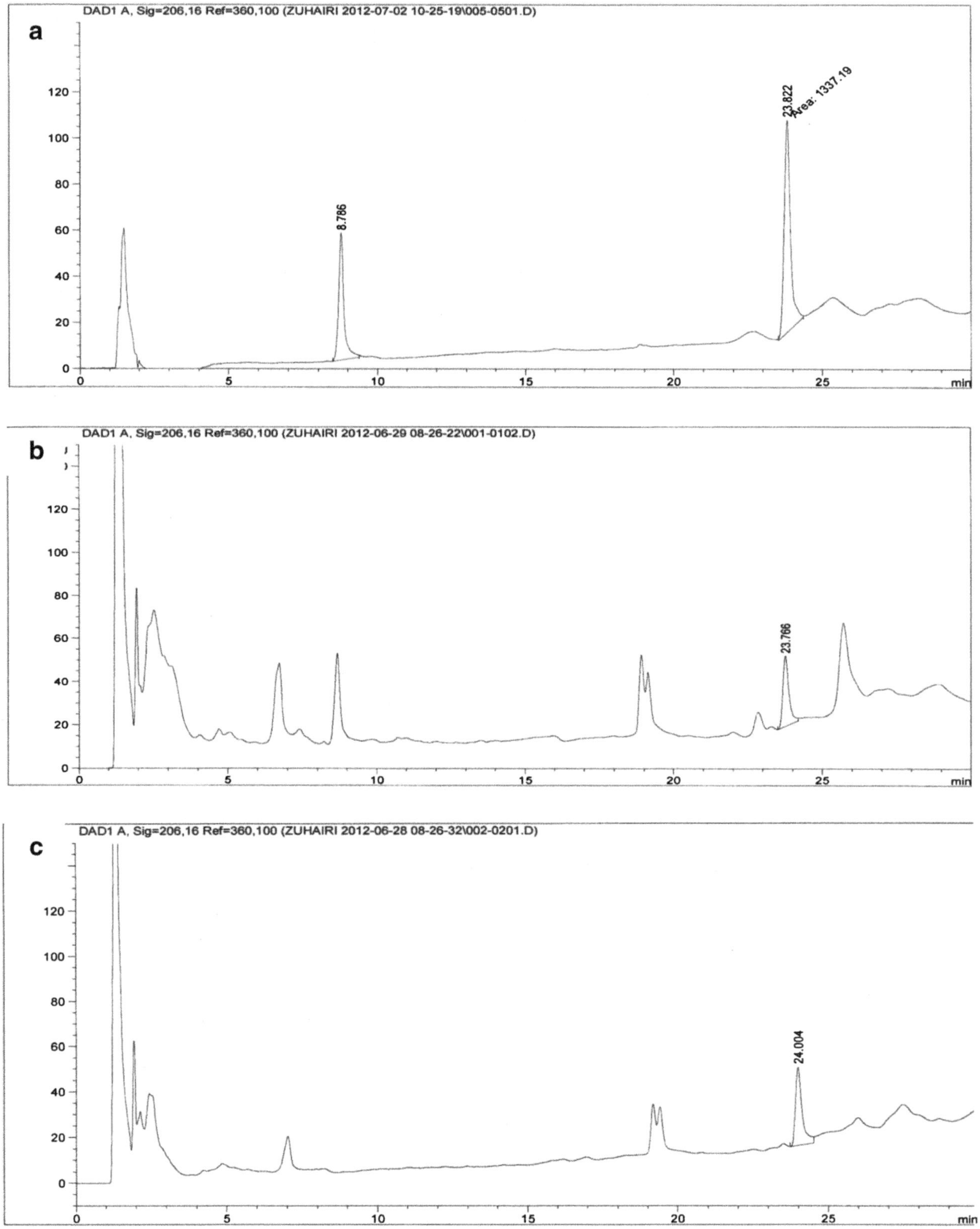

Figure 4 HPLC chromatograms of asiatic acid. (a) Standard sample, **(b)** asiatic acid in original powder, and **(c)** asiatic acid in nanopowders.

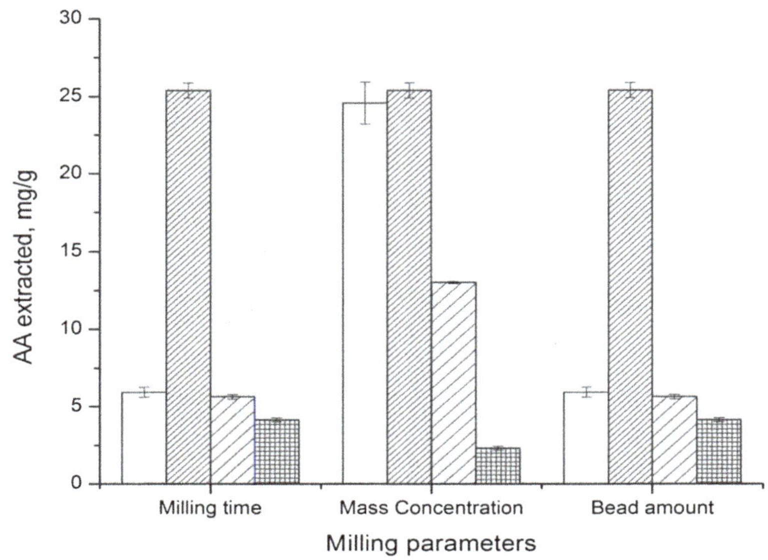

Figure 5 Variation of asiatic acid extracted at different milling conditions. Milling time (h), mass concentration (% (*w/v*)), and bead amount (g).

efficiency of extraction due to the difference in particle size of the powder and the ability of asiatic acid to retain their structure during the milling process. Each set of milling time (2, 4, 6, and 8 h) showed a reduction of particles to nanosize range. We further observed that a shorter milling time of 2 h produced a lower yield of asiatic acid content compared to 4 h of milling which could be explained by the difference in particle size of the powder as shown in Figure 2. Interestingly, after 6 h of milling, the content of asiatic acid reduced tremendously as compared to after 4 h of milling, and the amount is further reduced after 8 h of milling. At this time, further nanosize reduction has occurred, and the energy the milling process may have degraded the asiatic acid and consequently reduce the amount of asiatic acid extracted. Kormin also reported that the asiatic acid content in *C. asiatica* juice extract reduced significantly at higher temperatures [25]. Other findings have also showed that heat is an important factor that should be considered in extracting bioactive compound from plants. Effect of heat in the extraction of saponins from ginseng was clearly significant as the level of ginsenosides Rg1, Re, Rb1, Rc, R2, Rb3, and Rd decreased after the steaming process [26]. The same trend was also observed in experiments that investigate the effect of bead amounts on asiatic acid content. Effect of the nanopowders in

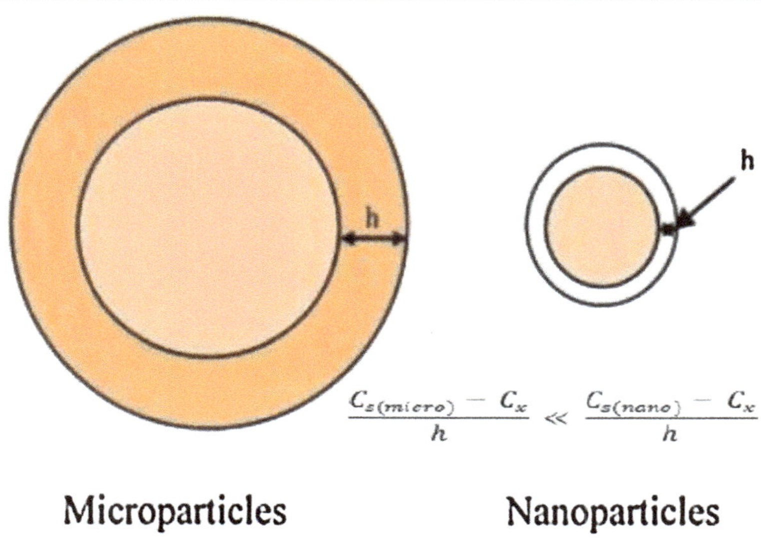

$$\frac{C_{s(micro)} - C_x}{h} \ll \frac{C_{s(nano)} - C_x}{h}$$

Figure 6 **Distance of diffusion layer in microparticles and nanoparticles.**

improving extraction of asiatic acid can be clearly observed by varying mass concentrations. The parameter values range from 0.4% to 2.8% (*w*/*v*) because our preliminary study showed that at this range, nanopowders range from 100 to 1,000 nm could be produced. Above these values, the milled product produced a greater size of particle (microrange). In this experiment, samples that were ground at 0.4% and 1.2% (*w*/*v*) showed a high amount of asiatic acid (24.5 and 25.4 mg/g) extracted due to increased extraction efficiency caused by smaller particle size that resulted from efficient energy distribution in the milling process. When milled at a higher concentration (2.0% (*w*/*v*)), the asiatic acid content was reduced to 13 mg/g, and further milling at 2.8% (*w*/*v*) caused a further drop in the extracted asiatic acid content (2.3 mg/g). The sudden drop in the amount of extracted asiatic acid amount was probably due a lower extraction efficiency caused by a larger particle size (500.7 and 581.7 nm). In this study, optimized milling parameters for the extraction of asiatic acid from *C. asiatica* is 4 h of milling time, 1.2% (*w*/*v*) mass concentration, and 25 g of bead load. To our knowledge, this method yields the highest amount of extracted asiatic acid from the plant compared to other methods previously reported which ranges from 7 to 14 mg/g [15,17].

Extraction mechanism in nanopowders

Generally, extraction of plant material involved two physical phenomena: diffusion through the cell wall and washing out the analyte into extraction solvent. Using nanopowders in the extraction of bioactive compound uses the same principles as using micron powders, but the only difference lies in the particle size used in the extraction. In this section, we discuss how the unique properties of nanopowders results in higher extraction yields.

Size reduction prior to extraction has been reported to be a crucial step to obtain from the plant matrix [15,19,27,28]. The reduction in size is reported to enhance the contact ratio between plant cell and the surrounding solvent [28]. In our study, the higher amount of asiatic acid extracted from milled samples can be explained by the amazing properties of nanoparticles contributed by an increase in surface area and the reduction of diffusional distance of the particles (Figure 6) [29]. Decreasing the particle size to the nanometer range caused an increase of surface area, leading to a higher number of cells directly exposed to the extraction by the solvent [28]. Based on Noyes-Whitney equation [30], the dissolution velocity of particles increases due to surface area enlargement as described in Equation 1:

$$\frac{\mathrm{dc}}{\mathrm{dt}} = DA\frac{C_s - C_x}{x}, \quad (1)$$

where dc/dt is the rate of dissolution velocity, D is the diffusion coefficient, A is the surface area of the solid, C_s is the concentration of the solid in the diffusion layer surrounding the solid, and C_x is the concentration of the solid in the bulk dissolution medium. The dissolution velocity dc/dt described in the above equation also depends on the concentration gradient $(C_s - C_x)/h$ where h, the diffusional distance is reduced for smaller particles [31]. Due to the reduced h, the concentration gradient of nanoparticles is large (Figure 5) [29,32] which causes the dissolution velocity to increase. This will enhance the diffusion of solvent into the cell wall, which in turn will allow the rapid release of active ingredients from the cell matrix of the plant. This is also supported by the phenomena where the disrupted cell wall structure of the plant material itself cause the active compound in the inner cytoplasm to be exposed directly while not being released from cell walls making effective ingredients dissolve out rapidly [33]. Past studies have shown that nanonization will allow the rapid release of active component from their cytoplasm matrix [34-36] and help improve the extraction of bioactive components of plants.

Conclusions

In general, we have provided further proof that nanonization of plant material played an important role in contributing to the extraction yield of bioactive compounds from plant materials. In this research, we have successfully produced nanopowders of *C. asiatica* by planetary ball milling. Our findings showed that a smaller particle size gave a higher extraction yield of asiatic acid compared to a bigger particle size. However, if the particle size is too small, the extraction yield of the compound will be reduced due to the effect of heat and pressure in the milling jar. As a conclusion, optimized milling parameters for high extraction yield of asiatic acid from *C. asiatica* was found to be 4 h of milling time, 1.2% (*w*/*v*) concentration of powder, and 25 g of bead load.

Methods

Chemicals and materials

Asiaticoside and asiatic acid standards (Biopurify Chegdu, Sichuan, China) were purchased from Sigma-Aldrich Corporation (St. Louis, MO, USA), acetonitrile (LC grade) was purchased from Fisher Chemicals, and methanol (HPLC grade) was purchased from Friedemann Schmidt (Berlin, Germany). Distilled water was purified using a Milli-Q water purification system (Millipore, Bedford, MA, USA).

Sample preparation

The plant material was washed with running tap water and rinsed with distilled water to remove dirt and contaminants. The clean plant material was stored in an oven at 50°C and was left to dry for 3 days. The dried plant material was ground using a conventional rotor mixer. The ground powder was then sieved through a 250-μm sieve and stored at 4°C prior to the ball milling process.

Optimization of ball milling parameters

Ball milling is a common mechanical process to produce superfine powders. In this research, a planetary ball mill was chosen as a grinding tool due to its simplicity. The sieved powder was first mixed with Pluronic F127 and milled at different mass concentrations (0.4%, 1.2%, 2.0%, and 2.8% (*w*/*v*)) at a constant milling time (4 h) and fixed bead amount (25 g). The resulting milling product was freeze-dried and analyzed for their physical characteristics. The best condition of milling (% (*w*/*v*)) was determined by the size of powder yielding the highest amount of asiatic acid extracted. This condition was then applied at different milling times (2, 4, 6, and 8 h) and different bead amounts (5, 25, 50, and 100 g), and the sample was analyzed for their physical characteristic and amount of asiatic acid extracted.

Particles size analysis

Prior to the measurement, nanosuspensions of each sample were taken and adjusted to 0.01% (*w*/*v*) and introduced into a disposable cuvette to be measured in triplicates. *z*-Average and polydispersity were recorded. Field emission scanning electron microscopy (FESEM; SMT SUPRA 40VP, Carl Zeiss AG, Oberkochen, Germany) imaging was carried out by placing 10 μL of the nanosuspension onto a glass slide and stored in an electronic desiccator (temperature 20°C, humidity 18% RH) for drying purposes. The dried samples were then coated with gold (approximately 10-nm thick) and were placed onto an adhesive tape on the FESEM stub.

HPLC analysis

Extraction

Coarse powder (0.5 g) was weighed and sonicated for 1 h using an ultrasonic processor (UP400S Ultrasonic Processor, Hielscher Ultrasonics GmbH, Teltow, Germany) at 60% amplitude and 1 cycle. The solids were removed by gravitational filtration, and the filtrate was collected. The nanopowders were also extracted in the same manner as above. The extract was dissolved in methanol (1 mg/mL) before being subjected to HPLC analysis.

HPLC conditions

The chromatographic separation was based on a modified method of Inamdar et al. [37] using a Zorbax Eclipse XDB-C18 4.6 × 150 mm (Agilent Technologies, Inc., Santa Clara, CA, USA), 5 μm at 25°C. The mobile phase, solvent A (ultrapure water) and solvent B (acetonitrile), was delivered at a flow rate of 1.0 mL min^{-1}. Gradient elution was employed with the ratio of A/B and varied as follows: 0 min, 80:20 and 35 min, 45:55. Standards or samples were introduced into the HPLC using an Agilent 1200 G1367B auto sampler, injection volume was 40 μL, and the detection wavelength was 206 nm. The amount of asiatic acid was determined using a standard calibration curves.

Conflict of interest

Hereby I, Muhammad Zuhairi Bin Borhan declare that as the author of the manuscript, have or do not have a direct financial relation with the commercial identities mentioned in my paper that might lead to a conflict of interest for any of the authors.

Authors' contributions

MZBB: Mainly involve in all laboratory works; Experimental Design, Sample preparation, Analysis sample; Mainly involve in preparation of manuscript. RA: Assisting in all laboratory works; Extraction of compound, HPLC Analysis; Assisting in preparation of manuscript. MR: Assisting in all laboratory works; FESEM Analysis, Particle Size Analysis; Assisting in preparation of manuscript. SA: Assisting in all laboratory works; FESEM Analysis, Particle Size Analysis, HPLC Analysis; Assisting in preparation of manuscript.

Acknowledgements

We would like to thank Universiti Teknologi MARA (UiTM) for the financial support.

Author details

[1]NANO-SciTech Centre (NST), Institute of Science, Shah Alam 40450, Selangor, Malaysia. [2]NANO-Innovation Centre (Nano-IC), Faculty of Applied Sciences, Shah Alam 40450, Selangor, Malaysia. [3]Faculty of Applied Sciences, Universiti Teknologi MARA (UiTM) Malaysia, Shah Alam 40450, Selangor, Malaysia.

References

1. Shukla, A, Rasik, AM, Jain, GK, Shankar, R, Kulshrestha, DK, Dhawan, BN: *In vitro* and *in vivo* wound healing activity of asiaticoside isolated from *Centella asiatica*. J Ethnopharmacol. **65**, 1–11 (1999)
2. Monti, D, Candido, A, Cruz Silva, MM, Křen, V, Riva, S, Danieli, B: Biocatalyzed generation of molecular diversity: selective modification of the saponin asiaticoside. Adv. Synth. Catal. **347**, 1168–1174 (2005)
3. Shim, PJ, Park, JH, Chang, MS, Lim, MJ, Kim, DH, Jung, YH, Jew, SS, Park, EH, Kim, HD: Asiaticoside mimetics as wound healing agent. Bioorg. Med. Chem. Let. **6**, 2937–2940 (1996)
4. Rush, WR, Murray, GR, Graham, DJM: The comparative steady-state bioavailability of the active ingredients of Madecassol. Eur. J. Drug Metab. Pharmacokinet. **18**, 323–326 (1993)
5. Zheng, XC, Wang, SH: Determination of asiatic acid in beagle dog plasma after oral administration of *Centella asiatica* extract by precolumn derivatization RP-HPLC. J. Chromatogr. B Analyt. Technol. Biomed. Life Sci. **877**, 477–481 (2009)
6. Tang, L, He, R, Yang, G, Tan, J, Zhou, L, Meng, X, Huang, XR, Lan, HY: Asiatic acid inhibits liver fibrosis by blocking TGF-beta/smad signaling *in vivo* and *in vitro*. PLoS ONE **7**(2), e31350 (2012)
7. Park, BC, Bosire, KO, Lee, ES, Lee, YS, Kim, JA: Asiatic acid induces apoptosis in SK-MEL-2 human melanoma cells. Cancer Lett. **218**, 81–90 (2005)

8. Nasir, MN, Abdullah, J, Habsah, M, Ghani, RI, Rammes, G: Inhibitory effect of asiatic acid on acetylcholinesterase, excitatory post synaptic potential and locomotor activity. Phytomedicine **19**, 311–316 (2012)
9. Jeong, BS, Mi, KL, Young, CK, Lee, ES: Modification of C2 functional group on asiatic acid and the evaluation of hepatoprotective effects. Arch. Pharm. Res. **30**, 282–289 (2007)
10. Jeong, BS, Young, CK, Lee, ES: Modification of C2,3,23,28 functional groups on asiatic acid and evaluation of hepatoprotective effects. Bull. Korean Chem. Soc. **28**, 977–982 (2007)
11. Sticher, O: Natural product isolation. Nat. Prod. Rep. **25**, 517 (2008)
12. Choi, JY, Yoo, JY, Kwak, HS, Nam, BU, Lee, J: Role of polymeric stabilizers for drug nanocrystal dispersions. Curr. Appl. Phys. **5**, 472–474 (2005)
13. Newman, DJ, Cragg, GM: Natural products as sources of new drugs over the 30 years from 1981 to 2010. J. Nat. Prod. **75**, 311–335 (2012)
14. Shirsath, SR, Sonawane, SH, Gogate, PR: Intensification of extraction of natural products using ultrasonic irradiations—a review of current status. Chem. Eng. Process. Process Intensif. **53**, 10–23 (2012)
15. Shen, Y, Liu, A, Ye, M, Wang, L, Chen, J, Wang, XR, Han, C: Analysis of biologically active constituents in *Centella asiatica* by microwave-assisted extraction combined with LC-MS. Chromatographia **70**, 431–438 (2009)
16. Teo, CC, Tan, SN, Yong, JWH, Hew, CS, Ong, ES: Pressurized hot water extraction (PHWE). J. Chromatogr. A **1217**, 2484–2494 (2010)
17. Kim, WJ, Kim, J, Veriansyah, B, Kim, JD, Lee, YW, Oh, SG, Tjandrawinata, RR: Extraction of bioactive components from *Centella asiatica* using subcritical water. J. Supercrit. Fluids **48**, 211–216 (2009)
18. Kim, W-J, Veriansyah, B, Lee, Y-W, Kim, J, Kim, J-D: Extraction of mangiferin from *Mahkota Dewa* (*Phaleria macrocarpa*) using subcritical water. J. Ind. Eng. Chem. **16**, 425–430 (2010)
19. Wu, J, Lin, L, Chau, FT: Ultrasound-assisted extraction of ginseng saponins from ginseng roots and cultured ginseng cells. Ultrason. Sonochem. **8**, 347–352 (2001)
20. Mustafa, A, Turner, C: Pressurized liquid extraction as a green approach in food and herbal plants extraction: a review. Anal. Chim. Acta. **703**, 8–18 (2011)
21. Rajkhowa, R, Wang, L, Kanwar, J, Wang, X: Fabrication of ultrafine powder from eri silk through attritor and jet milling. Powder Technol. **191**, 155–163 (2009)
22. He, M, Wang, Y, Forssberg, E: Parameter effects on wet ultrafine grinding of limestone through slurry rheology in a stirred media mill. Powder Technol. **161**, 10–21 (2006)
23. He, M, Forssberg, E: Influence of slurry rheology on stirred media milling of quartzite. Int. J. Miner. Process. **84**, 240–251 (2007)
24. Lee, J: Drug nano- and microparticles processed into solid dosage forms: physical properties. J. Pharm. Sci. **92**, 2057–2068 (2003)
25. Kormin, S: The effect of heat processing on triterpenes glycosides and antioxidant activity of herbal Pegaga (*Centella asiatica* L. Urban) drink. Master's thesis. Universiti Teknologi Malaysia (2005)
26. Sun, S, Wang, CZ, Tong, R, Li, XL, Fishbein, A, Wang, Q, He, TC, Du, W, Yuan, CS: Effects of steaming the root of *Panax notoginseng* on chemical composition and anticancer activities. Food Chem. **118**, 307–314 (2010)
27. Sun, Y, Bi, J, Zhang, L, Ye, B: Ultrasound-assisted extraction of three bufadienolides from Chinese medicine ChanSu. Ultrason. Sonochem. **19**, 1150–1154 (2012)
28. Vinatoru, M: An overview of the ultrasonically assisted extraction of bioactive principles from herbs. Ultrason. Sonochem. **8**, 303–313 (2001)
29. Mauludin, R, Müller, RH, Keck, CM: Development of an oral rutin nanocrystal formulation. Int. J. Pharm. **370**, 202–209 (2009)
30. Noyes, AA, Whitney, WR: The rate of solution of solid substances in their own solutions. J. Am. Chem. Soc. **19**, 930–934 (1897)
31. Müller, RH, Peters, K: Nanosuspensions for the formulation of poorly soluble drugs: I: preparation by a size-reduction technique. Int. J. Pharm. **160**, 229–237 (1998)
32. Müller, RH, Böhm, BHL, Grau, MJ: Formulations for poorly soluble drugs with poor bioavailability/2nd communication: stability, biopharmaceutical aspects, possible drug forms and registration aspects. Nanosuspensionen–Formulierungen fur schwerlosliche arzneistoffe mit geringer bioverfugbarkeit: 2 Mitteilung: Stabilitat, biopharmazeutische aspekte, mogliche arzneiformen und zulassungsfragen **61**, 175–178 (1999)
33. Su, YL, Fu, ZY, Wang, WM, Wang, H, Wang, YC, Zhang, JY, Ma, PY: **Ultrafine grinding of Radix salvia miltiorrhiza particles and the physicochemical properties by high speed centrifugal sheering.** In: *Book Ultrafine grinding of Radix salvia miltiorrhiza particles and the physicochemical properties by high speed centrifugal sheering* (Editor ed.^eds.), vol. 330-332 I. pp. 215-218. City; 2007:215-218
34. Su, YL, Fu, ZY, Quan, CJ, Wang, WM: Fabrication of nano *Rhizama Chuanxiong* particles and determination of tetramethylpyrazine. Trans. Nonferrous Metals Soc. China **16**, S393–S397 (2006)
35. Ma, PY, Fu, ZY, Su, YL, Ma, JJ: The nano pulverization of traditional Chinese medicine Liuwei Dihuang. J. Wuhan University of Tech-Mat. Sci. Ed. **21**, 105–108 (2006)
36. Ma, PY, Fu, ZY, Su, YL, Zhang, JY, Wang, WM, Wang, H, Wang, YC, Zhang, QJ: Modification of physicochemical and medicinal characterization of Liuwei Dihuang particles by ultrafine grinding. Powder Technol. **191**, 194–199 (2009)
37. Inamdar, PK, Yeole, RD, Ghogare, AB, De Souza, NJ: Determination of biologically active constituents in *Centella asiatica*. J. Chromatogr. A **742**, 127–130 (1996)

An experimental investigation of the enhanced oil recovery and improved performance of drilling fluids using titanium dioxide and fumed silica nanoparticles

Goshtasp Cheraghian[1], Mahmood Hemmati[2*], Mohsen Masihi[3] and Saeed Bazgir[4]

Abstract

Nanotechnology has contributed to the technological advances in various industrial biomaterials and renewable energy production over the last decade. Recently, a renewed interest arises in the application of nanotechnology for the upstream petroleum industry such as exploration, drilling and production. In particular, the adding of nanoparticles to injection fluids may drastically benefit enhanced oil recovery, such as changing the properties of the fluid, wettability alternation of rocks, advanced drag reduction, strengthening sand consolidation, reducing the interfacial tension and increasing the mobility of the capillary-trapped oil. The feasibility of these methods depends on many factors such as flow mechanisms in porous media and porous medium properties at microscopic and macroscopic scales. Previous studies have indicated that the oil recovery from porous media may be substantially increased by the injection of miscible fluids. This all sounds great and waterflooding has been used successfully for decades; however, it is important to carefully design and appropriately operate the waterflood. Using nanoparticles in all samples has resulted in recovery increase. Among these applications of the study is nanoenhanced oil recovery which can be applied in many water-wet reservoirs dominated by inhibition mechanism to extract more fluid through really small caliber pores. In these experiments, two nanoparticles dissolved in water are injected into simulated environment, and also, the effect of these nanoparticles in water-base drilling typical fluid have been investigated. Using nanoparticles in all samples has resulted in recovery increase. Finally, considering the experiments, it is demonstrated that flows with nanoparticles and, in particular, titanium dioxide (TiO_2) nanoparticles have the highest amount of recovery factors and thus using nanoparticles in waterflooding projects and even some in polymer flooding ones. Also, results of the other tests, regarding each typical drilling costs of each foot and importance of time in the operation, it is possible to replace technically and economically the ordinary addition (here, the widely used sodium hydroxide) with fumed silica nanoparticles in drilling fluid to prevent cement contamination of the drilling fluid. The advantages of nano-TiO_2 are possessing suitable thermal transition qualities in the drilling fluid.

Keywords: Nanoenhanced oil recovery (NEOR); Drilling fluid; Porous media; Inhibition mechanism

* Correspondence: hemmatim@ripi.ir
[2]Research Institute of Petroleum Industry (RIPI), Tehran, Iran
Full list of author information is available at the end of the article

Background

In the last few years, enhanced oil recovery (EOR) processes have regained interest from the research and development phases to the oil field EOR implementation. This renewed interest has been furthered by the currently high oil price, the increasing worldwide oil demand, the maturation of oil fields worldwide, and few new well discoveries (Aladasani and Bai 2010 in [1]). Hydrocarbon recovery occurs through two main processes: primary recovery and supplementary recovery. Primary recovery refers to the volume of hydrocarbon produced by the natural energy prevailing in the reservoir and/or artificial lift through a single wellbore, while supplementary or secondary recovery refers to the volume of hydrocarbon produced as a result of the addition of energy into the reservoir, such as fluid injection, to complement or increase the original energy within the reservoir (Dake 1978; Lyons and Plisga 2005 in [1]).

Customized nanoparticles have the ability to enhance oil recovery, improve exploration, and be useful in the formation scale control. Nanoparticles can be tailored to alter reservoir properties such as wettability, improve mobility ratio, or control formations migration. Nanofluids have successfully been developed in laboratories, and the upcoming challenge is to develop techniques for the cost of industrial-scale production of nanofluids. In nearly all cases, the thermal conductivity of conventional heat transfer of fluids is improved by the addition of small amounts of nanoparticles [2]. In addition, development of pressure and temperature-sensitive nanosensors will enable *in-situ* measurements within the reservoir [3]. Nanosensors can provide improved temperature and pressure ratings in deep wells and hostile environments [4].

Many novel enhanced oil recovery methods have been applied to increase oil recovery in oil fields [5,6]. The feasibility of these methods depends on many factors such as flow mechanisms in porous media and porous medium properties at microscopic and macroscopic scales have been reported the role of nanoparticles in EOR operations (Skauge et al. 2010 in [7]). As well, some researchers presented extensive literature reviews regarding the effect of wettability on microscopic displacements of fluid flow at pore scale (Zhao et al. 2010; Jamaloei and Kharrat, 2010 in [8-10]).

Suleimanov et al. (1995 in [11]) carried out experiments which showed how dispersed nanoparticels in an aqueous phase could modify the interfacial properties of a liquid/liquid system if their surface were modified by the presence of an ionic surfactant. The application of nanosuspension in their study permitted significant increase in the efficiency of oil displacement oil-water rate. In homogeneous porous media, oil recovery before water breakthrough was increased by 51% and 17% for surfactant aqueous solution with nanoparticle addition to water and surfactant aqueous solution, respectively [11].

Designed nanoparticles, especially nanocrystalline materials in combination with advanced drilling fluids, will probably improve the rate of penetration and decrease wear on drilling equipment significantly [12]. Nanobased mud additive is expected to improve the thermal conductivity of nanofluids, which consequently will provide more efficient cooling of drill bits, and longer operational cycles [13]. In addition, an additive in casing to increase compressive and exural strength, as well as light-weight, rugged structural materials was studied in the project thesis, The use of nanotechnology in the petroleum industry [4].

Polymer flooding is a tertiary recovery method by adding high-molecular-weight polyacrylamide into injected water to increase the viscosity of fluid, improve volumetric sweep efficiency, and further increase oil recovery factor. When oil is displaced by water, the oil/water mobility ratio is so high that the injected water fingers through the reservoirs. By injecting polymer solution into reservoirs, the oil/water mobility ratio can be much reduced, and the displacement front advances evenly to sweep a larger volume. The viscoelasticity of polymer solution can help the displacement of oil remaining in micropores that cannot be otherwise displaced by water flooding.

Previous studies have indicated that the oil recovery from porous media may be substantially increased by the injection of miscible fluids. This all sounds great and water flooding has been used successfully for decades; however, it is important to carefully design and appropriately operate waterflood. There are many factors to consider when designing a successful water flood including

- reservoir permeability (both absolute and relative)
- beginning and ending fluid saturations (oil, water and gas)
- reservoir heterogeneity
- oil gravity and viscosity
- water source and compatibility
- formation clay content
- depth and lifting costs

But if done right, a well run water flood will significantly nanoenhance oil recovery and produce attractive returns for many years. Waterfloods can also be improved by polymer flooding. Addition of polymer makes the water more viscous so that oil is produced faster. Obviously, this is not a good idea in a low-permeability reservoir or one with a high clay content that can adsorb the polymer. However, polymer-augmented waterfloods can be profitable. In this experiment, two nanoparticles titanium dioxide and fumed silica dissolved in water are

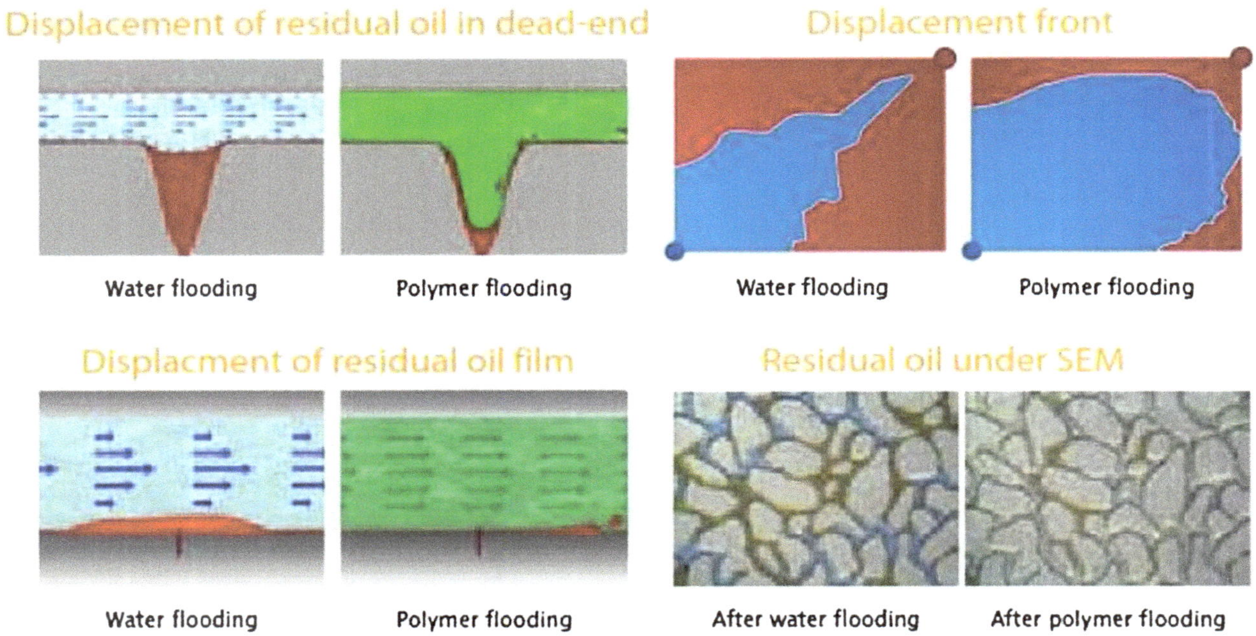

Figure 1 **The comparison of the displacement of oil by polymer flooding.**

injected into the simulated environment. Other additive matters that are used in making the compound are introduced in the next parts.

First, using a simulated reservoir, we prepare the container, from whose entrance we inject 100 psi air pressure and through the end there is an opening for flow exit. Here, also a metal net and a filter paper with the same cross sections are used. So, it is possible to simulate an environment like porous media.

Furthermore, it is tried in this paper to study the effect of two nanoparticles in water-base drilling-typical fluid. Specifically, the effect of this matter on the yield point (YP) and consistency index (**k**) and flow behavior index (**n**) was examined to improve cement-contaminated cement during drilling operation. Decreasing the flow behavior index by adding nanofumed silica in base fluid is among the results of the test. Cementing operation was of drilling steps employed several times to strengthen casing pipes. So, cement contaminations are avoidable in drilling. Silica (silicate mud) is a drilling fluid that matter used for specific purposes including being economic, lower biocontamination, better thermal

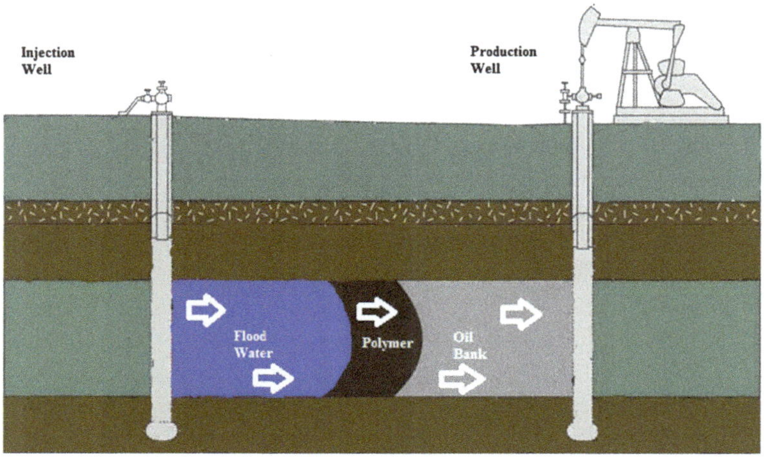

Figure 2 Polymer flooding enhanced oil recovery.

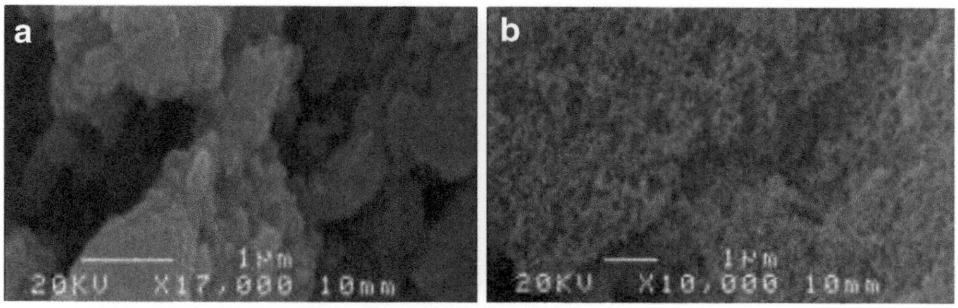

Figure 3 SEM micrographs of the pure oxides (a) TiO_2 and (b) fumed SiO_2.

and pressure stability, lubricating drilling pipes and shale prevention. Regarding each typical drilling costs of each foot and importance of time in the operation, it is possible to replace technically and economically the ordinary addition (here, the widely used sodium hydroxide) with fumed silica nanoparticles in drilling fluid to improve cement-contaminated drilling fluid. One of the nano-TiO_2 matter advantages is possessing suitable thermal transition qualities in drilling fluid. In drilling mud composition, there are a series of solid particles making electrochemical bounds with other liquid-phase matters. The resistance the matters show against cut is called yield point the unit; on the other hand, the point expresses viscosity and electrostatic forces between particles. In laminar flow, lines are parallel; mud moves in separate parallel layers with different velocities and passes through ring spaces or pipes. With the increase, flow velocities get closer in different layers. Displacement of mud is more difficult with slow flow because drilling particles tend to layers with low velocity and move in the layers and cause low velocity in the layers, their displacement gets harder. One way to avoid the problem is to increase the yield point.

The designating flow behavior index (**n**) indicates non-Newton behavior of a fluid, namely, Newton fluids assign n index of unity to themselves; the lower it is, the higher the mud capability in cleaning the well, namely, the wider the longitudinal profile of moving fluid and the lower the solids fall. Numerical reduction of n results in bending lapse of the moving fluid, and as a result, viscosity in the place of drill and drilling speed will increase. Consistency index indicates concentration

Table 1 Composition of the fluids

No. of fluid	Ingredient base fluid (wt.%)	Nanofume silica content (wt.%)	Nano-TiO_2 (wt.%)
1	Initial fluid (IF)	0	0
2	Initial fluid (IF)	0	1.4285
3	Initial fluid (IF)	1.4285	0
4	IF + Carboxymethyl cellulose (0.2857)	0	0
5	IF + Carboxymethyl cellulose (0.2857)	0	1.4285
6	IF + Carboxymethyl cellulose (0.2857)	1.4285	0
7	IF + Polyhachioides (0.2857)	0	0
8	IF + Polyhachioides (0.2857)	0	1.4285
9	IF + Polyhachioides (0.2857)	1.4285	0
11	IF + Sulfate solution (4)	0	0
12	IF + Sulfate solution (4)	0	1.4285
13	IF + Sulfate solution (4)	1.4285	0
14	IF + Cement (1.1428)	0	0
15	IF + Cement (1.1428)	0	1.4285
16	IF + Cement (1.1428)	1.4285	0
17	IF + Cement (1.1428%) + Sodium hydroxide (0.2857)	0	0
18	IF + Cement (1.1428%) + Sodium hydroxide (0.2857)	0	1.4285
19	IF + Cement (1.1428%) + Sodium hydroxide (0.2857)	1.4285	0

Table 2 Physical properties of fluids

No. of fluid	Viscosity (cp)	Conductivity (mV)	Mud cake diameter (mm)	Final gel strength	Initial gel strength	Temperature (°C)	Density of fluid (ppg)
1	9	−259	1.9	28	12	22	8.7
2	18	−213	2.71	31	25	22.5	8.7
3	14	−183	2.36	38	13	24	8.8
4	19	−219	1.6	38	18	23	8.6
5	24	−219	1.69	36	20	23	8.7
6	20.5	−161	1.39	52	12	23	8.7
7	11.5	−235	1.98	28	15	31	8.6
8	20	−212	2.56	45	29	22	8.7
9	16.5	−161	3.8	91	34	23	8.8
11	5.5	−177	0.64	6	3	20	8.7
12	5.5	−172	0.6	4	3	20	8.75
13	6	−165	1.31	9	3	20	8.88
14	43	−308	6.28	37	33	23	8.7
15	69	−323	12.48	59	42	23	8.8
16	39.5	−309	8.2	65	33	23	8.6
17	20	−315	9.74	18	12	25	8.7
18	61	−330	7.48	47	36	25	8.6
19	46.5	−337	9.12	73	51	26	8.6

or viscosity and pumping capability of drilling fluid; the higher the viscosity, the higher the k. In case the flow behavior index is low, cleaning the well gets harder and it is possible to increase viscosity gradually to enhance consistency in the index size. Consistency index (**k**) and flow behavior index change with flow chemical composition and temperature. However, considering relatively constant temperature of the flows under study, the only alternative is changing their chemical composition.

Results and discussion

The comparison of the displacement of oil by polymer flooding is shown in Figure 1, while enhanced oil recovery is shown in Figure 2.

In general, it is observed that for fluids in whose nanostructures or nanoparticles are used (Figure 3), YP is significantly higher than basic fluids. Regarding Tables 1 and 2, it can be seen that fluids YP having TiO_2 nanoparticles is higher than the other fluids, when sodium hydroxide is used in fluid. YP difference between nanofluids

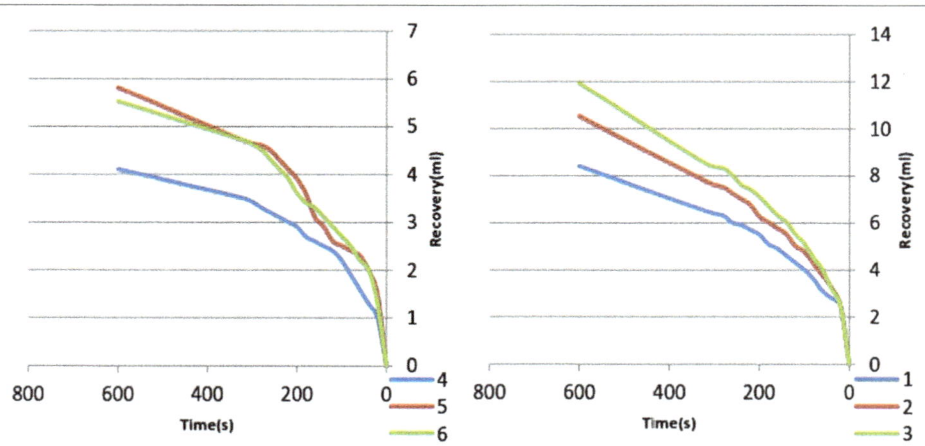

Figure 4 Recovery fluid vs. injection time. Regarding relatively fixed temperature and pH of all samples, we generally analyze them. Samples 1, 2 and 3 have basic fluid and samples 4, 5 and 6 have carboxymethyl cellulose.

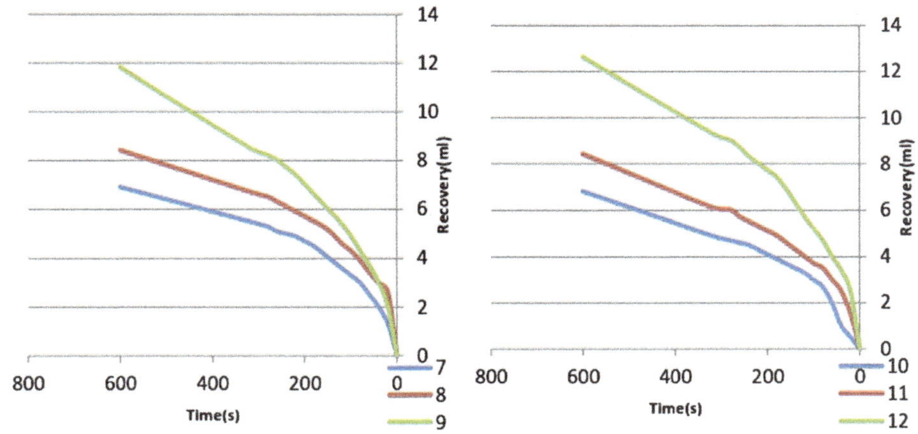

Figure 5 Recovery fluid vs. injection time. Regarding relatively fixed temperature and pH of all samples, we generally analyze that samples 7, 8 and 9 have polyrhachioides and samples 10, 11 and 12 also have 10% sulfate solution.

and basic fluids is considerably higher than the time the matter is not applied in the fluid structure. Also, sodium hydroxide results in higher YP fluid with fumed silica nanoparticles compared to TiO_2 nanoparticles with fluids. Among fluids, mud has a basic matter and TiO_2 nanoparticle has a maximum YP. Except fluid having fumed silica nanoparticle their bending tension grows with a constant rate, bending tension of fluids with sodium hydroxide increase lower than the time the matter present in the compositions. Meanwhile, basic fluid with nano-TiO_2 without sodium hydroxide has maximum slope of bending tension against bending rate and minimum amount is attributed to the basic fluid with sodium hydroxide relatively determined with a constant slope. Regarding fluids without sodium hydroxide, basic fluid with fumed silica nanoparticles and higher YP compared to the basic fluid,

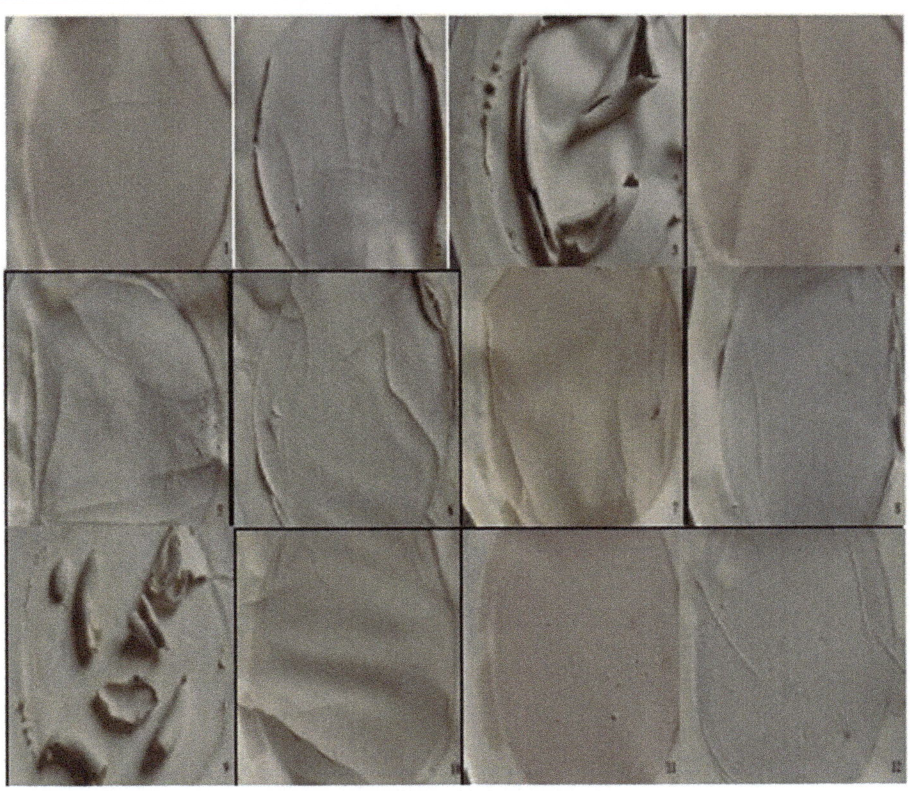

Figure 6 Effect of TiO_2 and fumed silica nanoparticles on the simulated porous media at depths.

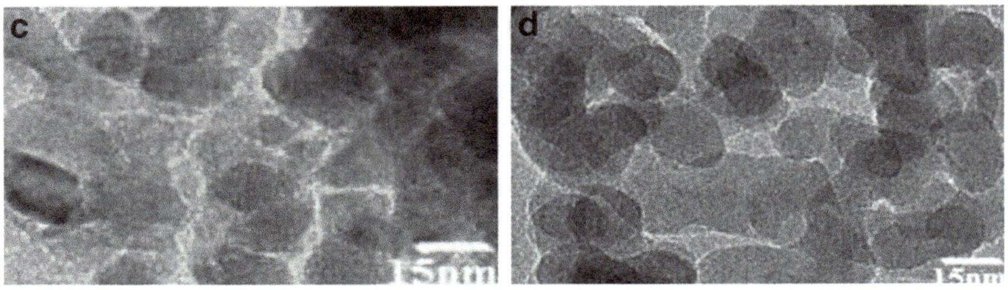

Figure 7 TEM bright field images of pure oxides (c) TiO_2 and (d) fumed SiO_2.

bending tension has no significant increase when bending rate increased, and in the end, its amount is lower than the basic fluid. On the other hand, by adding sodium hydroxide to basic fluid, bending tension increase regarding tension rate is much lower than its basic state. In fact, before two curves reach each other, at a fixed bending rate, mud has nanofumed silica and higher viscosity comparing to its basic state, while after increasing bending rate, the state changed for the fluids.

Like other industrial projects, feasibility of the samples are examined initially, then the results can be applied in field studies. Among the applications of this study is improvement in oil recovery which can be applied in many water wet reservoirs dominated by inhibition mechanism to extract more flow through really small caliber pores.

Conclusions

Using nanoparticles in all samples have resulted in recovery increase (Figures 4, 5, 6, 7). In all samples of flow with titanium dioxide nanoparticle, recovery is better except fluids with polyrhachioides and basic fluid, among which samples with fumed silica nanoparticles have the most recovery (Figure 8). Maximum recovery belongs to flows with sulfate and nano TiO_2. Considering the experiments, it is demonstrated that flows with nanoparticles and, in particular, the titanium dioxide nanoparticles have the highest amount of recovery and thus using nanoparticles in water flooding projects and even some of the polymer flooding ones (for IOR process).

In laminar flow, increasing particles motion tendency toward the lower velocity area and consequently their displacement needing YP increase is very significant; too much increase in YP in drilling fluids with TiO_2 and fumed silica nanoparticles is among positive points of adding the matters to basic fluids obviously seen in diagrams covering laminar flows problem in retrospective ring spaces greatly (Figures 9, 10).

Among all fluids under study, fluids with TiO_2 nanoparticles have more YP in both states compared to fluids with fumed silica nanoparticles, and in sum, YP was higher in TiO_2 nano-fluid.

It is observed that with increasing sodium hydroxide, numerical amount of **n** for TiO_2 nanoparticles drilling mud becomes higher than the other two. In sum, adding sodium hydroxide to the mud, **n** index decreases with a fixed, almost equal difference in nanoflows, and also maximum difference is seen in basic mud. The presence of fumed silica nanoparticles in particular with sodium hydroxide will have the minimum **n**. Increasing sodium hydroxide, **k** will also increase and in the mean time, fumed silica nanoparticles and sodium hydroxide flow

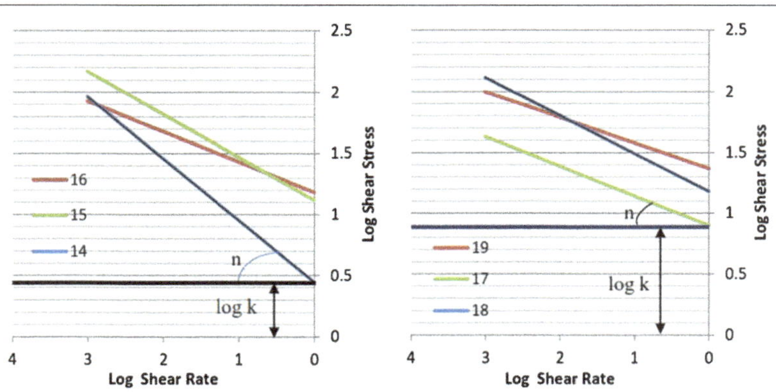

Figure 8 Flow behavior index and consistency index (improved cement-contaminated drilling fluid samples).

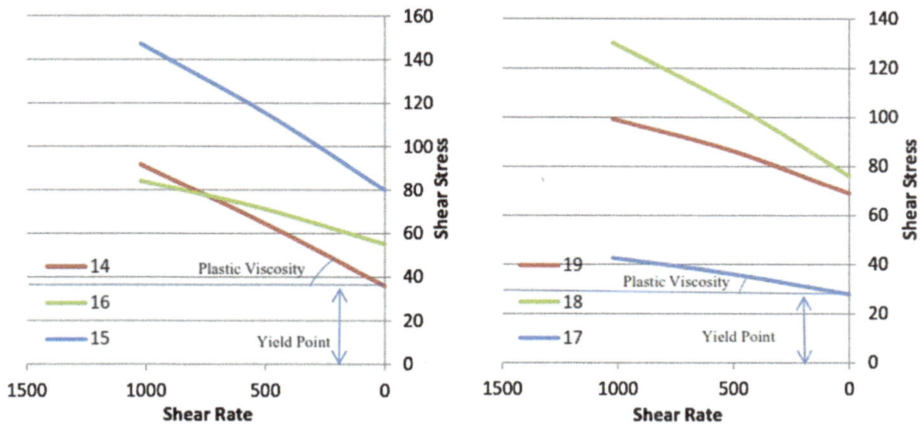

Figure 9 Yield Point and Plastic Viscosity (improved cement-contaminated drilling fluid samples).

will have higher n. Accordingly, it can cover low **n** in the flow (Figures 8). Additionally, although **n** is lower in fumed silica nanoparticle flows compared to flows with TiO_2 nanoparticles and sodium hydroxide, co-ordination index is equal. Using nanoparticles in drilling mud will result in coordination index increase in all samples (Figure 9).

Treating cement contamination of drilling flow by sodium hydroxide (Figure 10), it is observed that flows composed of fumed silica nanoparticles have minimum **n**, while when the particles are in cement-contaminated mud, numerical value of **n** equals the **n** value in flow treated by sodium hydroxide without any nanoparticles. In other words, it is possible to treat the n value of cement-contaminated mud by fumed silica nanoparticles as effective as with sodium hydroxide.

The flow with fumed silica nanoparticles and sodium hydroxide has minimum **n** and maximum **k** and maximum well-cleaning capability and results in high drilling rate by reducing viscosity at drill place.

Methods

In the first step, a certain amount of water is added to the compound and divided into eight plates. The first sample is considered as the base and left without change. To the second, fifth, eighth and eleventh samples, a certain amount of titanium dioxide is added and placed in the mixer. And fumed silica matter is added to samples 3, 6, 9 and 12 and then mixed to the extent that the mixture is homogenous. Then, each of the samples is separately located in pre-simulated environment. In 600s, the flow recovered from the device is registered in 20s intervals.

Figure 10 Effect of TiO_2 and fumed silica nano particles on the simulated drilling layers (drilling mud cake).

Competing interests

The authors declare that they have no competing interests.

Authors' contributions

GC has contributed to all experiments and has been supervised by MH who supervised the project and also he is corresponding author. GC has participated in the statistical analysis and in the preparation of the manuscript. All authors read and approved the final manuscript.

Acknowledgements

All planning, execution and discussion of laboratory experiments were done in cooperation with Sajjad Tardasti and supervisor Dr. M. Hemmati. I would therefore address a great thank you to S. Tardasti for the help, guidance and excellent collaboration throughout the whole Project.
Furthermore, the authors are grateful to the Department of Petroleum and Gas Engineering, Tehran Science and Research Branch, and Islamic Azad University (I.A.U.) of Iran for the financial support for this research.

Author details

[1]Department of Petroleum and Gas Engineering, Science and Research Branch, Islamic Azad University, Tehran, Iran. [2]Research Institute of Petroleum Industry (RIPI), Tehran, Iran. [3]Chemical and Petroleum Engineering Department, Sharif University of Technology, Tehran, Iran. [4]Department of Polymer Engineering, Science and Research Branch, Islamic Azad University, Tehran, Iran.

References

1. Romero-Zerón, L: Advances in enhanced oil recovery processes, introduction to enhanced oil recovery (EOR) processes and bioremediation of oil-contaminated sites. InTech, Canada. (2012). 978-953-51-0629-6
2. Kakac, S, Pramuanjaroenkij, A: Review of convective heat transfer enhancement with nanofluids. Int. J. Heat Mass Transf **52**, 3187–3196 (2009). 26 March
3. Alaskar, M, Ames, M, Connor, S, Liu, C, Cui, Y, Li, K, Horne, R: Nanoparticle and microparticle flow in porous and fractured media: an experimental study. SPE Annual Technical Conference and Exhibition held in Denver, Colorado, USA. (2011). 30 October - 2 November
4. Engeset, B: Project thesis - the use of nanotechnology in the petroleum industry. Thesis, Norwegian University of Science and Technology (2011)
5. Dong, M, Ma, S, Liu, Q: Enhanced heavy oil recovery through interfacial instability: a study of chemical flooding for Brintnell heavy oil. J. Fuel. **88**(6), 1049–1056 (2009)
6. Hongyan W, Xulong C, Jichao Z, Zhang A: Development and application of dilute surfactant-polymer flooding system for Shengli oilfield. J. Pet. Sci. Eng. **65**(1–2), 45–50 (2010)
7. Wu, J, He, J, Torsæter, O, Zhang, Z: Effect of nanoparticles on oil-water flow in a confined nanochannel: a molecular dynamics study. In: SPE International Oilfield Nanotechnology Conference, 12-14 June, Noordwijk. (2012). SPE 156995
8. Yadali Jamaloei, B, Kharrat, R: The performance evaluation of viscous-modified surfactant waterflooding in heavy oil reservoirs at varying salinity of injected polymer-contained surfactant solution. Iran. J. Chem. Chem. Eng **31**(1), 99–111 (2012)
9. Yadali Jamaloei, B, Kharrat, R: Analysis of microscopic displacement mechanisms of dilute surfactant flooding in oil-wet and water-wet porous media. Transp. Porous Media **81**(1), 1–19 (2010)
10. Yadali Jamaloei, B, Kharrat, R: The influence of pore geometry on flow instability and microscale displacement mechanisms of dilute surfactant flooding in mixed-wet porous media. J. Porous Media **14**(2), 91 (2011)
11. Suleimanov, B, Ismalov, F, Veliyev, E: Nanofluid for enhanced oil recovery. J. Pet. Sci. Eng. **78**, 431 (2011)
12. Krishnamoortu, R: Extracting the benefits of nanotechnology for the oil industry. JPT, Society of Petroleum Engineers. 11, 24–25 (2006). November 2006
13. Amanullah, M, Al-Arfaj, M, Al-Abdullatif, Z: Preliminary test results of nano-based drilling fluid for oil and gas field application, pp. 1–3. SPE/IADC Drilling Conference and Exhibition, Amsterdam, The Netherlands (2011)

Study of influential factors in synthesis and characterization of cobalt oxide nanoparticles

Ghazaleh Allaedini* and Abubakar Muhammad

Abstract

This study has investigated the different factors in obtaining cobalt oxide nanoparticles. The factors that have been studied are the effect of pH, synthesis temperature, capping agent, annealing temperature, and different usage of cobalt salts. Cobalt oxide nanoparticles were characterized by scanning electron microscopy, energy dispersive X-ray spectroscopy, UV, Fourier transform infrared spectroscopy, and X-ray diffraction. In order to find the effects of various factors, the optimum pH was achieved in the first experiment by the precipitation method; after finding out that in experiments with pH 8 to 9, the particles are homogeneous in shape and regular. The other experiments were repeated with the same pH but changing the factors. In the second experiment, the effect of synthesis temperature at 40°C, 60°C, 80°C, and 100°C was investigated. To find out the effect of capping agent, two experiments were done, one using the oleic acid and the other using 2-(dodecyloxy) acetic acid. In the case of investigating the annealing temperature, the black cobalt oxides were annealed at 400°C, 500°C, and 600°C. The last one was done using two different cobalt salts, cobalt nitrate and cobalt sulfate. Finally, cobalt oxide nanoparticles were characterized.

Keywords: Cobalt oxide nanoparticles; Co-precipitation method; pH effect; Synthesis temperature; Capping agent; Annealing temperature; Cobalt salt type

Background

Cobalt-based nanoparticles reside among the most promising materials for technological applications like information storage devices, magnetic fluids, and catalysts [1]. Co is a well-known ferromagnetic material which is commonly used as an alloying element in permanent magnets. It exists in two forms: hexagonal close packed (HCP) and face-centered cubic (FCC). HCP is the stable phase at room temperature, whereas FCC is stable at temperatures above 450°C [2]. Nanosized Co particles display a wide range of interesting size-dependent structural, electrical, magnetic, and catalytic properties [3]. In particular, because of their large surface area, Co nanoparticles showed high chemical reactivity, which makes them suitable for catalysis [4]. To have further applications of cobalt in different industries such as separation technology, information storage systems, catalysis, and biomedicine [5], nanoparticles are required to be discrete, identical in size and shape, and uniform in composition and crystal structure [6]. Thus, characterization should be done to have a broad knowledge on cobalt oxide nanoparticles. In this study, the different factors have been investigated to see their effects on the structure and morphology of cobalt oxide nanoparticles.

* Correspondence: jiny_ghazaleh@yahoo.com
Chemical Engineering Department, UKM, Bangi, Selangor, Malaysia

Results and discussion

Effect of pH

Scanning electron microscopy (SEM) images of Co_3O_4 synthesized by suspension at pH 8 to 10 in a co-precipitation reaction are shown in Figure 1. Comparing them, we see that nanocubic Co_3O_4 with the average particle size of 20 nm is formed when pH is 8 to 9, and when pH increases at 10 to 11, some irregular Co_3O_4 including the grains that are recombined in the products are formed. This fact can be explained as the condensation reaction of Co $(OH)_2$ precursors is able to happen at higher pH value, and agglomeration of the nanoparticles occurs. The average sizes of nanoparticles were calculated to be 20 nm at pH 8 to 9 and 30 nm at pH 10 to 11 using Debye-Scherrer's equation. At pH 9 to 10, some areas with distorted strings can be seen due to self-alignment orientation that has

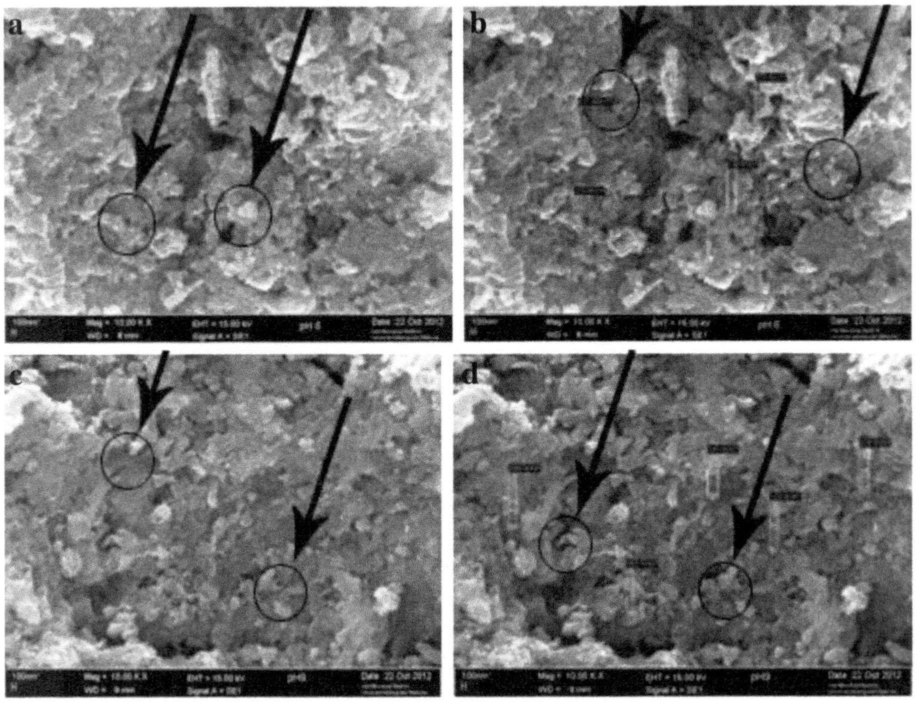

Figure 1 SEM images at pH 8 (a, b) and pH 9 (c, d).

happened, and this can be attributed to the presence of weak interactions. However, the crystalline Co_3O_4 nanoparticles can be observed as crystals in some areas of the micrographs. The pH range of 8 to 9 makes more uniform and smaller nanoparticles than pH 10 to 11. Large and agglomerated particles have been synthesized at higher pH since the nucleation rate is lower than the particle growth rate. Higher pH range increases synthesis rate; thus, the produced cobalt sample will be non-uniform and agglomerated. The obtained results show that the solution pH of 8 to 9 is a better one to synthesize uniform cobalt since at lower pH, condensation reaction was considerable so that the cobalt sample was completely agglomerated (Table 1). At higher pHs (10 to 11), reduction rate of Co ions was fast so that obtained samples were agglomerated less and a bit without any nanorods. In Figure 2, energy dispersive X-ray spectroscopy (EDX) spectra are shown.

Table 1 Weight and atomic percentages at pH 8 and 9

	Weight percentage (pH 8)	Atomic percentage (pH 8)	Weight percentage (pH 9)	Atomic percentage (pH 9)
CK	22.95	34.87	6.35	11.63
OK	45.73	52.15	48.16	66.22
CoK	22.49	10.24	29.43	16.15
NaK	8.83	2.73	16.07	6.00
Total	100	100	100	100

Identification was done by comparing the diffraction patterns with the standard X-ray diffraction pattern of the synthesized cobalt oxide that was analyzed to investigate the phase structure and the crystallinity which is shown in Figure 3. The patterns show sharp peaks relating to and according to standard Co_3O_4 X-ray diffraction (XRD) pattern (Joint Committee on Powder Diffraction Standards (JCPDS) no. 43–1003). All the peaks of cobalt oxide can be indexed to cubic phase.

Figure 3 shows a crystalline Co_3O_4 with small broadened profiles of all measured reflections and a crystalline powder which has sharp and a kind of intense diffraction peak. There are no impurity peaks, and this fact shows that the final product synthesized is Co_3O_4 with nanocubic structure by co-precipitation. In addition to the facts mentioned, the XRD patterns show the crystalline phase, and they are in agreement with reported values of selected area electron diffraction (SAED) pattern that shows the presence of well-defined clear spot. However, for pH 10 to 11, it has some broad diffraction lines which remind some inefficient crystallinity. Even if we look at the images of SEM (Figure 4), we can see some spherical agglomeration with some pores. The XRD plot of the cobalt powder at pH 8 to 9 shows the presence of both HCP and FCC phases. Also, the allotropic transition from HCP to FCC phase happened as the pH changes. The peak broadening at each reflection is also indicative of the formation of fine size cobalt powder.

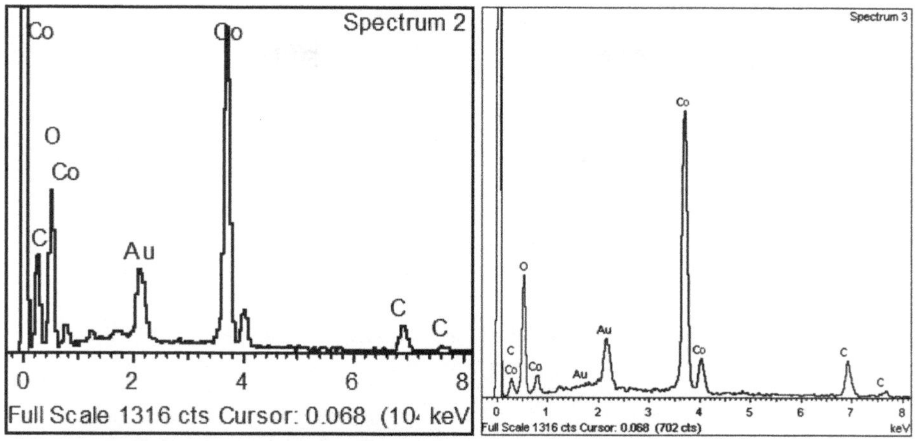

Figure 2 EDX spectra of nanoparticles at pH 8 (left) and pH 9 (right).

When the pH lies in the range of 8 to 9, Co_3O_4 dominates the product in the XRD pattern (Figure 5). At pH 10 to 11, a weak diffraction peak of CoO is observed. A small amount of cubic Co_3O_4 is found at pH 10 to 11 (Table 2). Based on Scherrer's formula, the average particle sizes of Co_3O_4 at pH 8 to 9 are 40 and 50 nm. Co_3O_4 gave the large crystallite size because it contains three Co ions which have large radii (Table 3). At pH 10 to 11, the range is 20 to 30 nm. The crystallite sizes were calculated using XRD data based on Debye-Scherrer's equation as follows:

$$d = \frac{0.89\,\lambda}{\beta \cos\theta}$$

where d, λ, θ and β are the crystallite size, X-ray wavelength (1.542 Å), Bragg diffraction angle, and full width at half maximum (FWHM) of the diffraction peak. After drying in an inert atmosphere at pH 8 to 9, a change from amorphous to crystalline state can be seen.

This shows narrow particle size distribution in a well-defined particle size, purity of the phase, and the structure and shape. A stretching frequency at 3,361.2 cm^{-1} and a weak asymmetric band at 2,919 cm^{-1} will prove the existence of OH group due to the absorption of water by nanoparticles during sample preparation (Figure 6). The presence of two strong M-O stretching and bending frequencies at 1,481 and 831 cm^{-1} shows phase purity of monodisperse one in the face-centered cubic structure. The peaks at 1,579 and 1,539 cm^{-1} correspond to Co_3O_4 asymmetric stretching vibrations. The peak at 3,610 cm^{-1} corresponds to CoO bending vibrations at pH 11.

The optical property of cobalt oxide nanoparticle has been found with high purity, and the crystallinity of the cobalt oxides were confirmed with the help of UV-visible absorption spectra by observing the absorption rate

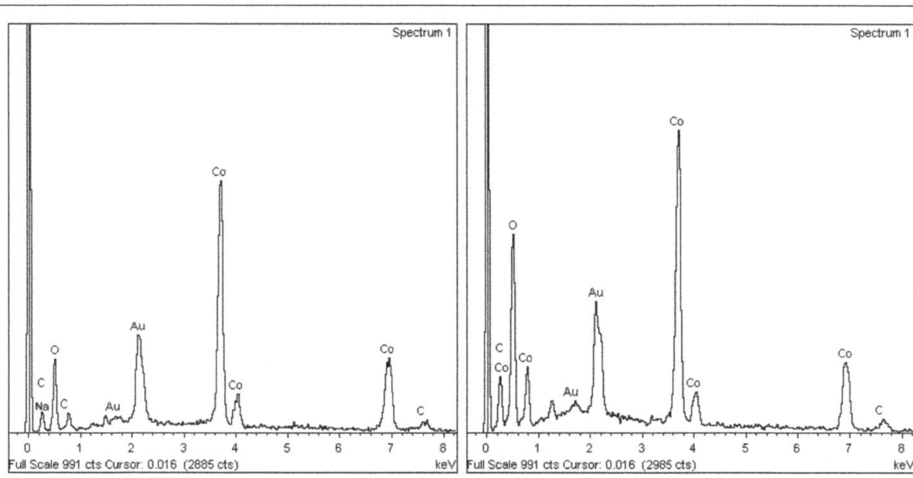

Figure 3 EDX spectra of nanoparticles at pH 10 (left) and pH 11 (right).

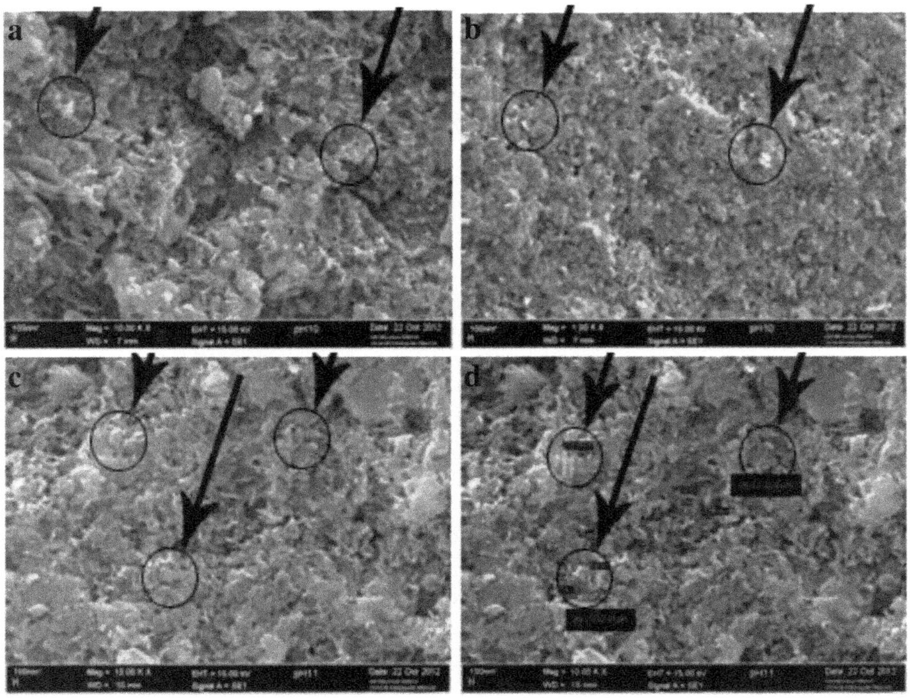

Figure 4 SEM images at pH 10 (a, b) and pH 11 (c, d).

band at 210 nm with a tail which is extended towards a longer wavelength (Figure 7), and the reason for that is their quantum size effects. In the extending tail, there is no significant absorption peak due to quantum confinement effects in the energy gap. The absorption peak is a bit broad peak due to the particle size. CoO nanoparticle are considered as stable, and this fact can be the result of symmetrical polarity structure which depends on the weak interaction of van der Waals forces within the particle regime.

Effect of synthesis temperature

Four samples were synthesized at different temperatures. The prepared samples were characterized by SEM.

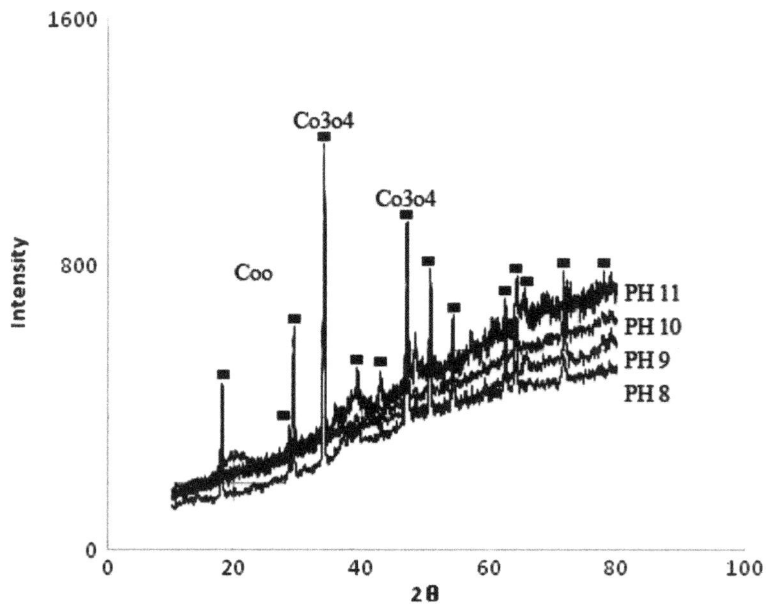

Figure 5 XRD patterns of cobalt oxide nanoparticles at different pHs.

Table 2 Weight and atomic percentages at pH 10 and 11

	Weight percentage (pH 10)	Atomic percentage (pH 10)	Weight percentage (pH 11)	Atomic percentage (pH 11)
CK	10.62	22.46	14.90	25.63
OK	29.14	46.29	43.97	56.77
CoK	26.05	16.51	19.30	9.95
NaK	34.20	14.74	21.83	7.65
Total	100	100	100	100

Figure 8a, b, c, d shows the SEM images of the cobalt samples which were synthesized at 40°C, 60°C, 80°C, and 100°C, respectively. When the reaction was performed at a temperature lower than 60°C, the reaction rate was very slow, and the prepared sample was amorphous and agglomerated. At temperatures higher than 60°C, the synthesis rate is fast, but the prepared samples showed that the higher temperature causes more agglomeration. Amorphous percentage of samples is increased because, at high reaction rate, nucleation and particle growth processes are not controllable. This fact can be concluded in figure 9 showing the EDX too. They were analyzed by XRD. Figure 10 shows the XRD patterns of these samples. As can be seen from the figure, the solution temperature has an important role in the composition and phase control of the synthesized sample. At 60°C and lower, pure metallic cobalt can be synthesized. At high temperatures, 80°C and more, cobalt will be formed in oxide forms such CoO, Co_2O_3, and CoO_2. Based on XRD results, the metallic cobalt content of samples is decreased when the synthesis temperature is increased from 40°C up to 100°C and more. However, the combination of SEM and XRD results for temperature studies shows that the sample

Table 3 Crystallite size and particle size found by Scherrer

pH	Crystallite size (nm)	Particle size (nm)
8	3.22	40
9	3.25	50
10	4.75	20
11	4.89	30

synthesized at 60°C has uniform morphology, small particles, and composition without any impurity. The comparison is even possible through the peaks at FTIR in Figure 11.

Effect of annealing temperature

Figure 12 shows X-ray diffraction patterns of Co_3O_4 at different annealing temperatures of 400°C, 500°C, and 600°C; small peaks were observed, and a poor crystallinity at 400°C is shown. The ones annealed at temperature more than 500°C exhibit sharp diffraction peak characteristics of the Co_3O_4 cubic spinel structure. From Figure 12, phases ($2\theta \approx 22.90°$), ($2\theta \approx 35.80°$), ($2\theta \approx 58.20°$), and ($2\theta \approx 78.10°$). XRD peaks were observed at the temperature of 600°C, and it is concluded that all of them were crystalline with a cubic spinel structure (JCPDS) and a random orientation, which generally occurs in the growth of Co_3O_4. The degree of orientation of the Co_3O_4 was strongly dependent on the annealing temperature. It increases as the annealing temperature increases. The average particle sizes of Co_3O_4 thin film were calculated using the full width at oxygen defects at the grain boundaries. As a result, these defects are favorable to the merging process to form larger Co_3O_4 grains while increasing the annealing temperature. As the annealing temperature increases from 400°C to 600°C, the intensity value of Co_3O_4 exhibits a tendency to decrease, which can be attributed to the coalescences of

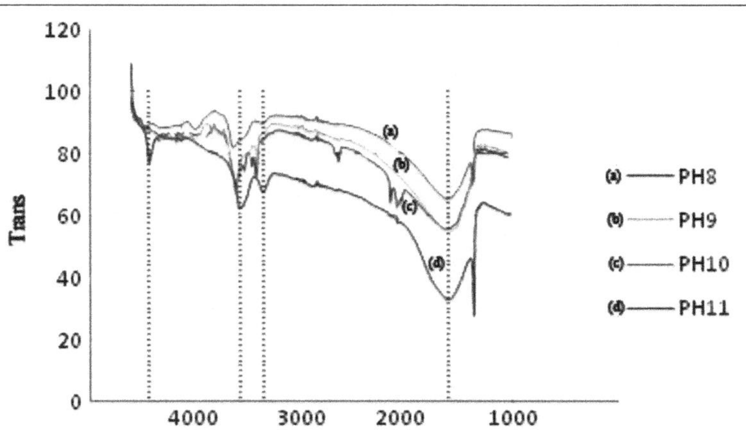

Figure 6 FTIR spectra of cobalt oxide nanoparticles at pH 8 to 11.

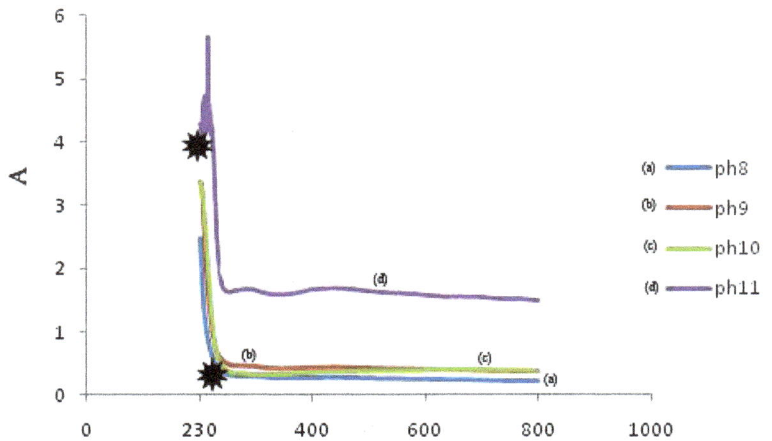

Figure 7 UV results of cobalt oxides at pH 8, 9, 10, and 11 from bottom to top.

grains at higher annealing temperature. As a result, it implies that the crystallinity of the Co_3O_4 is improved at higher annealing temperatures. This may be due to high annealing temperature providing energy to crystallites gaining enough energy to orient in proper equilibrium sites, resulting in the improvement of crystallinity and degree of orientation of the Co_3O_4.

Figure 13a, b, c shows SEM images of cobalt annealed at 400°C, 500°C, and 600°C. The figure shows nearly the same surface morphology with slight increase in grain size. The surface looks smooth and composed of very fine elongated particles smaller than 80 nm in length connected by two to three spherical grains of about 40 to 45 nm in diameters. From the SEM image, overgrowth of clusters is clearly seen. Initially grown. The percent of crystallization and related facts mentioned about SEM can be concluded from the images at Figure 14 and Figure 15 showing the EDX. Nanograins may have increased their size by further deposition and come closer to each other. The cobalt oxide surface is well

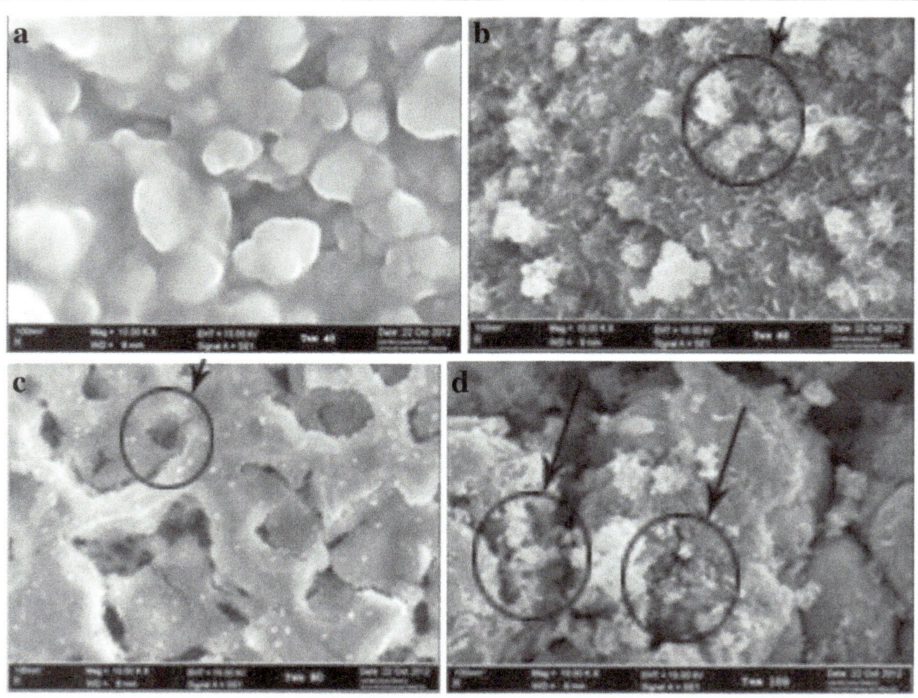

Figure 8 SEM images of cobalt oxides at temperatures of 40°C (a), 60°C (b), 80°C (c), and 100°C (d).

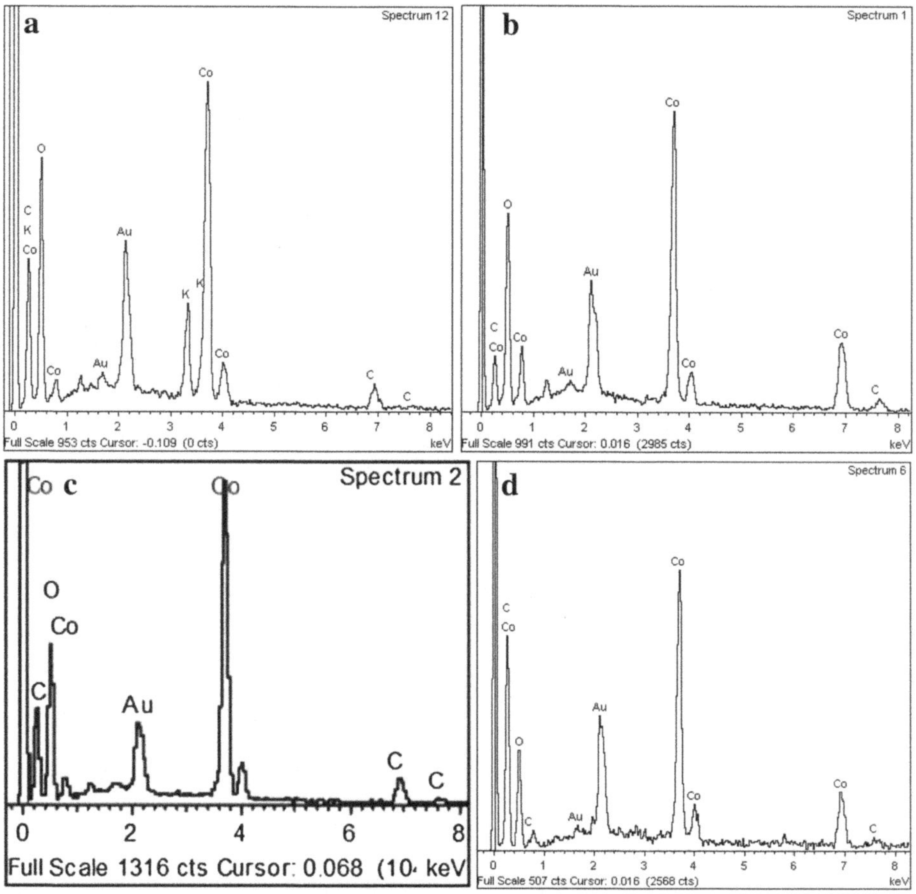

Figure 9 EDX spectra of cobalt oxides at temperatures of 40°C (a), 60°C (b), 80°C (c), and 100°C (d).

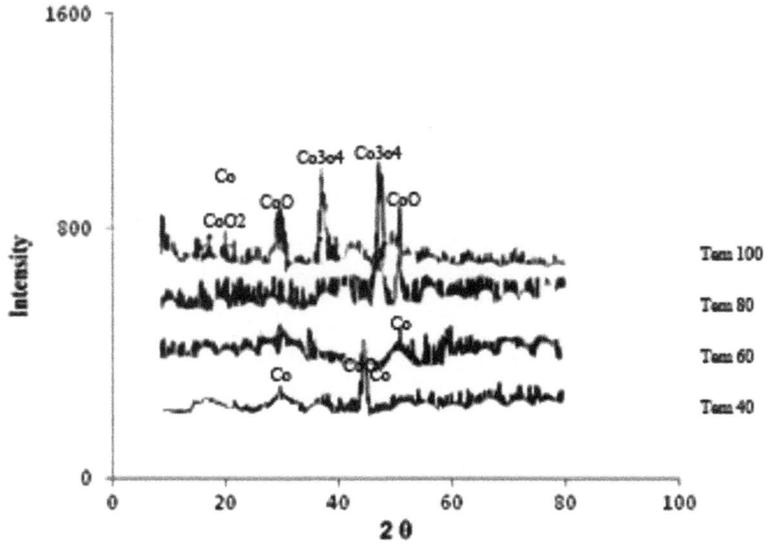

Figure 10 XRD patterns of samples synthesized at different temperatures: 40°C, 60°C, 80°C, and 100°C.

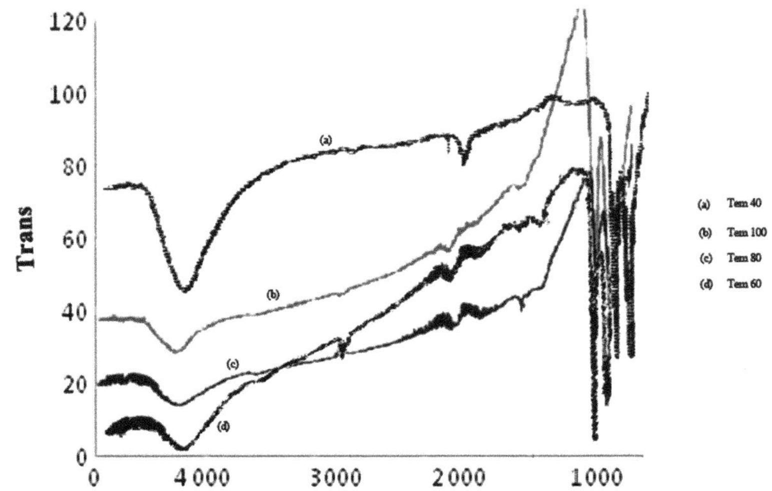

Figure 11 FTIR results of cobalt oxides at different synthesis temperatures.

covered without any pinholes and cracks. Such surface morphology may offer increased surface area.

Effect of capping agent

Fourier transform infrared spectroscopy (FTIR) is a powerful tool to study the capping of surfactants to the surface of the nanoparticles. Figure 16 shows the FTIR data for pure oleic acid and DDOA. The notable features of the spectrum in oleic acid are a band centered at 1,500 cm^{-1} and a peak occurring at 3,400 cm^{-1}. We also observe bands positioned at 2,700 and 1,070 cm^{-1} which are designated to symmetric and anti-symmetric stretching of CoO. The most important differences in this spectrum from that of oleic acid are the shift in the band from 1,500 to 1,100 cm^{-1} and the disappearance of the peak at 3,400 cm^{-1}.

The FTIR spectrum for purified DDOA molecule is shown in Figure 16. The band at 1,737 cm^{-1} is assigned to C=O stretch, while the peaks appearing at 2,732 and 3,000 cm^{-1} are assigned to symmetric and anti-symmetric stretching of CH_2.

Here, the most prominent difference when this molecule is present on Co nanoparticle surface is the shift observed

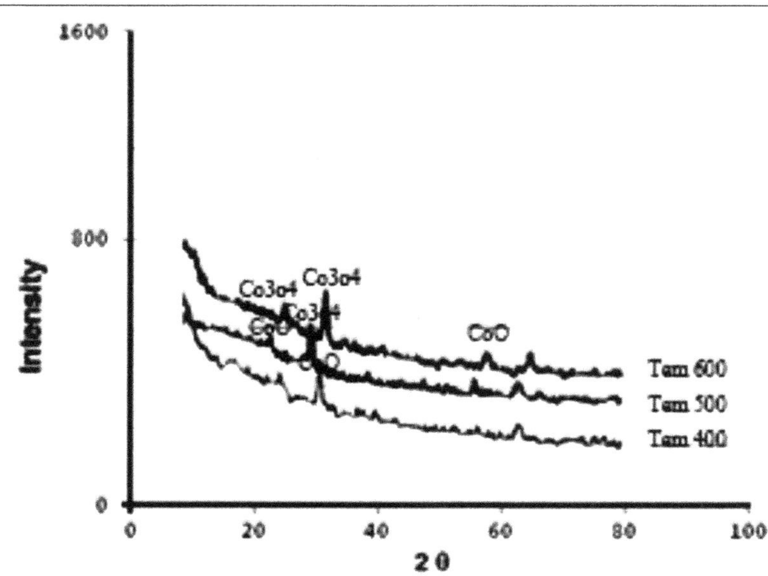

Figure 12 X-ray diffraction patterns of Co_3O_4 at different annealing temperatures.

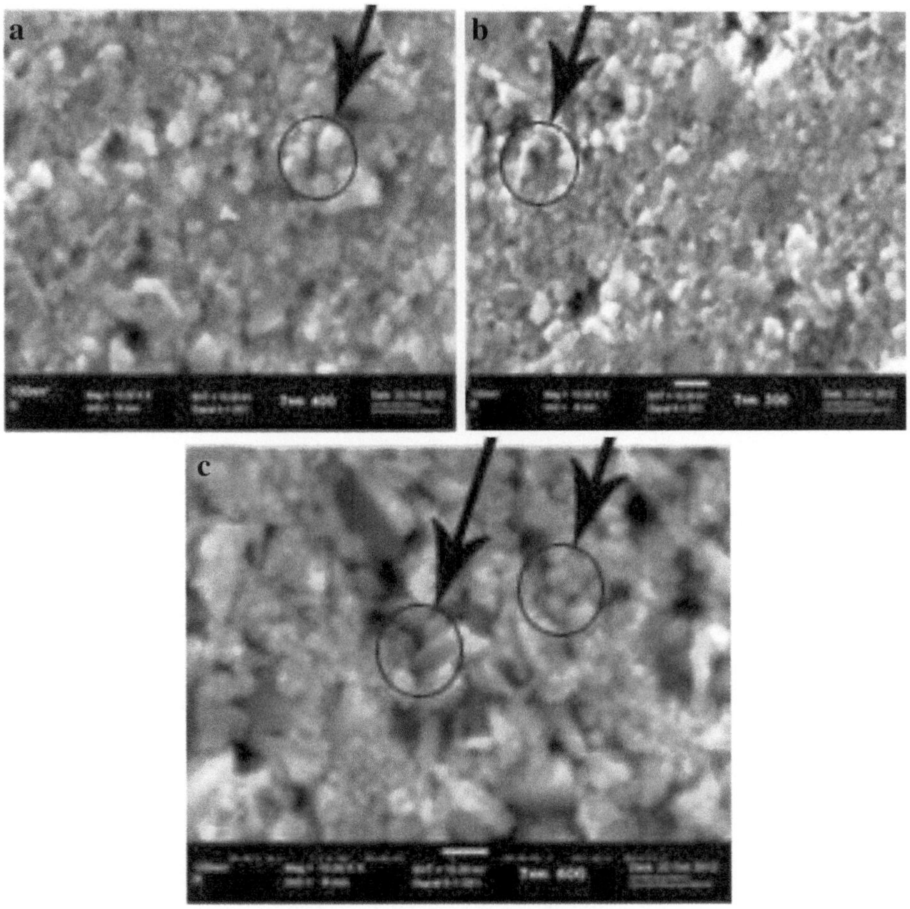

Figure 13 SEM images of cobalt oxides at annealing temperatures of 400°C (a), 500°C (b), and 600°C (c).

in the peak from 1,490 to 1,300 cm^{-1}. The pure oleic acid spectrum, peak positioned at 1,500 cm^{-1} is assigned to C=O stretch, while the peak occurring at 3,309 cm^{-1} represents the stretching of olefinic C-H. In the FTIR spectra of oleic acid-capped Co nanoparticles, a shift in peak corresponding to C=O is observed, while the peak corresponding to C=C-H stretch vanishes and also the peak occurring at 2,702 cm^{-1} corresponding to CoO. Disappearance of this peak indicates binding of double bond to the surface of nanoparticles, while the shift in C=O peak is due to carboxylate formation. Comparative FTIR study was carried out for OA-Co nanoparticles. We wish to highlight here that in the pure OA case, if the binding to the Co nanoparticles is through the double bond or carboxylic ends with Co nanoparticle surface, an unfavorable situation where -CH_3 end is exposed to aqueous environment would result. This could be the reason why oleic acid does not act as good capping agent in aqueous environment. In the case of DDOA, binding to the surface of nanoparticles occurs only through the -COOH group as it does not contain a double bond. This is indicated by the shift in peak. However, by changing appropriate precursors and organic transformation, changes to this ligand structure can be accomplished.

Figure 17 demonstrates the X-ray diffraction patterns for oleic acid. The as-prepared sample displays no features. Clear and distinguishable peaks are observed for Co_3O_4, while high peaks are observed for CoO. The figure also demonstrates the X-ray diffraction pattern for DDOA-capped Co nanoparticles. In this case, the diffraction pattern was recorded for the powder sample heated at 300°C as no diffraction was observed for the as-prepared sample. The peak observed corresponds to the Co_3O_4, and just two weak peaks correspond for CoO.

The particle size obtained with DDOA as capping agent was smaller than those obtained with oleic acid (Figure 18). The average sizes calculated using Scherrer's

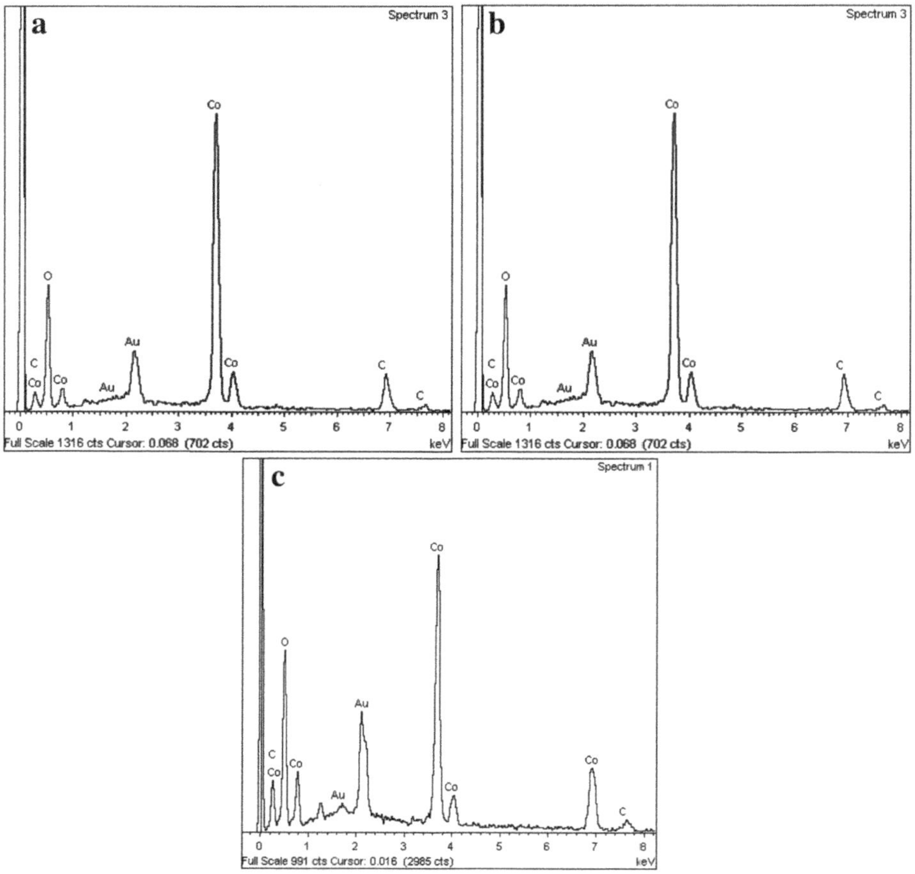

Figure 14 EDX spectra of cobalt oxides at annealing temperatures of 400°C (a), 500°C (b), and 600°C (c).

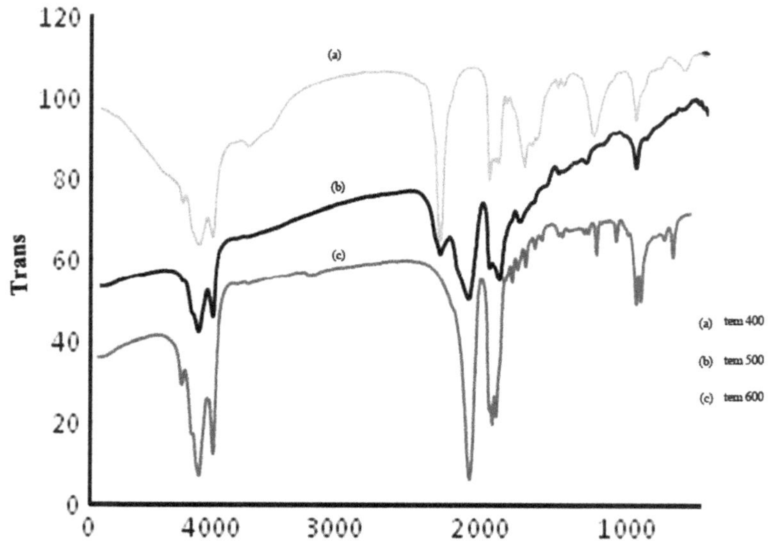

Figure 15 FTIR results of cobalt oxides at different annealing temperatures.

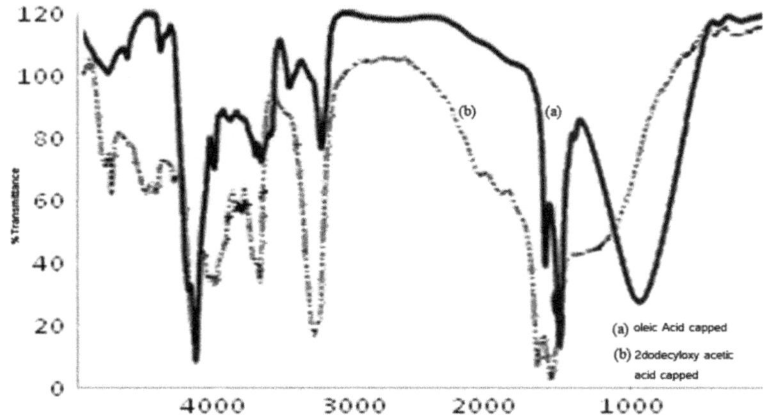

Figure 16 FTIR results of cobalt nanoparticles capped with oleic acid and DDOA.

formula were 35 and 45 nm for DDOA and oleic acid, respectively. SEM analysis carried out for oleic acid shows aggregated nanoparticles with larger particle size (Figure 19). The poor capping ability of oleic acid in water can be implicated for this observation as oleic acid cannot stabilize Co nanoparticles leading to coalescence of particles. From our studies, it is observed that ligands used in the syntheses play an important role in governing the crystal structure of cobalt nanoparticles.

Effect of cobalt salt type

Figure 20 shows the SEM images of Co_3O_4 using different cobalt salts. It can be seen from the figure that only nanocube morphologies of Co_3O_4 are shown using Co $(NO_3)2{\cdot}6H_2O$ as cobalt salt. For comparison, typical spherical Co_3O_4 and also nanocubic Co_3O_4 are obtained when the cobalt salts are $CoSO_4{\cdot}7H_2O$. It can be concluded that the morphologies of Co_3O_4 are closely dependent on the anion type in cobalt salts. In other words, the anion type in cobalt salt plays a key role in the morphology of Co_3O_4.

Several precursors (cobalt nitrate and cobalt sulfate) were used (Figure 21). It suggests that reduction of cobalt (II) nitrate to cobalt is requiring lower energy than the cobalt sulfate. This demonstrates that pure cobalt nanoparticles can be synthesized from cobalt (II) nitrate

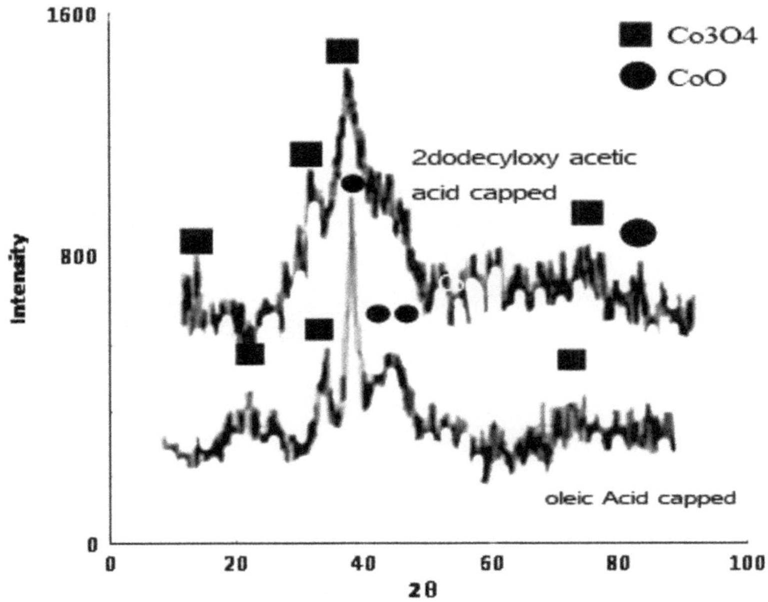

Figure 17 XRD spectra of oleic acid-capped cobalt nanoparticles and XRD patterns for as-prepared DDOA-capped cobalt nanoparticles.

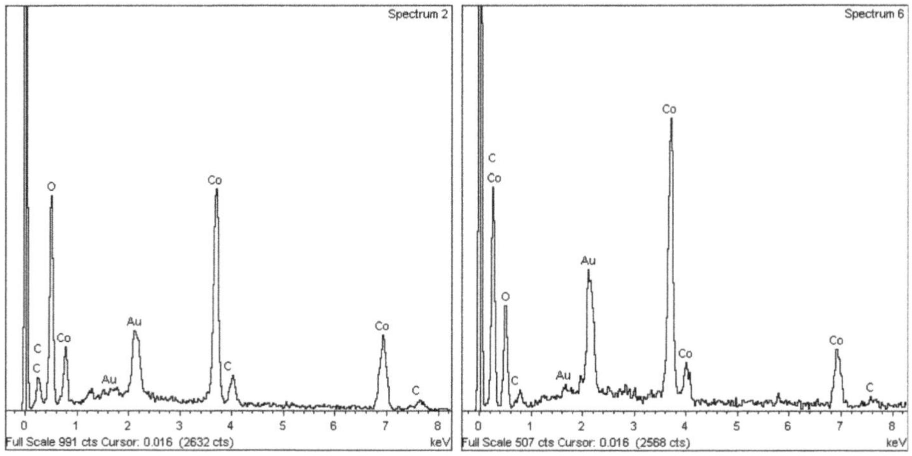

Figure 18 EDX spectra of cobalt nanoparticles with oleic acid (left) and DDOA (right) as capping agents.

as the removal of nitrates and water requires lower energy than the removal of water from cobalt sulfate. In addition, cobalt sulfate leaves carbon with the cobalt nanoparticles. XRD patterns of Co products prepared from nitrate and sulfate salts are shown in Figure 22. A broad peak at two angles between 40° and 50° (2θ = 42°, 48.50°, and 50°) corresponds to Co_3O_4 materials. Co_3O_4 is the dominant crystalline species. Peaks representing residual or impurity (formed during nitrate decomposition) are not seen, indicating complete decomposition or removal after the washing procedure, but for sulfate, there is just one weak peak for CoO (Figure 23).

Conclusion

As a conclusion, we can say that Co_3O_4 with the average particle size of 40 nm is formed when pH is 8 to 9, and when pH increases at 10 to 11, irregular Co_3O_4 with grains are seen and the size is averagely 30. Thus, pH has a great effect on the Co_3O_4 formation. To achieve the optimum pH, the other experiments were done. In the other experiment, two capping agents, DDOA and oleic acid, were used to see the result; particles with oleic acid are irregular in shape and are clustered. Particle size distribution indicates that the size of the particle is 35 nm. In addition to the mentioned factors, the cobalt salt also plays an important role. It can be concluded that the morphologies of Co_3O_4 are dependent on the anion type in cobalt salts. In other words, the anion type in cobalt salt plays a key role in the morphology of Co_3O_4. The morphologies of Co_3O_4 show irregular nanocubes, 45 nm in size, using Co $(NO_3)2{\cdot}6H_2O$ as cobalt salt. Also, in the experiment done to observe the effect of annealing temperature, it was concluded

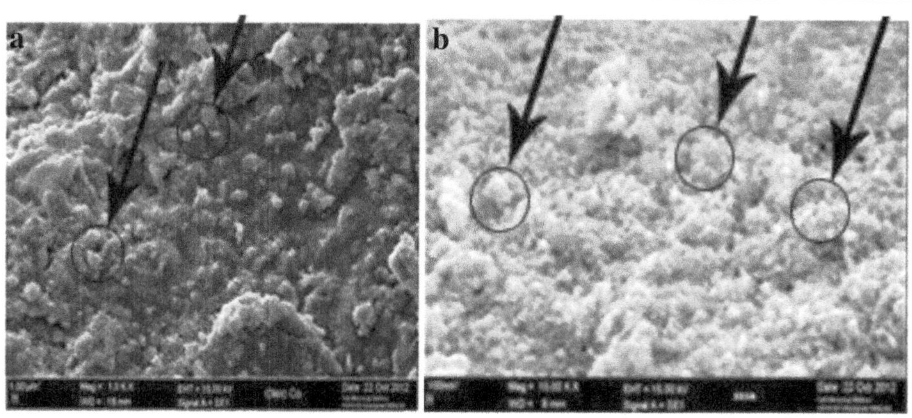

Figure 19 SEM images of cobalt nanoparticles with oleic acid (left) and DDOA (right) as capping agents.

Figure 20 SEM images of $CoSO_4$ (left) and Co $(NO_3)_2$ (right) as cobalt salts.

that as the annealing temperature increases from 400° C to 600°C, the intensity value of Co_3O_4 exhibits a tendency to decrease, which can be attributed to the coalescences of grains at higher annealing temperature. As a result, it implies that the crystallinity of the Co_3O_4 is improved at higher annealing temperatures. In conclusion of observing the effect of heat of reaction, it was seen that when the reaction was performed at temperature lower than 60°C, the reaction rate was very slow and the prepared sample was amorphous and agglomerated. At temperatures higher than 60°C, the synthesis rate is fast, but the prepared samples showed that the higher temperature causes more agglomeration and amorphous percentage of samples that is increased because, at high reaction rate, nucleation and particle growth processes are not controllable.

Methods

All the materials were purchased from Sigma Aldrich (St. Louis, MO, USA) and Acros Organics (Geel, Belgium). In order to have the optimum pH to control the size and morphology of cobalt nanoparticles, the experiment was done. Thus, after the optimum pH was found, the other experiments were done with the selected pH. A 15 mmol cobalt salt (Co $(NO_3)2{\cdot}6H_2O$) was dissolved into deionized water containing dispersant polyethylene glycol with relative molecular mass of about 20,000; then, excessive amount of NH_4OH was added with electromagnetic stirring at 50°C to form $Co(OH)_2$ gel. The pH value was monitored at 8, 9, 10, and 11 by dropping KOH. A certain volume of 40% (mass fraction) H_2O_2 was dropped into the above suspension. Finally, the suspension was transferred into a centrifuge. Then, it was placed inside an oven to be heated at 100°C then cooled

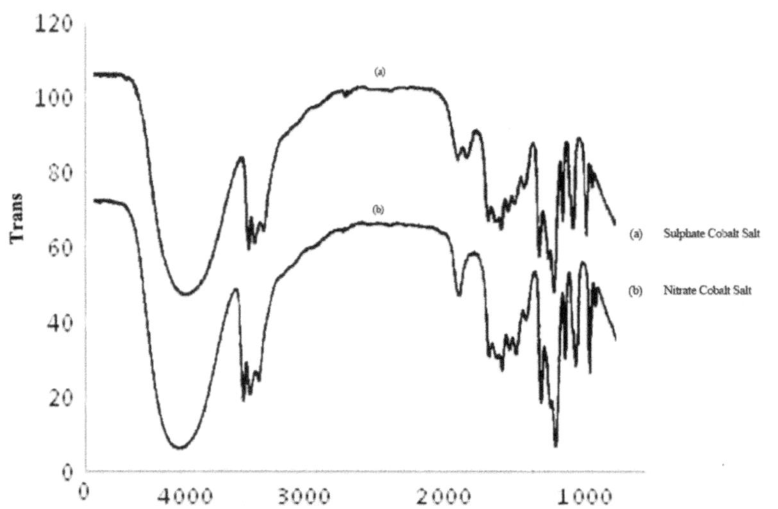

Figure 21 FTIR results of cobalt oxide with two cobalt types.

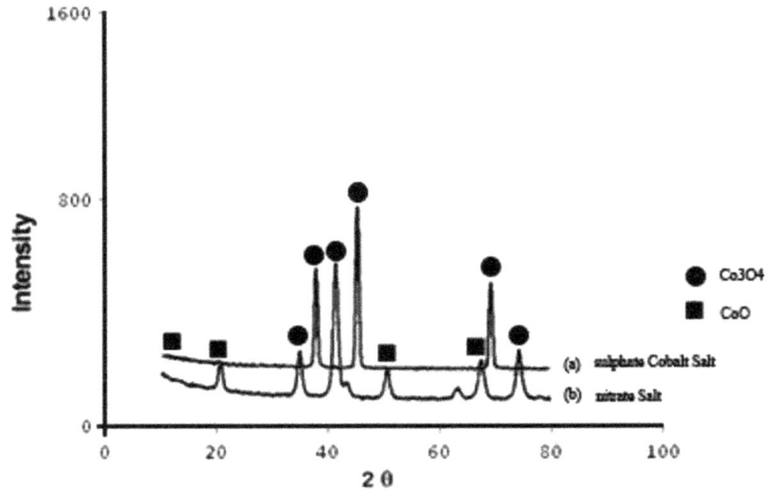

Figure 22 XRD results of cobalt oxides using sulfate (top) and nitrate (bottom) salts.

to room temperature in air naturally. The black products were centrifuged and washed with deionized water and ethanol for five times and then dried in a vacuum oven at 80°C for a day. The methods used to control pH are shown in Figure 24.

The methodology for other experiments is the same, but the factors have been changed such as synthesis temperature, annealing temperature, capping agent, and cobalt salts. In order to investigate the effect of synthesis temperature, the same procedure was done with pH 9 at four temperatures 40°C, 60°C, 80°C, and 100°C. In order to investigate the role of annealing temperature, the experiment was done as follows: Cobalt acetate tetrahydrate was added to 50 mL of methanol and stirred vigorously at 80°C for 1 h, leading to the formation of light pink colored powder. Then, H_2O_2 was added to remove the black precipitate. The as-prepared powder was sintered at various temperatures ranging from 400°C, 500°C, to 600°C with a fixed annealing time of 1 h in an ambient air to obtain Co_3O_4 with different crystallite sizes. Another experiment was carried out to see the effect of capping agent. Synthesis of cobalt nanoparticles was carried using sodium borohydride ($NaBH_4$) as reducing agent. In a typical experiment, 100 mL of oleic acid was taken, and cobalt nitrite was added to this. This mixture was stirred for a day. Twenty milligrams of sodium borohydride was dissolved in 1 mL of deionized water and ethanol; since the oleic acid is insoluble, it was added to the mixture dropwise. On addition of $NaBH_4$, the color of the solution changed from light pink to black indicating

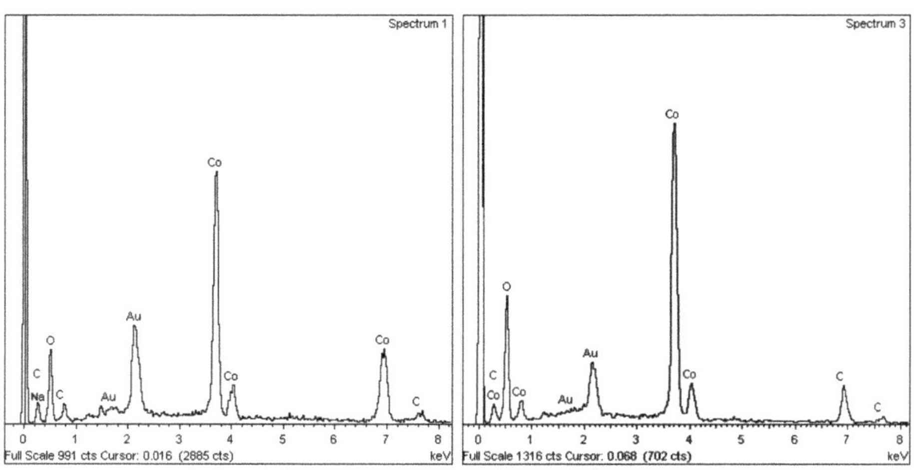

Figure 23 EDX spectra of cobalt oxides using sulfate (left) and nitrate (right) salts.

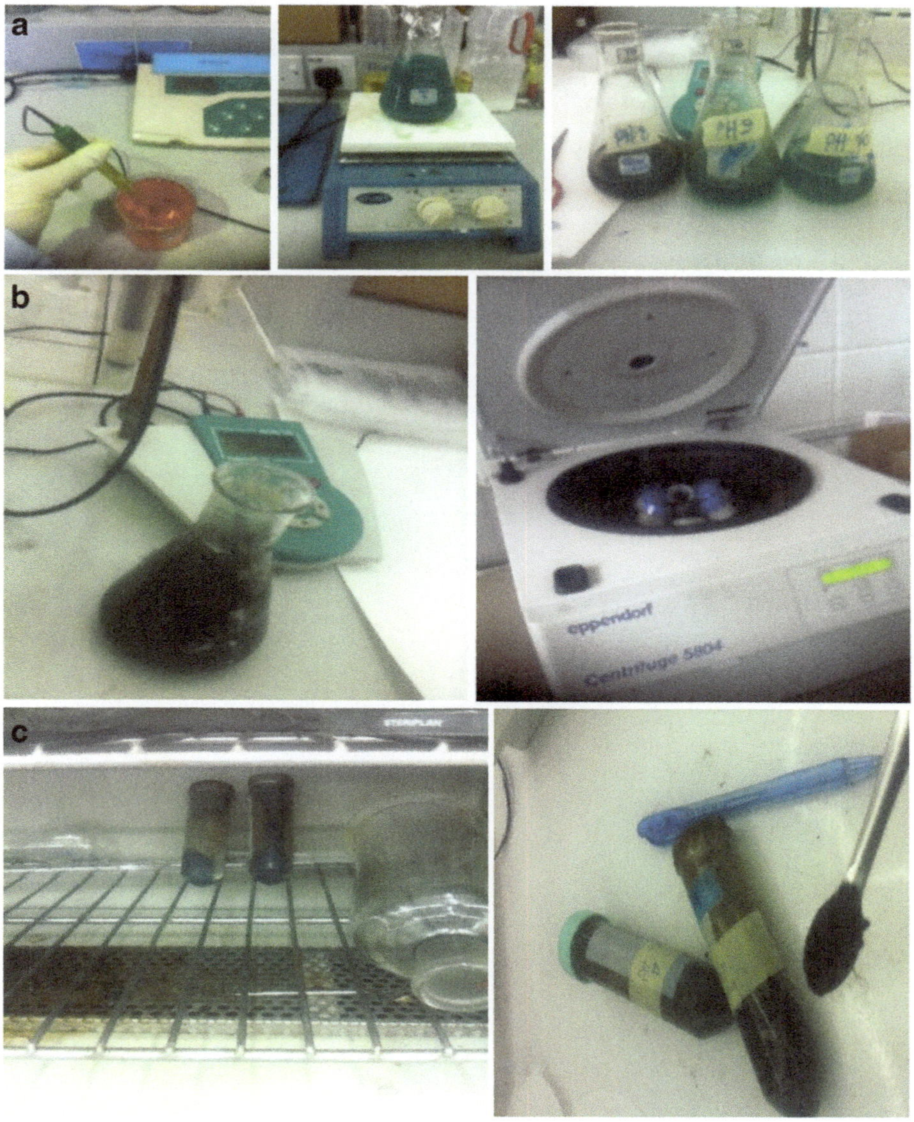

Figure 24 Methods used to control pH. (a) Stirring. **(b)** Centrifugation. **(c)** Heating and centrifugation.

the formation of cobalt nanoparticles. The nanoparticle solution was then centrifuged at 8,000 rpm for 20 min to remove uncoordinated surfactant and un-reacted $NaBH_4$. The same experiment was carried out but using 2-(dodecyloxy) acetic acid (DDOA) as capping agent.

In order to see the effect of cobalt salt for obtaining Co_3O_4 powders, two cobalt salts were used. The precipitation agents, 0.1 M Co $(NO_3)_2$ solution and 0.2 M NH_3 solution, were added together to a 150 mL of water at a constant pH = 9 and at room temperature. The green precipitate was filtrated, washed, and dried at 110°C overnight and then calcined in air; the same procedure was done except for using cobalt sulfate as cobalt salt source. The sample was obtained from 0.5 M $CoSO_4$ solution and 0.2 M NH_3 solution as precipitating agent at pH = 9.5.

Characterization

To study the obtained products, conventional characterization techniques were applied, such as XRD, SAED, SEM, IR spectroscopy, and UV–vis spectroscopy.

Abbreviations

DDOA: 2-(dodecyloxy) acetic acid; SAED: Selected area electron diffraction; SEM: Scanning electron microscope.

Competing interests

Both authors declare that they have no competing interests.

Authors' contributions

GA with the supervision of MA carried the studies. All authors read and approved the final manuscript.

Acknowledgments

We would like to appreciate the fund from UKM chemical Engineering department.

References

1. Dinega, DP, Bawendi, MG: Influence of iron oleate complex structure on iron oxide nanoparticle formation. J Appl Physiol **85**, 4325–4330 (1999)
2. Puntes, VF, Krishnan, KM, Alivisatos, AP: Synthesis, self-assembly, and magnetic behavior of a two-dimensional superlattice of single-crystal ε-Co nanoparticles. Appl Phys Lett **78**, 2187–2189 (2001)
3. Hyeon, T: Synthesis of highly crystalline and monodisperse magnetite nanocrystallites without a size-selection process. J Am Chem Soc **123**, 12798–12801 (2001)
4. Sergeev, C, Gubin, SP, Yurkov, GY, Kataeva, NA, Kurnakov: Microgranules and nanoparticles on their surface, pp. 1536–1540. Russian Academy of Sciences, Leninskii (2005)
5. Kodama, RH: Magnetic nanoparticles. J Magn Magn Mater **200**, 359–372 (1999)
6. Balela, MDL: Synthesis and characterization of cobalt nano particles by liquid-phase reduction. Master's thesis. RS201.N35 B183, pp. pp 1–4. Universiti Sains, Malaysia (2008). 8–32
7. Berger, D, Morfin, F, Matei, J, Volta, C: Studies on synthesis and characterization of Co_3O_4 powders for CO oxidation. J Optoelectron Adv Mat **9**, 1540–1545 (2007)
8. Yang, YP, Liu, RS, Huang, KL, Wang, LP, Liu, SQ, Zeng, WW: Preparation and electrochemical performance of nanosized Co_3O_4 via hydrothermal method. Trans Nonferrous Metals Society of China **17**, 1083–1085 (2007)

31

Simulation of nanodrug by theoretical approach

Saeideh Ghorbaninezhad* and Maryam Ghorbaninezhad

Abstract

In recent years, HIV-1 integrase (IN) has become an attractive target for designing antiretroviral agents. The development of raltegravir and other successful lead IN inhibitors has also influenced the IN inhibitor design strategy. This has led to the identification of several potent inhibitors in these last 2 years. The medicines which have the compound of C-centered and N-centered anti-HIV inhibitor with single-walled carbon nanotube are examined by density functional theory (DFT) method. In this paper, the end of the nanotubes, which was saturated with hydrogen atoms, was examined by DFT at the level of B3LYP and 6-31G(d) standard basis set. There are free Gibbs energy, free Helmholtz energy, enthalpy, bond length (Å), bond angle (in degrees), dihedral angle (in degrees), energy hyperconjugation, total energy (in Kcal mol^{-1}), moment dipole (in Debye), occupancy between nanotube (6, 6), and chalcone derivative. These cases and medicines show that complex 2 is more stable than the other complexes.

Keywords: NBO; Nanodrug; DFT; Chalcone; Inhibitor; Anti-HIV; Nanotube; SWCNT

Background

Since the discovery of carbon nanotubes by Iijima [1], extensive research has been devoted to their structural characterization [2]. Among the numerous delivery systems currently under investigations, carbon nanotubes (CNTs) seem to embody a promising option [3]. Pristine carbon nanotubes (pCNTs) are made up of carbon atoms arranged in a series of condensed benzene rings and wrapped into a tubular form (Figure 1). Concerning their use in biological systems, lack of solubility (both in organic solvents and aqueous solutions), formation of thick and inhomogeneous bundles, circulation half-life of 3 to 3.5 h [4], and biocompatibility and immunogenicity limitations raise great concerns. However, these observations hold only for pCNTs and, therefore, just indicate the need for further modifications in order to explore the feasibility of functionalized CNTs (f-CNTs) as safe bio-nanomaterial [5,6]. In particular, the application of f-CNTs as new nanovectors for drug delivery became doable soon after the demonstration of cellular uptake of this new material [7,8]. It is worth to mention that apart from a few cases of phagocytic incorporation inside macrophages [9,10] (which are known to be large cleaning cells able to remove foreign material including less soluble nanotubes), no uncoated pCNTs were reported to penetrate inside the cells without displaying remarkable effect. This last point should reinforce the use of f-CNTs as improved, less harmful nanovehicles, especially after our recent discovery regarding the lack of a direct correlation between the kind of functionalization on the surface of carbon nanotubes and the extent of their internalization [11-13]: either electrostatically neutral or charged f-CNTs could be taken up by cells with comparable amount, hence indicating that numerous different chemical procedures could be adapted to introduce several groups and functionalities. Further, an increased understanding of IN structural biology has opened up novel approaches to inhibit IN, such as targeting its multimerization or interaction with cellular cofactors [14]. In the paper, complex between chalcone and nanotube (6, 6) is investigated, and chalcone is used as anti-HIV drug. The chemistry of chalcones has generated intensive scientific studies throughout the world. Special interest has been focused on the synthesis and biodynamic activities of chalcones. The term 'chalcones' was given by Kostanecki and Tambor [15]. These compounds are also known as benzalacetophenone or benzylidene acetophenone. In chalcones, two aromatic rings are linked by an aliphatic three carbon chain. Chalcone bears a very good synthon, so a variety of novel heterocycles with good pharmaceutical profile can be designed. Chalcones are α- and β-unsaturated

* Correspondence: s.ghorbani.uni@gmail.com
Department of Chemistry, Islamic Azad University (Rasht Branch), Rasht, Iran

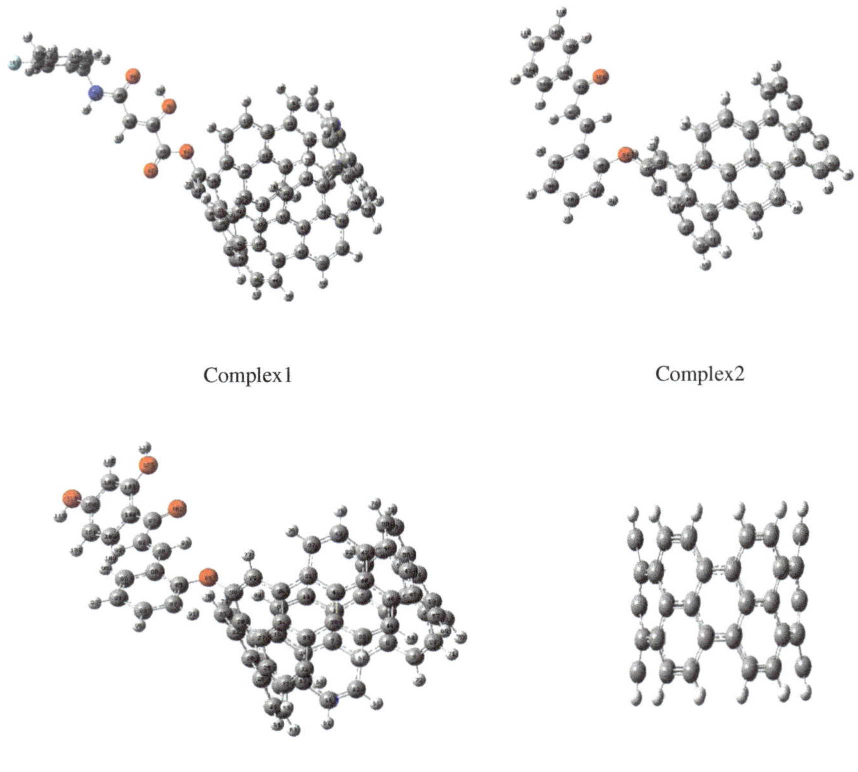

Figure 1 The optimized complex of chalcone derivative - SWCNT (6, 6) in B3LYP/6-31G(d) method at 298.15K.

ketones containing the reactive ketoethylenic group CO-CH=CH-. These are colored compounds because of the presence of the chromospheres -CO-CH=CH-, which depend in the presence of other auxochromes. Chalcones resemble the diketo acid functionality and are potential leads in designing potent IN inhibitors. Potent chalcones were identified through an NCI drug screening program [16]. The structures of chalcone are shown in Figure 2.

Results and discussion

We measured the parameters such as bond length (Å), natural bond orbital (NBO), and bond angle, dihedral angle, distances of analyzed models of the SWCNT (6, 6). The end of the nanotubes, which was saturated with hydrogen atoms, was examined by DFT at the level of B3LYP and 631G(d) standard basis set and is shown in Tables 1, 2, and 3. The electron that is given by the complexes in a reaction should make it as HOMO, while that which captured the complexes must be placed on its LUMO [17], so the atom on which the HOMO mainly scattered should be able to separate the electrons, while the atom, by holding the LUMO, should achieve electrons on this basis. HOMO-LUMO gap is usually associated with chemical stability against electronic excitation with a larger distance like greater stability [18]. The energy (kcal mol^{-1}) and dipole moments (Debye) indicate the consistency among the three complex calculations in the DFT method. The optimized configurations are shown in Figures 1 and 2. In Table 1, it becomes obvious that the complexes 2 and 3 have higher hyperconjugation energy than complex 1. The results also show that as the P increases through the sharing of hybrid atoms, the occupancy decreases. The hybrid *s* orbital shared in the oxygen atom in complex 3 is more than the hybrid *s* orbital shared in complexes 1 and 2. Most of the combined energy hyperconjugations are stable. Occupancy coefficient is smaller. Complex 2 is more stable than complexes 1 and 3. The energy hyperconjugation of complex 2 (21.25) is lesser than that of complex3. Reduced coupling energy is due to the resonance interference.

Based on the drug resonance structure, a phenolic ring is observed. In complex 2, the negative charge is located in orthoposition. However, in complex 3, the negative charge is in the oxygen group (OH). Thus,

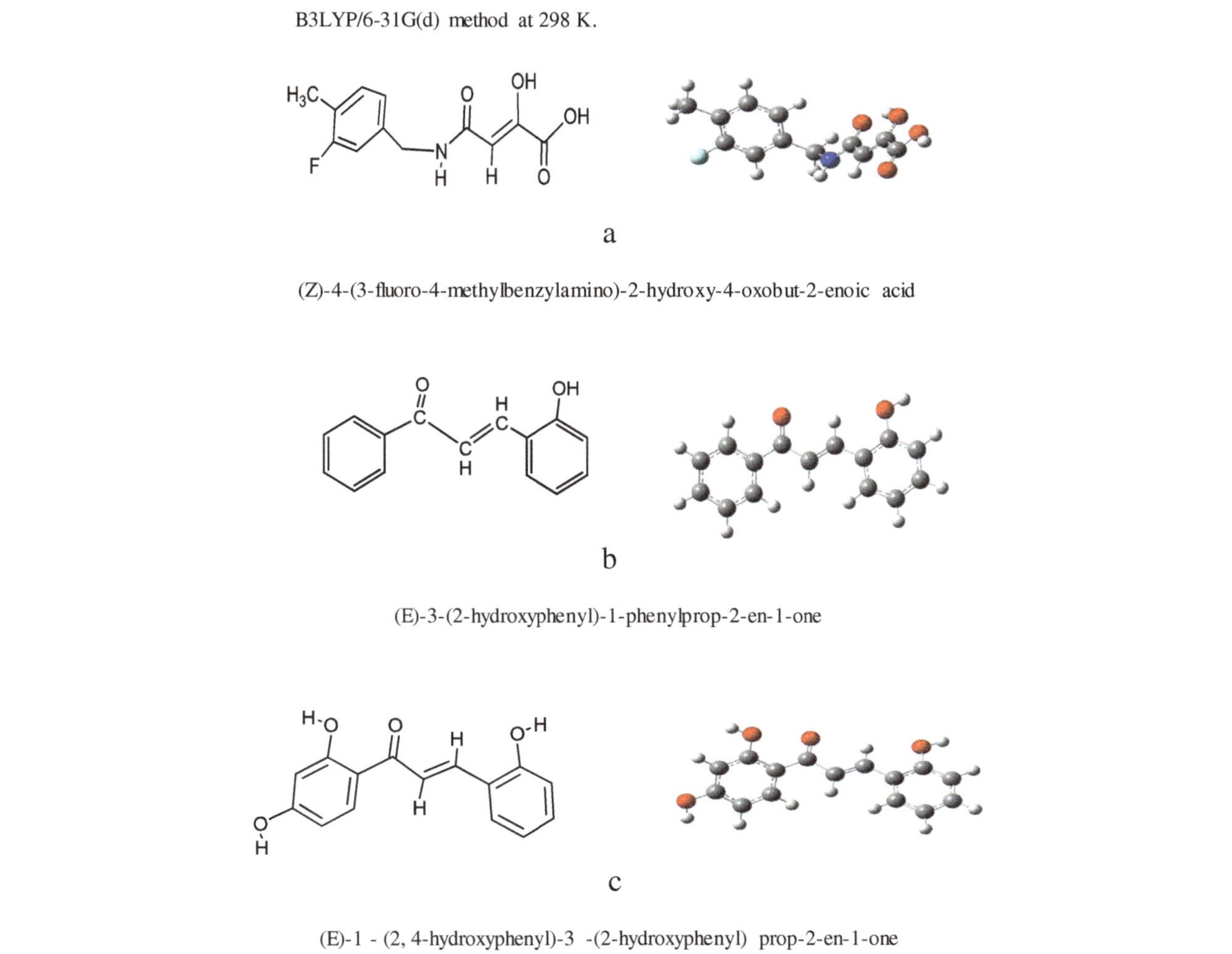

Figure 2 The structures of chalcone derivative. Optimizations were performed by B3LYP/6-31G(d) method at 298 K. **(a)** (Z)-4-(3-fluoro-4-methylbenzylamino)-2-hydroxy-4-oxobut-2-enoic acid. **(b)** (E)-3-(2-hydroxyphenyl)-1-phenylprop-2-en-1-one. **(c)** (E)-1-(2, 4-hydroxyphenyl)-3-(2-hydroxyphenyl)prop-2-en-1-one.

Table 1 The parameters of NBO complexes 1, 2, and 3 by B3LYP/631G(d) method at 298.15 K

Agent	Acceptor	Occupancy	Donor	Occupancy	Hybrid	ΣE2
Complex 1	σ* C28-30		LP(1)O84	0.03683		21.27
	π* C29-30	1.93552	LP(1)O84	0.23326	*sp*1.84	
	σ* C28-30	1.84587	LP(2)O84	0.03683	*sp*99.99	
	σ* C29-30		LP(2)O84	0.02234		
	π* C29-30		LP(2)O84	0.23326		
Complex 2	σ* C28-30		LP(1)O84	0.03694		21.25
	π* C29-30	1.93477	LP(1)O84	0.23167	*sp*1.99	
	σ* C28-30	1.84823	LP(2)O84	0.03694	*sp*12.91	
	σ* C29-30		LP(2)O84	0.02252		
	π* C29-30		LP(2)O84	0.23167		
Complex 3	σ* C28-30		LP(1)O84	0.03705		25.67
	π* C29-30	1.93368	LP(1)O84	0.24195	*sp*1.87	
	σ* C28-30	1.84353	LP(2)O84	0.03705	*sp*1.00	
	σ* C29-30		LP(2)O84	0.02139		
	π* C29-30		LP(2)O84	0.24195		

Table 2 The parameter of bond distances, bond angles, and torsional angles in B3LYP/631G(d) method optimized complexes 1, 2, and 3 at 298.15 K

Agent	Complex1	Complex 2	Complex3
$C_{30}O_{84}$	1.38360	1.38918	1.38844
$O_{84}C_{85}$	1.36355	1.37733	1.37905
$C_{85}...C_{87/86}$	1.50671	1.41639	1.41556
$C_{85}=C_{86}$ or $C_{85}...C_{87}$	1.34399	-	1.39631
$C_{88}...N_{92}$	1.37423	-	-
$C_{88}=C_{92}$	-	1.35015	1.34867
$C_{88}=O_{93/92}$	1.22635	1.33211	-
$C_{95/92}\text{-}C_{97}$	1.51805	1.48257	1.49038
$C_{97}=O_{101}$	-	1.23025	1.22693
$C_{89/88}H_{93}$	-	1.08671	1.08697
$C_{97}C_{101/100}$	1.40144	1.50356	1.49705
$C_{102}F_{107}$	1.35539	-	-
$C_{108}O_{112}$	-	-	1.3632
$N_{92}C_{95}$	1.46093	-	-
$C_{30}O_{84}C_{85}$	119.49672	119.98567	119.72205
$O_{84}C_{85/95}...C_{86}$	122.94446	115.99633	116.30968
$C_{85}...C_{86}...H_{89}/C_{89}$	115.63440	117.35796	117.31516
$C_{86}C_{89}=N_{92}/C_{92}$	112.89707	127.05408	126.82625
$C_{92}C_{97}=C_{101}/O_{101}$	120.54112	121.23147	121.14706
$C_{100}...C_{103}F/O/H^{1}_{107}$	118.48520	117.99263	118.85317
$C_{30}O_{84}C_{85}...C^{2}_{87}$	65.15798	10.06339	10.44740
$C_{85}=C_{86}C_{89}N_{92}/H^{3}_{95}$	−171.74691	178.34956	178.08447
$C_{28}...C_{30}O_{84}C^{4}_{85}$	−160.04093	74.54557	75.32017
$C_{106}...C_{108}O_{112}H^{5}_{113}$	-	-	179.82075
$C_{88}...N_{92}C_{95}F^{6}_{107}$	179.93485	-	1.32961

complex 2 is stable. The hybrid orbital S of a compound is lower. The occupancy factor is larger. In complex 2, the hybrid and occupancy are $sp^{1.88}$ and 1.9355, respectively.

The calculations of the total energies and energy hyperconjugation (E^2) of the optimized structures, dipole moments (μ), occupancy, and hybrid orbital at B3LYP/631G(d) levels are presented in Tables 2 and 3. The DFT calculated geometric parameters for complexes 1, 2, and 3 are compared in Table 2. The bond lengths of $C_{30}\text{-}O_{84}$ calculated for complexes 1, 2, and 3 at the DFT level are 1.383, 1.389, and 1.388 Å, respectively at the B3LYP/631G(d) level. The bond length of $C_{85}...C_{87}$ for complex 1 is 1.50671 Å; for complexes 2 and 3, 1.41639 and 1.41556 Å, respectively. The bond length of $C_{88}...N_{92}$ (1.37423 Å) in complex 1 is lower than the bond length of $C_{88}=C_{92}$ in complexes 2 (1.35015 Å) and 3 (1.34867 Å). This is because the electronegative nitrogen is bigger than carbon. The bond length of $C_{88}=O_{93/92}$ in complexes 1 and 2 are

Table 3 The parameter of Mulliken charge, energy HOMO/LUMO and gap energy in B3LYP/631G(d) method at 298.15 K

Agent	C_{30} Mulliken charge	Energy of HOMO (Kcal mol^{-1})	O_{84} Mulican charge	Energy of LUMO (Kcal mol^{-1})	Gap of energy
Complex 1	0.282424	−0.1579	−0.586927	−0.0889	0.06906
Complex 2	0.282510	−0.1567	−0.586736	−0.0876	0.06911
Complex 3	0.291184	−0.1538	−0.581214	−0.0855	0.06837

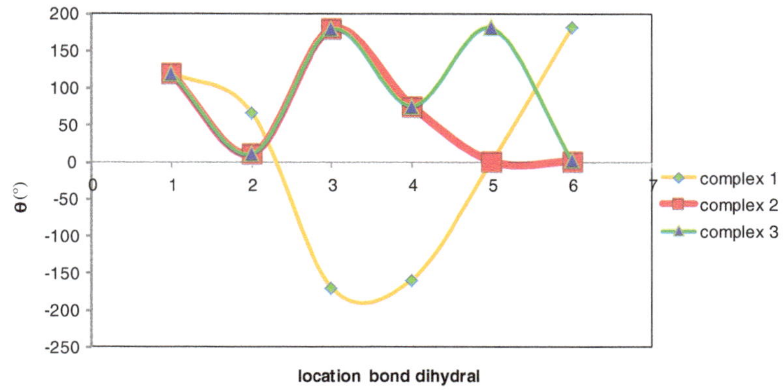

Figure 3 Parameter of dihedral angle change to locations 1, 2, 3, 4, 5, and 6 of complexes 1, 2, and 3 formed in B3LYP/631g(*d*) method at 298.15 K.

1.22636 and 1.33211 Å respectively lower than those of C=C, C...N in complexes 2, 3, and 1. The dihedral angles of $C_{100}...C_{103}F/O/H^{1}_{107}$ for complexes 1, 2, and 3 are 118.48520°, 117.99263°, and 118.85317° (in F, O, and H atoms). In locations 1, 2, 3, 4, 5, and 6, the hydra angle differences are observed in Table 1 and Figure 3. In Table 3, the Mulliken charges in donor electronegative atom O_{84} and acceptor C_{30} are negative and positive, respectively. The gap of energy of complex 2 is larger than that of complexes 1 and 3. Therefore, complex 2 is stable (Table 4).

Conclusions

Based on the above discussion, the following conclusions were made:

- The Mulliken charges of the donor atoms are negative.
- The formed bond between oxygen and carbon is stronger and has more hybrid S orbital sharing, so the bond length becomes shorter.
- There is no reaction between the nanotube and drugs in the normal body temperature.
- NBO analysis shows high hyperconjugations in all compounds.
- By increasing the hybrid *P* orbital sharing, the occupancy decreases.

Table 4 Comparison between stability energy and moment dipole of complexes 1, 2, and 3 formed in B3LYP/631g(d) method at 298.15 K

Agent	Stability energy (Kcal mol^{-1})	Dipole moment (Debye)
Complex 1	−1,900,685.711	−0.2357
Complex 2	−1,995,192.983	1.09133
Complex 3	−1,972,541.490	0.5679

Methods

A computer was used to account all calculations, which has Intel® Core™ 2 quad CPU 8400 with 4 GB RAM Gaussian 98 plan package [19] Gauss view, and nanotube model [20] maker at DFT of theory, the B3LYP useful hybrid [21] with the standard of 631G(d) basis set [22,23]. Schemes were used to display the geometric optimization of the nanotube (6, 6) with 84 atoms and compound chalcones (HIV-1 inhibitors). Separately (Figures 1, 2, and 3) then, complexes 1,2, and 3 were formed. The reaction of the nanotube and chalcone is shown in Equation 1.

$$\text{Chalcone} + \text{Nanotube}(6,6)\text{Chalcone–Nanotube}(6,6) + H_2 \quad (1)$$

Competing interests

The authors declare that they have no competing interests.

Authors' contributions

SG and MG participated in writing the manuscript. Both authors read and approved the final manuscript.

Acknowledgments

The authors would like to thank the financial support by Islamic Azad University, Rasht, Iran.

References

1. Iijima, S: Helical microtubules of graphitic carbon. Nature **354**, 56 (1991)
2. Colomer, JF, Henrard, L, Lambin, P, Van Tendeloo, G: Electron diffraction study of small bundles of single-wall carbon nanotubes with unique helicity. Physic Rev B **64**, 125425 (2001)
3. Pouton, CW, Seymour, LW: Key issues in non-viral gene delivery. Adv. Drug Deliv. Rev. **46**, 187–203 (2001)
4. Singh, R, Pantarotto, D, Lacerda, L, Pastorin, G, Klumpp, C, Prato, M, Bianco, A, Kostarelos, K: Tissue biodistribution and blood clearance rates of

intravenously administered carbon nanotube radiotracers. Proc. Natl. Acad. Sci. USA **103**, 3357–3362 (2006)
5. Wong Shi Kam, N, Jessop, TC, Wender, PA, Dai, H: Nanotube molecular transporters: internalization of carbon nanotube-protein conjugates into mammalian cells. J. Am. Chem. Soc **126**, 6850–6851 (2004)
6. Pastorin, G: Crucial functionalizations of carbon nanotubes for improved drug delivery: a valuable option? Pharm. Res. **26**, 746–769 (2009)
7. Taft, BJ, Lazareck, AD, Withey, GD, Yin, A, Xu, JM, Kelley, SO: Site-specific assembly of DNA and appended cargo on arrayed carbon nanotubes. J. Am. Chem. Soc. **126**, 12750–12751 (2004)
8. Pantarotto, D, Singh, R, McCarthy, D, Erhardt, M, Briand, JP, Prato, M, Kostarelos, K, Bianco, A: Functionalized carbon nanotubes for plasmid DNA gene delivery. Angew. Chem. Int. Ed. **43**, 5242–5236 (2004)
9. Cato, MH, D'Annibale, F, Mills, DM, Cerignoli, F, Dawson, MI, Bergamaschi, E, Bottini, N, Magrini, A, Bergamaschi, A, Rosato, N, Rickert, RC, Mustelin, T, Bottini, M: Cell-type specific and cytoplasmic targeting of PEGylated carbon nanotube-based nanoassemblies. J. Nanosci. Nanotechnol. **8**, 2259–2269 (2008)
10. Gannon, CJ, Cherukuri, P: Carbon nanotube-enhanced thermal destruction of cancer cells in a noninvasive radiofrequency field. Cancer **110**, 2654–2665 (2007)
11. Cherukuri, P, Bachilo, SM, Litovsky, SH, Weisman, RB: Near-infrared fluorescence microscopy of single-walled carbon nanotubes in phagocytic cells. J. Am. Chem. Soc. **126**, 15638–15639 (2004)
12. Dumortier, H, Lacotte, S, Pastorin, G, Marega, R, Wu, W, Bonifazi, D, Briand, JP, Prato, M, Muller, S, Bianco, A: Functionalized carbon nanotubes are non-cytoxic and preserve the functionality of primary immune cells. Nano. Lett. **6**, 1522–1528 (2006)
13. Kostarelos, K, Lacerda, L, Pastorin, G, Wu, W, Wieckowski, S, Luangsivilay, J, Godefroy, S, Pantarotto, D, Briand, JP, Muller, S, Prato, M, Bianco, A: Cellular uptake of functionalized carbon nanotubes is independent of functional group and cell type. Nat. Nanotechnol. **2**, 108–113 (2007)
14. Ramkumar, K, Serrao, E, Odde, S, Neamati, N: HIV-1 integrase inhibitors: 2007–2008 update. Med. Res. Rev. **30**, 890–954 (2010)
15. Kostanecki, SV, Tambor, J: Ueber die sechs isomeren monooxybenzalacetophenone (monooxychalkone). Ber. Dtsch. Chem. Ges. **32**, 1921–1926 (1899)
16. Moyle, G, Gatell, J, Perno, CF, Ratanasuwan, W, Schechter, M, Tsoukas, C: Potential for new antiretrovirals to address unmet needs in the management of HIV-1 infection. AIDS Patient Care STDS **22**, 459–471 (2008)
17. Wrodnigg, TM, Eder, B: Topics in Current Chemistry, p. 215. Springer, Berlin (2001)
18. Stephens, PJ, Devlin, FJ, Chabalowski, CF, Frisch, MJ: Ab initio calculation of vibrational absorption and circular dichroism spectra using density functional force fields. J. Phys. Chem. **98**, 11623–11624 (1994)
19. Frisch, MJ, Trucks, GW, Schlegel, HB, Scuseria, GE, Robb, MA, Cheeseman, JR, Montgomery Jr, JA, Stratmann, RE, Burant, JC, Dapprich, S, Millam, JM, Daniels, AD, Kudin, KN, Strain, MC, Farkas, O, Tomasi, J, Barone, V, Cossi, M, Cammi, R, Mennucci, B, Pomelli, C, Adamo, C, Clifford, SO, Chterski, J, Petersson, GA, Ayala, PY, Cui, Q, Morokuma, K, Malick, DK, Rabuck, AD, et al.: Gaussian 98. Gaussian Inc., Pittsburgh, PA (1998)
20. J Crystal Soft: JCrystal. www.jcrystal.com (2013). Accessed 28 Feb 2013
21. Becke, AD: A new mixing of Hartree–Fock and local density-functional theories. J. Chem. Phys. **98**, 1372–1377 (1993)
22. Frisch, AE, Foresman, JB: Gaussian 98 User's Reference. Gaussian, Wallingford (1998)
23. Foresman, JB, Frisch, A: Exploring Chemistry with Electronic Structure Methods, 2nd edn. Gaussian, Wallingford (1996)

Functionalized multi-walled carbon nanotubes for enhanced photocurrent in dye-sensitized solar cells

Chandra Sekharan Nair Omana Amma Sreekala[1*], Jinchu Indiramma[1], Kandala Bala Subramanya Pavan Kumar[1], Karyaveettil Savithriamma Sreelatha[1,2] and Mahesh Saran Roy[3]

Abstract

The influence of the incorporation of nitric acid-treated (functionalized) multi-walled carbon nanotubes (f-MWCNTs) and unmodified MWCNTs in TiO_2 films are investigated by observing the photocurrent-voltage characteristics of dye-sensitized solar cells (DSSCs) with and without MWCNTs. The short-circuit photocurrent (J_{sc}) of the modified DSSC increased by 46% compared with that of a cell with plain TiO_2 film. The open-circuit voltage remained the same for all cases. The enhanced J_{sc} is explained by the increased surface area of the film, enhanced cluster formation of TiO_2 particles around f-MWCNTs, and improved interconnectivity of TiO_2 particles in the presence of f-MWCNTs. The efficiency of the cell increased by 45% due to J_{sc} enhancement.

Keywords: Multi-walled carbon nanotubes, Dye-sensitized solar cell, Short-circuit photocurrent, Open-circuit voltage, Cluster formation

Background

Dye-sensitized solar cells (DSSCs) have attracted much attention in the last two decades since reported for the first time by Gratzel [1]. Attempts are ongoing to increase the adsorption of dyes across the solar spectrum, smoothen the function of the transport of photoexcited electrons, and encourage the diffusion of electrolyte ions. The process involves the injection of electrons from the photoexcited dye into the conduction band of the semiconductor oxide (TiO_2), from where they pass through the transparent conducting fluorine-doped tin oxide (FTO) current collector and finally into the external circuit. Furthermore, the sensitizer is regenerated by electron transfer from a donor, typically iodide ions, which are dissolved in the electrolyte present in the porous semiconductor. The triiodide ions formed during the reaction diffuse to the counter electrode, where they are reduced back to iodide by the conduction band electrons that have passed though the external circuit that performs the electrical work.

At present, poly(3,4-ethylenedioxythiophene):poly (styrenesulfonate)(PEDOT:PSS)-coated FTO glass is used as a counter electrode instead of a costly platinum-sputtered FTO glass to catalyze the reduction of $I^{3-}$to I^- in the redox electrolyte [2-4]. Recently, it was reported that the conductivity of PEDOT:PSS film is enhanced if a liquid organic compound such as dimethyl sulfoxide (DMSO), dimethyl formamide (DMF), or tetrahydrofuran is added to the PEDOT:PSS aqueous solution [5,6]. The choice of preparation method for depositing conducting polymers on the conducting glass is also significant in determining the efficiency of DSSCs. Several methods, such as spin coating [7] and chemical polymerization [3], have been employed to deposit conductive polymers on the counter electrode, both of which provide good electrochemical stability. Lee et al. [8] further modified the counter electrode. Using multi-walled carbon nanotubes, they were able to replace the costly platinum electrode, with little change in efficiency of the cell.

Modification of the working electrode is also important. Vincent et al. [9] reported the inclusion of carbon nanotubes in an inorganic TiO_2 sol–gel matrix. Huang and Gao studied the growth behavior of multi-walled

* Correspondence: cosreekala@gmail.com
[1]School of Biotechnology, Amrita Vishwa Vidyapeetham, Amritapuri Campus, Kollam, Kerala 690525, India
Full list of author information is available at the end of the article

carbon nanotubes (MWCNTs) by incorporating them on immobilized rutile TiO_2 (r-TiO_2) [10]. Photovoltaic devices were also fabricated by Ago and colleagues using composites of MWCNTs and conjugated polymers [11]. Ionic gel electrolytes were prepared by Usui et al. by dispersing MWCNTs into ionic liquid electrolytes. They achieved higher energy conversion efficiency with corresponding DSSCs, relative to those using untreated ionic liquid electrolytes [12]. It has been reported that the DSSCs fabricated with anatase and r-TiO_2 are comparable in their photovoltaic properties at one-sun intensity (100 mW cm^{-2}) [13,14]. It has been shown that r-TiO_2 has superior light-scattering properties compared to anatase TiO_2 because of its higher refractive index. Other benefits of r-TiO_2 are that it is cheaper to produce and is chemically more stable [15]. However, the electron transport is slower in the rutile layer than in the anatase layer because of the interparticle connectivity associated with particle packing density and is evident from the data suggested by intensity-modulated photocurrent spectroscopy and scanning electron microscopy [13]. Sung et al. reported that the introduction of MWCNTs in poly(methyl methacrylate) polymers increased the conductivity of the composite [16]. On treating carbon nanotubes with a concentrated acid mixture of H_2SO_4 and HNO_3 [17-20] or of H_2SO_4 and $KMnO_4$ [21], carboxylic acid groups can be introduced into the nanotubes in addition to shortening them [22]. It is assumed that MWCNTs attached with carboxylic acid groups can adhere more strongly to TiO_2 particles than untreated MWCNTs due to the strong covalent bond formed between the dye molecule and the TiO_2, resulting in more efficient DSSCs.

In the present study, Rose bengal, an organic dye whose molecular structure is shown in Figure 1, is used as the photosensitizer, and a graphite-coated DMSO-treated PEDOT:PSS is used as the counter electrode for our dye-sensitized solar cell. In the present research, we have preferred MWCNTs and studied their beneficial influence, with and without modification, when incorporated into P25 TiO_2 films and their overall impact on the photovoltaic properties of our DSSCs. Three types of investigations are presented in this paper. First, MWCNTs are incorporated in P25 TiO_2 films and sensitized with Rose bengal dye. In the second investigation, P25 TiO_2 films are sensitized with a solution of Rose bengal dye and f-MWCNTs (MWCNTs functionalized with -COOH), and in the third one, f-MWCNTs are incorporated in the P25 TiO_2 films and then sensitized with Rose bengal dye.

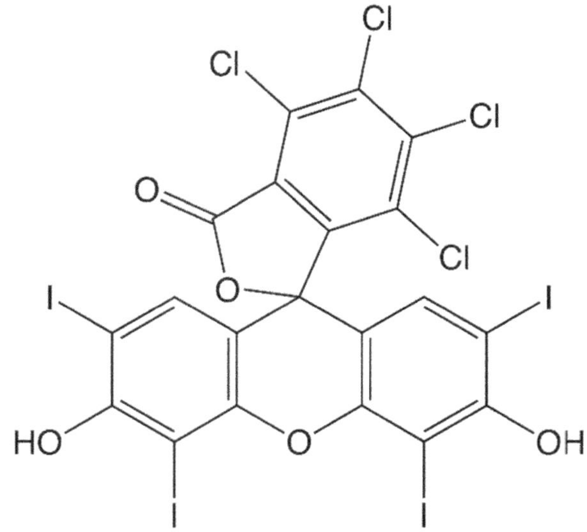

Figure 1 Molecular structure of Rose bengal.

Results and discussion

An FTIR spectrum of MWCNTs is shown in Figure 2a. It was then compared with the FTIR spectrum of f-MWCNTs shown in Figure 2b. The attachment of carboxylic acid groups to acid-treated MWCNTs (f-MWCNTs) is identified. FTIR spectra of the f-MWCNTs showed a C=O peak of the COOH groups at 1,730 cm^{-1} and a broad OH peak around 3,400 cm^{-1} (not shown); an additional C=C peak of MWCNTs occurred near 1,600 cm^{-1}. On the basis of these results, the presence of COOH groups in the f-MWCNTs was confirmed.

As a result of the shortening and simultaneous functionalization with COOH groups by nitric acid, f-MWCNTs dispersed well in water [23] and DMF as shown in Figure 3. Their good miscibility with water is beneficial for the fabrication of TiO_2 film electrodes.

The AFM images of the prepared films, unmodified TiO_2 film(Figure 4a), MWCNT-incorporated TiO_2 film(Figure 4b), and f-MWCNT-incorporated TiO_2 film(Figure 4c), were taken and analyzed. Interestingly, the calculated root-mean-square (RMS) roughness values were 117, 153, and 103 nm, respectively. By the introduction of MWCNTs, the roughness value was found to have increased due to the MWCNTs not fitting well in the gaps between the TiO_2 nanoparticles. However, when the MWCNTs were functionalized, the RMS roughness value was found to have decreased. This reduction in roughness was attributed to the smoothing effect induced by the f-MWCNTs as they better fit in the gaps between the TiO_2 nanoparticles. This indicates that there is more favorable cluster formation in P25 TiO_2 films in the presence of f-MWCNTs. Also, there is an increased surface area of the films fabricated in the presence of f-MWCNTs.

Figure 5 shows the Raman spectra of the TiO_2 film which is compared with the TiO_2 film in the

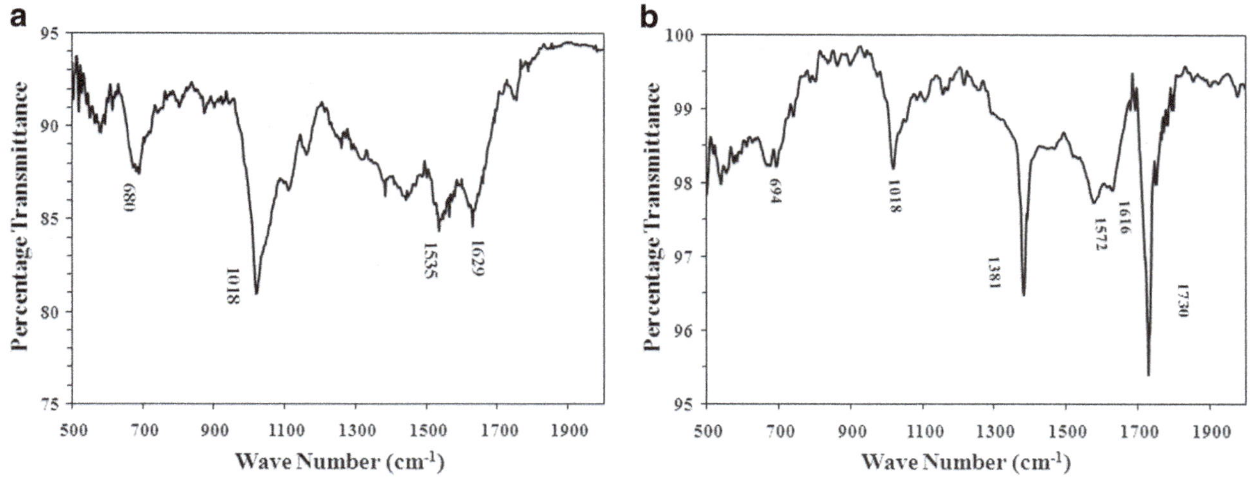

Figure 2 FTIR spectra of (a) MWCNTs and (b) f-MWCNTs.

presence of MWCNTs and f-MWCNTs. It is found that the MWCNT-incorporated film shows blueshifts by 4 cm^{-1} and f-MWCNTs shifts by 7 cm^{-1}. These blueshifts can be attributed to the strain effects [24] at the f-MWCNT/TiO_2 interfaces, which may influence the vibration frequencies. The differences in the Raman spectra between f-MWCNTs versus MWCNTs are due to the increased adherence of TiO_2 particles to f-MWCNTs.

The optical absorption spectra of the P25 (the standard electrode), PE1, PE2, and PE3 electrodes are shown in Figure 6. It is seen in the figure that for all the photoelectrodes, the absorption band extends from 320 to 600 nm, but the intensity of absorption is different for different electrodes. The graphs show that the dye-enhanced absorption in PE3 is more than that of other electrodes. This is due to the increased surface area of the films fabricated in the presence of f-MWCNTs.

Figure 7 shows the current density-voltage (J-V) curves of DSSCs made by incorporating MWCNTs, with and without functionalization, in the TiO_2 films, compared with that of a cell fabricated with P25 film only. The results are summarized in Table 1. The illumination intensity used was 100 mW cm^{-2}. Compared to a cell fabricated with P25 film only, the cell with PE1 shows a higher J_{sc} by about 5%. Similarly, the cells with PE2 and PE3 show J_{sc} improvements of about 17% and 46%, respectively. However, there are hardly any changes in the open-circuit voltage (V_{oc}) of the four cells. The essential reason for the enhanced J_{sc} is attributed to the increased surface area of the TiO_2 films in the presence of f-MWCNTs. This can be verified by measuring absorbance of desorbed dye from the TiO_2 film. Enhanced quantity of dye adsorption is indicative of the enhanced surface area of TiO_2 film for a monolayer of dye on the TiO_2 particles.

Absorbance measurements of the desorbed dye from TiO_2 films with and without f-MWCNTs in Figure 8 indicate increased dye adsorption on the surface. The J_{sc} increases are consistent with this order of increase of surface area. The increased surface area of TiO_2 films with the incorporation of f-MWCNTs may be due to additional adsorption sites gained from the adherence of TiO_2 particles to f-MWCNTs that are otherwise not available when TiO_2 particles aggregate themselves in the P25 film. As a result of the J_{sc} enhancements, the overall energy conversion efficiency (η) increases by 7%, 32%, and 60% for the cells prepared with films PE1, PE2, and PE3, respectively, when compared to that of the cell made with P25 film only. The surface area of the TiO_2 films is increased in the presence of f-MWCNTs which

Figure 3 MWCNTs (1) and f-MWCNTs (2) in DMF.

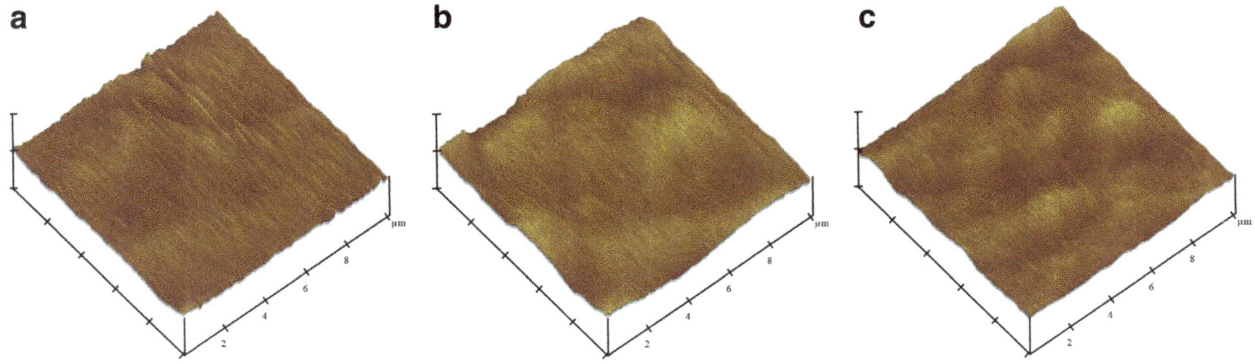

Figure 4 AFM images of the prepared films. (**a**) Unmodified TiO_2film, (**b**) MWCNT-incorporated TiO_2film, and (**c**) f-MWCNT-incorporated TiO_2 film.

is the reason for the enhanced J_{sc}. This is evident from the absorption spectra of the photoelectrodes (Figure 6).

Enhanced quantity of dye adsorption in the presence of f-MWCNTs is indicative of the enhanced surface area of the TiO_2 film. This suggests that the surface area of TiO_2 films for dye adsorption has increased in the order of surface area of P25<PE1<PE2<PE3. It may be due to additional adsorption sites gained from the adherence of TiO_2 particles to f-MWCNTs. The J_{sc} increases are consistent with this order of increase of surface area. The increased surface area of TiO_2 films and the incorporation of f-MWCNTs provide for better dispersion of TiO_2 particles compared to TiO_2 particle aggregation in unmodified P25 film. Because of the fine distribution of f-MWCNTs in the TiO_2 films, as can be understood from the AFM images (Figure 5), a better dispersion of TiO_2 particles in the PE3 can be visualized. The more even dispersion of TiO_2 particles provides additional adsorption sites for the dye. The favorable TiO_2 cluster formation induced by f-MWCNTs in the films on the order of that of P25<PE1<PE2<PE3 is in agreement with the order of the J_{sc} increase. AFM studies of the photoelectrodes also verify the same. The J_{sc} enhancement can also be explained by the increase in interconnection among the TiO_2 particles in the TiO_2 films prepared with f-MWCNTs. Formation of f-MWCNT-induced clusters, as mentioned already, enables an increase in interconnection among the TiO_2 particles in the film, relative to that among particles in films without f-MWCNTs. The increased interconnectivity in turn increases the electrical conductivity of the film in the presence of f-MWCNTs. In addition to improved interconnectivity among TiO_2 particles in the films in the presence of MWCNTs, the anchoring of TiO_2 particles to nanotubes can promote charge separation, owing to the fact that carboxylic acid groups of f-MWCNTs are also able to attach themselves to TiO_2 particles. Since absorbance is proportional to the free electron density in the conduction band of TiO_2, the implication is a higher electron transport in films PE1, PE2, and PE3. The absorbance pattern of films P25, PE1, PE2, and PE3 are consistent with their respective J_{sc} enhancements. Electrons trapped in the surface states appear to be responsible for the slower photocurrent response in the case of the DSSC with bare P25 film [25,26]. It is to be emphasized here that the performance of the cell has not been optimized with regard to film thickness, dye purification, and device architecture. This study focused on the influence of MWCNTs and TiO_2/f-MWCNTs, when incorporated

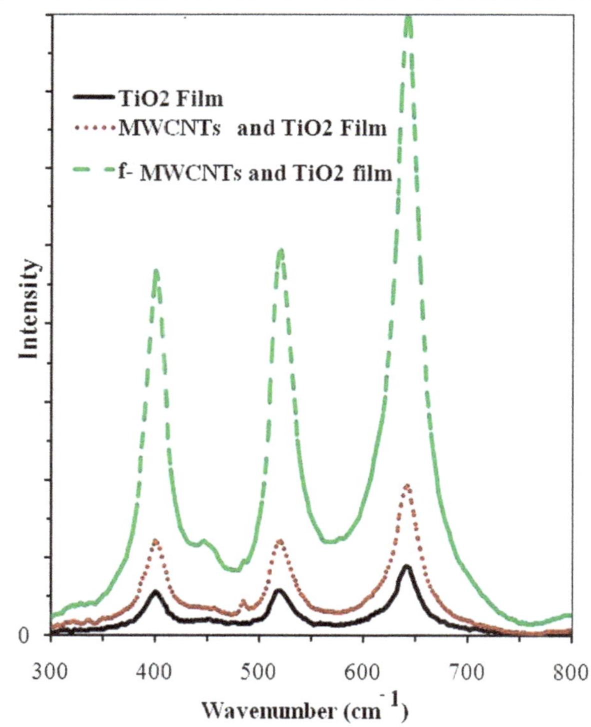

Figure 5 Raman spectra of the TiO_2 film compared with TiO_2 film with MWCNTs and f-MWCNTs.

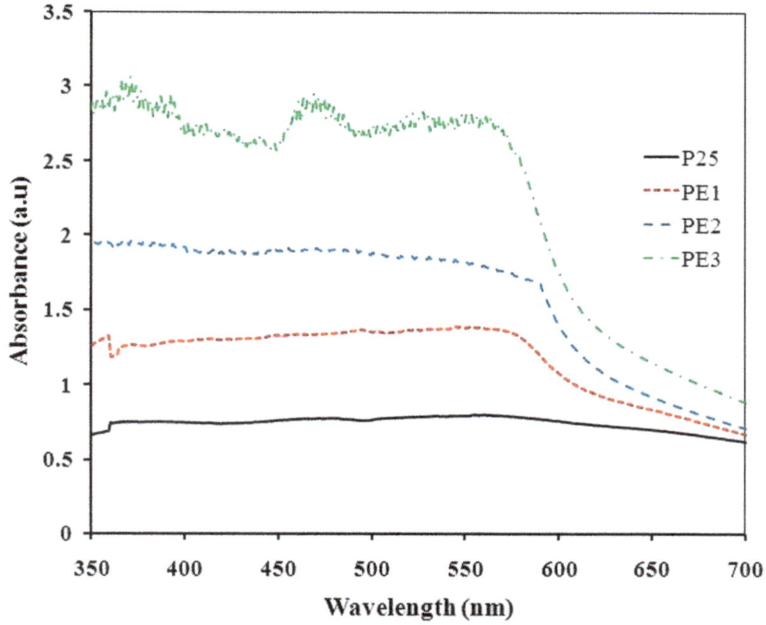

Figure 6 Optical absorption spectra of P25, PE1, PE2, and PE3 electrodes.

in P25 TiO_2 film, on the photovoltaic properties of the corresponding DSSCs. Comparative measurements are made in this regard with reference to a cell without f-MWCNTs.

Conclusions

Incorporation of acid-treated f-MWCNTs in P25 TiO_2 films yields a considerable increase in J_{sc} of about 46% over unmodified P25 TiO_2 films with the same open-circuit voltage for the pertinent dye-sensitized solar cells, with respect to the unmodified cell. The enhanced essential J_{sc} is attributed to the increased surface area of the films fabricated in the presence of f-MWCNTs. The J_{sc} enhancement also arises from more favorable cluster formation in P25 TiO_2 films in the presence of f-MWCNTs than that in the absence of them. Improved

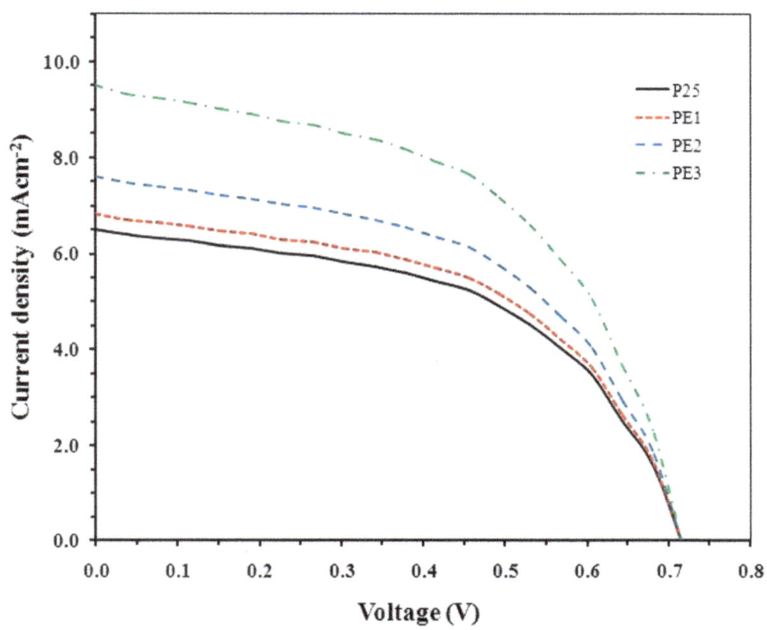

Figure 7 *J-V* curves of DSSCsmade by incorporating MWCNTs in the TiO_2 films.

Table 1 Photovoltaic parameters of the DSSCs with different photoelectrodes

Photoelectrode	J_{sc}(mAcm^{-2})	V_{oc} (V)	FF	Efficiency(%)
P25	6.51	0.68	0.54	2.41
PE1	6.83	0.68	0.54	2.53
PE2	7.62	0.68	0.54	2.82
PE3	9.50	0.68	0.54	3.52

interconnectivity among TiO_2 particles in the presence of the f-MWCNTs is seen as another reason for the J_{sc} enhancements. The J_{sc} enhancement is also consistent with the increase in dye adsorption of the respective photoelectrodes in the cells.

Methods

Materials

MWCNTs with a length of 1 to 2 μm and a diameter of 10 to 30 nm were prepared by CVD and purchased from Intelligent Materials Pvt. Ltd., Chandigarh, India. Fluoride-doped tin oxide-coated glass substrate with a resistance of 10 Ω/cm^2 was purchased from Solaronix, Aubonne, Switzerland. Titanium dioxide (Degussa P25) and Rose bengal (RB) dye were procured from Orion Chem. Pvt. Ltd., Navi Mumbai, India, and HiMedia Laboratories Pvt. Ltd., Mumbai, India, respectively. For electrolyte preparation, polyethylene glycol (PEG;MW 400), potassium iodide, and iodine were procured from Aldrich, St. Louis, MO, USA. PEDOT:PSS is also from Aldrich. Other chemicals such as DMSO, nitric acid, DMF, 2-propanol, acetylacetone, Triton X-100, and acetonitrile are from Merck, Mumbai, India.

Preparation of photoelectrodes

The conducting glass substrates (FTO) are cleaned and rinsed with deionized water and 2-propanol and then soaked in propanol for 12 h. The FTO substrates are air-dried prior to film preparation. The TiO_2 colloidal solution is prepared by grinding the TiO_2 (Degussa P25) powder purchased from Aldrich with 2 ml of distilled water and acetylacetone for 30 min. Finally, 8.0 ml of distilled water and 0.1 ml of Triton X-100 are added with continuous mixing for 10 min, and the FTO is coated using the doctor blade technique. Functionalization of MWCNTs is done as follows: 10 mg of MWCNTs is mixed with 70% concentrated HNO_3 and is refluxed for 8 h at 140°C. The resulting product is cooled and centrifuged at 4,500 rpm for 30 min, followed by washing with distilled water at 3,000 rpm for 45 min. MWCNTs and f-MWCNTs in DMF are shown in Figure 3. It can be seen that f-MWCNTs disperse well in organic liquids. The final product is dried at 80°C, designated as f-MWCNTs, and characterized by Fourier transform infrared (FTIR) spectrophotometry and Raman spectroscopy.

The standard photoelectrode, P25, was prepared as follows: TiO_2 was coated on FTO conducting glassand fired at 450°C for 30 min, and Rose bengal dye was adsorbed through 12 h of sensitization. The other electrodes, PE1, PE2, and PE3, were prepared as follows: PE1: MWCNTs (0.5mg) were incorporated in P25 TiO_2 (1.2 g) by direct mixing to form a colloidal paste, coated

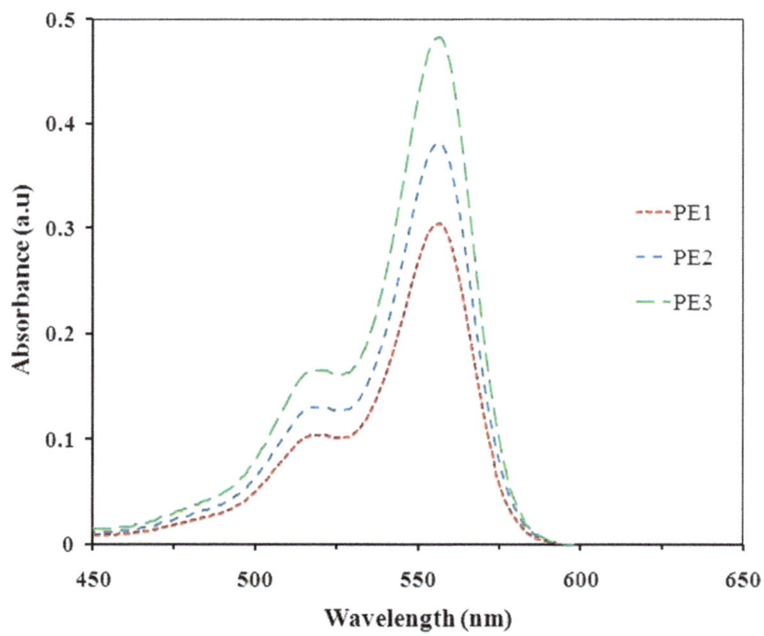

Figure 8 Absorbance measurements of the desorbed dye from TiO_2 films with and without f-MWCNTs.

on FTO conducting glass, and fired at 450°C and Rose bengal dye was adsorbed through 12 h of sensitization. PE2 was made by sensitizing fired P25 TiO_2-coated FTO glass with a mixture of Rose bengal dye and f-MWCNTs. PE3 was made by sensitizing the Rose bengal dye on fired FTO glass coated with a mixture of P25 TiO_2 and f-MWCNTs. Nearly the same TiO_2 film thickness is maintained in all four cases, this being 16 μm. The reason for these thick films is to prevent MWCNTs from protruding from the films, which could have caused short circuits in the DSSCs.

Preparation of counter electrode and assembling of DSSC

The PEDOT:PSS-coated counter electrode is prepared as explained in [27]. The dye-sensitized solar cell is assembled by clamping the dye-sensitized photoelectrode with a DMSO-treated PEDOT:PSS-coated FTO counter electrode to form a sandwich-type structure. Electrolyte solution was prepared as reported in [27], by taking a proportionate quantity of 0.5mol of KI,0.05mol of iodine, and 0.14 mol of PEG (MW 400) in 50 ml of acetonitrile solvent. The conductivity of the electrolyte solution is recorded with a conductivity bridge meter, and it was found to be 13.6 mS cm^{-1}. One drop of this electrolyte solution was injected into the interspace between the photoelectrode and the counter electrode. The schematic diagram of the device is shown in Figure one of our previous paper [27].

The absorption spectrum of the fabricated device was recorded using an Analytik Jena photodiode array (New Delhi, India) or a PerkinElmer Lambda 35 spectrophotometer (Waltham, MA, USA). FTIR spectra were obtained using a Shimadzu IRAffinity-1 FTIR spectrophotometer (Kyoto, Japan) and were used to identify the formation of carboxylic acid groups on the f-MWCNTs. Raman spectra were obtained using an Avalon Instruments Raman Station R3 (Boston, MA, USA) in which a green laser of λ=532 nm was used. The scanning range was from 200 to 4,000 cm^{-1} to identify the attachment of TiO_2 to f-MWCNTs. Atomic force microscopy (AFM) images were taken using a Veeco Digital Instruments AFM Nanoscope 111A, software version 5 (Plainview, NY, USA). The tips used were contact mode tips. Photocurrent-voltage curves were measured using a Keithley Electrometer 6517A (Cleveland, OH, USA). A 300-W Xe lamp with an AM 1.5 spectrum and an output power of 100 mW cm^{-2} was used to illuminate the active area, 1 cm^2 of the TiO_2 electrode.

Competing interests

The authors declare that they have no competing interests.

Authors' contributions

JI and KBSP fabricated and characterized the device, and COS, KSS, and MSR performed the technical analysis. COS, KSS, and MSR participated in the design of the study and coordination. All authors read and approved the final manuscript.

Acknowledgement

We are grateful to the Department of Science and Technology (DST), Government of India, New Delhi, for the financial support in the form of a grant under WOS-A Scheme (SR/WOS-A/PS-04/2008).

Author details

[1]School of Biotechnology, Amrita Vishwa Vidyapeetham, Amritapuri Campus, Kollam, Kerala 690525, India. [2]Government Polytechnic College, Kottayam, Kerala 686013, India. [3]Defence Laboratory, Jodhpur, Rajasthan 342011, India.

References

1. Gratzel, M: Dye-sensitized solar cells. J. Photochem. Photobiol. C Photochem. Rev. **4**(2), 145–153 (2003)
2. Yohannes, T, Inganas, O: Photoelectrochemical studies of the junction between poly[3-(4-octylphenyl)thiophene] and a redox polymer electrolyte. Solar Energy Materials Solar Cells **51**(2), 193–202 (1998)
3. Bay, L, West, K, Jenssen, BW, Jacobsen, T: Electrochemical reaction rates in a dye-sensitised solar cell—the iodide/tri-iodide redox system. Solar Energy Materials Solar Cells **90**(3), 341–351 (2006)
4. Haywang, G, Jonnes, F: Poly(alkylenedioxythiophene)s—new, very stable conducting polymers. Adv. Mater. **4**(2), 116–118 (1992)
5. Kim, JY, Jung, JH, Lee, DE, Joo, J: Enhancement of electrical conductivity of poly(3,4-ethylenedioxythiophene)/poly(4-styrenesulfonate) by a change of solvents. Synth. Met. **126**(2–3), 311–316 (2002)
6. Quyang, J, Xu, Q, Chu, CW, Yang, Y, Li, G, Shinas, J: On the mechanism of conductivity enhancement in poly(3,4-ethylenedioxythiophene):poly(styrene sulfonate) film through solvent treatment. Polymer **45**(25), 8443–8450 (2004)
7. Satio, Y, Kube, W, Kitamura, T, Wade, Y, Yanagida, S: I–/I3–redox reaction behavior on poly(3,4-thylenedioxythiophene) counter electrode in dye-sensitized solar cells. J. Photochemi. Photobiol. A Chem. **164**(1–3), 153–157 (2004)
8. Lee, WJ, Ramasamy, E, Lee, DY, Song, JS: Efficient dye-sensitized solar cells with catalytic multiwall carbon nanotube counter electrodes. ACS Appl. Mater. Interfaces **1**(6), 1145–1149 (2009)
9. Vincent, P, Brioude, A, Journet, C, Rabaste, S, Purcell, ST, Le Brusq, J, Plenet, JC: Inclusion of carbon nanotubes in a TiO2 sol–gel matrix. J. Non-Cryst. Solids **311**(2), 130–137 (2002)
10. Huang, Q, Gao, L: Immobilization of rutile TiO2 on multiwalled carbon nanotubes. J. Mater. Chem. **13**(7), 1517–1519 (2003)
11. Ago, H, Petritsch, K, Shaffer, MSP, Windle, AH, Friend, RH: Composites of carbon nanotubes and conjugated polymers for photovoltaic devices. Adv. Mater. **11**(15), 1281–1285 (1999)
12. Usui, H, Matsui, H, Tanabe, N, Yanagida, S: Improved dye-sensitized solar cells using ionic nanocomposite gel electrolytes. J. Photochem. Photobiol. A Chem. **164**(1–3), 97–101 (2004)
13. Park, NG, van de Lagemaat, J, Frank, AJ: Comparison of dye-sensitized rutile- and anatase-based TiO_2 solar cells. J. Phys. Chem. B **104**(38), 8989–8994 (2000)
14. NG Park, G, Schlichthorl, J, Van De Lagemaat, HM, Cheong, A, Mascarenhas AJ, F: Dye-sensitized TiO2 solar cells: structural and photoelectrochemical characterization of nanocrystalline electrodes formed from the hydrolysis of TiCl4. J. Phys. Chem. B **103**(17), 3308–3314 (1999)
15. Kim, KJ, Benkstein, KD, van de Lagemaat, J, Frank, AJ: Characteristics of low-temperature annealed TiO2 films deposited by precipitation from hydrolyzed TiCl4 solutions. Chem. Mater. **14**(3), 1042–1047 (2002)
16. Sung, JH, Kim, HS, Jin, HJ, Choi, HJ, Chin, IJ: Nanofibrous membranes prepared by multiwalled carbon nanotube/poly(methyl methacrylate) composites. Macromolecules **37**(26), 9899–9902 (2004)
17. Liu, Z, Shen, Z, Zhu, T, Hou, S, Ying, L, Shi, Z, Gu, Z: Organizing single-walled carbon nanotubes on gold using a wet chemical self-assembling technique. Langmuir **16**(8), 3569–3573 (2000)
18. Nan, X, Gu, Z, Liu, Z: Immobilizing shortened single-walled carbon nanotubes (SWNTs) on gold using a surface condensation method. J. Colloid Interface Sci. **245**(2), 311–318 (2002)

19. Liu, J, Rinzler, AG, Dai, H, Hafner, JH, Bradley, RK, Boul, PJ, Lu, A, Iverson, T, Shelimov, K, Huffman, CB, Rodriguez-Macias, F, Colbert, DT, Smalley, RE: Fullerene pipes. Science **280**(5367), 1253–1257 (1998)
20. Esumi, K, Ishigami, M, Nakajima, A, Sawada, K, Honda, H: Chemical treatment of carbon nanotubes. Carbon **34**(2), 279–281 (1996)
21. Hiura, H, Ebbesen, TW, Tanigaki, K: Opening and purification of carbon nanotubes in high yields. Adv. Mater. **7**(3), 275–276 (1995)
22. Saito, T, Matsushige, K, Tanaka, K: Chemical treatment and modification of multi-walled carbon nanotubes. Physica B Condensed Matter **323**(1–4), 280–283 (2002)
23. Shaffer, MSP, Fan, X, Windle, AH: Dispersion and packing of carbon nanotubes. Carbon **36**(11), 1603–1612 (1998)
24. Xu, CY, Zhang, PX, Yan, L: Blue shift of Raman peak from coated TiO2 nanoparticles. J. Raman Spectroscopy **32**(10), 862–865 (2001)
25. Schwartsburg, K, Willig, F: Influence of trap filling on photocurrent transients poly crystalline TiO2. Appl. Phys. Lett. **58**(3), 2520–2522 (1991)
26. Qian, X, Qin, D, Song, Q, Bai, Y, Li, T, Tang, X, Wang, E, Dong, S: Surface photovoltage spectra and photoelectrochemical properties of semiconductor-sensitized nanostructured TiO_2 electrodes. Thin Solid Films **385**(1–2), 152–161 (2001)
27. Roy, MS, Balraju, P, Manish, K, Sharma, GD: Dye-sensitized solar cell based on Rose bengal dye and nanocrystalline TiO2. Solar Energy Mater. Solar Cells **92**(8), 909–913 (2008)

33

Bioengineered silver nanobowls using *Trichoderma viride* and its antibacterial activity against gram-positive and gram-negative bacteria

Kethirabalan Chitra and Gurusamy Annadurai*

Abstract

Myconanotechnology is a study of nanoparticle synthesis using fungi and their application mainly in medicine. Myconanotechnology has great advantage due to wide range and diversity of the fungi. Recently, fungal-mediated synthesis of nanoparticles is a reliable and ecofriendly method. The present work investigates the synthesis of silver nanoparticles using *Trichoderma viride* which is a non-pathogenic fungus. The cell filtrate of *T. viride* was used for the reduction of silver nitrate to silver nanoparticles. The pH of the cell filtrate was changed, and the effect of the pH was monitored on the synthesis of the silver nanoparticles. While changing the pH of the medium, the shape and size of the nanoparticles were controlled. The silver nanobowls were synthesized at first time using *T. viride*. The synthesized silver nanoparticles were characterized using UV spectrophotometer, X-ray diffractometer, Fourier transform infrared spectrophotometer, scanning Electron microscope (SEM), and transmission electron microscope (TEM). The SEM and TEM images showed the shape and size of the nanoparticles, and the synthesized nanoparticles were nanobowl in shape and polydispersed in size. The antibacterial activity of silver nanoparticles was carried out against gram-positive and gram-negative organisms. The maximum zone of inhibition occurred at 50 μl, while the concentration of the nanoparticles increased the inhibition activity also increased.

Keywords: *Trichoderma viride*, Silver nanoparticles, Extracellular synthesis, Transmission electron microscope, Antibacterial activity

Background

Myconanotechnology is the connecting term of mycology and nanotechnology, and it has a significant potential in the production of nanoparticles. The synthesis of nanoparticles using fungi has gained much importance because they are easy to culture in bulk, and the extracellular secretion of enzymes has an added advantage in the downstream processing. The fungi produce higher amount of protein compared to bacteria, thus resulting in the higher production of nanoparticles [1-3]. Because of their properties, fungi might be extensively used for the rapid and eco-friendly biosynthesis of metal nanoparticles. It has been monitored that the fungal-mediated synthesized nanoparticles have good monodispersity and good dimensions [2,4]. The intracellular and extracellular methods are being used for the synthesis of nanoparticles via a biological route. The extracellular method is more advantageous than intracellular method because the intracellular method needs an additional step to obtain the purified nanoparticles [5,6].

Synthesis of noble metals has become mandatory issue in the present decade due to the abnormal increase of their market values [7]. Metals such as gold, silver, and copper have been widely used for the synthesis of stable dispersion of nanoparticles, which are being useful in the area of photography, biological labeling, photonics, optoelectronics, and surface-enhanced Raman scattering detection [8-10]. Various physical and chemical methods are used for the synthesis of nanoparticles; these methods are intended at controlling the physical properties of particles. Still, most of these methods are in the development stage, and problems are often solved through the stability of the nanoparticle preparation, control of the crystal growth, and aggregation of the

* Correspondence: gannadurai@hotmail.com
Environmental Nanotechnology Division, Sri Paramakalyani Centre for Environmental Sciences, Manonmaniam Sundaranar University, Alwarkurichi, Tamil Nadu 627412, India

particles [11]. As a result of some drawbacks in physical and chemical methods, biological synthesis method has been developed to obtain biocompatible, inexpensive, eco-friendly, and size-controlled nanoparticles [12]. Nowadays, noble metal nanoparticles play a vital role in the fields of medicine, biology, physics, chemistry, and materials science. Among noble metal nanoparticles, silver is of particular interest because of its distinctive properties, such as good electrical conductivity, chemical stability, and catalytic and antibacterial activities [6,13,14].

Trichoderma viride is a rhizosphere competent strain found in all type of soils; it is a non-pathogenic fast-growing organism which is used as a biocontrol agent [15,16]. The present work demonstrated the synthesis of silver nanoparticles using *T. viride*. The synthesized silver nanoparticles were characterized using UV spectrophotometer analysis, X-ray diffractometer to analyze the nature of the nanoparticles, and scanning electron microscope to analyze the surface morphology of the nanoparticles, transmission electron microscope to analyze the size of the nanoparticles, and selected area electron diffraction (SAED) pattern to confirm the crystalline nature of the nanoparticles. Fourier transform infrared spectroscopy (FTIR) is used to identify the functional groups of the silver nanoparticles. The bactericidal activity of silver nanoparticles was carried out using disc diffusion method.

Results and discussion

The UV absorption spectral studies were carried out to confirm the formation of silver nanoparticles using *T. viride* (Figure 1). The absorption peak in UV spectrum equivalent to the surface plasmon resonance, the peak found at 400 nm, and the maximum synthesis of silver nanoparticles occurred at 96 h of incubation. The UV spectral absorption reading confirmed that the well-dispersed, but not aggregated silver nanoparticles, were formed using *T. viride* cell filtrate [5]. The silver nanoparticles were formed very rapidly within 30 min. Fayaz et al. [16] have reported the extracellular synthesis of silver nanoparticles using *T. viride* after 4 h of incubation. The inset in Figure 1 shows the conical flasks containing *T. viride* filtrate (a) before and (b) after reaction with silver ions. The aqueous solution of the biomass is green in color before the reaction with silver ions, which turned to brown after the reaction with silver ions. The change of color indicates the formation of silver nanoparticles using *T. viride*. The color change occurs due to the excitation of surface plasmon resonance vibration of silver; hence, it confirmed the reduction of silver ions extracellulary [17,18].

Figure 2 shows the UV spectrum of pH effect on the synthesis of silver nanoparticles using *T. viride*. The sharp peak was observed at 400 nm, and the maximum production of silver nanoparticles occurred at pH 9. The inset picture (Figure 2) showed that the color intensity of the aqueous medium increased, while the pH of the fungal cell filtrate increased. The color of the solution varied from yellow to dark brown. The pH of the medium has a great effect on the control of the shape and size of the nanoparticles. The color variation from yellow to dark brown of the solution occurred because of the increase amount of nanoparticle production in the fungal cell filtrate medium [17,19]. Veerasamy et al. [20] explained that the particles are larger at lower pH, and a large amount of smaller diameter particles were formed at higher pH due to the large number of functional groups obtainable for nanoparticle binding. The

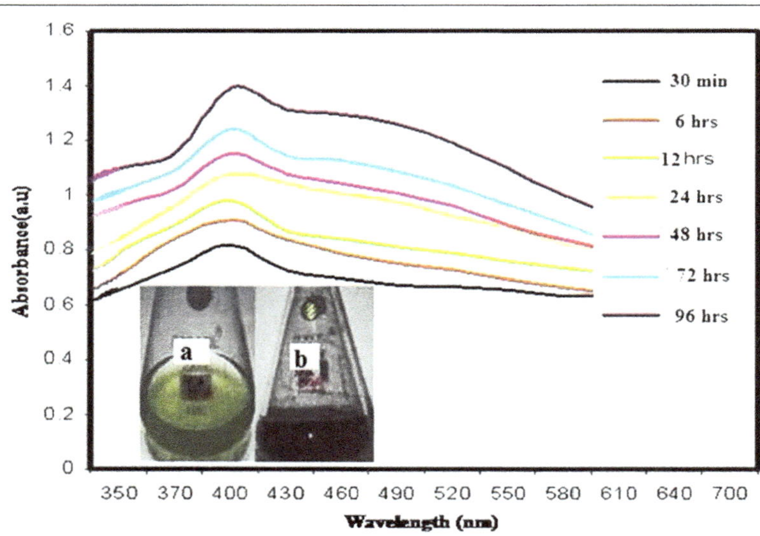

Figure 1 The absorption spectra of silver nanoparticles using *T. viride* cell filtrate at various time intervals. Inset shows (**a**) *T. viride* cell filtrate without silver nitrate and (**b**) 96 h of reaction between silver nitrate and fungal cell filtrate.

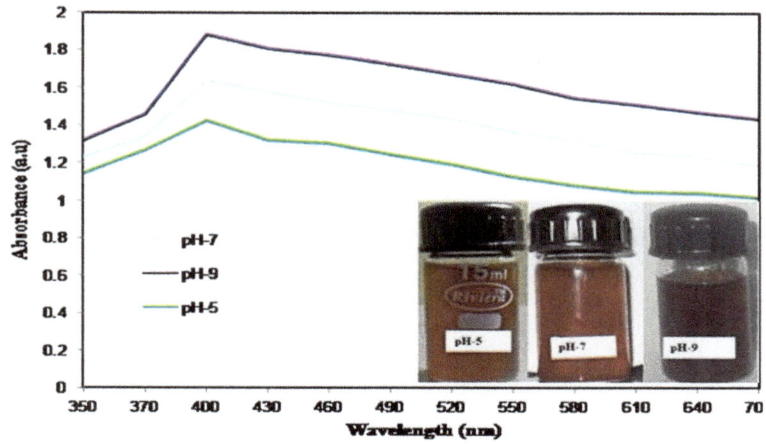

Figure 2 Effect of pH on the synthesis of silver nanoparticles using *T. viride*. The inset shows the color variation of the medium depending on the pH.

nanoparticles which are synthesized using pH 9 were used for further characterization studies.

The X-ray diffraction studies were carried out to confirm the crystalline nature of the nanoparticles, and the XRD pattern (Figure 3) of the synthesized silver nanoparticles shows the crystalline nature of silver nanoparticles synthesized using *T. viride*. The XRD pattern shows strong peaks in the entire spectrum of 2θ values ranging from 20 to 80. The silver nanoparticles synthesized in this experiment were in nanocrystal form as evidenced by the peaks at 2θ values of 38.24°, 44.11°, 64.64°, and 77.47° analogous to (111), (200), (220), and (311) planes for silver. The unassigned peaks present in spectrum indicate the presence of organic matters which are present in the fungal cell filtrate.

FTIR spectrum is used to identify the possible interactions among silver salts and proteins (present in the fungal cell filtrate) which are responsible for the reduction of silver ions and stabilization of nanoparticles. The FTIR spectrum of the biosynthesized silver nanoparticles using *T. viride* (Figure 4) shows the absorption peaks at 3,396, 2,925, 2,855, 1,740, 1,651, 1,548, and 1,380 cm^{-1}. The peak at 3,396 and 2,925 cm^{-1} reveals the presence of N-H bend, indicating the primary and secondary amine groups of protein. Likewise, the bands at 1,651 and 1,548 cm^{-1} correspond to the primary and secondary amine groups of N-H bending and carbonyl stretching vibrations of protein, respectively [21]. Therefore, the FTIR study has shown that the carbonyl group of amino acid residues and peptides of proteins has a stronger metal-binding capability. Most possibly, the proteins could form a coat to cover the nanoparticles and act as capping agents for silver nanoparticle formation to avert the agglomeration of the nanoparticles; therefore, the particles are stabilized [22]. The two bands at 2,855 and 1,380 cm^{-1} represent the C-H of symmetric and asymmetric stretching vibrations of aromatic and aliphatic modes, and the C-H stretching vibrations of

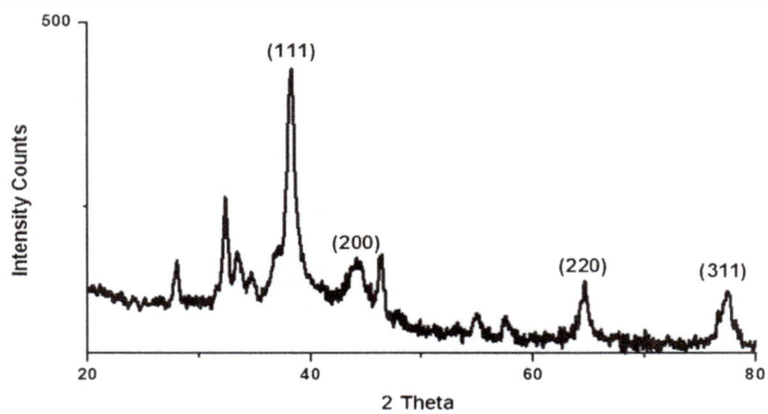

Figure 3 XRD pattern of biosynthesized silver nanoparticles using *T. viride*.

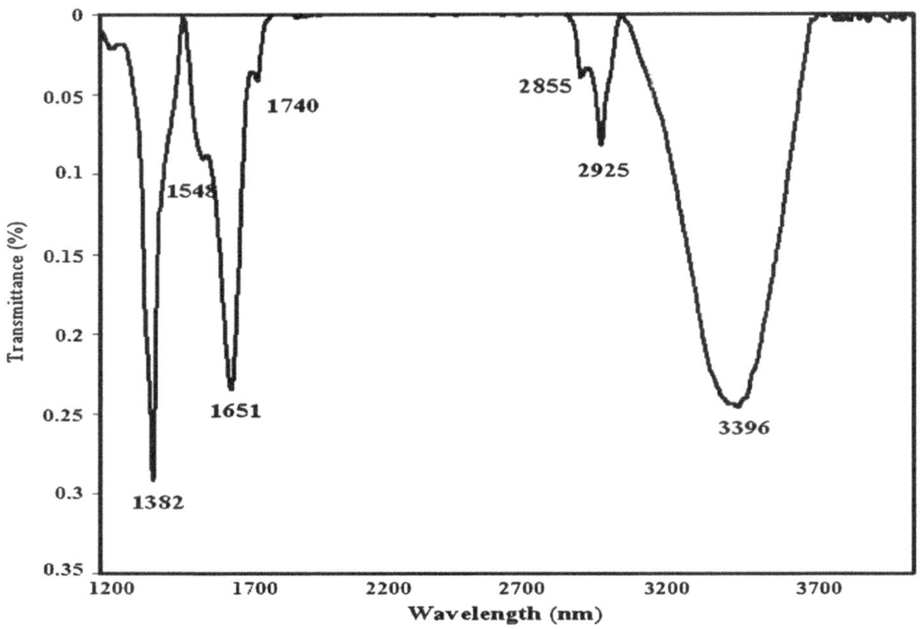

Figure 4 FTIR spectrum of silver nanoparticles.

aromatic and aliphatic amines, respectively. The band at 1,740 cm^{-1} corresponds to C=O stretching vibrations of aldehyde group. Hence, the observations indicate that the bioreductions of silver nitrate are tied to the oxidation of hydroxyl groups in fungal cell filtrate [23]. The whole observation suggests that the biological molecules might possibly be the reason for the formation and stabilization of metal nanoparticles in the fungal cell filtrate.

After 120 h of incubation, the biosynthesized silver nanoparticles were air dried and examined under scanning electron and transmission electron microscopes. The SEM image (Figure 5a,b) of the silver nanoparticles shows the well-dispersed nanoparticles, and the biosynthesized silver nanoparticles have bowl-like shapes. The TEM image (Figure 6a,b) revealed the size of the synthesized silver nanoparticles. These nanoparticles were polydispersed, and their sizes were in the range from 28 to 59.17 nm. The selected area of electron diffraction pattern (inset in Figure 6) of the silver nanoparticles shows their face centered cubic structure, and the synthesized silver nanoparticles are in crystalline nature. This is the first study to report on silver nanoparticle synthesis using *T. viride* resulting in nanobowl-shaped nanoparticles. The definite reaction mechanism for the synthesis of silver nanoparticles using fungal cell filtrate is not yet to be elucidated. The fungal biomass secretes proteinic compounds in liquid medium which may be accountable for the synthesis of nanoparticles, and consequently, the proteins

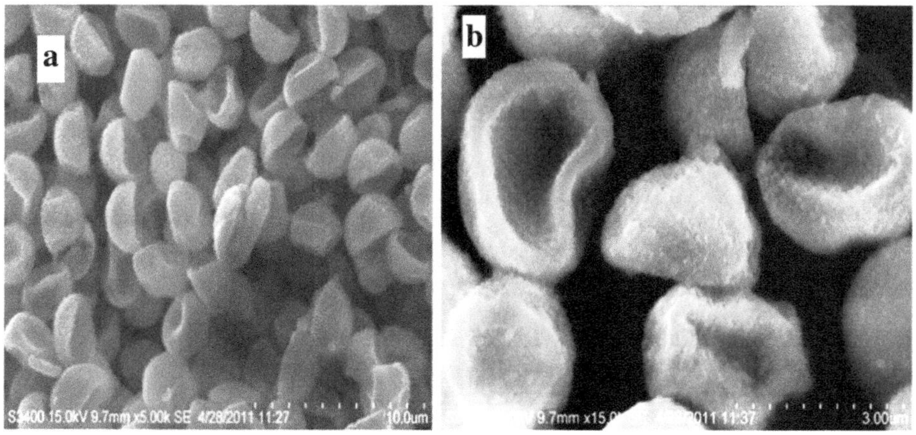

Figure 5 SEM images show the bowl shape of the biosynthesized nanoparticles using *T. viride* (a-b). ×15,000 magnification.

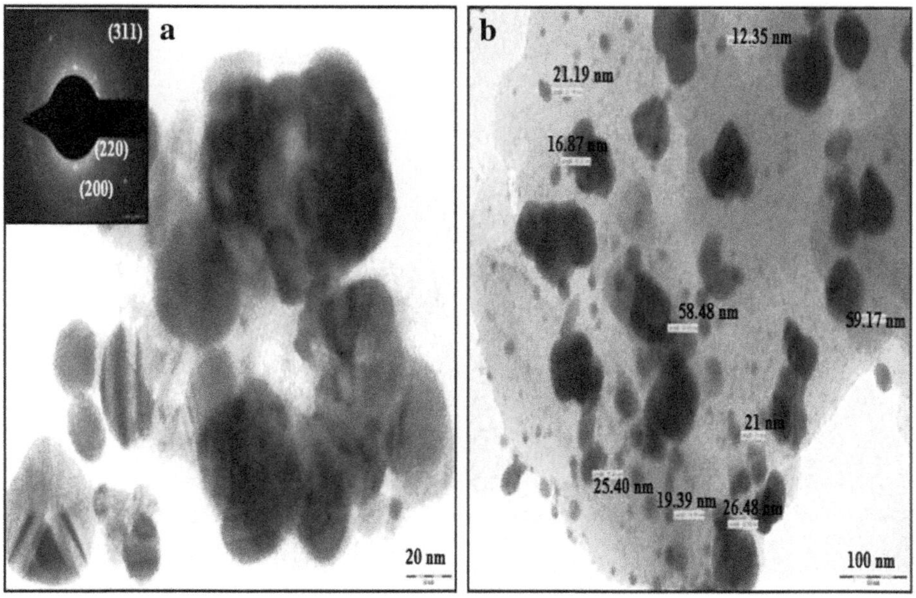

Figure 6 TEM images of the biosynthesized nanoparticles. Scale bars (**a**) 20 nm and (**b**) 100 nm. Inset shows the selected area of diffractometer pattern of the synthesized silver nanoparticles.

attach and improve the stability of the nanoparticles [5]. Two protein bands were identified with molecular weights of 45 and 39 KDa, respectively, in silver nanoparticle sample, which may be responsible for the synthesis of silver nanoparticles (Figure 7). The extracellular proteins have great advantage in the synthesis of nanoparticles because they provide large amount of nanoparticles in pure form, free from other cell components [3].

Well diffusion method was used to provide the evidence for the antibacterial activity of biosynthesized silver nanoparticles against *Bacillus subtilis* and *Klebsiella planticola*. Different concentrations (10 to 50 μl) of silver nanoparticles were used to confirm the antibacterial efficiency of the silver nanoparticles. The maximum zone of inhibition was observed in *K. planticola* followed by *B. subtilis* (Figure 8). The antibacterial activity of silver nanoparticles was indicated by the formation of the zone. The diameter of the inhibition zone was measured in millimeters. The maximum zone of inhibition occurred at 50 μl concentration of silver nanoparticles. The possible mechanism for the bactericidal activity of silver nanoparticles may be the penetration of the nanoparticles into the cell wall and entry inside the cells, compressing the DNA which simultaneously inactivates the cellular proteins [24,25].

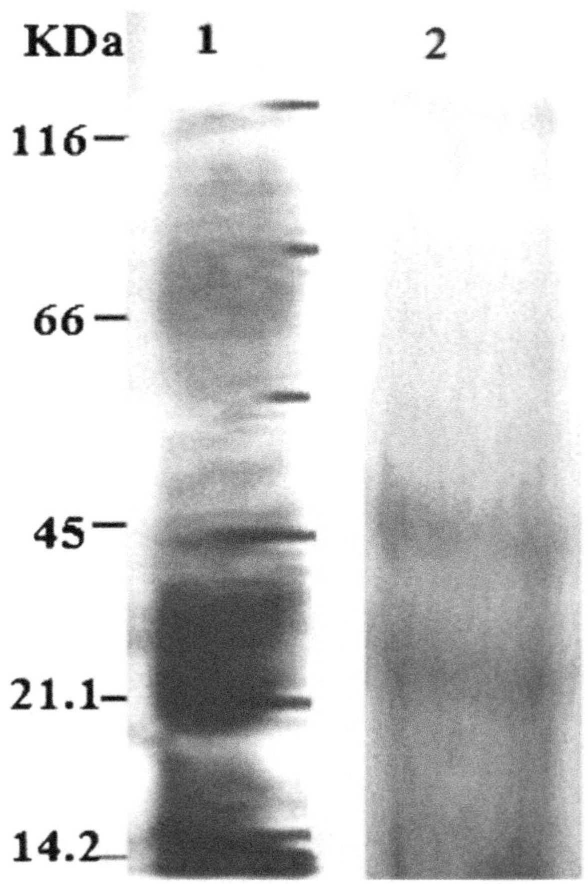

Figure 7 SDS-PAGE gel electrophoresis (1) marker protein and (2) synthesized silver nanoparticles.

Conclusion

An ecological, cost-effective method was employed for the extracellular synthesis of silver nanoparticles using *T. viride*. X-ray diffractometry revealed the crystalline nature of the nanoparticles. Scanning electron microscopy

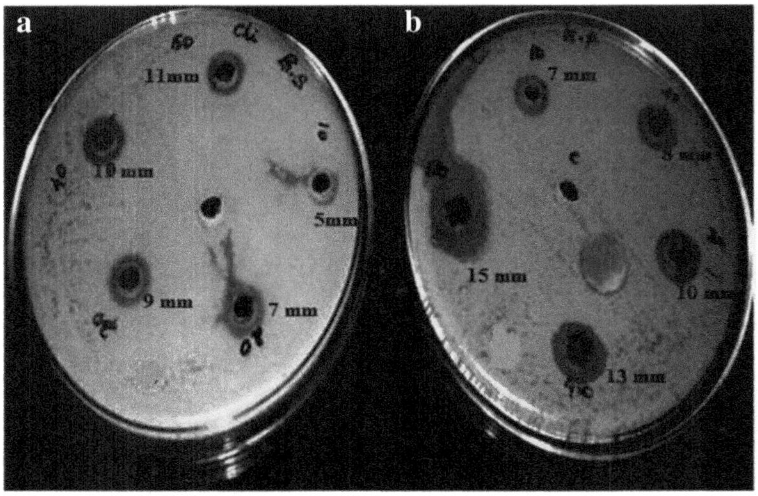

Figure 8 Antimicrobial activity of biosynthesized silver nanoparticles against (a) *B. subtilis* and (b) *K. planticola*.

revealed the bowl-shaped nanoparticles. Transmission electron microscope shows the average size of the nanoparticles to be around 28 nm, and the SAED confirms the crystalline nature of the nanoparticles. The antibacterial activity occurred due to the penetration of the silver nanoparticles into the cell walls, reducing the function of the organisms.

Methods

Preparation of silver nanoparticles

T. viride was obtained from MTCC-800 Mumbai, India. The culture was sub-cultured in SDA media. Then, the *T. viride* was inoculated in a liquid SDA media and grown aerobically at 35°C and agitated at 150 rpm on orbital shaker. The biomass was harvested after 72 h of growth by filtration using Whatman No. 1 filter paper followed by extensive washing with distilled water to remove any medium component from the biomass. The 10 g of biomass was brought in contact with 100 ml of double distilled water for 48 h at 35°C and agitated as described earlier. After 48 h of incubation, the cell filtrate was obtained by passing it through Whatman filter paper No. 1. The silver nitrate (1 mM) was added in 100 ml of cell filtrate, and the reaction was carried out at room temperature. The UV spectroscopic absorbance was taken at various time intervals.

Effect of pH on the synthesis of nanoparticles

The cell filtrate was obtained using the above-stated process, and then the pH (5, 7, and 9) of the cell filtrate was adjusted using 0.1 N of NaOH and O.1 N of HCl followed by the addition of 1 mM of silver nitrate to the cell filtrate. After 24 h of reaction, the UV spectrophotometric reading was taken at different wavelengths.

Characterization of the synthesized silver nanoparticles

The synthesized silver nanoparticles were characterized using UV (Perkin Elmer Inc., MA, USA) visible absorbance spectra, absorbance taken at various time intervals at different wavelength, powder X-ray diffractometer (Bruker D8 Advance using CuKα radiation, at 40 keV in the range of 10 to 80, Bruker Corporation, Ettlingen, Germany), scanning electron microscope (S-3400N, Hitachi, Tokyo, Japan), Fourier transform infrared spectrometer (6700, Thermo Nicolet, MA, USA), transmission electron microscope (CM200, Koninklijke Philips Electronics N.V., Amsterdam, Netherlands), and selected area electron diffractometer to characterize the nature, shape, functional groups, and size of the synthesized silver nanoparticles.

SDS-PAGE gel electrophoresis

The proteins which might be the reason for the synthesis of nanoparticles were determined using SDS-PAGE gel electrophoresis. Silver nitrate (1 mM) was added to the *T. viride* cell filtrate. After 24 h of incubation, it was precipitated with ammonium sulfate (80% saturation). Then, the pellet was dialyzed against phosphate buffer (0.05 M) using dialysis bag. The sample was analyzed on SDS-Page, and the molecular weight of the protein was determined using marker protein.

Antibacterial activity of silver nanoparticles

The bactericidal activity of the synthesized silver nanoparticles was observed against *B. subtilis* (3053) and *K. planticola* (2277). The bactericidal activity was carried out using well diffusion method. Mueller-Hinton agar medium (MHA) was prepared and poured into plates, and wells were made in the agar medium. Then, 24-h-old test bacterial cultures were swabbed in MHA medium to form a confluent lawn of bacterial cultures. Various concentrations (10 to 50 µl) of silver nanoparticles were loaded into the wells. Then, the plates were incubated at 35°C for 24 h. After 24 h of incubation, the plates were observed for zone of inhibition.

Competing interests
The authors declare that they have no competing interests.

Authors' contributions
KC and GGA provided the same contributions in this article. Both authors read and approved the final manuscript.

Acknowledgments
Authors gratefully acknowledge the DST-FIS-sponsored program from the Department of Science Technology, New Delhi, India for funding this research (ref. no. S/FST/ESI-101/2010). Authors convey their thanks to the Central Instrumentation Facility Pondicherry for providing the FTIR and SEM facilities, the Sophisticated Analytical Instrumentation Facility IIT Bombay for providing TEM, and the School of Advanced Sciences (Characterization Facility Wing) VIT University Vellore for providing the XRD facilities.

References
1. Rai, M, Yadav, A, Bridge, P, Gade, A, Rai, MK, Bridge, PD: Myconanotechnology: a new and emerging science. In: Rai, M., Bridge, P.D. (eds.) Applied Mycology, pp. 258–267. CAB International, New York (2009)
2. Rai, M, Yadav, A, Gade, A: Mycofabrication, mechanistic aspect and multifunctionality of metal nanoparticls – where are we? And where should we go. In: Mendez-Vilas, A. (ed.) Current Research, Technology and Education Topics in Applied Microbiology and Microbial Biotechnology, pp. 1343–1354. Formatex Research Center, Badajoz, Spain (2010)
3. Sastry, M, Ahmad, A, Khan, MI, Kumar, R: Biosynthesis of metal nanoparticles using fungi and actinomycetes. Current Sci **85**, 162–170 (2003)
4. Mandal, D, Bolander, ME, Mukhopadhyay, D, Sarkar, G, Mukherjee, P: The use of microorganisms for the formation of metal nanoparticles and their application. Appl. Microbiol. Biotechnol. **69**, 485–492 (2008). doi:10.1007/s00253-005-0179-3
5. Kuber, C, Bhainsa, SF, Souza, D: Extracellular biosynthesis of silver nanoparticles using the fungus *Aspergillus fumigatus*. Colloids Surf B: Biointerfaces **47**, 160–164 (2006). doi:10.1016/j.colsurfb.2005.11.026
6. Prathna, TC, Chandrasekaran, N, Raichur, AM, Mukherjee, A: Biomimetic synthesis of silver nanoparticles by *Citrus limon* (lemon) aqueous extract and theoretical prediction of particle size. Colloids Surf B: Biointerfaces **82**, 152–159 (2011). doi:10.1016/j.colsurfb.2010.08.036
7. Binupriya, AR, Sathiskumar, M, Yun, SI: Biocrystallization of silver and gold ions by inactive cell filtrate of *Rhizopus stolonifer*. Colloids Surf B: Biointerfaces **79**, 531–534 (2010). doi:10.1016/j.colsurfb. 2010.05.021
8. Sharma, VK, Yingard, RA, Lin, Y: Silver nanoparticles: green synthesis and their antimicrobial activities. Adv. Colloid. Interface. Sci. **145**, 83–96 (2009). doi:10.1016/j.cis.2008.09.002
9. Simth, AM, Duan, H, Rhyner, MN, Ruan, G, Nie, SA: A systematic examination of surface coatings on the optical and chemical properties of semiconductor quantum dots. Phys. Chem. Chem. Phys. **8**, 3895–3903 (2006). doi:10.1039/b606572b
10. Kearns, GJ, Foster, EW, Hutchison, JE: Substrates for direct imaging of chemically functionalized SiO_2 surfaces by transmission electron microscopy. Anal. Che. **78**, 298–303 (2006). doi:10.1021/ac051459k
11. Gericke, M, Pinches, A: Biological synthesis of metal nanoparticles. Hydrometallurgy **83**, 132–140 (2006). doi:10.1016/j.hydromet.2006.03.019
12. Sadhasivum, S, Shanmugam, P, Yun, K: Biosynthesis of silver nanoparticles by *Streptomyces hygroscopicus* and antimicrobial activity against medically important pathogenic microorganisms. Colloids and Surface B: Biointerfaces **81**, 358–362 (2010). doi:10.1016/j.colsurfb.2010.07.036
13. Wang, J, Wang, Z: Rapid synthesis of hexagon-shaped gold nanoplates by microwave assistant method. Mater. Lett. **61**, 4149–4151 (2007). doi:10.1016/j.matlet.2007.01.043
14. Yokoyama, K, Welchons, K: The conjugation of amyloid beta protein on the gold colloidal nanoparticles surfaces. Nanotechnology **18**, 105101 (2007). doi:10.1088/0957-4484/18/10/105101
15. Sathiyaseelan, K, Sivasakthivelan, P, Lenin, G: Evaluation of antagonistic activity and shelf life study of *Trichoderma viride*. Bontany. Res. Int. **2**, 195–197 (2009)
16. Fayaz, M, Tiwary, CS, Kalaichelvan, PT, Venkatesan, R: Blue orange light emission from biogenic synthesized silver nanoparticles using *Trichoderma viride*. Colloids Surf B: Biointerfaces **75**, 175–178 (2010). doi:10.1016/j.colsurfb.2009.08.028
17. Ahmad, A, Mukherjee, P, Senapathi, S, Mandal, D, Khan, MI, Kumar, R, Sastry, M: Extracellular biosynthesis of silver nanoparticles using the fungus *Fusarium oxysporum*. Colloids Surf B: Biointerfaces **28**, 313–318 (2003). doi:10.1016/SO927-7765 (02)001741
18. Balaji, DS, Basavaraja, S, Deshpande, R, Mahesh, DB: Extracellular biosynthesis of functionalized silver nanoparticles by strains of *Cladosporium cladosporioides* fungus. Colloids Surf B: Biointerfaces **68**, 88–92 (2009). doi:10.1016/j.colsurfb.2008.09.022
19. Kathiresan, K, Manivannan, S, Nabeel, MA, Dhivya, B: Studies on silver nanoparticles synthesized by a marine fungus Penicillium fellutanum isolated from coastal mangrove sediment. Colloids and Surface B: Biointerf **71**, 133–137 (2009). doi:10.1016/j.colsurfb.2009.01.016
20. Veerasamy, R, Xin, TZ., Gunasagaran, S, Xiang TFW, Yang, EFC, Jeyakumar, N, Dhanaraj, SA: Biosynthesis of silver nanoparticles using *mangosteen* leaf extract and evaluation of their antibacterial activities. J. Saudi. Chem. Soc. **15**, 113–120 (2010). doi:10.1016/j.jscs.2010.06.004
21. Sawle, BD, Salimath, B, Deshpande, R, Bedre, MD, Prabhakar, BK, Venkataraman, A: Biosynthesis and stabilization of Au and Au-Ag alloy nanoparticles by fungus *Fusarium semitectum*. Sci. Technol. Adv. Mater. **9**, 035012 (2008). doi:10.1088/1468-6996/9/3/035012
22. Vigneshwaran, N, Ashtapure, M, Varadarajan, PV, Nachane, RP, Paralikar, KM, Balasubramanya, RH: Biological synthesis of silver nanoparticles using the fungus *Aspergillus flaves*. Mater. Lett. **61**, 1413–1418 (2007). doi:10.1016/j.matlet.2006.07.042
23. Sangi, R, Verma, P: Biomimetic synthesis and characterization of protein capped silver nanoparticles. Biores. Technol. **100**, 501–504 (2009). doi:10.1016/j.biortech.2008.05.048
24. Feng, QL, Wu, J, Chen, GQ, Cui, FZ, Kim, TN, Kim, JO: A mechanistic study of the antibacterial effect of silver ions on *Escherichia coli* and *Staphylococcus aureus*. J. Biomed. Mater. Res. **52**, 662–668 (2000). doi:10.1002/1097-4636(20001215)
25. Sondi, I, Sondi, BS: Silver nanoparticles as antimicrobial agent: a case study on *E. coli* as a model for Gram-negative bacteria. J. Colloid. Interface. Sci **275**, 177–182 (2004). doi:10.1016/j.jcis.2004.02.012

The interaction between carbon nanotube and skin anti-cancer drugs: a DFT and NBO approach

Maryam Hesabi[1*] and Masoumeh Hesabi[2]

Abstract

Nowadays, nanotubes are used widely as container agents to transport bioactive peptides, proteins, nucleic acids, and drugs, and used to deliver their cargos to cells and organs in human bodies. In this work, the possibility of the formation of stable composite between carbon nanotubes and skin anti-cancer drugs has been investigated. The nanotube used in this study includes 60 C atoms of (6,6) type. Density functional theory-based methods (B3LYP/6-31G) show that the composites between drugs (aminolevulinic acid and tretinoin) and nanotube are more stable than the single agents. The difference in the hybridization of C and O atoms can cause to a difference between bond lengths, angles, and charges. The natural bond orbital calculations indicate that some donor atoms (lone pair of oxygen or nitrogen atoms) can transfer electron to acceptor atoms (σ * or π * in carbon nanotube) and the occupancy of the oxygen lone pair with increasing *p* orbital share of the lone pair electrons of oxygen decreases. Then, the highest occupied molecular orbital (HOMO) and the lowest unoccupied molecular orbital (LUMO), the HOMO-LUMO bandgap, and the electronic chemical potential (μ) for the lowest energy derived to estimate the structural stability of the composites have been investigated. Results show that nanotube-tretinoin is more stable than nanotube - aminolevulinic acid.

Keywords: Nanotube, Aminolevulinic acid, Tretinoin, DFT, HOMO-LUMO gap

Background

Skin cancer is the most common form of human cancer. It is estimated that over one million new cases occur annually. The annual rates of all forms of skin cancer are increasing each year, representing a growing public concern. As a result, many drugs have been developed to treat skin cancer. The available anti-cancer drugs have distinct mechanisms of action which may vary in their effect on different types of normal and cancer cells. In addition, there are very few demonstrable biochemical differences between cancerous cells and normal cells. For this reason, the effectiveness of many anti-cancer drugs is limited by their toxicity to normal rapidly growing cells in the intestinal and bone marrow areas. A final problem is that cancerous cells which are initially suppressed using a specific drug may develop a resistance to that drug. For this reason, cancer treatments may consist of using several drugs in combination for varying lengths of time [1-3]. The anti-cancer drugs that were selected in this work are tretinoin and aminolevulinic acid.

Tretinoin, or all-*trans*-retinoic acid, is a naturally occurring derivative of vitamin A (retinol). Retinoids such as tretinoin are important regulators of cell reproduction, proliferation, and differentiation and are used to treat acne and photodamaged skin and to manage keratinization disorders such as ichthyosis and keratosis follicularis.

Topical administration of tretinoin has proved to be effective in treating the clinical signs of photodamaged skin. Tretinoin, a natural vitamin A metabolite, is an anti-cancer drug used in the treatment of acute promyelocytic leukemia. It is more commonly used to treat skin disorders such as acne, warts, hyperpigmentation, and reactions to sunlight [4-6]. The other drug is aminolevulinic acid (4-oxopentanoic acid) which is to be used in photodynamic therapy to treat a skin condition called actinic keratosis [7].

The interesting properties of nanotubes have caused researchers and companies to consider using them in several fields, because carbon nanotubes have the highest

* Correspondence: marhesabi@gmail.com
[1]Department of Chemistry, Faculty of Science, Islamic Azad University, Rasht Branch, Rasht 41476, Iran
Full list of author information is available at the end of the article

strength to weight ratio of any known material [8,9]. The highly regular atomic structure of carbon nanotubes (CNTs) and the large degree of the structure purity make it accessible to accurate computer modeling by a variety of theoretical techniques. CNTs have emerged as a new alternative and efficient tool for transporting and translocating therapeutic molecules [10,11]. It is well recognized that the transport and bioavailability of drugs are significant factors in improving their distribution, therapeutics, selectivity, and in ameliorating their toxic effects [12-14]. Single-wall carbon nanotubes (SWNTs) with very high specific surface areas can be easily derivatized with biomolecules either through chemical attachment, adsorption, or encapsulation. Such bioconjugates on SWNTs have the ability to deliver bioactive molecules across cell membranes and even into the cell nuclei [15,16]. Scientists have shown that nanotubes can release drugs in cells without damaging the healthy cells of the tissue. The advantages of using a carbon nanotube-assisted drug delivery system include efficient targeting and amplification of tumor targeting due to an enhanced permeability and retention effect of the carbon nanotube which can be efficiently loaded with the drug. The use of a non-toxic drug which is activated to its cytotoxic form in the tumor cells helps preserve the non-targeted normal tissue of the patient, thereby potentially reducing the side effects resulting from the therapy [17-19].

In this present work, the interactions between carbon nanotubes and the composites of these drugs with nanotube (1, nanotube - aminolevulinic acid; 2, nanotube - tretinoin) based on the density functional theory (DFT) methods and natural bond orbital (NBO) analysis have been investigated.

DFT, in particular, the exchange-correlation B3LYP hybrid density functional, is widely used in molecular modeling studies to predict structure, spectroscopic parameters, and energy changes of small-, middle-, and large-size molecules [20,21].

Finally, the lowering highest occupied molecular orbital-lowest unoccupied molecular orbital (HOMO-LUMO) energy gaps have also been discussed.

Results and discussion

The hybrid values, Mulliken charges, bond lengths, and angles are shown in Table 1.

The results show that by increasing the *p* orbital share in the hybridization, the bond length increases. So, this hybrid can affect the bond angle, too. The Mulliken charges for the oxygen atoms in each compound are presented. The charge on the N in nanotube - tretinoin is smaller than that of nanotube - aminolevulinic acid. Therefore, the interaction in nanotube - tretinoin is stronger than that of nanotube - aminolevulinic acid that causes large negative charge on it (Table 2).

From the optimized structures, the relative and formation energies, the HOMO-LUMO bandgap as differences between the HOMO and LUMO, and the electronic chemical potential (μ) as half of the energy of the HOMO and LUMO have been found as a measure of the structural stability properties. Table 2

Table 1 Calculated parameters at B3LYP/6-31G level

Agent	C Hybrid	O Hybrid	C-O bond length/Å	C-C-O bond angle	C Mulliken charge	O Mulliken charge
Nanotube - aminolevulinic acid	$SP^{3.45}$	$SP^{1.97}$	1.420	114.17	0.2255	−0.5810
Nanotube - tretinoin	$SP^{3.36}$	$SP^{1.99}$	1.412	114.37	0.2379	−0.5937

SP, *s* orbital/*p* orbital.

Table 2 Obtained parameters for two composites at B3LYP/6-31G level

Agent	Energy/kcal mol^{-1}	ΔE/kcal mol^{-1}	HOMO (ev)	LUMO (ev)	HOMO-LUMO gap (ev)	μ (ev)
Nanotube	582,302.6915	-	−3.65	−2.07	1.58	−2.86
Aminolevulinic acid	1,726,959.5211	-	−4.99	−0.83	5.82	−2.91
Tretinoin	1,442,730.5983	-	−4.36	−1.89	2.47	−3.13
Nanotube - aminolevulinic acid	284,228.5235	−2,025,033.6892	−3.69	−2.10	1.59	−2.90
Nanotube - tretinoin	0	−2,025,033.2898	−3.57	−2.04	1.53	−2.81

Table 3 Calculated natural hybrids, occupancies, and the second-order perturbation energy E_2 at B3LYP/6-31G level

Agent	Lewis-type NBOs			Non-Lewis NBOs		E_2/kcal mol^{-1}	ΣE_2/kcal mol^{-1}
	Type	Hybrid	Occupancy	Type	Occupancy		
Nanotube - aminolevulinic acid	$LP(1)_O$	$SP^{1.95}$	1.95702	σ^* C53 to C58	0.02860	0.89	
	$LP(2)_O$	$SP^{99.99}$	1.76124	σ^* C47 to C53	0.02860	5.47	28.49
				π^* C47 to C53	0.03096	22.13	
Nanotube - tretinoin	$LP(1)_O$	$SP^{1.87}$	1.95742	σ^* C35 to C37	0.02922	0.82	
				σ^* C37 to C43	0.02229	1.32	
				π^* C37 to C43	0.24365	5.23	
	$LP(2)_O$	$SP^{99.99}$	1.77605	σ^* C35 to C37	0.02922	0.51	29.23
				σ^* C37 to C43	0.02229	0.71	
				π^* C37 to C43	0.24365	20.64	

LP_O, oxygen lone pair.

indicates that the energy values obtained are negative, signifying that such functionalization is of favorable stability. These parameters indicate that nanotube - tretinoin is more stable than nanotube - aminolevulinic acid and both of them are more stable than the single agents. Energy data show that the electron correlation increases the relative energy of less stable composite. The nanotube - tretinoin has the lowest HOMO-LUMO bandgaps; therefore, they confirm the stability of this compound.

As can be observed in Table 3, the composites between drugs and nanotube are more stable than other drugs.

The delocalization of electron density between the filled Lewis-type NBOs and empty anti-bonding non-Lewis NBOs leads to the loss of occupancy from the localized NBOs of the idealized Lewis structure into the empty non-Lewis orbitals. For each donor NBO and acceptor NBO, the stabilization energy is presented as the second-order perturbation interaction energy (E_2) [27,28]. The E_2 energy values confirm the stability of nanotube - tretinoin, too.

These results show that the hyperconjugation between the lone pair orbitals of oxygen as donors and some σ^* or π^* orbitals as acceptors can occur. The most important common interaction in the nanotube -tretinoin composite is $LP(2)_O \rightarrow \pi^*_{C37\ to\ C43}$ and in the nanotube - aminolevulinic acid composite is $LP(2)_O \rightarrow \pi^*_{C47\ to\ C53}$. Tretinoin has the highest energy and can make the structure more stable than the other drug. Also, the occupancy decreases with increasing the*p* orbital share of the lone pair electrons of oxygen.

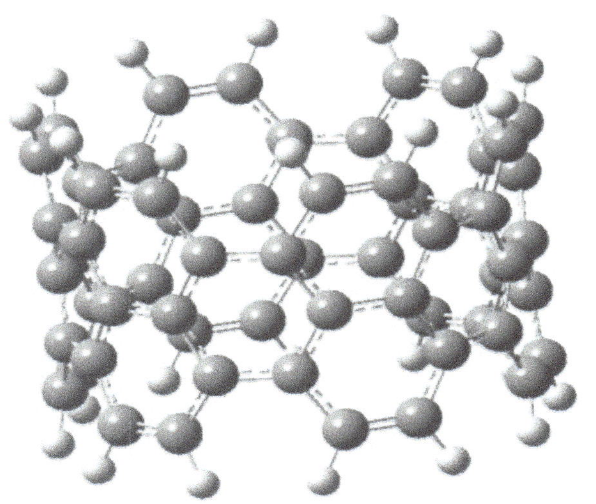

Figure 1 Nanotube (6,6).

Conclusions

Based on the B3LYP/6-31G-optimized ground-state geometries, the NBO analysis of composites has demonstrated a charge transfer and effective energy interaction between the oxygen lone pair (LP_O) and the anti-bonding

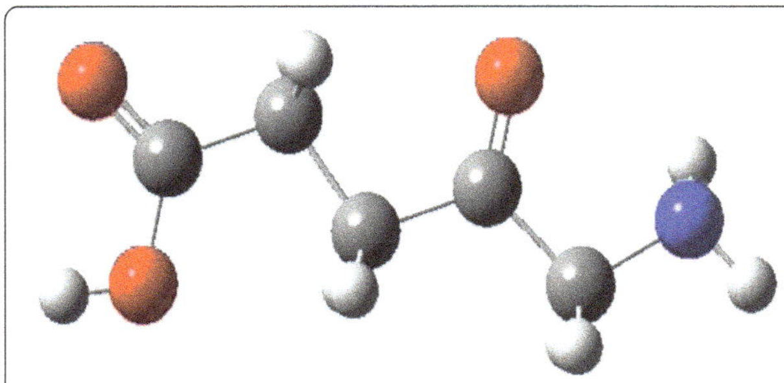

Figure 2 Aminolevulinic acid.

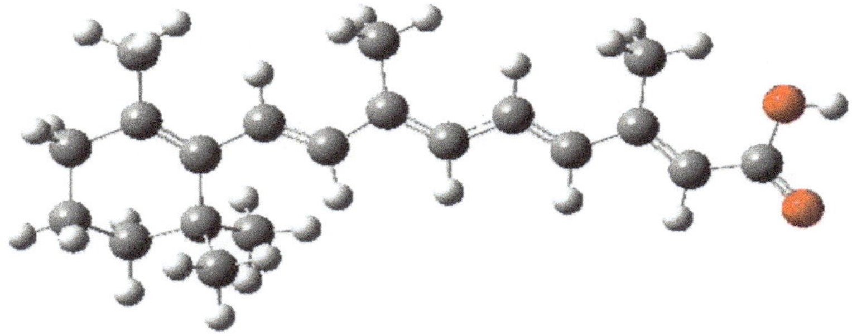

Figure 3 Tretinoin.

orbitals (σ^* or π^*). As can be observed, the most important common interaction is $LP_O \rightarrow \pi^*_{C37\ to\ C43}$ for nanotube - tretinoin. This factor provides more stability in this drug. Also, the NBO analysis show that by an increase of the*p* orbital share, the occupancy of the LP_O decreases.

To justify the structural stability of composites, the HOMO-LUMO bandgap, the electronic chemical potential (μ), the second-order perturbation energy E_2, the Mulliken charges, and the occupancy have been found as a measure of the structural stability properties for the purpose.

Results indicate that nanotube - tretinoin has higher E_2 and μ energies and lower HOMO-LUMO gap and Mulliken charge. Therefore, this composite is more stable than nanotube- aminolevulinic acid, and both of them are more stable than the single drug.

Methods

Computational method

All of the calculations were carried out using a personal computer which has an Intel® Pentium® dual CPU with 2-GB RAM (Intel®, Santa Clara, CA, USA). At first, a nanotube including 60 C atoms (6,6) is formed using a nanotube modeler package [22]. Then, this nanotube is optimized using GaussView [23] and Gaussian 03 [24] softwares(Gaussian Inc., Wroclaw, Poland) by DFT/B3LYP method and 6-31G basis set (Figure 1). The selected drugs (aminolevulinic acid and tretinoin) were made using GaussView and optimized using Gaussian 03 by B3LYP functional with the basis set 6-31G (Figures 2 and 3) [25]. Then, the composites between nanotube and drugs were formed and optimized by B3LYP/6-31G method (Figures 4 and 5). Molecule hybridizations, bond angles, Mulliken charges, and bond lengths were calculated. The delocalization

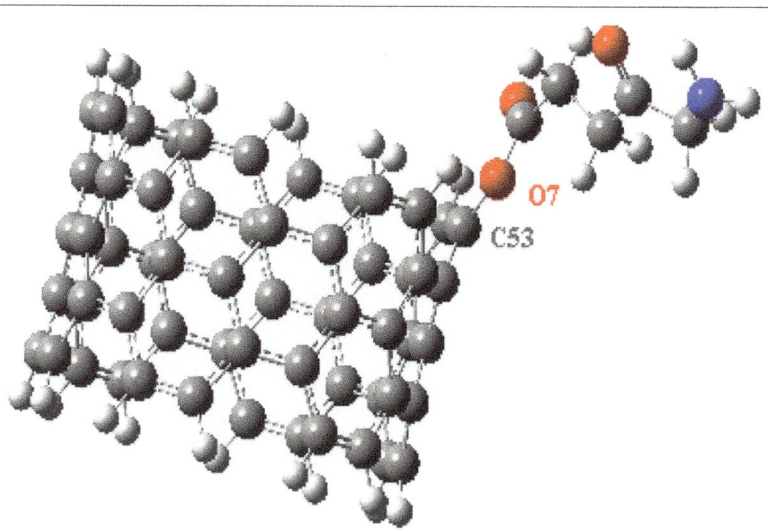

Figure 4 Nanotube - aminolevulinic acid.

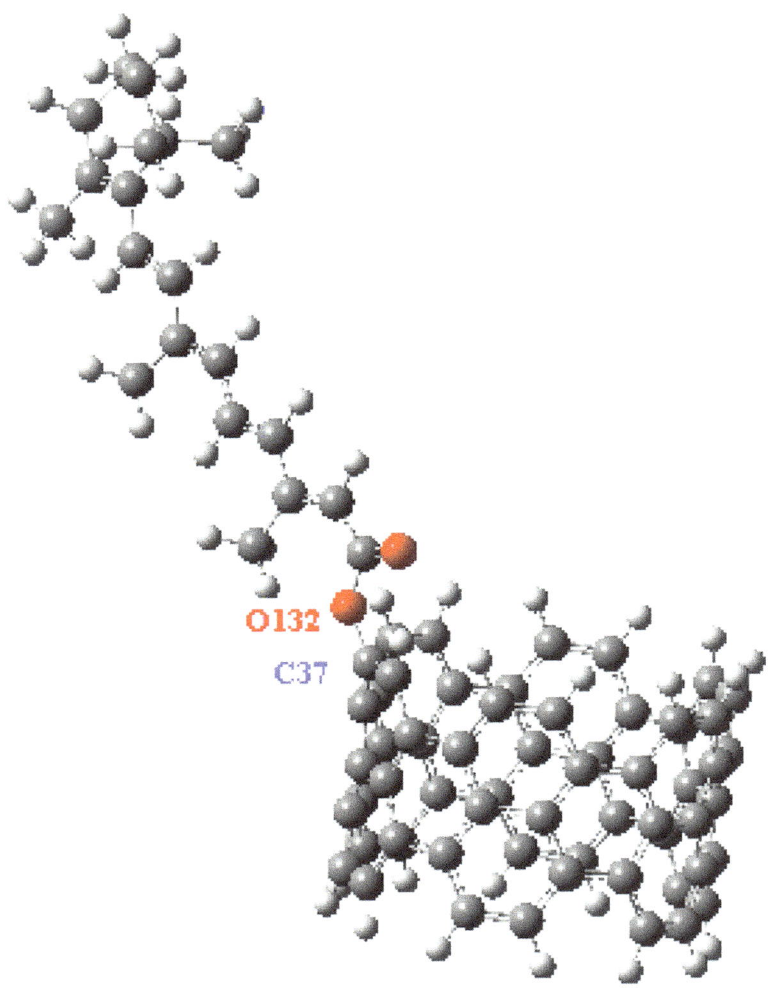

Figure 5 Nanotube - tretinoin.

of electron density between the filled lone pair of Lewis-type NBOs and empty anti-bonding non-Lewis NBOs was obtained by NBO analyzing at B3LYP/6-31G level [26].

Competing interests

The authors declare that they have no competing interests.

Authors' contributions

Both authors participated in writing the manuscript. Both authors read and approved the final manuscript.

Acknowledgements

The authors would like to thank the financial support by Islamic Azad University, Rasht, Iran.

Author details

[1]Department of Chemistry, Faculty of Science, Islamic Azad University, Rasht Branch, Rasht 41476, Iran. [2]Faculty of Nursing, Shahid Beheshti University of Medical Sciences, Rasht Branch, Rasht 41625, Iran.

References

1. Lin, J, Eder, M, Weinmann, S: Behavioral counseling to prevent skin cancer: a systematic review for the U.S. Preventive Services Task Force. Ann. Intern. Med **154**, 190–201 (2011)
2. Narayanan, D, Saladi, R, Fox, J: Ultraviolet radiation and skin cancer. Int. J. Dermatol. **49**, 978–986 (2010)
3. Sersa, G, Miklavcic, D, Cemazar, M, Rudolf, Z, Pucihar, G, Snoj, M: Electrochemotherapy in treatment of tumours. Eur. J. Surg. Oncol. **34**, 232–240 (2008)
4. Joensuu, H: Systemic chemotherapy for cancer: from weapon to treatment. Lancet Oncol **9**, 304 (2008)
5. Sanz, M: Treatment of acute promyelocytic leukemia. Hematol. Am. Soc. Hematol. Educ. Prcgr, 147–155 (2006)
6. Loureiro, K, Kao, K, Jones, K, Alvarado, S, Chavez, C, Dick, L, Felix, R, Johnson, D, Chambers, C: Minor malformations characteristic of the retinoic acid embryopathy and other birth outcomes in children of women exposed to topical tretinoin during early pregnancy. Am. J. Med. Genet. A **136**, 117–121 (2005)
7. Stummer, W, Pichlmeier, U, Meinel, T, Wiestler, OD, Zanella, F, Reulen, HJ: Fluorescence-guided surgery with 5-aminolevulinic acid for resection of malignant glioma: a randomised controlled multicentre phase III trial. Lancet Oncol. **7**, 392–401 (2006)
8. Liang Zang, J, Yuan, Q, Chao Wang, F, Pu Zhao, Y: A comparative study of Young's modulus of single-walled carbon nanotube. Comput.Mater. Sci **46**, 621–625 (2009)

9. Favila, A, Gallo, M, Glossman-Mitnik, D: DFT studies of functionalized carbon nanotubes and fullerenes as nanovectors for drug delivery of antitubercular compounds. J. Mol. Mod. **13**, 505 (2007)
10. Maiti, A: Adsorption of carbon monoxide on a (6,6) armchair carbon nanotube: *Ab initio*study. MicroelecJ (2006). Article in press
11. Tserpesa, KI, Papanikos, P: Finite element modeling of single-walled carbon nanotubes. Composites B **36**, 468–477 (2005)
12. Kostarelos, K, Bianco, A, Prato, M: Promises, facts and challenges for carbon nanotubes in imaging and therapeutic. Nature **2**, 108 (2007)
13. Bottini, M, Cerignoli, F, Dawson, MI, Magrini, A, Rosato, N, Mustelin, T: Noncovalentlysilylated carbon nanotubes decorated with quantum dots. Biomacromolecules **7**, 2259–2263 (2006)
14. Nakayama-Ratchford, N, Bangsaruntip, S, Sun, X, Welsher, K, Dai, H: DNA wrapping and cleavage of a carbon nanotube–polypseudorotaxane conjugate. J. Am. Chem. Soc. **129**, 2448–2449 (2007)
15. Bianco, A, Kostarelos, K, Prato, M: Opportunities and challenges of carbon-based nanomaterials for cancer therapy. Expert Opin. Drug Deliv. **5**, 331–342 (2008)
16. Kam, NWS, Liu, Z, Dai, HJ: Functionalization of carbon nanotubes via cleavable disulfide bonds for efficient intracellular delivery of siRNA and potent gene silencing. Am. Chem. Soc. **127**, 6021–6026 (2005)
17. Reed, AE, Curtiss, LA, Weinhold, F: Intermolecular interactions from a natural bond orbital, donor-acceptor viewpoint. J. Chem. Rev. **88**, 899–926 (1988)
18. Yunpeng, Y, Xiaoyuan, C: Integrin targeting for tumor optical imaging. Theranostics **1**, 102–126 (2011)
19. Zhuang, L, Tabakman, S, Welsher, K: Carbon nanotube in biology and medicine. Nano Res. **2**, 85–120 (2009)
20. Wongchoosuk, C, Udomvech, A, Kerdcharoen, T: The geometrical and electronic structures of open-end fully functionalized single-walled carbon nanotubes. J. Curr. Appl. Phys **9**, 352–358 (2009)
21. Chelmecka, E, Pasterny, K, Kupka, T, Stobinski, L: Density functional theory studies of OH-modified open ended single wall zigzag carbon nanotube (SWCNTs). J. Mol. Struct. (THEOCHEM) **948**, 93–98 (2010)
22. Fernandez, L, Viruela-Martin, P, Latorre, J, Guillem, C, Baltran, A, Amoros, P: Molecular precursors of mesostructured silica materials in the atrane route: A DFT/GIAO/NBO theoretical study. J. THEOCHEM. **822**, 89–102 (2007)
23. Zahedi, E, Aghaie, M, Zare, K: Density functional study of NBO, NICS and 14N NQR parameters of 5-methylcytosine tautomers in the gas phase. J. Mol. Struct. (THEOCHEM) **905**, 101–105 (2009)
24. Frisch, MJ, Trucks, GW, Schlegel, HB, Scuseria, GE, Robb, MA, Cheeseman, JR, Montgomery Jr, JA, Vreven, T, Kudin, KN, Burant, JC, Millam, JM, Iyengar, SS, Tomasi, J, Barone, V, Mennucci, B, Cossi, M, Scalmani, G, Rega, N, Petersson, GA, Nakatsuji, H, Hada, M, Ehara, M, Toyota, K, Fukuda, R, Hasegawa, J, Ishida, M, Nakajima, T, Honda, Y, Kitao, O, Nakai, H, et al: Gaussian 03, Revision C. 02. Gaussian Inc, Wallingford, CT (2004)
25. Becke, AD: A new mixing of Hartree-Fock and local density-functional theories. J. Chem. Phys. **98**, 1372–1377 (1993)
26. Padmaja, L, Amalanathan, M, Ravikumar, C, Hubert Joe, I: Density functional theory calculations and vibrational spectral analysis of 3, 5-(dinitrobenzoic acid). Spectrochim. Acta **A74**, 349–356 (2009)
27. Roohi, H, Ebrahimi, A, Habibi, SM, Jarahi, E: NBO and AIM analyses of the anomeric effect in fluoromethanthiol. J. Mol. Struct. (THEOCHEM) **772**, 65–73 (2006)
28. Nori-shargh, D, Roohi, F, Deyhimi, F, Naeem-Abyaneh, R: Conformational behaviors of trans-2,3-bis(methylthio)-1,4-dioxane, -dithiane and -diselenane. A hybrid-DFT study and NBO interpretations. J. Mol. Struct. (THEOCHEM) **763**, 21–28 (2006)

Biosynthesis of silver nanoparticles using *Salmonella typhirium*

Hamid Reza Ghorbani

Abstract

The use of microorganisms in the synthesis of nanoparticles emerges as an eco-friendly and exciting approach. In the present investigation, we report the biosynthesis of silver nanoparticles employing the bacterium *Salmonella typhirium*. Silver nanoparticles were synthesized through the reduction of aqueous Ag^+ ion using the growth culture supernatants in the bright conditions. The synthetic process was fast, and silver nanoparticles were formed within 4 min of silver ion coming in contact with the cell filtrate. An absorption peak at 427 nm in a Uv-vis spectrophotometer was detected indicating the presence of Ag nanoparticles. The morphology of nanoparticles was observed by transmission electron microscopy. The size of the nanoparticles was determined to be 87 ± 30 nm, applying dynamic light scattering. Reduction degree of Ag^+ was measured as 75% by atomic absorption spectrophotometry. The silver nanoparticles have been prepared from silver sulfate during this investigation for the first time.

Keywords: Ag nanoparticles, *Salmonella*, Biosynthesis, DLS analysis

Background

Nanotechnology plays a very important role in many key technologies of the new millennium [1]. The application of nanoscale and nanostructure materials within the range of 1 to 100 nm is an emerging area of nanoscience and nanotechnology. Nanomaterials may provide solutions to technological and environmental challenges in the areas of solar energy conversion, catalysis, medicine, and water treatment [2].

Nanoparticles of silver have many important applications that include spectrally selective coating for solar energy absorption and intercalation material for electrical batteries, as optical receptors, polarizing filters, catalysts in chemical reaction, biolabelling, and as antimicrobial agents [3].

For the production of nanoparticles, one needs to know the physical and chemical principles of nanoscale materials and also know how to commercialize them. Generally, metal nanoparticles can be prepared and stabilized by chemical, physical, and biological methods; the chemical approach, such as chemical reduction, electrochemical techniques, photochemical reduction [2,4], and pyrolysis [5], and physical methods, such as Arc-discharge and physical vapor condensation [6], are used. Living organisms have huge potentials for the production of nanoparticles of wide applications.

Studies show that the size, morphology, stability, and properties (chemical and physical) of the metal nanoparticles are strongly influenced by the experimental conditions, the kinetics of interaction of metal ions with reducing agents, and the adsorption processes of stabilizing agent with metal nanoparticles. Therefore, the design of a synthesis method in which the size, morphology, stability, and properties are under control is a major field of interest [2,7].

Researchers prefer a biological synthesis because the distribution control of particles obtained from this method is better than other methods [8]. In addition, this method involves no environmental toxicity which is usually accompanied with other chemical methods [9].

The first of the synthesis of silver nanoparticles by a bacterium has been reported in 2000. Joerger et al. used *P.stutzeri* bacterium (AG259) for silver nanoparticle synthesis with a size smaller than 200 nm [10]. In another study, silver nanocrystal biosynthesis was performed using the bacterium *Bacillus licheniformis*. This was indicated by the change in color from whitish-yellow to brown [11].

Correspondence: hamidghorbani6@gmail.com
Department of Chemical engineering, Qaemshahr Branch, Islamic Azad University, Qaemshahr, Iran

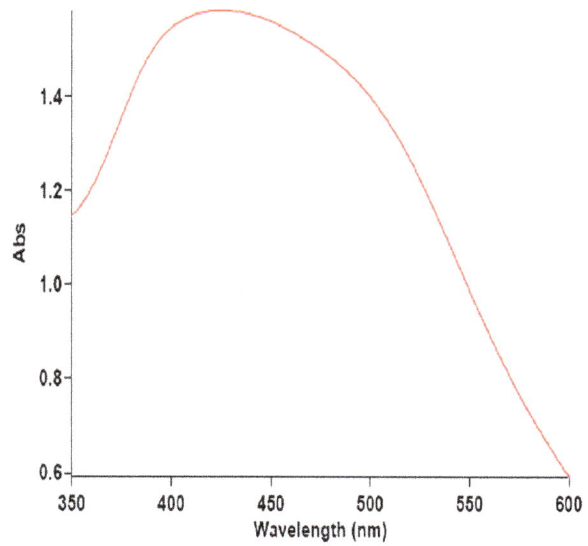

Figure 1 Uv-vis spectra of Ag colloids spectra. Recorded after the addition of supernatant (1 ml) to 100 ml of silver sulfate solution (0.0005 M).

In 2009, it was investigated that there was a different visible light irradiation effect on the formation of silver nanoparticles from silver nitrate using the culture supernatant of Klebsiella pneumonia. Also, the study experimentally investigated the liquid mixing process effect on silver nanoparticle synthesis by visible light irradiation [12].

Research shows that one of the most important factors in the production of silver nanoparticles is the time of its formation. In addition, biosynthesis of silver nanoparticles using microorganisms is rather slow. However, finding microorganisms to rapidly synthesize Ag nanoparticles is an important aspect. In this study, we used *Salmonella typhirium* as the bacterium to reduce Ag+ ions in an aqueous solution.

Results and discussion

For the analytical study of the prepared sample, the amount of absorption within the wavelength of 350 to 600 nm was observed by UV-vis spectroscopy. This technique has proven to be very useful for analyzing nanoparticles [3,7]. As illustrated in Figure 1, a strong surface plasmon resonance was centered at ca. 427 nm. Observation of this strong but broad surface plasmon peak has been well documented for various metal nanoparticles, with sizes ranging widely from 2 to100 nm [13,14].

In order to carefully study the size of the produced silver nanoparticles, dynamic light scattering analysis was used. As seen in Figure 2, the average size of nanoparticles produced in the bioreactor is about 129.3 nm, and majority of the nanoparticles have the size of 87.18 nm.

The difference between the smallest and largest sizes of nanoparticles is 30.92 nm which indicates narrow distribution of the produced nanoparticles. By means of atomic absorption spectrometry, the

Results

			Diam. (nm)	% Number	Width (nm)
Z-Average (d.nm):	129.3	Peak 1:	87.18	100.0	30.92
PdI:	0.128	Peak 2:	0.000	0.0	0.000
Intercept:	0.921	Peak 3:	0.000	0.0	0.000

Result quality : Good

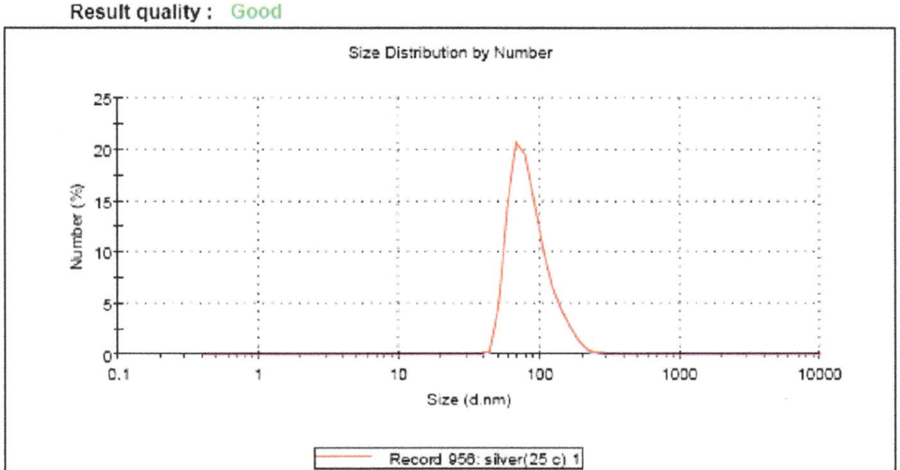

Figure 2 The curve of size distribution (Ag nanoparticles) by number.

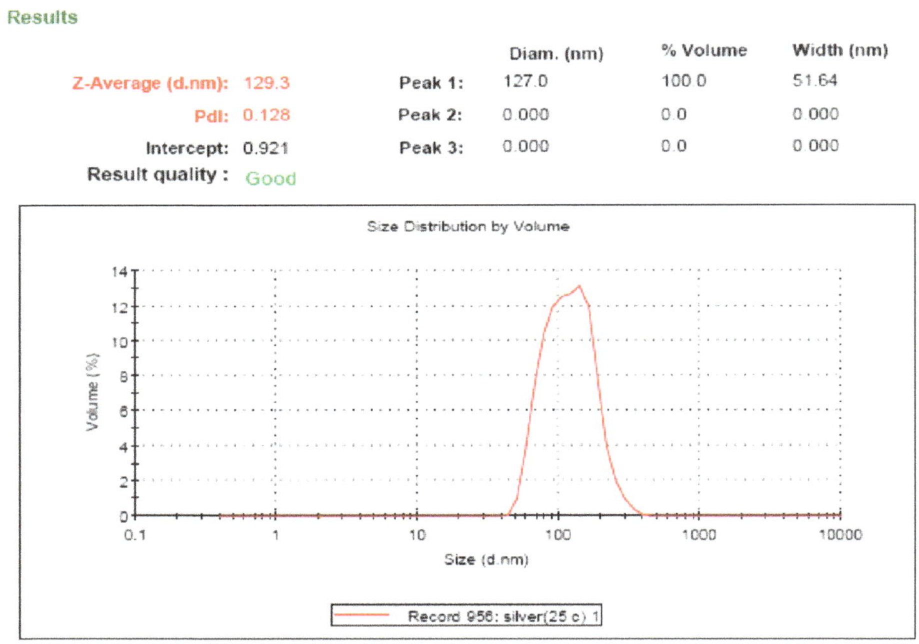

Figure 3 The curve of size distribution (Ag nanoparticles) by volume.

amount of Ag+ reduction was measured and calculated. The amount of Ag+ reduction in the solution after 4 min was 75% which indicates high efficiency. From this amount of reduction, the largest nanoparticle volume (weight) was obtained at the size of 127 nm. In fact, silver nanoparticles with the size of 127 nm had the largest amount of reduction. Dynamic light scattering (DLS) volume analysis shows this fact (Figure 3).

The silver nanoparticles synthesized by *Salmonella* were studied by transmission electron microscopy (TEM), and images show and confirm silver nanoparticle production at nanosize. TEM images of the produced nanoparticles are shown in Figure 4.

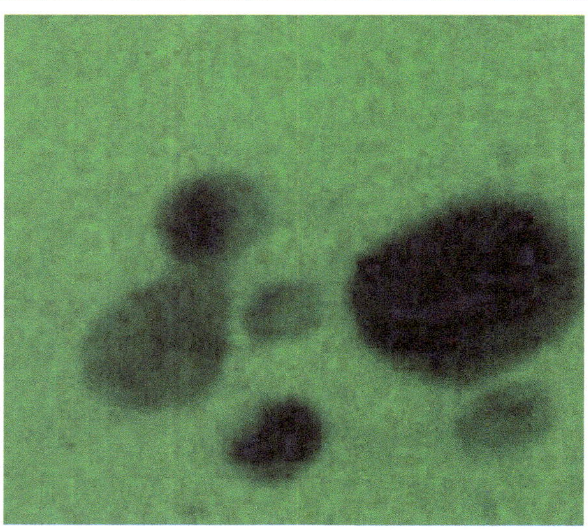

Figure 4 TEM micrograph of silver particles synthesized by *Salmonella* (scale bare: 100 nm).

Conclusions

Here, we described an inexpensive approach for reducing silver sulfate solution to produce nanoparticles. Silver nanoparticles in the range of approximately 50 to 150 nm are synthesized by the supernatant of *Salmonella typhirium* when it is added to silver sulfate. It seems that this method is the fastest biological method to form Ag nanoparticles. The time required to complete the silver ion reduction was obtained about 4 min. According to previous reports, the reductase enzymes released by the bacteria play the most important role in the

Figure 5 Solutions of silver sulfate (0.0005M) before (right) and after (left) exposure to the supernatant of *Salmonella*.

reduction of silver ions. It seems the reductase enzymes released by *Salmonella typhirium* can reduce silver ions quickly; although, more research should take place in this area. This methodology could be used for synthesizing a number of metallic nanoparticles involving other metals with good size and shape morphology. This study would therefore lead to an easy (and rapid) procedure for producing silver nanoparticles with the added advantage of biosafety.

Methods

Organism

Salmonella typhirium was provided by the Department of Veterinary Medicine, Tehran University of Iran. The bacterium after growth was maintained on an agar plate (Macconkey agar) at 4°C temperature.

Growth medium

The Trypti casein soy broth medium (Company Merck, Whitehouse Station, NJ, USA) was used for bacterial culture in this study. The medium was prepared, and after autoclaving, it was maintained in order to be used at later stages.

Preparation of inoculum

A loopful of colony from an agar plate was picked up and aseptically transferred into a 100-cm^3 medium in a 250-cm^3 conical flask. The flask was cultured overnight at 37°C in order to prepare a suspension of the *Salmonella* bacterium.

Ag nanoparticle synthesis

The culture was centrifuged at 5,000 rpm for 15 min, and the supernatant was used for the synthesis of silver nanoparticles. Distilled water was used as solvent in the synthesis of silver nanoparticles. One milliliter of supernatant was added separately to the reaction vessel containing 100 ml Ag_2SO_4 at a concentration of 0.0005M. The reaction between this supernatant and Ag+ ions were carried out in bright conditions for 4 min. After 4 min, the colorless solution of silver sulfate in the container turns into brown color (Figure 5). This color change indicates a possibility of silver nanoparticle production. In order to prove the existence of nanoparticles, product efficiency, and their size, Uv-vis spectroscopy, atomic absorption spectroscopy, transmission electron microscopy, and DLS analysis were used.

Competing interests

The author has no competing interests.

Acknowledgements

We are thankful to the University of Qaemshahr Research Council for the partial support of this work.

References

1. Mandal, D, Bolander, ME, Mukhopadhyay, D, Sarkar, G, Mukherjee, P: The use of microorganisms for the formation of metal nanoparticles and their application. Appl. Microbiol. Biotechnol. **69**, 485–492 (2006)
2. Sharma, VK, Yngard, RA, Lin, Y: Silver nanoparticles: Green synthesis and their antimicrobial activities. Adv. Colloid Interfac. **145**, 83–96 (2009)
3. Ahmad, A, Mukherjee, P, Senapati, S, Mandal, D, Islam Khan, M, Kumar, R, Sastry, M: Extracellular biosynthesis of silver nanoparticles using the fungus Fusariumoxysporum. Colloid Surfaces B. **28**, 313–318 (2003)
4. Jacob, J, Kapoor, S, Biswas, N, Mukherjee, T: Size tunable synthesis of silver nanoparticles in water–ethylene glycol mixtures. Colloid Surfaces A. **301**, 329–334 (2007)
5. Qiaoxin, Z, Hao, L, Xiaohui, W, Xiaoliang, S, Xinglong, D: Fabrication and characterization of nano silver powder prepared by Spray Pyrolysis. J. Wuhan Univ. Technol. **24**, 871–874 (2009)
6. Tavakoli, A, Sohrabi, M, Kargari, A: A review of methods for synthesis of nanostructured metals with emphasis on iron compounds. Chem. Pap. **61**, 151–170 (2007)
7. Vigneshwaran, N, Ashtaputre, NM, Varadarajan, PV, Nachane, RP, Paralikar, KM, Balasubramanya, RH: Biological synthesis of silver nanoparticles using the fungus Aspergillusflavus. Mater. Lett. **61**, 1413–1418 (2007)
8. Durán, N, Marcato, PD, De Souza, G, Alves, O, Esposito, E: Antibacterial effect of silver nanoparticles produced by fungal process on textile fabrics and their effluent treatment. J. Biomed. Nanotechnol. **3**, 203–208 (2007)
9. Kumar, SA, Abyaneh, MK, Gosavi, SW, Kulkarni, SK, Pasricha, R, Ahmad, A, Khan, MI: Nitrate reductase-mediated synthesis of silver nanoparticles from $AgNO_3$. Biotechnol. Lett. **29**, 439–445 (2007)
10. Joerger, R, Klaus, T, Granqvist, CG: Biologically produced silver-carbon composite materials for optically functional thin film coatings. Adv. Mater. **12**, 407–409 (2000)
11. Kalimuthu, K, Babu, RS, Venkataraman, D, Bilal, M, Gurunathan, S: Biosynthesis of silver nanocrystals by Bacillus licheniformis. Colloid Surface B. **65**, 150–153 (2008)
12. Mokhtari, N, Daneshpajouh, S, Seyedbagheri, S, Atashdehghan, R, Abdi, K, Sarkar, S, Minaian, S, Shahverdi, HR, Shahverdi, AR: Biological synthesis of very small silver nanoparticles by culture supernatant of Klebsiella pneumonia: The effects of visible-light irradiation and the liquid mixing process. Mater. Res. Bull. **44**, 1415–1421 (2009)
13. He, S, Guo, Z, Zhang, Y, Zhang, S, Wang, J, Gu, N: Biosynthesis of gold nanoparticles using the bacteria Rhodopseudomonas capsulate. Mater. Lett. **61**, 3984–3987 (2007)
14. Pörtner, R, Giese, C: An Overview on Bioreactor Design, Prototyping and Process Control for Reproducible Three-Dimensional Tissue Culture. In: Marx and Sandig: Drug Testing In Vitro – Breakthroughs and Trends in Cell Culture Technology. Wiley-VCH Verlag GmbH & Co. KgaA, Weinheim (2007)

Antibacterial properties of nano silver finish cellulose fabric

Mohammad Mirjalili[1*], Niloofar Yaghmaei[1] and Marjan Mirjalili[2]

Abstract

In the present study, an attempt has been made to impart antimicrobial finishing to the cellulose fabric using nano silver solution at various concentrations (5, 10, 15, 20 and 25 ml) and an eco-friendly crosslinking agent by the exhaustion technique. Curing conditions were varied, keeping curing temperatures at 110°C and 100°C and curing times to 1 and 2 min. To assess the quality of the finished fabric, various properties such as washing fastness and zone of inhibition were studied. The zones of inhibition have been studied using *Escherichia coli* bacteria to determine antimicrobial activity of the fabric; the surface characteristics of these fabrics have been studied by scanning electron microscopy (SEM).

In the case of the treated commercial cotton fabric product, the zones of inhibition are at a minimum of 14 mm and a maximum of 18 mm for Gram-negative bacteria. SEM study and antibacterial tests of the silver-finished fabrics indicated that, generally, silver nanoparticles were well dispersed on the fabric surface. Antibacterial test was used to estimate the biological activity of the treated fabrics, and Gram-negative bacteria (*Escherichia coli*) were used for this purpose. The washing fastness of finished textiles was investigated in terms of ISO 105 CO3-1982 standard. The results obtained proved the good and long-lasting bacteriostatic efficacy of silver nanoparticles applied during the finishing of cotton and viscose.

Keywords: Antimicrobial, Nano silver, SEM, Cotton, *Escherichia coli*

Background

In the last few years, the market for antimicrobial textiles has shown double-digit growth. This growth has been fuelled by the increased need of consumers for fresh, clean and hygienic clothing. Extensive research is taking place to develop new antimicrobial finishes. This paper reports, in detail, the mechanism of antimicrobial activity and the principles of antimicrobial finishing of textiles. Antimicrobial finishes add value to textiles and garments by providing protection in different ways.

There are different types of fungicides or bactericides such as metal salt, organic metallics, quaternary ammonium salts, formaldehyde, amines, urea, phenols and antibiotics, which have the ability to interrupt the usual metabolism of microorganism and inhibit their growth thereby imparting antibacterial and antifungal activity to cellulosic fibers. Since antibacterial method to textiles is the chemical method, antibacterial finishing via nanotechnology is used instead [1].

Nanoscale particles provide a narrow size distribution, which is required to obtain a uniform material response. Materials such as paints, pigments, electronic inks and ferrofluids, as well as advanced functional and structural ceramics, require that the particles be uniform in size and stable against agglomeration. Fine particles, particularly nanoscale particles with significant surface areas, often agglomerate to minimize the total surface or interfacial energy of the system. Although the process of using solution chemistry can be a practical route for the synthesis of both sub-micrometer and nanoscale particles of many materials, issues such as the control of size, distribution of particles, morphology, crystallinity, particle agglomeration during and after synthesis and separation of these particles from the reactant need further investigation.

* Correspondence: mir_textile@yahoo.com
[1]Department of Textile Engineering, Yazd Branch, Islamic Azad University, Yazd, Iran
Full list of author information is available at the end of the article

Druids used silver to preserve food. American settlers put silver dollars in milk to stop spoilage. Silver leaf was used during World War I to combat infection in wounds. Human skin has many surface bacteria present at any time; that is not a bad thing [1,2].

Antimicrobial activity of silver

The antimicrobial activity of silver has been recognised by clinicians for over 100 years. However, it is only in the last few decades that the mode of action of silver as an antimicrobial agent has been studied with any rigour.

Metallic silver is relatively unreactive; however, when exposed to aqueous environments, some ionic silver (Ag+) is released. Certain salts (e.g. silver nitrate) are readily soluble in water and have been exploited as antiseptic agents for many decades. The generation of silver ions can also be achieved through ion exchange using complexes of silver with other inorganic materials (e.g. silver-zeolite complexes; silver nanoparticles have also been demonstrated to exhibit antimicrobial properties against both bacteria and viruses with close attachment to the microbial cell/virus particles being demonstrated with activity being size dependent. Despite this, the principle activity of silver is as a result of the production of silver ions within an aqueous matrix. This therefore implies that for silver to have an antimicrobial effect, free water must be present. Silver ions interact with a number of components of both bacterial, protozoal and fungal cells. Toxicity to microbial cells is exhibited at very low concentrations with masses within the range of a few fg cell-1s being associated with bactericidal activity. The kinetics of kill vary depending on the source of silver ions with silver derived from ion-exchange processes demonstrating delayed activity compared with that derived from soluble salts. Activity appears to increase with temperature and pH. Studies have demonstrated that silver ions interact with sulfydryl (–SH) groups of proteins as well as the bases of DNA leading either to the inhibition of respiratory processes or DNA unwinding. Inhibition of cell division and damage to bacterial cell envelopes are also recorded, and interaction with hydrogen-bonding processes has been demonstrated to occur. Interruption of cell wall synthesis resulting in the loss of essential nutrients has been shown to occur in yeasts and may well occur in other fungi.

Antiviral activity of silver ions has been recorded, and interaction with -SH groups has been implicated in the mode of action. The association of silver nanoparticles with the envelope of certain viruses has been suggested to prevent them from being infective. Much of the research into the mechanism of action of silver ions has been associated with its use as a therapeutic agent especially as a topical dressing on burns. The concentration employed in and released from treated articles is significantly lower than in these applications. Under such conditions, it has been suggested that in many cases, the concentration of silver ions available following hydration of the surface of a treated article is too low to produce antimicrobial activity associated with many of the mechanisms described above. However, silver ions have been demonstrated to interact with the proteins and possibly phospholipids associated with the proton pump of bacterial membranes. This results in a collapse of the membrane proton gradient causing a disruption of many of the mechanisms of cellular metabolism and hence cell death. Silver ions clearly do not possess a single mode of action. They interact with a wide range of molecular processes within microorganisms resulting in a range of effects from the inhibition of growth, loss of infectivity to cell death. The mechanism depends on both the concentration of silver ions present and the sensitivity of the microbial species to silver. Contact time, temperature, pH and the presence of free water all impact on both the rate and extent of antimicrobial activity [3-16].

Bound type antimicrobial agents

Another set of antimicrobials with a completely different mode of action is one that bonds molecularly to the textile. This product makes the substrate surface antimicrobially active and works by rupturing the cell membrane of the microorganism when it comes into direct contact. These give durable antimicrobial properties to textiles.

Antimicrobial agent based on silver

Silver kills bacteria by strangling them in a warm and moist environment. Highly bioactive silver ions bind with proteins inside and outside bacterial cell membranes thus inhibiting cell respiration and reproduction. Silver is three to four times more active at pH 8 than at pH 6. Silver products are effective against bacteria but not as effective against other organisms like fungi, mould and mildew; they can be used with polyester where many other products cannot. Alginate and chitosan have also been used to make novel antimicrobial materials in combination with silver. Various techniques have been explored to attach silver to textile materials. To prepare antimicrobial fabrics suitable for sterilisation of air, cellulose was grafted with acrylic acid and treated with silver nitrate to bind the silver ions to the COOH group of graft copolymer. To develop a durable finish on wool, it was treated with a complexing agent such as tannic acid or ethylene diaminetetraaceticdianhydride. Thus, treated wool can

Table 1 Bacteria reduction of fabrics treated with nano silver with different concentrations of nano silver against *E.coli*

Nano silver (ml)	Temperature of curing (°C)	Curing time (min)	Reduction percent (%)
5	100	1	99.9
		2	99.89
	110	1	100
		2	100
10	100	1	99.99
		2	99.83
	110	1	100
		2	100
15	100	1	99.87
		2	99.98
	110	1	99.90
		2	100
25	100	1	99.91
		2	99.87
	110	1	100
		2	100

react easily with copper and silver and inhibit the propagation of *S. aureus* and intestinal bacteria effectively. Deposition or interstitial precipitation of tetrasilver tetroxide crystals within the interstices of fibres, yarns and/or fabrics has also been reported in a US patent [17-21].

Results and discussion

Antimicrobial efficiency

From Table 1 and Figure 1, it is clear that the higher the concentration of silver and the higher the curing temperature and time, the better the zone of inhibition. Here, for Gram-negative bacteria (*E. coli*), the minimum zone of inhibition is 14 mm and the maximum is 18.6 mm.

Scanning electron microscopy (SEM)

The antimicrobial finished samples were observed visually, and the topography or morphology of the fabric samples was analysed by high-resolution SEM. The photographs are shown in Figure 2. From this figure, it is thoroughly possible to recognize the silver particles on cotton.

Washing fastness

Textiles are subjected to frequent washing during their use, and the requirement of durability is a very important parameter. According to Table 2, after washing, bacteria reduction of fabrics treated with nano silver gives good results.

Conclusion

The results of both fabrics applied in this work to determine the silver content in fabrics for microbiological estimation of bactericidal efficacy show that silver nanoparticles are well coupled with the fabric, indicating the long-lasting durability of such a finish against washing. The results obtained with the SEM image of cotton fabrics, and the results derived before and after

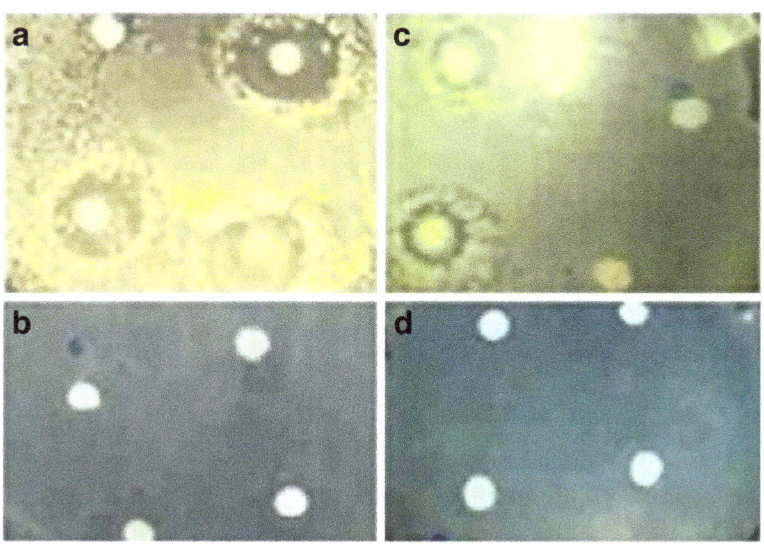

Figure 1 The antibacterial test for treated cotton (a,c), treated viscose (b,d),untreated samples (a,b) and treated samples (c,d).

Figure 2 SEM image of the silver particles on fabric.

finishing demonstrate that both fabrics are protected from *Escherichia coli*. The silver nanoparticles used in this work were well dispersed in the finishing bath nanoparticles, even in very small amounts, can provide the final product with bacteriostatic properties due to the fact that nanoscaled materials have a high surface area-to-volume ratio.

Methods

Materials

Preparation of fabric

Viscose and woven cotton fabrics of plain weave with a surface mass of 100 g/m^2 underwent

a) *Desizing.* The grey fabric is desized with 5 gpl enzyme and 10 gpl common salt at 60°C for 2 h.
b) *Scouring.* The desized fabric is treated with 2.5% sodium hydroxide, 1.5% sodium carbonate and 0.5% NP - at 100°C boiling point for 6 h.
c) *Bleaching.* The scoured fabric is bleached with 1% H_2O_2 (50%) at 80°C for 1 h. Silver nanoparticles, produced by Sigma-Aldrich (St. Louis, MO, USA), for crosslink nano silver with viscose and cotton 20.5 cm^3 were added to the finishing bath according to the fabric treatment graph in Figure 3.

Table 2 Comparison of bacteria reduction of fabrics treated with nano silver (*E. coli*)

Fabrics	Before washing	After five washing cycles	After 10 washing cycles	After 20 washing cycles
Viscose	100%	100%	99.99%	99.97%
Cotton	100%	99.99%	99.95%	99.90%

After home laundry of 0, 5, 10 and 20 times.

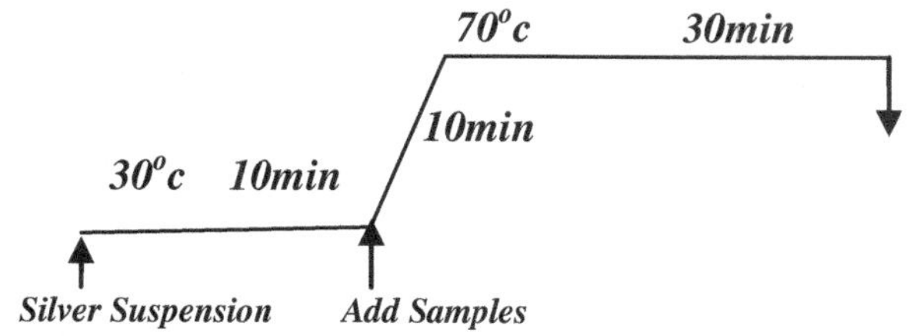

Figure 3 Exhaustion graph.

Antibacterial finishing of textiles

Escherichia coli (Gram-negative bacterium internal name: LOCK 0836 1998) was used for the evaluation of antimicrobial properties. The agar diffusion test is a method of testing the effectiveness of estimated antimicrobial activity of AATCC6538 (Figure 1). Colonies of bacteria recovered on the agar plate were counted, and the reduction percent of bacteria (*R*) was calculated using the following equation:

$$R(\%) = (B–A) \times 100/B, \qquad (1)$$

where *A* is the number of bacteria colonies from the treated specimen after inoculation over a 24-h contact period and *B* is the number of bacteria colonies from untreated control specimen after inoculation at 0 contact time.

Scanning electron microscopy (SEM)

The antimicrobial finished samples were observed visually, and the topography or morphology of the fabric samples was analysed by high-resolution SEM.

Washing fastness test

To evaluate the durability of the antibacterial effect after washing, the treated fabrics were cut into the size of 7.5 × 13.5 cm and washed according to the standard ISO 105 CO3-1982 test method.

Competing interests
The authors declare that they have no competing interests.

Authors' contributions
MM (Mohammad) designed the experiments and revised the manuscript. NY studied the literature and helped MM (Marjan) to fabricate the specimens and perform the tests. Both of them drafted the manuscript. All authors read and approved the final manuscript.

Acknowledgements
The authors want to thank Prof. Omid Moradi and the editor Sherlyn C. Machica for correcting the paper.

Author details
[1]Department of Textile Engineering, Yazd Branch, Islamic Azad University, Yazd, Iran. [2]Department of Chemistry, Yazd Branch, Islamic Azad University, Yazd, Iran.

References

1. Burnision, N, Bygott, C, Stratton, J: Nano Technology Meets TiO_2. Surface Coating International Part A , 179–814 (2004)
2. Fei, B, Deng, Z, Xin, JH, Xhang, Y, Pang, G: Room temperature synthesis of rutile nanorods and their applications on cloth. Nanotechnol **17**, 1927–1932 (2006)
3. Brown, MRW, Anderson, RA: The bactericidal effect of silver ions on *Pseudmonas aeruginosa*. J. Pharm. Pharacol. **20**(Supplement), 1–3 (1968)
4. Im, K, Takasaki, Y, Endo, A, Kuriyama, M: Antibacterial activity of A-type zeolite supporting silver ions in distilled water. J. Antibacterial. Antifungal. Agents **24**, 269–274 (1996)
5. Sondi, I, Salopek-Sondi, B: Silver nanoparticles as antimicrobial agent: a case study on *E coli* as a model for gram-negative bacteria. J. Colloid Interface Sci. **275**, 177–182 (2004)
6. Elchiguerra, JL, Burt, JL, Morones, JR, Camacho-Bragado, A, Gao, X, Lara, HH, Yacaman, MJ: Interaction of silver nanoparticles with HIV-I. J. Nanobiotechnology **3**, 6 (2005)
7. Morones, JR, Elechiguerra, JL, Camacho, A, Holt, K, Kouri, JB, Ramírez, JT, Yacaman, MJ: The bactericidal effect of silver nanoparticles. Nanotechnology **16**, 2346–2353 (2005)
8. McDonnell, G, Denver, R: Antiseptics and disinfectants: activity, action and resistance. Clinical. Microbiol. Reviews. **12**(1), 147–179 (1999)
9. Matsumura, Y, Yoshikata, K, Kunisaki, S-I, Tsuschido, T: Mode of bactericidal action of silver zeolite and its comparison with that of silver nitrate. App. Env. Micro. **69**(7), 4278–4281 (2003)
10. Bragg, PD, Rannie, DJ: The effect of silver ions on the respiratory chain of *E coli*. Can J Microbiol **20**, 883–889 (1974)
11. Batarseh, KI: Anomaly and correlation of killing in the therapeutic properties of silver (I) chelation with glutamic and tartaric acids. J. Antimicrobial Chemotherapy **54**, 546–548 (2004)
12. Richards, RME, Taylor, RB, Xing, DKL: Effect of silver on whole cells and speroplasts of a silver resistant *Pseudomonas aeruginosa*. Microbios **39**, 151–158 (1984)
13. Russell, AD, Hugo, WB: Antimicrobial activity and action of silver. Prog. Med. Chem. **31**, 351–371 (1994)
14. Wells, TN, Scully, P, Paravicini, G, Proudfoot, AE, Payton, MA: Mechanism of irreversible inactivation of phosphomannose isomerases by silver ions and flamazine. Biochemistry **34**(24), 896–903 (1995)
15. Thurmann, RB, Gerba, CP: The molecular mechanisms of copper and silver ion disinfection of bacteria and viruses. Crit. Rev. Environ. Control. **18**, 295–315 (1989)
16. Dibrov, P, Dzioba, J, Gosink, KK, Hase, CC: Chemiosmotic mechanism of antimicrobial activity of Ag+ in *Vibrio cholerae*. Antimicrob. Agents Chemother. **46**(8), 2668–2670 (2002)
17. Achwal, WB: Antimicrobial finishes and their modifications. Colourage, **50**(1), 58–59 (2003)
18. Gulrajani, ML: Cold plasma an environment friendly textile processing medium. Asian Dyer , 39–41 (2004)
19. Qin, Y: Novel antimicrobial fibres. Textile Magazine, **31**(2), 14–17 (2004)
20. Freddi, G, Arai, T, Colonna, GM, Boschi, A, Tsukada, M: Binding of metal cations to chemically modified wool and antimicrobial properties of the wool-metal complexes. J. Appl. Polym. Sci. **82**, 3513–3519 (2001)
21. Antelman, MS: High performance silver (I,III) oxide antimicrob al textile articles.. US Patent 6,436,420, 20 August 2002

Permissions

All chapters in this book were first published in JNC, by Springer; hereby published with permission under the Creative Commons Attribution License or equivalent. Every chapter published in this book has been scrutinized by our experts. Their significance has been extensively debated. The topics covered herein carry significant findings which will fuel the growth of the discipline. They may even be implemented as practical applications or may be referred to as a beginning point for another development.

The contributors of this book come from diverse backgrounds, making this book a truly international effort. This book will bring forth new frontiers with its revolutionizing research information and detailed analysis of the nascent developments around the world.

We would like to thank all the contributing authors for lending their expertise to make the book truly unique. They have played a crucial role in the development of this book. Without their invaluable contributions this book wouldn't have been possible. They have made vital efforts to compile up to date information on the varied aspects of this subject to make this book a valuable addition to the collection of many professionals and students.

This book was conceptualized with the vision of imparting up-to-date information and advanced data in this field. To ensure the same, a matchless editorial board was set up. Every individual on the board went through rigorous rounds of assessment to prove their worth. After which they invested a large part of their time researching and compiling the most relevant data for our readers.

The editorial board has been involved in producing this book since its inception. They have spent rigorous hours researching and exploring the diverse topics which have resulted in the successful publishing of this book. They have passed on their knowledge of decades through this book. To expedite this challenging task, the publisher supported the team at every step. A small team of assistant editors was also appointed to further simplify the editing procedure and attain best results for the readers.

Apart from the editorial board, the designing team has also invested a significant amount of their time in understanding the subject and creating the most relevant covers. They scrutinized every image to scout for the most suitable representation of the subject and create an appropriate cover for the book.

The publishing team has been an ardent support to the editorial, designing and production team. Their endless efforts to recruit the best for this project, has resulted in the accomplishment of this book. They are a veteran in the field of academics and their pool of knowledge is as vast as their experience in printing. Their expertise and guidance has proved useful at every step. Their uncompromising quality standards have made this book an exceptional effort. Their encouragement from time to time has been an inspiration for everyone.

The publisher and the editorial board hope that this book will prove to be a valuable piece of knowledge for researchers, students, practitioners and scholars across the globe.

List of Contributors

Leila Mahdavian
Department of Chemistry, Islamic Azad University, Doroud Branch, P.O. Box: 133, Doroud, Iran

Muhammad Salleh Shamsudin
Centre of Nanoscience and Nanotechnology (NANO-SciTech Centre), Institute of Science, Universiti Teknologi MARA (UiTM), Shah Alam, Selangor Darul Ehsan 40450, Malaysia
School of Physics and Material Studies, Faculty of Applied Sciences, Universiti Teknologi MARA (UiTM), Shah Alam, Selangor Darul Ehsan 40450, Malaysia
University of Southampton Malaysia Campus (USMC), No. 3, Persiaran Canselor 1, Kota Ilmu Educity@Iskandar, Nusajaya, Johor Darul Takzim 79200, Malaysia

Maryam Mohammad
Centre of Nanoscience and Nanotechnology (NANO-SciTech Centre), Institute of Science, Universiti Teknologi MARA (UiTM), Shah Alam, Selangor Darul Ehsan 40450, Malaysia
School of Physics and Material Studies, Faculty of Applied Sciences, Universiti Teknologi MARA (UiTM), Shah Alam, Selangor Darul Ehsan 40450, Malaysia

Syazwan Afif Mohd Zobir
Centre of Nanoscience and Nanotechnology (NANO-SciTech Centre), Institute of Science, Universiti Teknologi MARA (UiTM), Shah Alam, Selangor Darul Ehsan 40450, Malaysia
School of Physics and Material Studies, Faculty of Applied Sciences, Universiti Teknologi MARA (UiTM), Shah Alam, Selangor Darul Ehsan 40450, Malaysia

Noor Asnida Asli
Centre of Nanoscience and Nanotechnology (NANO-SciTech Centre), Institute of Science, Universiti Teknologi MARA (UiTM), Shah Alam, Selangor Darul Ehsan 40450, Malaysia
School of Physics and Material Studies, Faculty of Applied Sciences, Universiti Teknologi MARA (UiTM), Shah Alam, Selangor Darul Ehsan 40450, Malaysia

Suriani Abu Bakar
Department of Physics, Faculty of Science and Mathematics, Universiti Pendidikan Sultan Idris (UPSI), Tanjung Malim, Perak Darul Ridzuan 35900, Malaysia

Saifollah Abdullah
School of Physics and Material Studies, Faculty of Applied Sciences, Universiti Teknologi MARA (UiTM), Shah Alam, Selangor Darul Ehsan 40450, Malaysia

Syed Yusainee Syed Yahya
School of Physics and Material Studies, Faculty of Applied Sciences, Universiti Teknologi MARA (UiTM), Shah Alam, Selangor Darul Ehsan 40450, Malaysia

Mohamad Rusop Mahmood
Centre of Nanoscience and Nanotechnology (NANO-SciTech Centre), Institute of Science, Universiti Teknologi MARA (UiTM), Shah Alam, Selangor Darul Ehsan 40450, Malaysia
NANO-Electronic Centre, Faculty of Electrical Engineering, Universiti Teknologi MARA (UiTM), Shah Alam, Selangor Darul Ehsan 40450, Malaysia

Karim Zare
Department of Chemistry, Science and Research Campus, Islamic Azad University, Tehran, Iran
Department of Chemistry, Shahid Beheshti University, Evin, Tehran, Iran
Young Researchers and Elites Club of Science and Research Campus, Islamic Azad University, Tehran, Iran

Morasa Amani Malkeshi
Department of Chemistry, Science and Research Campus, Islamic Azad University, Tehran, Iran
Young Researchers and Elites Club of Science and Research Campus, Islamic Azad University, Tehran, Iran

Mahmood Moradi
Department of Physics, Science College, Shiraz University, Shiraz, Iran
Nanotechnology Research Center, Shiraz University, Shiraz, Iran

Mina Zamanian
Department of Physics, Science College, Shiraz University, Shiraz, Iran

Mohammad Noormohammadi
Department of Physics, Science College, Shiraz University, Shiraz, Iran

Azadeh Tadjarodi
Research Laboratory of Inorganic Materials Synthesis, Department of Chemistry, Iran University of Science and Technology, Tehran 16846-13114, Iran

Mina Imani
Research Laboratory of Inorganic Materials Synthesis, Department of Chemistry, Iran University of Science and Technology, Tehran 16846-13114, Iran

Hamed Kerdari
Department of Chemistry, Saveh Branch, Islamic Azad University, Saveh 39187-366, Iran

Mostafa Mahinroosta
No. 8, Negarestan 4, Nazarabad, Alborz 3331779638, Iran

Reza Ansari
Chemistry Department, Faculty of Science, University of Guilan, PO Box: 41635-1914, Rasht, Iran

Zahra Mosayebzadeh
Chemistry Department, Faculty of Science, University of Guilan, PO Box: 41635-1914, Rasht, Iran

Majid Arvand
Chemistry Department, Faculty of Science, University of Guilan, PO Box: 41635-1914, Rasht, Iran

Ali Mohammad-khah
Chemistry Department, Faculty of Science, University of Guilan, PO Box: 41635-1914, Rasht, Iran

Loghman Karimi
Young Researchers and Elites Club, Yazd Branch, Islamic Azad University, Yazd, Iran

Salar Zohoori
Department of Textile Engineering, Yazd Branch, Islamic Azad University, Yazd, Iran

Said Benramache
Material Sciences Department, Faculty of Science, University of Biskra, 07000, Algeria

Boubaker Benhaoua
VTRS Laboratory, Institute of Technology, University of El-Oued, 39000, Algeria

Hamza Bentrah
Mechanical Department, Faculty of Technology, University of Biskra, 07000, Algeria

Ahmed A Farghali
Nanotechnology Department, Faculty of Postgraduate Studies for Advanced Sciences, Beni-Suef University, Beni-Suef 62111, Egypt

Mohamed Bahgat
Minerals Technology Department Central Metallurgical R & D Institute (CMRDI), Helwan 11421, Egypt

Waleed M A ElRouby
Nanotechnology Department, Faculty of Postgraduate Studies for Advanced Sciences, Beni-Suef University, Beni-Suef 62111, Egypt

Mohamed H Khedr
Nanotechnology Department, Faculty of Postgraduate Studies for Advanced Sciences, Beni-Suef University, Beni-Suef 62111, Egypt

Sajjad Sedaghat
Department of Chemistry, Islamic Azad University, Shahr-eQods Branch, Tehran, Iran

Narges Ajami
Department of Chemistry, Payame Noor University, PO BOX, 19395-3697, Tehran, IRAN

Niloufar Bahrami Panah
Department of Chemistry, Payame Noor University, PO BOX, 19395-3697, Tehran, IRAN

Keloth Basavaiah
Department of Inorganic and Analytical Chemistry, Andhra University, Visakhapatnam, Andhra Pradesh 530003, India

Yarramsetti Pavankumar
Department of Inorganic and Analytical Chemistry, Andhra University, Visakhapatnam, Andhra Pradesh 530003, India

Alamanda Vara Prasada Rao
Department of Inorganic and Analytical Chemistry, Andhra University, Visakhapatnam, Andhra Pradesh 530003, India

Nasim Shadmani
Young Researchers and Elites Club, Rasht Branch, Islamic Azad University, Rasht, Iran

Elham Pournamdari
Department of Chemistry, Islamshahr Branch, Islamic Azad University, Islamshahr, Iran

Kaykhosrow Khojier
Department of Physics, Chalous Branch, Islamic Azad University, Chalous 46615-397, Iran

Mohammad Reza Karami Mehr
Department of Physics, Faculty of Science, Central Tehran Branch, Islamic Azad University, Tehran 13185-786, Iran

Hadi Savaloni
Department of Physics, University of Tehran, North Kargar Street, Tehran 14395-515, Iran

Rajesh Barve
Department of Repertory, Virar Homoeopathic Medical College, Virar, 401305, Maharashtra, India

Ramesh Chaughule
Department of Chemistry, Ramnarain Ruia College, Matunga, Mumbai 400019, India

Fahimeh Najafi
Department of Chemistry, Roudehen Branch, Islamic Azad University, Roudehen, Iran

Seyyed Vahid Atghia
Young Researchers Club, Rasht Branch, Islamic Azad University, Rasht, Iran

Somayyeh Sarvi Beigbaghlou
Young Researchers Club, Rasht Branch, Islamic Azad University, Rasht, Iran

Emilye Rosas Landa Loustau
Centro de Investigación en Energía, Universidad Nacional Autónoma de México (UNAM), Temixco, CP. 62580, México
Centro de Ciencias de la Complejidad (UNAM), Mexico, CP 04510, Mexico

Jesús A del Río
Centro de Investigación en Energía, Universidad Nacional Autónoma de México (UNAM), Temixco, CP. 62580, México
Centro de Ciencias de la Complejidad (UNAM), Mexico, CP 04510, Mexico

Julia Tagüeña-Martínez
Centro de Investigación en Energía, Universidad Nacional Autónoma de México (UNAM), Temixco, CP. 62580, México

Luis E Sansores
Instituto de Investigaciones en Materiales (UNAM), Mexico, CP 04510, Mexico

Rocío Nava
Centro de Investigación en Energía, Universidad Nacional Autónoma de México (UNAM), Temixco, CP. 62580, México

Kasi Gopinath
Department of Nanoscience and Technology, Alagappa University, Karaikudi, Tamil Nadu 630 004, India

Shanmugam Gowri
Department of Nanoscience and Technology, Alagappa University, Karaikudi, Tamil Nadu 630 004, India

Ayyakannu Arumugam
Department of Nanoscience and Technology, Alagappa University, Karaikudi, Tamil Nadu 630 004, India

Solmaz Maleki Dizaj
Faculty of Pharmacy, Tabriz University of Medical Science, Tabriz, Iran
Student Research Committee, Tabriz University of Medical Science, Tabriz, Iran

Rimpy Gupta
Post-Graduate Department of Chemistry, University of Jammu, Jammu Tawi 180006, India

Sumit Sanotra
Post-Graduate Department of Chemistry, University of Jammu, Jammu Tawi 180006, India

Haq Nawaz Sheikh
Post-Graduate Department of Chemistry, University of Jammu, Jammu Tawi 180006, India

Bansi Lal Kalsotra
Post-Graduate Department of Chemistry, University of Jammu, Jammu Tawi 180006, India

Fatemeh Sadeghi
Department of Physics, Payam Noor University, Tehran 19395-4697, Iran

Farzad Khani
Department of Mathematics, Razi University, Kermanshah 6715685438, Iran

Abbas Hemati Azandaryani
Department of Applied Chemistry, Faculty of Chemistry, Razi University, Kermanshah 6715685438, Iran
Nano Drug Delivery Research Center, Kermanshah University of Medical Sciences, Kermanshah 6714676573, Iran

Yadollah Mansouri
Environmental Epidemiological Research Center, Kermanshah University of Medical Sciences, Kermanshah 6714676573, Iran
Department of Analytical Chemistry, Faculty of Chemistry, Razi University, Kermanshah 6715685438, Iran

Zahra Shaykhi Mehrabadi
Department of Applied Chemistry, Faculty of Chemistry, Razi University, Kermanshah 6715685438, Iran

Amir Nikjou
Department of Physics, Payam Noor University, Tehran 19395-4697, Iran

Fahimeh Najafi
Department of Chemistry, Roudehen Branch, Islamic Azad University, Roudehen, Iran

Hamidreza Sadegh
Department of Chemistry, Shahr-e- Qods Branch, Islamic Azad University, Tehran, Iran

Ramin Shahryari ghoshekandi
Department of Chemistry, Shahr-e- Qods Branch, Islamic Azad University, Tehran, Iran

Yoshio Kobayashi
Department of Biomolecular Functional Engineering, College of Engineering, Ibaraki University, 4-12-1 Naka-narusawa-cho, Hitachi, Ibaraki 316-8511, Japan

Hiromitsu Inose
Department of Biomolecular Functional Engineering, College of Engineering, Ibaraki University, 4-12-1 Naka-narusawa-cho, Hitachi, Ibaraki 316-8511, Japan

Tomohiko Nakagawa
Division of Surgical Oncology, Graduate School of Medicine, Tohoku University, Seiryo-machi, Aoba-ku, Sendai, Miyagi 980-8574, Japan

Yohsuke Kubota
Division of Surgical Oncology, Graduate School of Medicine, Tohoku University, Seiryo-machi, Aoba-ku, Sendai, Miyagi 980-8574, Japan

Kohsuke Gonda
Division of Surgical Oncology, Graduate School of Medicine, Tohoku University, Seiryo-machi, Aoba-ku, Sendai, Miyagi 980-8574, Japan

Noriaki Ohuchi
Division of Surgical Oncology, Graduate School of Medicine, Tohoku University, Seiryo-machi, Aoba-ku, Sendai, Miyagi 980-8574, Japan

Mohammad Amdad Ali
Department of Agricultural Chemistry, Bangladesh Agricultural University, Mymensingh 2202, Bangladesh

Mahmudur Rahman Idris
Department of Chemistry, University of Dhaka, Dhaka 1000, Bangladesh

Md Emran Quayum
Department of Chemistry, University of Dhaka, Dhaka 1000, Bangladesh

Muhammad Zuhairi Borhan
NANO-SciTech Centre (NST), Institute of Science, Shah Alam 40450, Selangor, Malaysia
NANO-Innovation Centre (Nano-IC), Faculty of Applied Sciences, Shah Alam 40450, Selangor, Malaysia
Faculty of Applied Sciences, Universiti Teknologi MARA (UiTM) Malaysia, Shah Alam 40450, Selangor, Malaysia

Rohaya Ahmad
Faculty of Applied Sciences, Universiti Teknologi MARA (UiTM) Malaysia, Shah Alam 40450, Selangor, Malaysia

Mohamad Rusop
NANO-SciTech Centre (NST), Institute of Science, Shah Alam 40450, Selangor, Malaysia
NANO-Innovation Centre (Nano-IC), Faculty of Applied Sciences, Shah Alam 40450, Selangor, Malaysia

Saifollah Abdullah
NANO-Innovation Centre (Nano-IC), Faculty of Applied Sciences, Shah Alam 40450, Selangor, Malaysia

Goshtasp Cheraghian
Department of Petroleum and Gas Engineering, Science and Research Branch, Islamic Azad University, Tehran, Iran

Mahmood Hemmati
Research Institute of Petroleum Industry (RIPI), Tehran, Iran

Mohsen Masihi
Chemical and Petroleum Engineering Department, Sharif University of Technology, Tehran, Iran

Saeed Bazgir
Department of Polymer Engineering, Science and Research Branch, Islamic Azad University, Tehran, Iran

Ghazaleh Allaedini
Chemical Engineering Department, UKM, Bangi, Selangor, Malaysia

Abubakar Muhammad
Chemical Engineering Department, UKM, Bangi, Selangor, Malaysia

Saeideh Ghorbaninezhad
Department of Chemistry, Islamic Azad University (Rasht Branch), Rasht, Iran

Maryam Ghorbaninezhad
Department of Chemistry, Islamic Azad University (Rasht Branch), Rasht, Iran

Chandra Sekharan Nair Omana Amma Sreekala
School of Biotechnology, Amrita Vishwa Vidyapeetham, Amritapuri Campus, Kollam, Kerala 690525, India

Jinchu Indiramma
School of Biotechnology, Amrita Vishwa Vidyapeetham, Amritapuri Campus, Kollam, Kerala 690525, India

Kandala Bala Subramanya Pavan Kumar
School of Biotechnology, Amrita Vishwa Vidyapeetham, Amritapuri Campus, Kollam, Kerala 690525, India

Karyaveettil Savithriamma Sreelatha
School of Biotechnology, Amrita Vishwa Vidyapeetham, Amritapuri Campus, Kollam, Kerala 690525, India
Government Polytechnic College, Kottayam, Kerala 686013, India

Mahesh Saran Roy
Defence Laboratory, Jodhpur, Rajasthan 342011, India

Kethirabalan Chitra
Environmental Nanotechnology Division, Sri Paramakalyani Centre for Environmental Sciences, Manonmaniam Sundaranar University, Alwarkurichi, Tamil Nadu 627412, India

Gurusamy Annadurai
Environmental Nanotechnology Division, Sri Paramakalyani Centre for Environmental Sciences, Manonmaniam Sundaranar University, Alwarkurichi, Tamil Nadu 627412, India

Maryam Hesabi
Department of Chemistry, Faculty of Science, Islamic Azad University, Rasht Branch, Rasht 41476, Iran

Masoumeh Hesabi
Faculty of Nursing, Shahid Beheshti University of Medical Sciences, Rasht Branch, Rasht 41625, Iran

Hamid Reza Ghorbani
Department of Chemical engineering, Qaemshahr Branch, Islamic Azad University, Qaemshahr, Iran

Mohammad Mirjalili
Department of Textile Engineering, Yazd Branch, Islamic Azad University, Yazd, Iran

Niloofar Yaghmaei
Department of Textile Engineering, Yazd Branch, Islamic Azad University, Yazd, Iran

Marjan Mirjalili
Department of Chemistry, Yazd Branch, Islamic Azad University, Yazd, Iran